BARRON'S

THE TRUSTED NAME IN TEST PREP

2025

AP®

Environmental Science

PREMIUM

T0308446

Gary S. Thorpe, M.S.

Acknowledgments

I would like to thank my wife Patti for her understanding and patience while updating this 2025 edition and Jennifer Giammusso (Content Editor) and Aarthi Padmanaban (Production Editor) at Barron's (a division of Kaplan) for their advice and support.

Published by Kaplan North America, LLC d/b/a Barron's Educational Series
1515 West Cypress Creek Road
Fort Lauderdale, Florida 33309
www.barronseduc.com

ISBN: 978-1-5062-9189-5

10 9 8 7 6 5 4 3 2 1

Kaplan North America, LLC d/b/a Barron's Educational Series print books are available at special quantity discounts to use for sales promotions, employee premiums, or educational purposes. For more information or to purchase books, please call the Simon & Schuster special sales department at 866-506-1949.

About the Author

Gary S. Thorpe taught A.P. Chemistry at Beverly Hills High School and Environmental Science at various colleges in Los Angeles for 33 years. In his "spare time," Gary was a reserve Los Angeles Police Department patrol officer for 28 years. Gary lives in Los Angeles and Idyllwild, California, with his wife, Patti. Gary and Patti have two daughters: Lt. Commander Erin Thorpe, U.S. Navy, and Kris Weissman, Program Design and Implementation Director – Broderick-Haight Consulting.

Table of Contents

PRACTICE TESTS

Visit Barron's Online Learning Hub for more full-length practice tests and labs.

How to Use This Book

This book offers a comprehensive review of the Enduring Understandings, Key Concepts, Learning Objectives, and Essential Knowledge taught in the AP Environmental Science course, with many content examples and practice questions. Each element of this book is designed to prepare you for the exam.

Review and Practice

This book includes nine content review units that align with the nine units of the AP Environmental Science course curriculum. In addition, each chapter contains end-of-unit practice questions that highlight stimulus material, such as graphs and tables, food webs, environmental diagrams, and maps. All multiple-choice and free-response questions include answer explanations and sample scoring.

Practice Tests

The final section of this book offers the opportunity to take two full-length practice tests that include all question types found on the actual exam. There are 80 multiple-choice questions on the exam with only four answer choices. You will have 90 minutes for the multiple-choice questions. There are three FRQs free-response questions, or essay questions, on the exam, which you must answer in 70 minutes. A comprehensive answer explanation is provided for each multiple-choice question, and a rubric is provided for all free-response questions.

Online Practice

There are three full-length practice exams available at online.barronsbooks.com. You may take these exams in practice (untimed) mode or in timed mode. All questions include answer explanations. This book also includes online virtual labs for each chapter. Reinforce your understanding of environmental science by reviewing simulated laboratory experiments. All lab questions are fully explained.

For Students

By using this book, you have the opportunity to prepare for the exam by concentrating on the areas that the College Board has identified as "essential knowledge." This book highlights the key concepts in each unit and explains them by giving content examples. How you use the book depends on how your school offers the course. Nevertheless, by answering the review questions at the end of each unit and by taking the practice tests, you will have an indication of how well you will do on the actual exam.

For Teachers

Suggest to your students that they use this book along with their textbook. If you develop your own assignments and tests around the Enduring Understandings, Key Concepts, Learning Objectives, and Essential Knowledge found in the College Board's recently revised AP Environmental Science Curriculum Framework, this book should offer your students the opportunity to use the material for review.

This book and its online features were designed to be used beginning on Day 1. Advise your students to not wait until a few weeks before the APES testing day to begin reviewing. Many teachers use this book along with the textbook during the course of the year. Be sure to have students mark the practice questions in this book that they missed so that it is easier for them to go back later to review them before the exam.

Please Note: Barron's AP Environmental Science Flashcards can also be used in tandem with this guide for additional practice. The flashcards are organized by units and are aligned with the College Board AP Environmental Science Course. It is great for a whole class review or individual study.

BARRON'S ESSENTIAL 5

As you review the content in this book and work toward earning that **5** on your AP ENVIRONMENTAL SCIENCE exam, here are five things that you **MUST** do:

1. **Practice writing your own answers to the free-response questions (FRQs)** found in the two full-length practice exams at the end of this book ***and*** the three full-length online practice exams. After you review each FRQ, close this book and write your own response to the question. Then, compare your answers to those in this book. The more you practice writing *your own* responses, the higher your score will be on the actual APES exam.

2. **Be sure to review the Case Studies presented in this book.** You will find several questions on the exam that focus on specific events that have occurred throughout recent history and involve environmental science principles and issues. When possible, reference these Case Studies in your responses to the FRQs.

3. **The AP Environmental Science exam focuses on having you analyze data and draw logical conclusions.** All practice tests in this fully aligned edition are written to reinforce those required skills.

4. **Become comfortable with doing the types of math problems found in this book, especially those involving calculations.** These types of problems are found in each practice exam.

5. **Review the relevant acts, laws, and treaties detailed in each chapter.** You will find several questions on the APES exam that test your knowledge on them; you can also reference them in your answers to the FRQs when appropriate.

Questions Commonly Asked About the APES Exam

Q: WHAT'S THE BEST WAY TO PREPARE FOR THE APES EXAM?

A: Barron's *AP Environmental Science Premium* is NOT a cram book to be used at the last minute, but rather was designed to be used throughout your AP Environmental Science course to reinforce the most important concepts that you will find on the actual APES exam. Also, be sure to use Barron's fourth edition flashcards in tandem with this book, especially in the last few weeks before the exam, to review and reinforce major concepts and principles.

Q: HOW ARE THE MULTIPLE-CHOICE QUESTIONS SCORED?

A: The AP Environmental Science exam has 80 multiple-choice questions, which you need to complete in 90 minutes, with four answers to choose from. The format of the questions requires you to read and analyze text and graphs in order to answer the questions properly. The two practice exams found in this book and the three online practice exams are each written in this same format.

Q: HOW ARE THE FREE-RESPONSE QUESTIONS (FRQs) GRADED?

A: The AP Environmental Science exam has only three essay questions, also known as free-response questions (FRQs), which you need to complete within 70 minutes. Barron's uses FRQs submitted by top AP Environmental Science teachers from across the United States, along with complete scoring guides (rubrics).

Q: WILL I FIND QUESTIONS ABOUT LABS?

A: Yes. In the most recent test format, the emphasis is on reading and interpreting data and analyzing graphs. All of the units in this book have at least one online APES lab investigation. You will be provided the background to the lab, the directions on how to perform the lab, and the data had you actually done the lab. You will then be asked to analyze and interpret the data, draw graphs, etc. No other APES test prep book includes this important feature.

Q: WHAT DO THE SCORES MEAN?

A: Scores range from 1, being the lowest, to 5, being the highest. For many colleges, scores of 3, 4, or 5 are generally considered passing and are given full college credit. Go to *https://apstudent.collegeboard.org/creditandplacement/search-credit-policies* to find out which colleges accept AP credit and what scores are required. Historically, about half of the students who take the APES exam "pass" and about half do not. The following are current APES exam results:

5	4	3	2	1	>3	<3
8.3%	28.4%	17%	26.4%	19.9%	53.7%	46.3%

Q: WHAT MATERIALS DO I TAKE WITH ME TO THE EXAM?

A:
- Admission ticket
- Black or blue erasable pens for the FRQs
- Calculator—students will be permitted to use a scientific or four-function calculator on both sections of the exam.
- Official photo (e.g., student ID), including signature

- Several sharpened #2 pencils with non-smudging erasers
- Watch

DO NOT BRING

- Cell phones
- Colored pencils
- Food or drink
- Highlighters
- Rulers

Q: SHOULD I GUESS?

A: Yes. There is no penalty for guessing.

Q: CAN I CANCEL MY SCORES?

A: Yes; however, your request must be received by June 15 of the year you take the test. You may also request one or more of your scores NOT be sent to colleges.

Q: CAN I WRITE ON THE TEST BOOKLET?

A: Yes. Use the test booklet to make brief outlines of how you want to organize your FRQs (essays) or do math calculations. Several examples of this technique are presented in this book.

Q: HOW DO I GET MORE INFORMATION?

A: Log on to *apcentral.collegeboard.com* or *collegeboard.com/apstudents*. You can also join other APES students and teachers, along with the author, at *facebook.com/barronsapes*.

1

The Living World: Ecosystems

Learning Objectives

In this unit, you will learn:

- ➔ 1.1 Introduction to Ecosystems
- ➔ 1.2 Terrestrial Biomes
- ➔ 1.3 Aquatic Biomes
- ➔ 1.4 The Carbon Cycle
- ➔ 1.5 The Nitrogen Cycle
- ➔ 1.6 The Phosphorus Cycle
- ➔ 1.7 The Hydrologic (Water) Cycle
- ➔ 1.8 Primary Productivity
- ➔ 1.9 Trophic Levels
- ➔ 1.10 Energy Flow and the 10% Rule
- ➔ 1.11 Food Chains and Food Webs

For added practice with this unit and topics, log onto Online Labs 1A–1D at online.barronsbooks.com

1.1 Introduction to Ecosystems

An ecosystem is a community of living (biotic) organisms interacting with the nonliving (abiotic) components of their environment as a system through various nutrient and energy cycles.

Biological Populations and Communities

- **Organism**—a living thing that can function on its own

↓

- **Species**—organisms that resemble each other; are similar in genetic makeup, chemistry, and behavior; and are able to interbreed and produce fertile offspring

  Intraspecific = within the same species; *interspecific* means between different species.

↓

- **Population**—organisms of the same species that interact with each other and occupy a specific area

↓

- **Community**—populations of different species

Ecological Niches

An ecological niche is the particular area within a habitat occupied by an organism, as well as the function of that organism within its ecological community. The physical environment influences how organisms affect and are affected by resources and competitors. The niche reflects the specific adaptations that a species has acquired through evolution.

Characteristics of a niche include:

- habitat
- interactions with living (biotic) and nonliving (abiotic) factors
- place/role in the food web
- types and amounts of resources available

Generalist vs. Specialist Species

Generalists	Specialists
Able to survive on a wide variety of food resources	Specific/limited number of prey
Able to withstand a wide range of environmental conditions	Prone to extinction, sensitive to environmental change
Live in broad niches	Live in narrow niches; e.g., pandas
Examples: cockroaches, humans, mice	Examples: giant pandas, koalas, mountain gorillas

When environmental conditions are stable, specialist species have an advantage since there are few competitors, as each species occupies its own unique niche (Competitive Exclusion Principle). However, when habitats are subjected to rapid changes, the generalist species usually fare better since they are generally more adaptable.

Interactions Among Species

Symbiosis is a term used to describe any type of close and long-term biological interaction between two different biological organisms of the same or different species. Symbiosis can be obligatory (one or both of the organisms entirely depend(s) on the other for survival) or facultative (the organisms can generally live independently).

Listed below are examples of various types of symbiotic interactions.

Interaction	Description
Amensalism 	The interaction between two species whereby one species suffers and the other species is not affected. Example: The black walnut tree releases a chemical that kills neighboring plants.
Commensalism 	The interaction between two species whereby one organism benefits and the other species is not affected. Forms of commensalism include: (1) using another organism for transportation, e.g., the remora on a shark or mites on dung beetles; (2) using another organism for housing, e.g., epiphytic plants like orchids growing on trees or birds living in the holes of trees; and (3) using something that another organism created, e.g., hermit crabs using the shells of marine snails for protection.
Competition 	Competition can be either intraspecific (competition between members of the same species) or interspecific (competition between members of different species). Competition is the driving force of evolution whether it is for food, mating partners, or territory. Intraspecific competition results in organisms best suited for surviving in a changing environment. Competition is prominent in predator–prey relationships, with the predator seeking food and the prey seeking survival. The Competitive Exclusion Principal states "Two species that compete for the same limited resource cannot coexist at constant population values. When one species has even the slightest advantage over another, the one with the advantage will dominate in the long term which results either to the extinction of the weaker competitor or to an evolutionary or behavioral shift toward a different ecological niche.
Mutualism 	Mutualism is the interaction between two species whereby both species benefit. Example: bees fly from flower to flower gathering nectar, which they make into honey, benefiting the bees. When bees land in a flower, some of the pollen sticks to their bodies. When they land in the next flower, some of the pollen from the first flower rubs off, pollinating the second plant. The bees get to eat, and the flowering plants get to reproduce.

Interaction	Description
Parasitism 	Parasitism is an interaction between two species whereby one species is benefited, and the other species is harmed. Ectoparasites live on the surface or exterior of the host (e.g., ticks or fleas), whereas endoparasites live within their hosts (e.g., tapeworms). Epiparasites feed on other parasites (e.g., a protozoan living in the digestive tract of a flea living on a dog). Social parasites involve behaviors that benefit the parasite and harm the host (e.g., cuckoo birds that use other birds to raise their young). Hosts have evolved defense mechanisms (e.g., immune systems or plant toxins) to diminish parasitism, whereas parasites have evolved mechanisms to keep their hosts alive.
Predation 	Predators hunt and kill prey. Opportunistic predators (e.g., alligators, dogs, humans, leopards) kill and eat almost anything, whereas specialist predators (e.g., anteaters) only prey upon certain organisms.
Saprotrophism 	Saprotrophs obtain their nutrients from dead or decaying plants or animals through absorption of soluble organic compounds and include many bacteria, fungi, and protozoa. Vultures and dung beetles are also saprotrophs.

Law of Tolerance

Earth's ecosystems are affected by both biotic (living) and abiotic (nonliving) factors, and are regulated by the Law of Tolerance, which states that the existence, abundance, and distribution of species depends on the tolerance level of each species to both physical and chemical factors (Figure 1.1). The abundance or distribution of an organism can be controlled by certain factors (e.g., the climatic, topographic, and biological requirements of plants and animals) where levels of these exceed the maximum or minimum limits of tolerance of that organism.

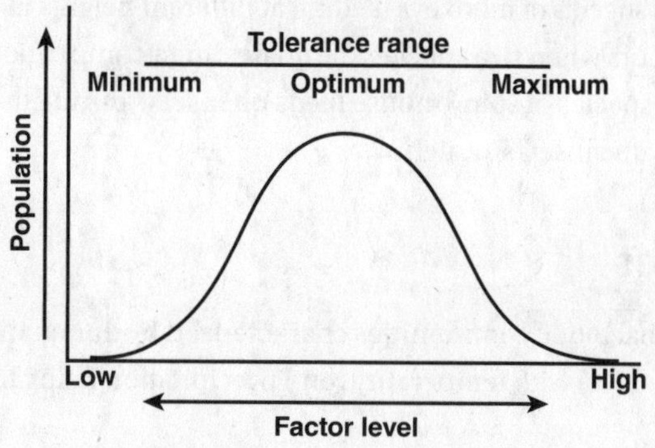

Figure 1.1 Law of Tolerance

Limiting Factors

A limiting factor is any abiotic factor that limits or prevents the growth of a population. Limiting factors in terrestrial ecosystems may include the level of soil nutrients, the available amount of water and light, and temperature. In aquatic ecosystems, major limiting factors may include the pH of the water, the amount of dissolved oxygen, light, or the degree of salinity.

Predator–Prey Population Relationships

Figure 1.2 shows the relationship (over time) of population cycles of both predators and their prey. Predator–prey cycles are based on a feeding relationship between two species, for example, if the prey species rapidly multiplies, the number of predators increases until the predators eventually eat so many of the prey that the prey population dwindles again. Soon afterward, predator numbers also decrease due to starvation. This leads to a rapid increase in the prey population, and a new cycle begins.

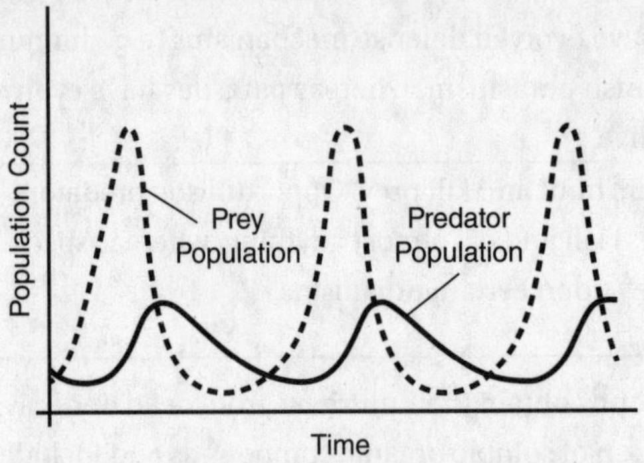

Figure 1.2 Predator–prey population cycle

Resource Partitioning

Resources in an environment are generally limited, and some species have evolved to "share" a specific resource. This sharing may take several forms:

- **Morphological partitioning** occurs when two species share the same resource but have evolved slightly different structures to utilize the same resource (e.g., two different species of bees have evolved different proboscises [tubular mouthpart used for feeding] lengths to utilize various size flowers of the same species).
- **Spatial partitioning** occurs when competing species use the same resource by occupying different areas or habitats within the range of occurrence of the resource (e.g., different species of fish feeding at different depths in a lake or different species of monkeys feeding at different heights in a tree).
- **Temporal partitioning** occurs when two species eliminate direct competition by utilizing the same resource at different times (e.g., one species of spiny mouse feeds on insects during the day while a second species of spiny mouse feeds on the same insects at night).

1.2 Terrestrial Biomes

Biomes are major regional or global biotic communities characterized by dominant forms of plant life and the prevailing climate (Figures 1.3 and 1.4) with temperature and precipitation being the most important determinants of biomes (Figure 1.5).

Biomes are classified by the types of dominant plant and animal life. The climate determines plants, and the plants determine animals. Species diversity within a biome is directly related to net productivity or the rate of generation of biomass (typically measured in $g/m^2/day$), availability of moisture, and temperature.

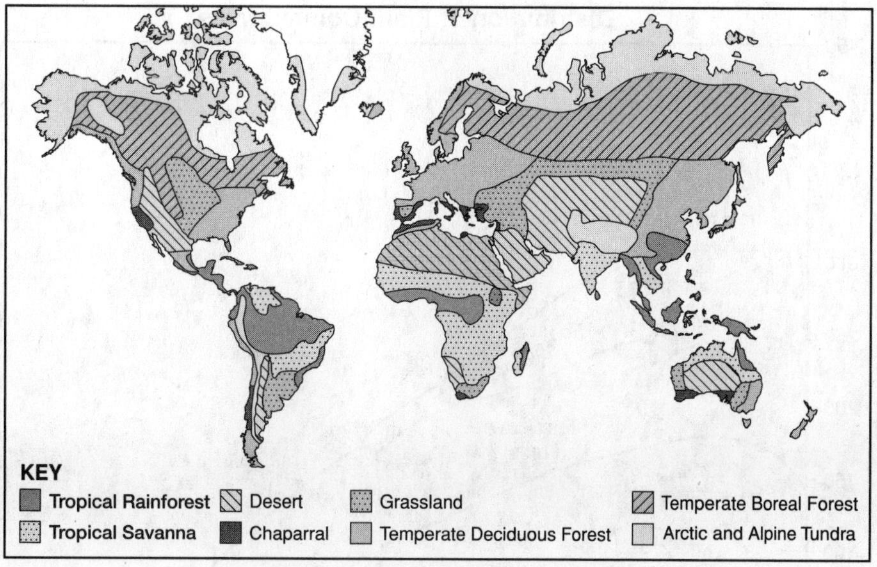

Figure 1.3 Major biomes of the world

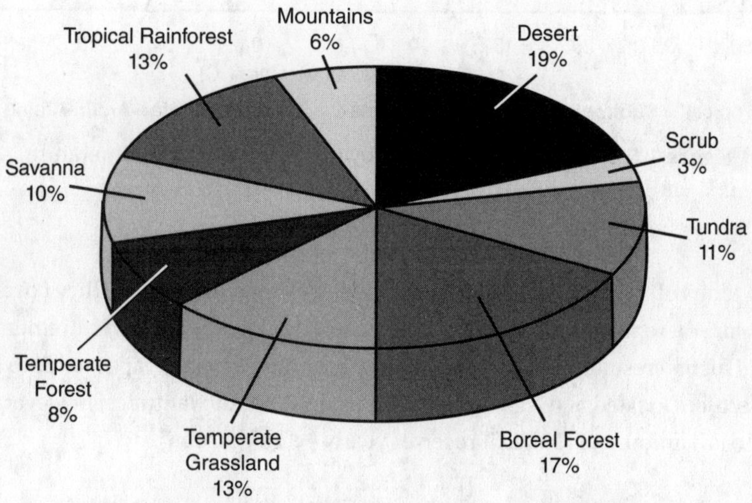

Figure 1.4 Percentage of total Earth land surface

Be aware of the following general information on terrestrial biomes:

- Many places on Earth share similar climatic conditions despite being located in different areas (e.g., Death Valley, California, and the Sahara Desert in northern Africa). As a result of natural selection, comparable ecosystems, including the adaptations of both plants and animals, have developed in these geographically distinct areas.
- Most terrestrial biomes are identified by the plant life primarily found within them.
- The geographical distribution (and productivity) of the various terrestrial biomes are controlled primarily by the average air temperature and the amount of rainfall the biome receives.

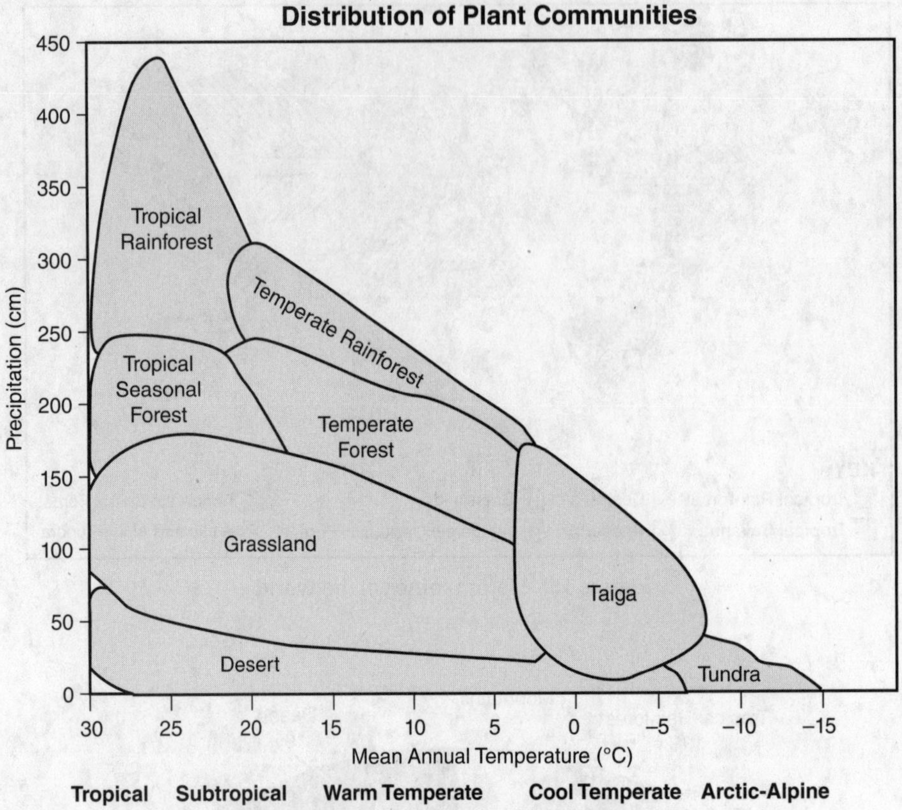

Figure 1.5 Distribution of biomes by precipitation and temperature

Deserts

Deserts are defined in terms of the amount of rainfall they receive, not temperature. They cover about 20% of Earth's surface and occur where rainfall is less than 20 inches (50 cm) per year. Daily extremes in temperature (very hot days with very cold nights) result from exceptionally low humidity as water vapor tends to block solar radiation (Figure 1.6). Most deserts are located between 15° and 35° north and south latitudes; however, the Arctic tundra is a cold desert due to the low amount of rainfall it receives yearly (< 10″ [25 cm]).

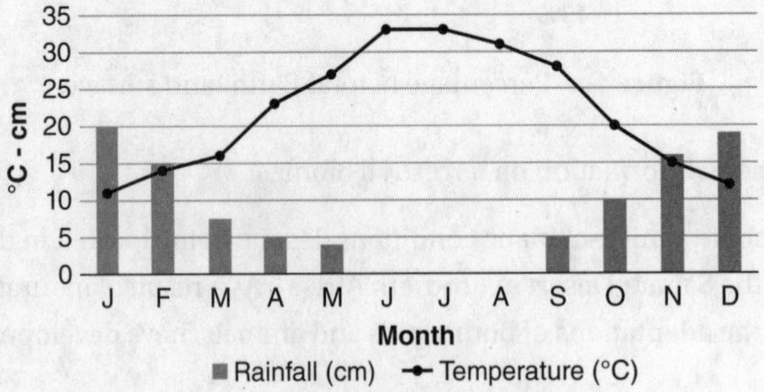

Figure 1.6 Desert climatograph

Desert soils often have abundant inorganic nutrients, but little to no organic matter. Desert plants are spaced apart due to limiting factors and primarily comprise plants that have fleshy leaves or stems that store water and are known as succulents (e.g., cacti).

Succulents have

- deep roots to tap groundwater;
- open stomata at night;
- shallow roots to collect and store water after short rainfalls;
- small surface areas exposed to sunlight;
- vertical orientation to minimize exposure to the sun; and
- waxy leaves to minimize transpiration.

Sharp spines on cacti create shade, reduce drying airflow, discourage herbivores, and reflect sunlight. Cacti may also secrete toxins into the soil to prevent interspecific completion (allelopathy).

Desert plants such as wildflowers

- are dependent on water for germination;
- have short life spans;
- perform their entire life cycle from seed to flower to seed within a single growing season (e.g., wildflowers); and
- store biomass in seeds.

Desert animals

- are generally small;
- are often nocturnal;
- have small surface areas; and
- spend time in underground burrows where it is cooler.

Aestivation (summer hibernation) is common, although some animals are able to metabolize dry seeds; e.g., kangaroo rats are metabolically able to produce their own water and secrete concentrated urine, and insects and reptiles have thick outer coverings to minimize water loss.

Natural disturbances are common in the form of occasional fires and sudden, infrequent, but intense rains that cause flooding.

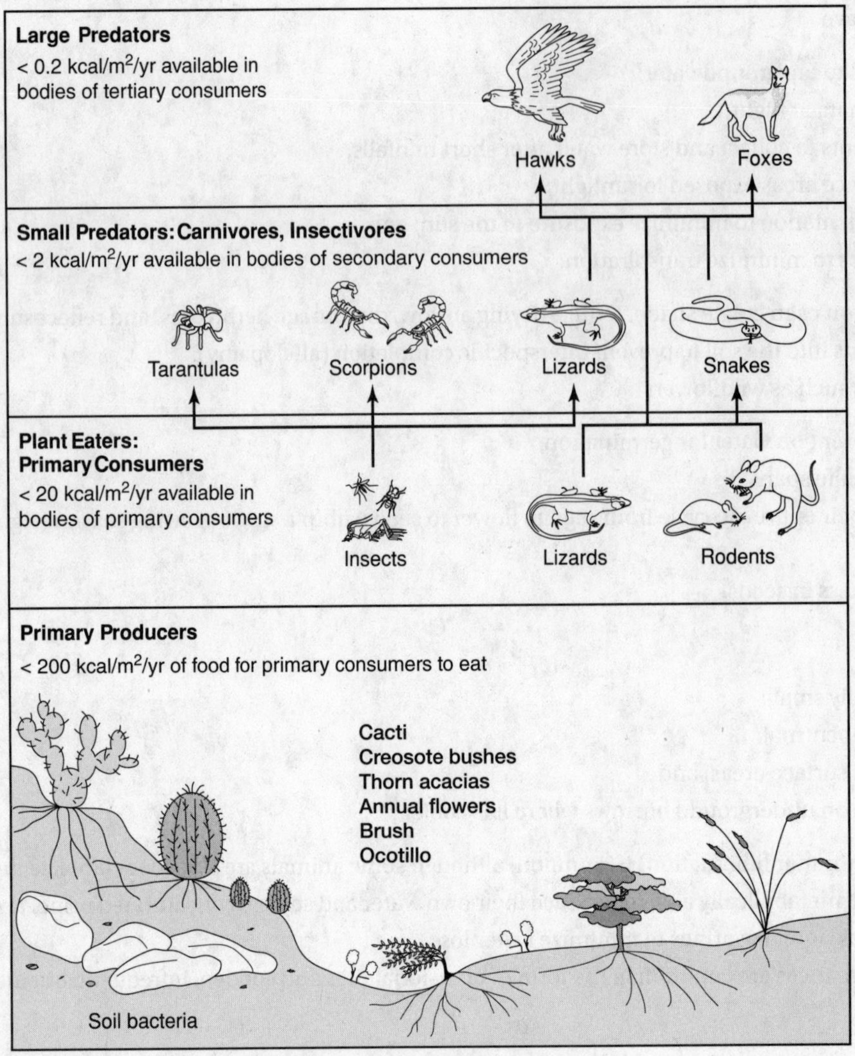

Figure 1.7 Desert food web

Major Environmental Threats

- Global warming is increasing the incidence of drought, which dries up water holes.
- Grazing livestock can destroy many desert plants and animals.
- Higher temperatures may produce an increasing number of wildfires that alter desert landscapes by eliminating slow-growing trees and shrubs and replacing them with fast-growing grasses.
- Irrigation that is used for agriculture may, in the long term, lead to salt levels in the soil that become too high to support plants.
- Nuclear waste may be dumped in deserts, which have also been used as nuclear testing grounds.
- Off-road recreational activities destroy habitats.
- Residential development.

TIP

According to the Second Law of Thermodynamics, as energy flows through systems, more of it becomes unusable at each step or transformation.

Solutions

- Dig artificial grooves in the ground to retain rainfall and trap windblown seeds.
- Find new ways to rotate crops to protect fragile soil.
- Only use off-road vehicles on designated trails and roadways.

- Plant leguminous plants, which extract nitrogen from the air and fix it in the ground, to restore soil fertility.
- Plant sand-fixing bushes and trees.
- Use existing water resources more efficiently and better control salinization.

Forests

Forests are the most dominant terrestrial ecosystem and cover about one-third of Earth's land surface, mostly in North America, the Russian Federation, and South America, and account for 75% of all gross primary productivity and plant biomass on Earth. Forests at different latitudes and elevations form distinctly different ecozones, such as boreal forests near the poles and tropical forests near the equator (Figures 1.8 and 1.9).

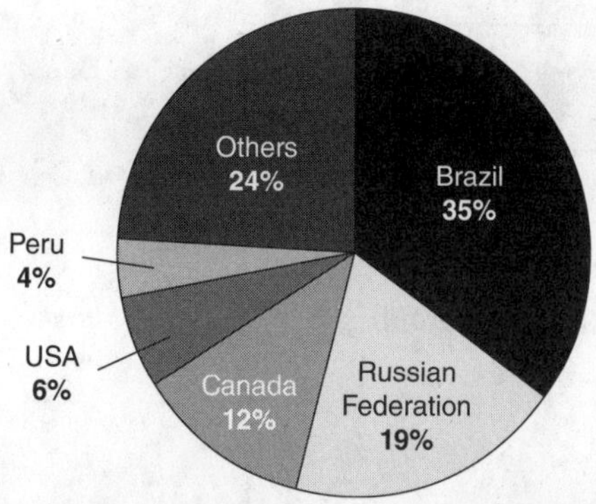

Figure 1.8 World forest distribution

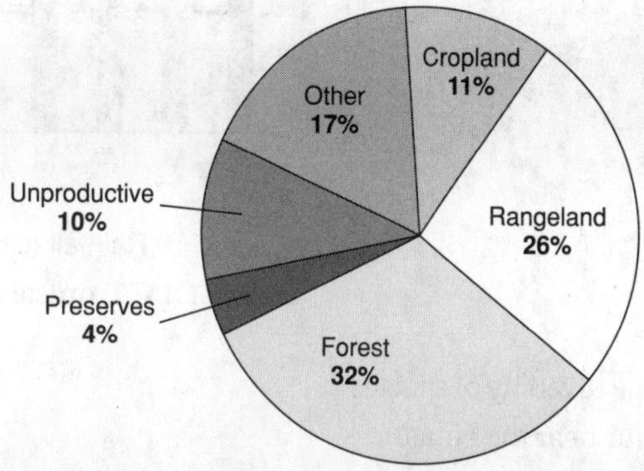

Figure 1.9 World land use

Forest Layers

There are two types of forest layers:

- **Closed canopy**—tree crowns (including branches, leaves, and reproductive structures extending from the trunk) cover more than 20% of the ground's surface. The majority (80%) of the forest biome is classified as closed canopy.
- **Open canopy**—tree crowns cover less than 20% of the ground surface.

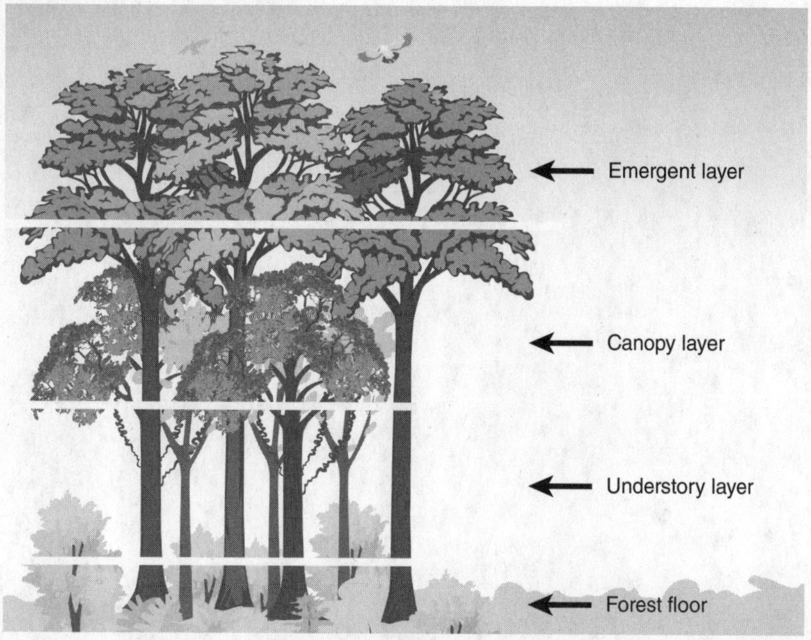

Figure 1.10 Forest layers

Tropical Rainforests

Characteristics

- Animals include numerous birds, bats, small mammals, and insects.
- Decomposition is rapid and soils are subject to heavy leaching.
- Distinct seasonality where winter is absent and only two seasons are present (rainy and dry). The length of daylight is 12 hours and varies little year-round (Figure 1.11).

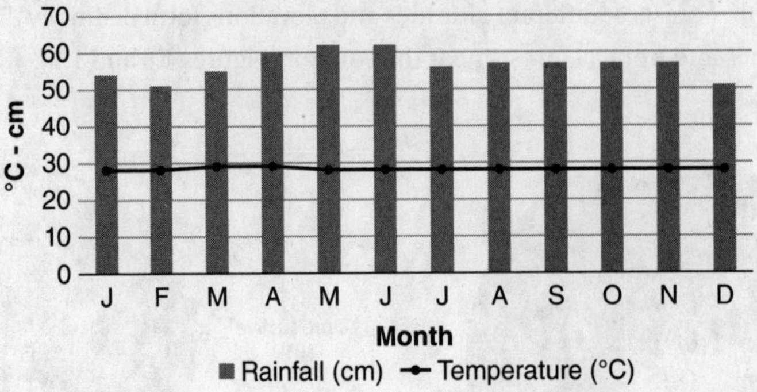

Figure 1.11 Tropical rainforest climatograph

- Large diversity of species.
- Occur near the equator.
- Plants are highly diverse; e.g., orchids, bromeliads, vines (lianas), ferns, mosses, and palms are present in tropical forests.
- Precipitation is evenly distributed throughout the year, with annual rainfall exceeding 80 inches (200 cm).
- Soil is nutrient-poor because competition is intense for nutrients, with most nutrients being quickly assimilated and stored in plant tissue.
- Temperature is warm to hot and varies little throughout the year.
- Tree canopy is multilayered and continuous, allowing little light penetration.
- Trees are tall, with buttressed trunks and shallow roots, and are mostly evergreen, with large, dark green leaves.

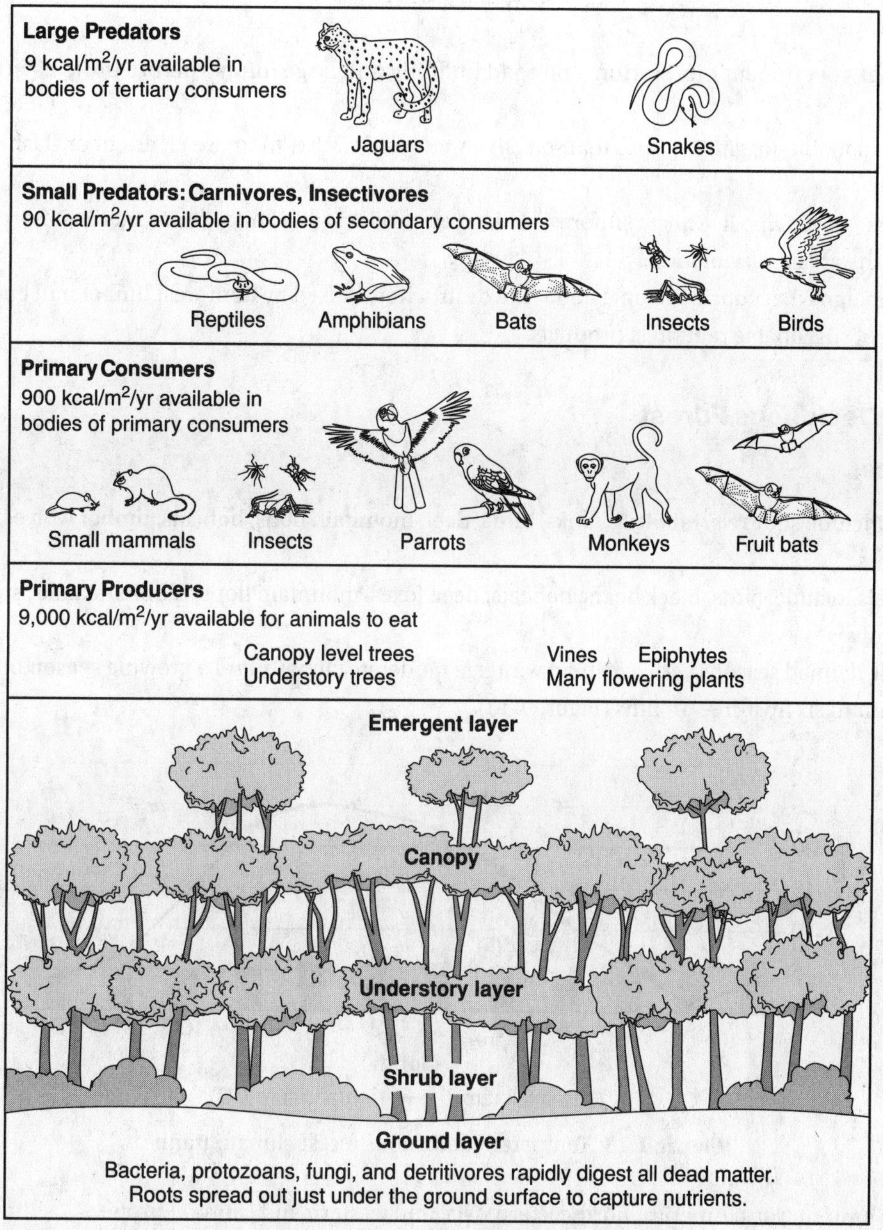

Figure 1.12 Tropical rainforest food web

Major Environmental Threats

- Acid rain damages leaves and causes trees and other plants to produce fewer seeds.
- Governments and industries clear-cut forests to make way for roads.
- Human intrusion into natural areas causes disruption to wildlife and flora (e.g., pollution, light, sound, chemicals, etc.).
- Hydroelectric projects flood acres of rainforests.
- Mining operations clear forests to build roads and dig mines.
- More than half of all tropical forests have already been destroyed.
- Non-native organisms are introduced that compete for food and nutrients with native wildlife and flora.
- Slash-and-burn techniques are used to clear the land for raising cattle and for cropland.
- Trees are cut for timber and pulp (used for manufacturing paper), and land is cleared for agriculture.
- Wildfires destroy millions of acres of forest worldwide every year.

Solutions

- Administer government moratoriums on road building and large infrastructure projects in the rainforest.
- Create sustainable-logging regimes that selectively cull trees rather than use clear-cut or slash-and-burn methods.
- Encourage people who live near rainforests to harvest products (e.g., nuts, fruits, medicines) rather than clear-cut the area for farmland.
- Start campaigns that educate people about the destruction caused by rainforest timber and encourage the purchase of sustainable rainforest products.

Temperate Deciduous Forest

Characteristics

- Animals include squirrels, rabbits, skunks, birds, deer, mountain lions, bobcats, timber wolves, foxes, and black bears.

  Animals include: birds, black bears, bobcats, deer, foxes, mountain lions, rabbits, skunks, squirrels, and timber wolves.
- Have well-defined seasons with a distinct winter, a moderate climate, and a growing season of 140–200 days during four to six frost-free months (Figure 1.13).

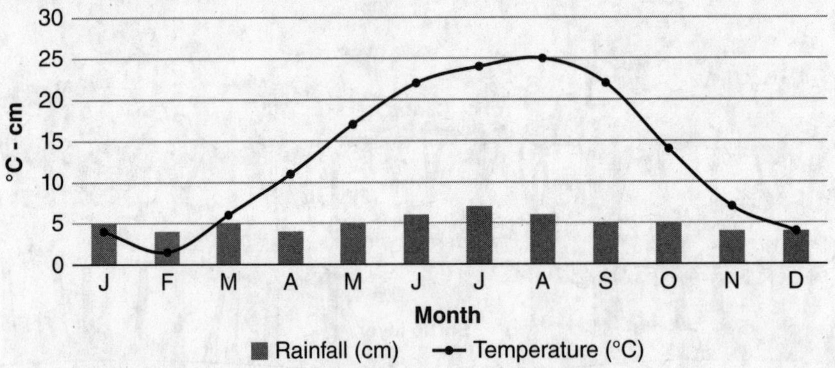

Figure 1.13 Temperate deciduous forest climatograph

- Occur in eastern North America, northeastern Asia, and western and central Europe.
- Only scattered remnants of original temperate forests remain due to development, land clearing, and timbering.
- Precipitation ranges from 30 to 60 inches (75–150 cm) and is distributed fairly evenly throughout the year.
- Soil is fertile and enriched with decaying litter made from leaves and other plant material.
- Temperature varies from −20°F to 85°F (−30°C to 30°C).
- Tree canopy is moderately dense and allows light to penetrate, resulting in well-developed and richly diversified understory vegetation and stratification of animals.
- Trees are distinguished by broad leaves that are lost annually (deciduous) and include such species as oak, hickory, beech, hemlock, maple, cottonwood, elm, willow, and spring-flowering herbs.

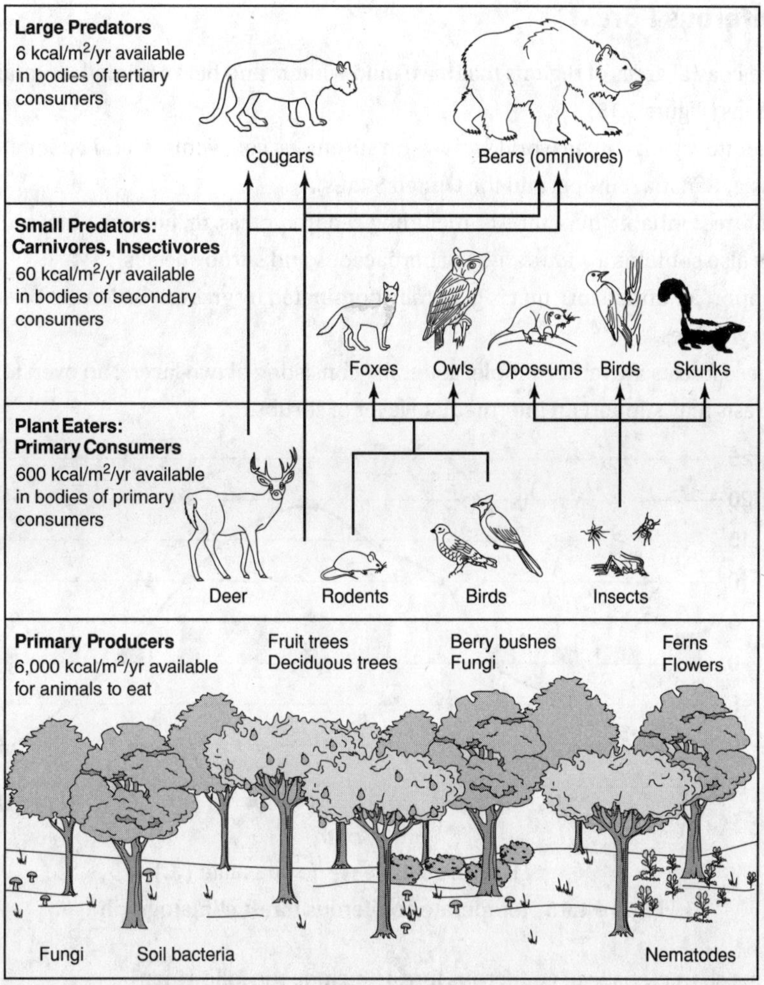

Figure 1.14 Temperate deciduous forest food web

Major Environmental Threats

- Acid rain caused by industrial and vehicular emissions poses the biggest threat to temperate deciduous forests. Over time, acid rain damages tree leaves, causes trees to produce fewer and smaller seeds, and reduces resistance to disease.
- Climate change
- Spread of invasive, non-native species that compete for space and food
- Strip-mining and required clear-cutting
- Unsustainable forestry practices

Solutions

- Increase opportunities for people to switch to renewable sources of energy.
- Initiate a "user tax" for homes and businesses that use nonrenewable energy sources.
- Lobby legislators to require areas where strip-mining is practiced to restore the land to its original condition at the conclusion of mining activities.
- Provide tax incentives for installing solar collectors on roofs.
- Take aggressive steps to remove invasive, non-native species when they are first discovered.

Temperate Coniferous Forest

- Common in the coastal areas of regions that have mild winters and heavy rainfall, or inland in drier climates or mountain areas (Figure 1.15).
- Found in temperate regions of the world with warm summers, cool winters, and adequate rainfall to sustain a forest (e.g., Asia, Canada, Europe, and the United States).
- Many species of trees inhabit these forests, including cedar, cypress, fir, juniper, pine, redwood, and spruce. The understory also contains a wide variety of herbaceous and shrub species.
- Pine forests support an understory that is generally dominated by grasses, which are often subject to ecologically important wildfires.
- Structurally, these forests are rather simple, generally consisting of two layers: an overstory and an understory. Some forests may support an intermediate layer of shrubs.

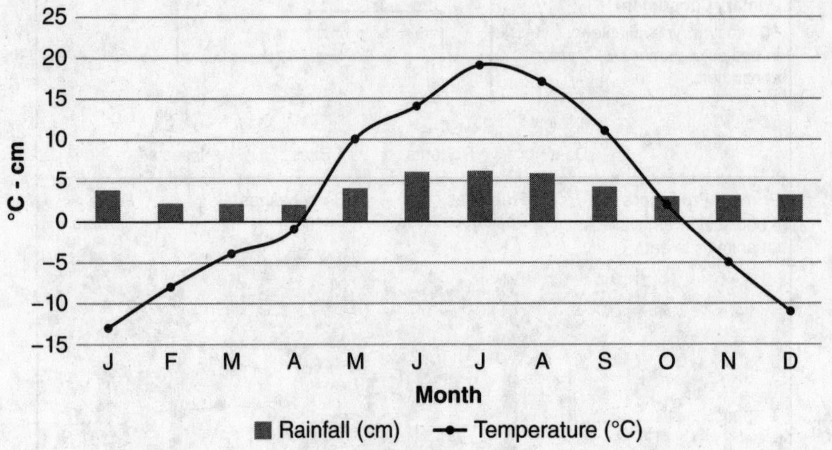

Figure 1.15 Temperate coniferous forest climatograph

Specific adaptations characteristic of coniferous forests include the following:

- Birds have layers of feathers, and many animals have thick fur to protect them from cold weather extremes (e.g., bears, beavers, foxes, and mink).
- Conical-shaped trees enhance shedding of snow and prevent loss of branches.
- Dark green color of the needles helps to absorb maximum light for photosynthesis.
- Many animals hibernate during the winter to conserve energy, as food resources are scarce, and build up fat during the warmer months when food is available.
- Needles are retained throughout the year and allow trees to start photosynthesis as soon as temperatures become favorable. Needles have thick, waxy coatings, waterproof cuticles, and sunken stomates. Needles also reduce the surface area compared to a leaf and minimize water loss due to transpiration.
- Some animals migrate to warmer climates during the winter months (e.g., caribou, Canadian geese, and elk).

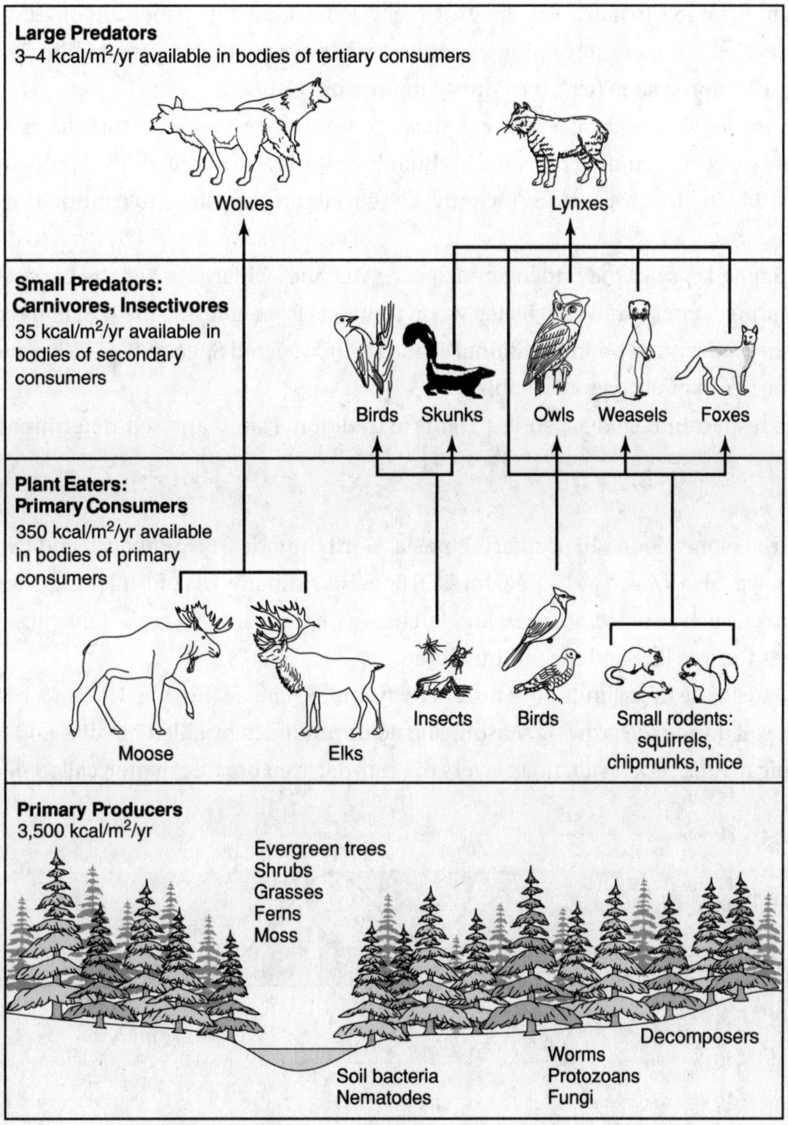

Figure 1.16 Temperate coniferous forest food web

Major Environmental Threats

- Acid rain, which weakens trees by damaging their leaves (needles) and limiting the nutrients available to them
- Clear-cutting driven by agricultural or timber interests
- Excessive degree of trapping and hunting, which creates an imbalance within the ecosystem, causing a negative impact on biodiversity
- Global warming and associated changes in precipitation patterns, causing climatic zones to shift toward the poles
- The construction of highways and the disruptions caused by the conversion of land for agricultural, industrial, and/or residential purposes that act as barriers and isolate populations of the same species from feeding grounds, migration routes, and breeding opportunities, thus limiting and reducing gene pool diversity

Solutions

- Companies can introduce "zero deforestation" policies that clean up their supply chains (e.g., holding their suppliers accountable for producing commodities like timber, beef, soy, palm oil, and paper in a way that has a minimal impact on climate).

- Companies can set targets to maximize the use of recycled wood, pulp, paper, and fiber in their products. For the non-recycled products they buy, companies should ensure that any virgin fiber used is certified by a third-party certification system (e.g., the Forest Stewardship Council).
- Developing countries with tropical forests can make commitments to protect their forests in exchange for the opportunity to receive funding for capacity-building efforts and national-level reductions in deforestation emissions, which provides a strong incentive for developing countries to continually improve their forest protection programs.
- In the United States, laws like the Endangered Species Act, the Wilderness Act, the Lacey Act, and the Roadless Rule help protect forests and stop illegal wood products from entering the U.S. marketplace. Other laws, such as the Convention on International Trade in Endangered Species (CITES), protect forests and the endangered species that rely on forest habitats.
- People need to respect indigenous peoples' rights to traditional lands and self-determination.

Taiga

- Largest terrestrial biome; found in northern Eurasia, North America, Scandinavia, and two-thirds of Siberia.
- The southern taiga (also known as "boreal forest") consists primarily of cold-tolerant evergreen conifers with needle-like leaves, such as pines, spruces, and larches, with the northern area of the taiga being more barren as it approaches the tree line and the tundra biome.
- The harsh climate in the taiga limits both productivity and resilience (Figures 1.17 and 1.18). Cold temperatures, very wet soil during the growing season, and acids produced by fallen needles and moss inhibit the full decay of organic matter. As a result, thick layers of semi-decayed organic matter, called peat, form.

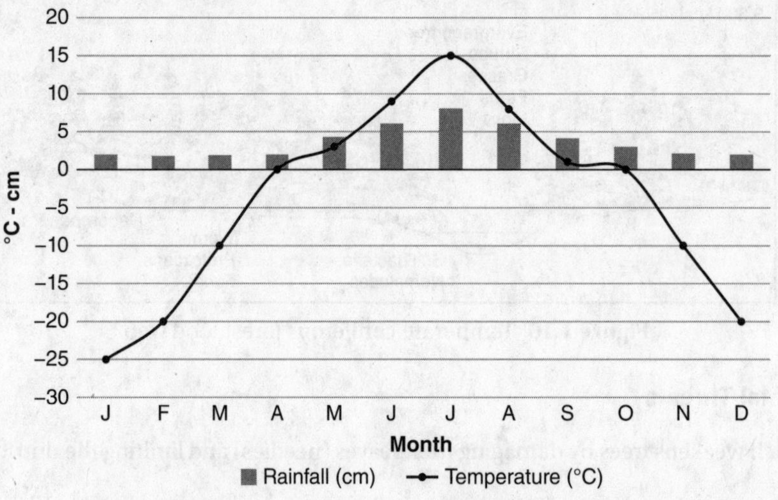

Figure 1.17 Southern taiga (boreal forest) climatograph

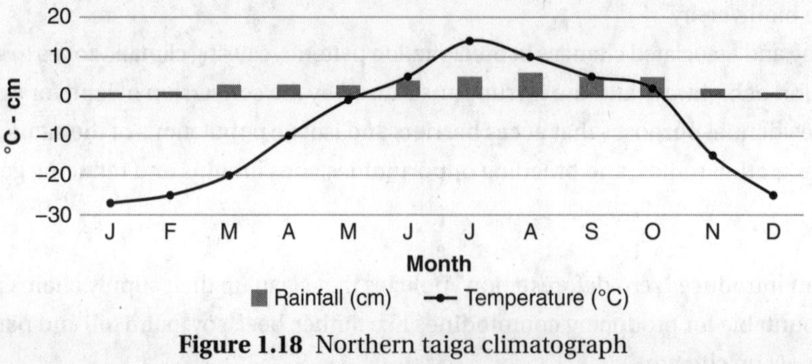

Figure 1.18 Northern taiga climatograph

- Seasons are divided into short, moist, moderately warm summers and long, dry, very cold winters.
- Soil is thin, nutrient poor, and acidic. In the boreal forests, the tree canopy permits low light penetration, and as a result the understory is limited.
- Animals consist of woodpeckers, hawks, moose, bears, weasels, lynxes, deer, hares, chipmunks, shrews, and bats (Figure 1.19).

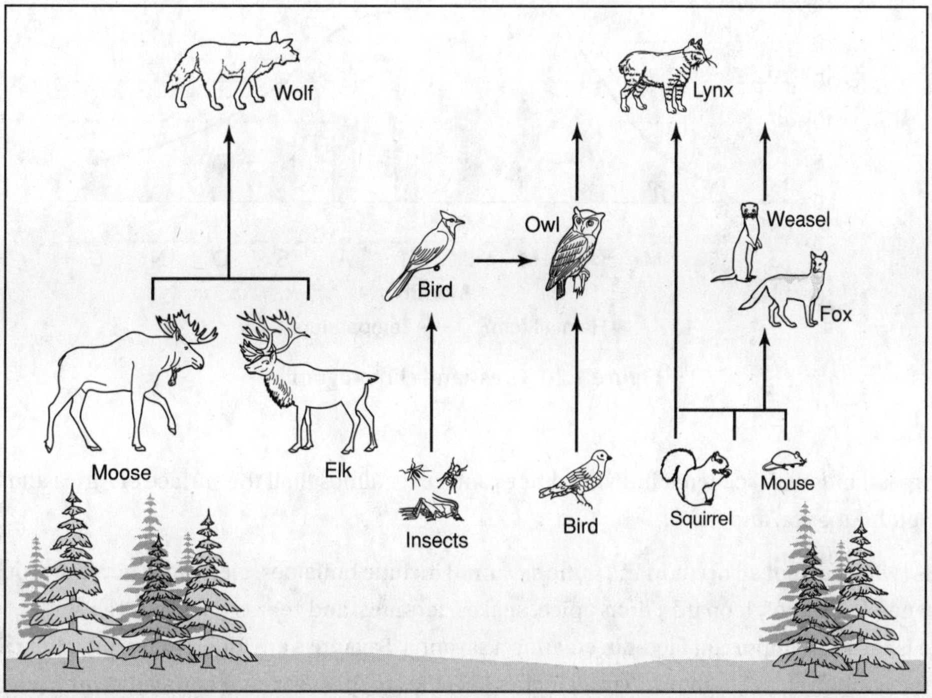

Figure 1.19 Southern taiga (boreal forest) food web

Major Environmental Threats

- Animals being hunted for fur
- An increase in hydroelectric power plants
- An increase in wildlife in some areas because of drier conditions as a result of global warming
- Gas and oil exploration
- Plantation forestry with the introduction of foreign tree species
- Road building
- Some areas are experiencing a decrease in wildfires due to large-scale clear-cutting, which reduces the amount of nutrients returning to the soil.
- Trees diseased by parasites

Solutions

- Adopt more sustainable forestry practices.
- Adopt safer and more sustainable methods to control the number of parasites (e.g., crop rotation, cover crops, soil enrichment, the use of natural pest predators, and bio-intensive integrated pest management).
- Increase alternative energy options to minimize global warming.
- Increase the number of protected areas and park reserves to restrict human influence.
- Limit road construction, mining activities, and the building of pipelines.
- Limit tourism and respect local cultures.

Grasslands

Grasslands are characterized as lands dominated by grasses rather than by large shrubs or trees. There are two main divisions of grasslands: (1) savannas or tropical grasslands; and (2) temperate grasslands.

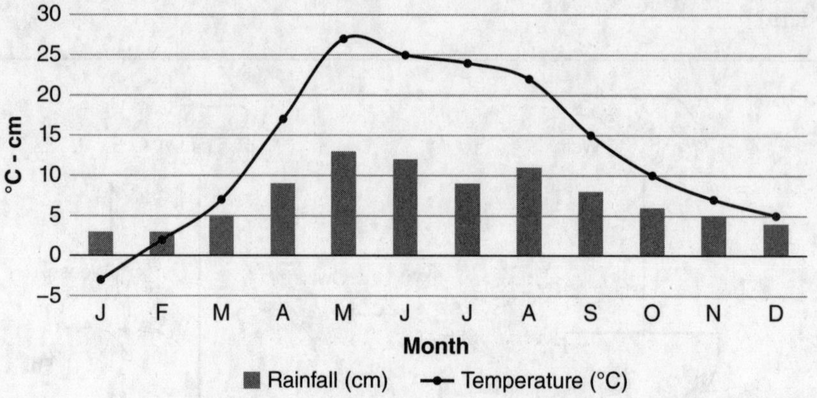

Figure 1.20 Grassland climatograph

Savannas

Savannas are grasslands with scattered individual trees and cover almost half the surface of Africa and large areas of Australia, South America, and India.

- Animals (which do not all occur in the same savanna) include buffaloes, elephants, giraffes, ground squirrels, hyenas, kangaroos, leopards, lions, mice, snakes, termites, and zebras (Figure 1.21).
- Climate is the most important factor in creating a savanna. Savannas are found in warm or hot climates where the annual rainfall is about 20 to 50 inches (50–130 cm) per year concentrated in six to eight months of the year, followed by a long period of drought when fires can occur.
- The soil of a savanna is porous, with rapid drainage of water and a thin layer of humus, which provides the vegetation with nutrients. The predominant vegetation consists of grasses and small broad-leafed plants that grow with grasses. Deciduous trees and shrubs are scattered across the open landscape. Savannas have both a dry and a rainy season, and seasonal fires play a vital role in the savanna's biodiversity.

Figure 1.21 Savanna food web

Temperate Grassland

Grasses are the dominant vegetation, while trees and large shrubs are absent (due to limited precipitation). Examples of temperate grasslands include the veldts of South Africa, the pampas of Argentina, the steppes (short grasses) of Russia, and the plains and prairies (tall grasses) of central North America.

- Animals (which do not all occur in the same temperate grassland) include gazelles, zebras, rhinoceroses, lions, wolves, prairie dogs, rabbits, deer, mice, coyotes, foxes, skunks, badgers, blackbirds, grouses, meadowlarks, quails, sparrows, hawks, owls, snakes, grasshoppers, and spiders (Figure 1.22).
- Climate is characterized by hot summers and cold winters, and rainfall is moderate. The amount of annual rainfall influences the height of grassland vegetation, with taller grasses in wetter regions. As in the savanna, seasonal drought and occasional fires are factors that affect biodiversity.
- Seasonal drought, occasional fires, and grazing by large mammals prevent woody shrubs and trees from invading and becoming established. However, a few trees, such as cottonwoods, oaks, and willows, grow in river valleys, with some species of flowers growing among the grasses.
- The soil is deep and dark, with fertile upper layers, and it is nutrient-rich from the growth and decay of deep, many-branched grass roots. The rotted roots hold the soil together and provide a food source for living plants.

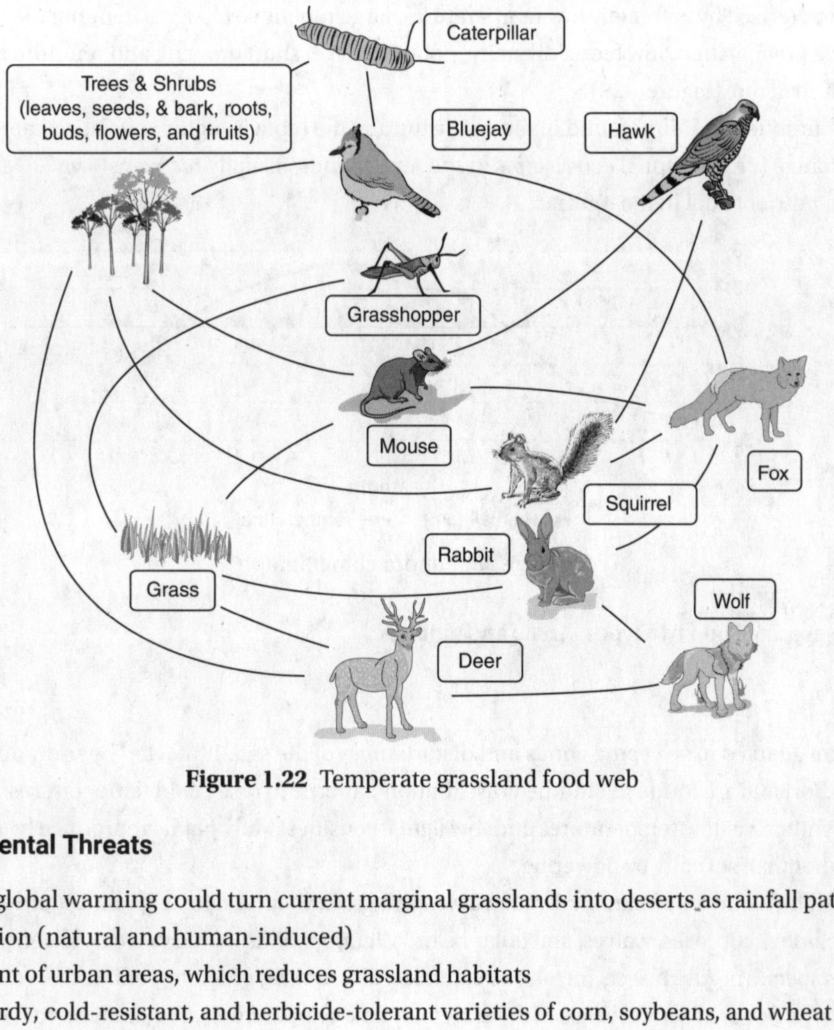

Figure 1.22 Temperate grassland food web

Major Environmental Threats

- Continued global warming could turn current marginal grasslands into deserts as rainfall patterns change.
- Desertification (natural and human-induced)
- Development of urban areas, which reduces grassland habitats
- Drought-hardy, cold-resistant, and herbicide-tolerant varieties of corn, soybeans, and wheat allow crops to expand into native grassland.
- Overgrazing
- Poaching of endangered species

- Soil compaction produced by overgrazing
- Soil salinization
- The conversion of natural grassland for agricultural purposes (e.g., plowing, which results in the wind blowing away topsoil)
- Where only one crop is grown, pests and disease can spread easily, creating the need for potentially toxic pesticides.

Environmental Solutions

- Conduct dry-season burning to obtain fresh growth and to restore calcium to the soil that had built up in the dry grasses.
- Continue education efforts on how to protect the soil and prevent soil erosion.
- Plant trees as windbreaks.
- Protect and restore wetlands, which are an important part of grassland ecology.
- Rotate agricultural crops to prevent the extraction of nutrients.

Tundra

The tundra is characterized by extremely low temperatures, large repetitive changes in population size, limited soil nutrients, little precipitation, low biotic diversity, poor drainage, short growing and reproductive seasons, and simple vegetation structure (Figure 1.23).

Because of the unique conditions found in the Arctic tundra, the biota is highly specialized and very sensitive to environmental change (i.e., disrupted ecosystems in the Arctic tundra usually recover slowly). Dead organic material functions as a nutrient pool in the tundra.

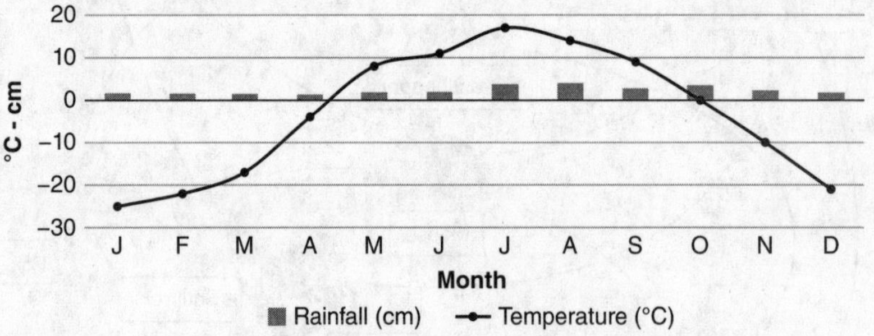

Figure 1.23 Tundra climatograph

The tundra is separated into two types: Arctic and alpine.

Arctic Tundra

- All plants are adapted to sweeping winds and disturbances of the soil. Protected by snow during the winter, plants are short and are found in clumped distribution patterns to resist cold temperatures. They can carry out photosynthesis at low temperatures and low light intensities. Most plants reproduce by budding and division rather than sexually by flowering.
- Animals include herbivorous mammals, such as lemmings, caribou, Arctic hares, and squirrels. Carnivorous animals include Arctic foxes, wolves, and polar bears. Migratory birds include ravens, falcons, terns, snowbirds, and various species of gull. Insects include mosquitoes, flies, moths, grasshoppers, and bees. Reptiles and amphibians are few or absent because of the extremely cold temperatures. Fish include cod, salmon, and trout.
- Food webs are simple and characterized by low biodiversity (Figure 1.24). Animals are adapted and highly specialized for long, cold winters and to breed and raise young quickly in the summer. Mammals and birds also have additional insulation from fat. Many animals hibernate or migrate south during the winter because food is not abundant.

- The Arctic tundra is located in the Northern Hemisphere, encircling the North Pole and extending south to the coniferous forests of the taiga, and is known for its cold, dry, desert-like conditions.
- The average winter temperature is about −30°F (−34°C), while summer temperatures range from 37°F to 54°F (3°C to 12°C), which enables this biome to sustain life. Yearly precipitation, including melting snow, is 6 to 10 inches (15 to 25 cm).
- The cold, dry conditions create slow growth rates and slow decomposition rates for both organic matter and pollutants. The very short growing season averages around 50 days per year.
- The thin, shallow, easily compacted, nutrient-poor soil forms slowly. A layer of permanently frozen subsoil called permafrost exists, consisting mostly of gravel and finer material. When water saturates the upper surface, bogs and ponds may form, providing moisture for plants that are able to resist the cold climate, such as low shrubs, mosses, grasses, approximately 400 varieties of flower, and lichen.

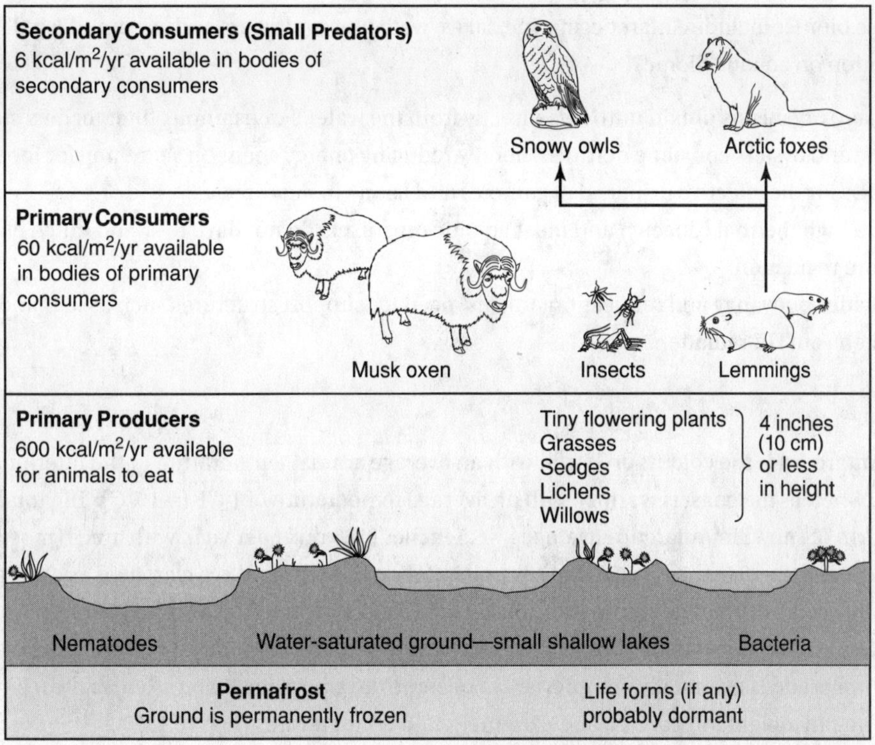

Figure 1.24 Tundra food web

Alpine Tundra

- Alpine tundra is located on mountains throughout the world at high altitude where trees cannot grow. The growing season is approximately 180 days, with nighttime temperatures usually falling below freezing.
- Unlike the Arctic tundra, the soil in the alpine tundra is well drained. Plants are very similar to those of the Arctic tundra and include grasses, dwarf trees, and small-leafed shrubs. Animals living in the alpine tundra include mountain goats, sheep, elk, birds, beetles, grasshoppers, and butterflies.

Major Environmental Threats

- Air pollution contaminates and kills lichen, a significant food source for many animals.
- Exploration of oil, gas, and minerals and the construction of pipelines and roads can cause physical disturbances of the permafrost and habitat fragmentation.
- Invasive species outcompete native vegetation and reduce the diversity of plant cover.
- Melting permafrost, as a result of global warming, is radically changing the biome and what species are able to live there.

- Oil spills can kill wildlife and significantly damage tundra ecosystems.
- Ozone depletion at the North and South Poles means stronger ultraviolet rays that harm wildlife; with some restrictions on ozone-depleting chemicals, this problem is diminishing to some extent.

Environmental Solutions

- Cutting harmful, planet-warming pollution by moving away from fossil fuels is key to safeguarding Earth's tundra habitats.
- Other measures include creating a refuge to protect certain species and regions while limiting or banning industrial activity.

1.3 Aquatic Biomes

The main aquatic biomes include Antarctic, marine, lakes, wetlands, and rivers and streams. Listed here is some general information on aquatic biomes:

- Many aquatic organisms obtain nutrients directly from the water. For example, filter feeders such as barnacles, clams, and oysters consume detritus, thereby reducing energy spent on searching for food.
- Water allows for the effective dispersal of gametes and larvae to new areas.
- Water has a high thermal capacity, and most aquatic organisms do not have to spend much energy on temperature regulation.
- Water provides buoyancy and reduces organisms' need for support structures such as legs and trunks.
- Water screens out UV radiation.

Antarctic

The climate of Antarctica is the coldest on Earth, with an average annual temperature in the interior of around −70°F (−57°C), whereas the coast is warmer with an average temperature of 14°F (−10°C). The total precipitation (mostly in the form of snow) in Antarctica averages ~6.5 inches (166 mm) per year, with areas (mostly in the interior) that receive less than 10 inches (~250 mm) or precipitation per year and are classified as deserts. Rainfall is rare and generally occurs during the summer in coastal areas and surrounding islands. The air in Antarctica is also very dry and, when combined with low temperatures, results in very low humidity. On most parts of the continent, the snow rarely melts and is eventually compressed to become the glacial ice that makes up the ice sheet. Winters have little light, no phytoplankton growth, and extremely cold temperatures (Figure 1.25).

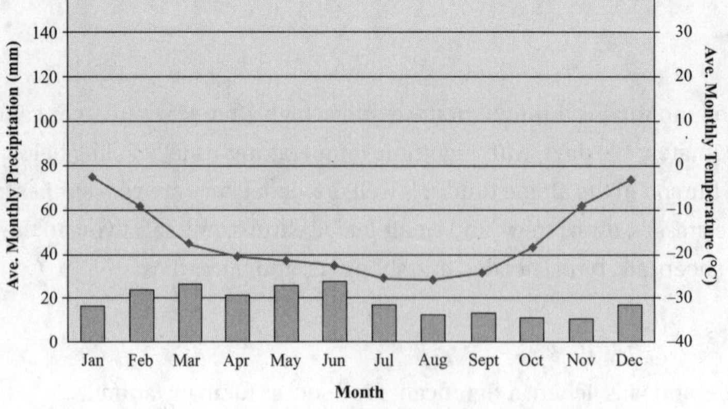

Figure 1.25 Antarctic climatograph

Though terrestrial biodiversity is low, Antarctic seas are extremely productive because phytoplankton grow abundantly during the extended daylight of summer. This massive primary-producer population supports large populations of krill. Krill (shrimp-like crustaceans) are a key food source in this ecosystem and serve as food for many predators (e.g., penguins, seals, and whales) (Figure 1.26).

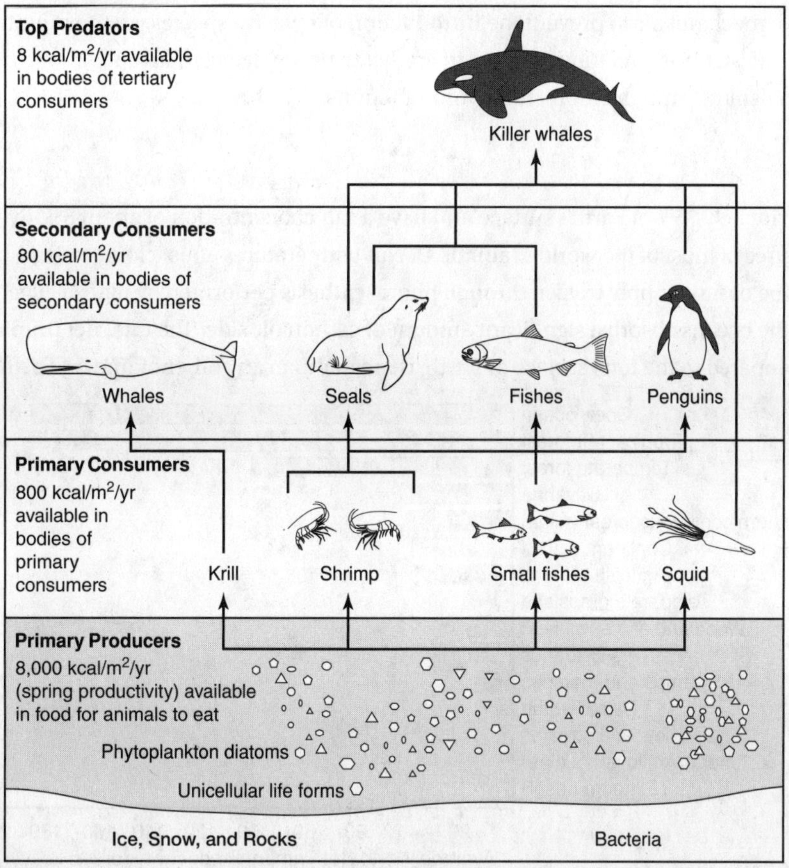

Figure 1.26 Antarctic food web

Major Environmental Threats

- Climate change and global warming, resulting in a warming of the sea and the loss of sea- and land-based ice
- Future exploration and exploitation of mineral reserves, oil, and gas. The current Antarctic Treaty bans all mining and mineral exploitation.
- Future fishing, both legal and illegal. Fishing for krill could be particularly significant as krill are at the bottom of many Antarctic food chains.
- Invasive species. Organisms that are not native to Antarctica are being transported there on ships. Some are being released in ship ballast water, and seeds and spores are becoming attached to boots, clothing, and equipment. Some of these invasive species are now able to survive in the Antarctic as a consequence of global warming.

  Rats, in particular, are a threat to Antarctica's ground-nesting birds. These birds are particularly vulnerable as there were never any ground-based predators prior to the introduction of these rats. Therefore, the birds have not developed any defense mechanism behaviors.
- Oceanic acidification (from extra dissolved carbon dioxide) and its long-term effects on the oceanic carbon cycle
- Pollution and CFCs (chlorofluorocarbons) are responsible for the ozone hole that has appeared over Antarctica for over 30 years.
- Tourism, with the pollutants that accompany ships and aircraft, and the effects of people and infrastructure on wildlife and the environment

Solutions

- Address the issue of worldwide climate change.
- Create effective mechanisms to prevent the introduction of invasive species, create monitoring systems for detecting new infestations, and move rapidly to eradicate newly detected invaders.
- Place limits on fishing through enforcement and sanctions.

Marine

Oceans cover approximately 75% of Earth's surface and have a salt concentration of about 3%. Evaporation of seawater is the primary source of most of the world's rainfall. Ocean temperatures affect cloud cover, surface temperature, and wind patterns. The oceans supply oxygen through photosynthesis performed by marine algae and photosynthetic bacteria, and the oceans absorb a significant amount of carbon dioxide. The total net primary productivity of the oceans, when compared to the total surface of Earth, is greater than any other biome on Earth (Figure 1.27).

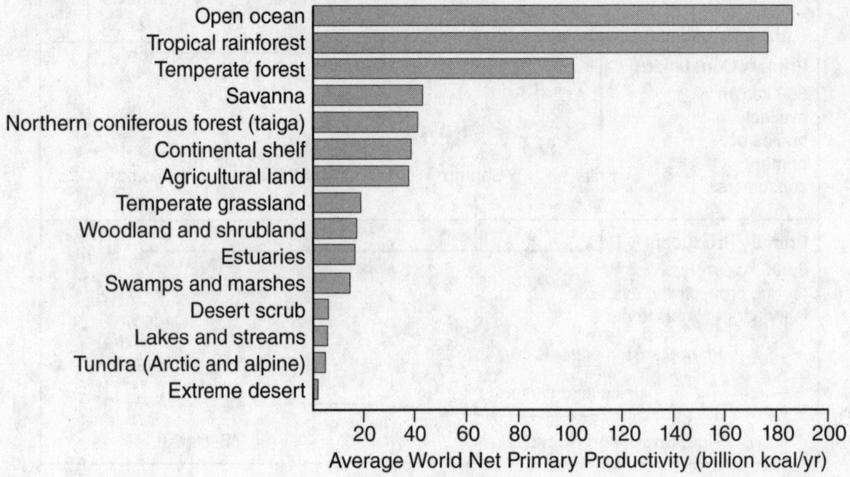

Figure 1.27 Net primary productivity (NPP) compared by total surface area on Earth

Ocean Circulation

The Northern Hemisphere is dominated by land, and the Southern Hemisphere is dominated by oceans. Air temperature differences between summer and winter are more extreme in the Northern Hemisphere because the land warms and cools more quickly than water does. Heat is transported from the equator to the poles mostly by atmospheric air currents but also by oceanic water currents. The warm waters near the surface and colder waters at deeper levels move by convection, which is the circular motion that happens when warmer air or liquid—which has faster-moving molecules, making it less dense—rises, while the cooler air or liquid sinks; i.e., changes in ocean temperatures have a direct bearing on ocean currents.

Surface ocean currents are driven by wind patterns that result from the flow of warmer, higher-pressure air generated at the tropics to colder, lower-pressure areas at the polar regions. Deep-water, density-driven currents are primarily controlled by differences in water temperature and density. Deeper ocean waters are colder and more dense than near-surface waters. In the Northern Hemisphere, north-flowing ocean currents originating near the equator are warm, while south-flowing ocean currents are colder.

The Great Ocean Conveyor Belt

There is constant ocean-water motion in the form of a "conveyor belt" driven by thermohaline currents, which are themselves driven by the density differences caused by temperature and water salinity, in contrast to most surface currents, which are driven primarily by winds (Figure 1.28).

Cold, salty water is dense and sinks to the bottom of the ocean, while warmer water is less dense and rises to the surface. Warm water from the Gulf Stream enters the Norwegian Sea and provides heat to the atmosphere in the northern latitudes. The loss of heat by the water in this area makes the water cooler and denser, causing it to sink. This cold bottom water then flows south to Antarctica, and eventually the cold bottom waters warm and rise to the surface in the Pacific and Indian Oceans.

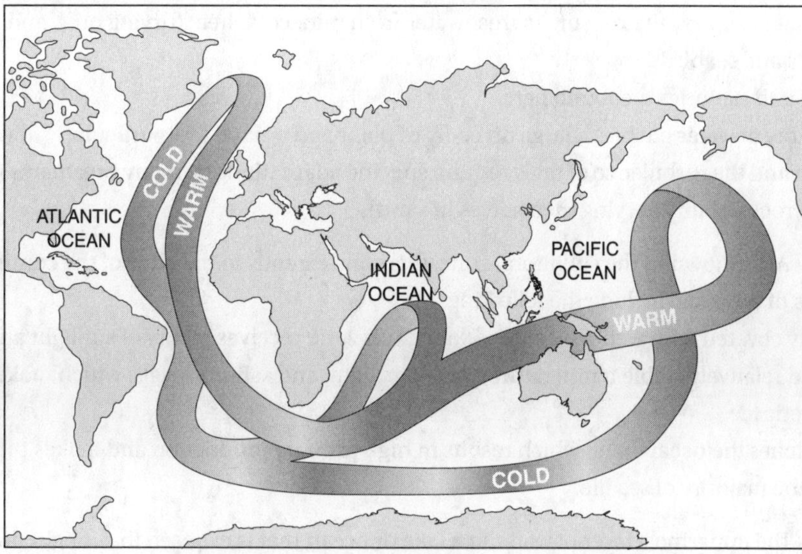

Figure 1.28 The Great Ocean Conveyor Belt

It takes around 1,600 years for seawater to make a round trip around the earth. The Great Ocean Conveyor Belt plays an important role in supplying heat to the polar regions and in regulating the amount of sea ice there.

For more information on increasing temperatures in the polar regions and the effects of warmer ocean temperatures on atmospheric CO_2 concentrations, refer to Unit 9.

Ocean Zones

Marine communities are distributed through several zones based upon the depth of the water, the degree of light penetration, and the distance from the shore (Figure 1.29).

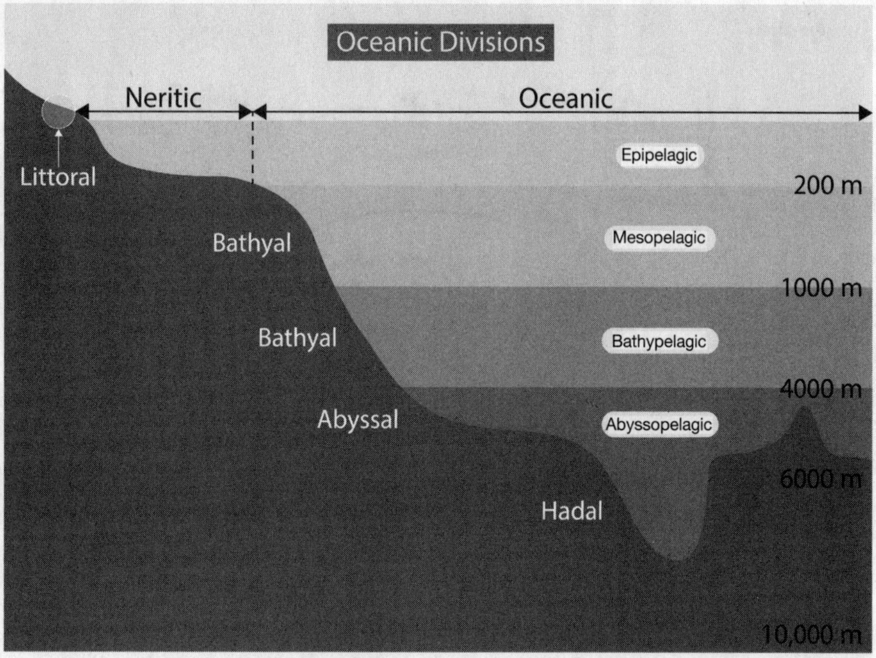

Figure 1.29 Ocean zones

The zones that you should be most familiar with are listed here:

1. **Littoral Zone**—The littoral (or intertidal) zone is the part of the ocean that is closest to the shore. Characteristics of this zone include the following:
   - Organisms must cope with exposure to freshwater from rain, cold, heat (drying out), and predation by land animals and seabirds.
   - Sand dunes and estuaries are found here.
   - The availability of water enables a large diversity of plant and animal life, particularly in wetlands.
   - Wave action and the turbulence of tides require specific adaptations by many organisms (e.g., barnacles attaching to rocks, clams burying themselves in sand).
2. **Neritic Zone**—Also known as the sublittoral zone, this zone extends to the edge of the continental shelf. Characteristics of this zone include the following:
   - Permanently covered with well-oxygenated water, this zone receives plenty of sunlight and has low water pressure and relatively stable temperature, pressure, light, and salinity levels, which makes it suitable for photosynthetic life.
   - Sunlight reaches the ocean floor, which results in high primary production and makes the neritic zone the location of the majority of sea life.
3. **Photic Zone**—The uppermost layer of water in a lake or ocean that is exposed to sunlight down to the depth where 1% of surface sunlight is available, and the layer just above the depth where the rate of carbon dioxide uptake by plants is equal to the rate of carbon dioxide production by animals. Characteristics of this zone include the following:
   - Approximately 90% of all aquatic marine life, which consists of copepods, phytoplankton (e.g., dinoflagellates and diatoms), zooplankton (e.g., drifters such as jellyfish and small crustaceans such as krill), and nekton or active swimmers (e.g., fish, seals, and whales), live in the photic (light) zone. The depth of the photic zone is dependent on how clear the water is (~50 feet [15 m] in murky water to ~700 feet [215 m] in clear water).
   - The photic zone allows for photosynthesis, which forms the base of most energy and food pyramids on Earth (Figure 1.30).

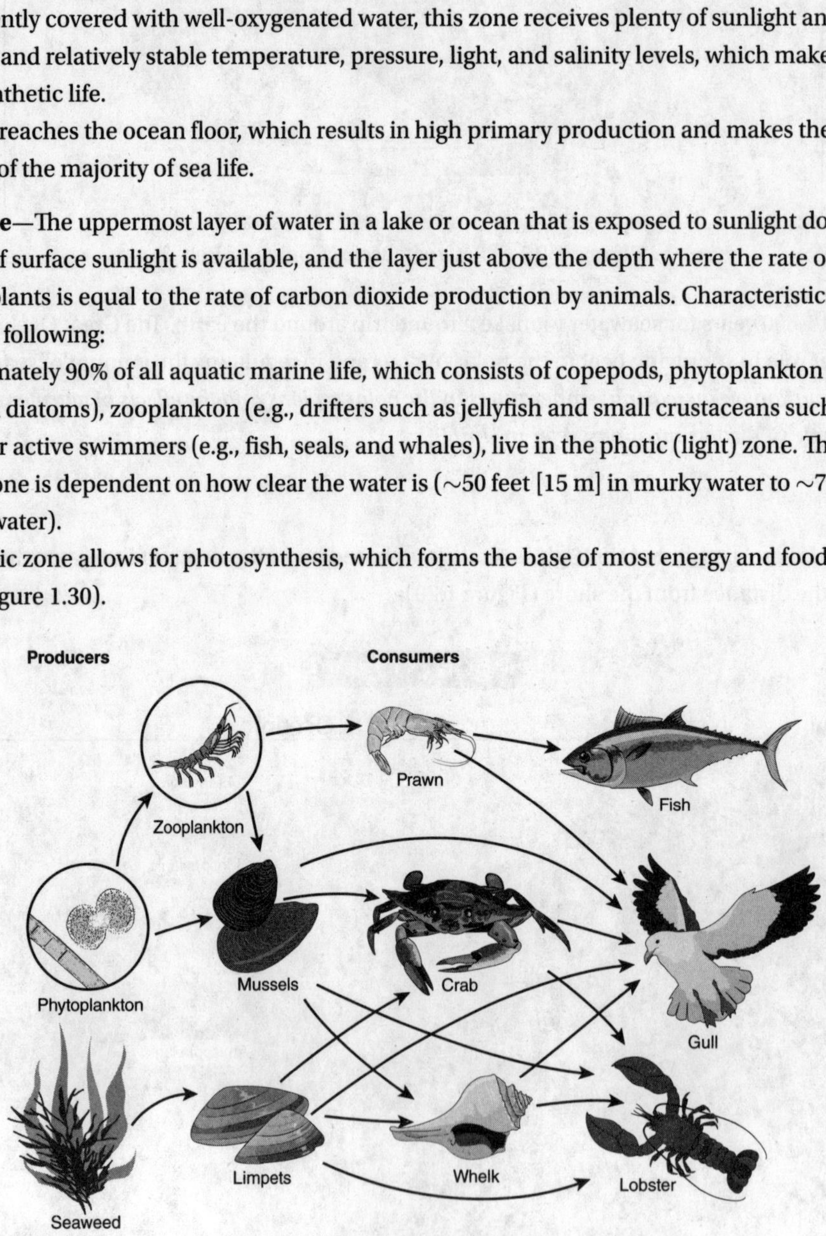

Figure 1.30 Marine food web

Corals

Corals are marine invertebrates that typically live in compact colonies of many identical individual polyps (small, sac-like animals with a set of tentacles surrounding a central mouth opening and an exoskeleton made of calcium carbonate at the base). The group of polyps includes the important reef builders that inhabit tropical oceans. Corals can reproduce asexually by budding or dividing. They can also breed sexually by releasing gametes.

Although some corals are able to catch small fish and plankton using stinging cells on their tentacles, most corals obtain the majority of their energy and nutrients from photosynthetic unicellular dinoflagellates, commonly known as zooxanthellae, that live within their tissues. Such corals require sunlight and grow in clear, shallow water, typically at depths less than 200 feet (60 m). Corals are major contributors to the physical structure of the coral reefs that develop in tropical and subtropical waters.

There are three major types of coral reef (Figure 1.31):

1. **Fringing reefs** grow near the coastline around islands and continents and are separated from the shore by narrow, shallow lagoons. Fringing reefs are the most common type of reef.
2. **Barrier reefs**, such as Australia's Great Barrier Reef, are also parallel to the coastline but are separated by deeper, wider lagoons. At their shallowest point, they can reach the water's surface, forming a "barrier" to navigation.
3. **Atolls** are rings of coral that create protected lagoons and are usually located in the middle of the sea. Atolls usually form when islands, often the tops of underwater volcanoes, surrounded by fringing reefs, sink into the sea, or the sea level rises around them. The fringing reefs continue to grow and eventually form circles with lagoons inside.

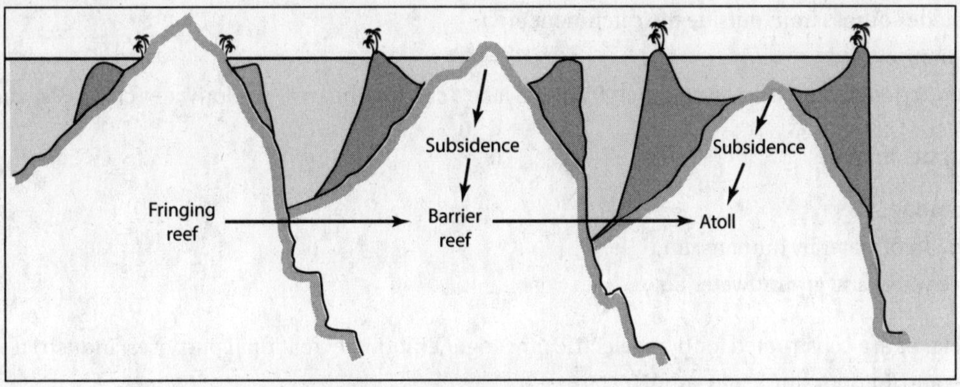

Figure 1.31 Types of coral reefs

Major Environmental Threats

- Coastal pollution from oil drilling and larger oil spills (e.g., *Exxon Valdez* and the *Deepwater Horizon*).
- Excessive ocean acidity due to increased levels of atmospheric CO_2 that is taken up by seawater. The higher levels of CO_2 increase the acidity of the ocean water (lowering the pH), resulting in the death (bleaching) of coral.
- Overfishing using cyanide and/or dynamite, pollution from sewage and agriculture (which lowers the oxygen content of seawater), outbreaks of predatory sea stars (starfish), invasive species, and sedimentation from poor land-use practices.
- Rising ocean temperatures, which result from global warming. Warmer than normal seawater breaks down the relationship between corals and symbiotic microalgae.

Solutions

- Ban or reduce non-biodegradable plastic.
- Enforce international treaties that ban the sale of products derived from endangered species (e.g., shark fins (soup), coral jewelry, and whale meat).
- Take steps to reduce the effects of climate change (e.g., rising ocean water temperatures, reduction of marine biodiversity, and coastal inundation of seawater).

Lakes

Lakes are large natural bodies of standing freshwater formed when precipitation, runoff, or groundwater seepage fills depressions in Earth's surface. Most lakes on Earth are located in the Northern Hemisphere at higher latitudes (between 30 and 60 degrees) and are generally found in mountainous areas, rift zones, areas with ongoing or recent glaciations, or along the courses of mature rivers.

Processes that form lakes include the following:

- Advance and retreat of glaciers that scrape depressions in Earth's surface where water can accumulate (e.g., Great Lakes)
- Crater lakes formed in volcanic craters and calderas (e.g., Crater Lake)
- Oxbow lakes formed by erosion in river valleys
- Salt or saline lakes that form where there is no natural outlet or where the water evaporates rapidly (e.g., Dead Sea, Great Salt Lake, and the Aral Sea)
- Tectonic uplift of a mountain range that creates a depression that accumulates water

Lake inputs include

- manmade sources from outside the catchment area;
- precipitation; and
- runoff carried by streams and channels from the lake's catchment area, groundwater channels, and aquifers.

Lake outputs include

- evaporation;
- extraction of water by humans; and
- surface water and groundwater flow.

Artificial lakes are constructed for hydroelectric power generation, recreational purposes, industrial and agricultural use, and/or domestic water supply.

The depth to which light can reach in lakes depends on turbidity, or the amount and type of suspended particles in the water. These particles can be either sedimentary (e.g., silt) or biological (e.g., algae or detritus) in origin.

The material at the bottom of a lake can be composed of a wide variety of inorganic materials, such as silt or sand, and/or organic materials, such as decaying plant or animal matter. The composition of the lake bed has a significant impact on the flora and fauna found near the lake, as it contributes to the amount and the types of nutrients available (Figure 1.32).

Because of the high specific heat capacity of water, lakes moderate the surrounding region's temperature and climate.

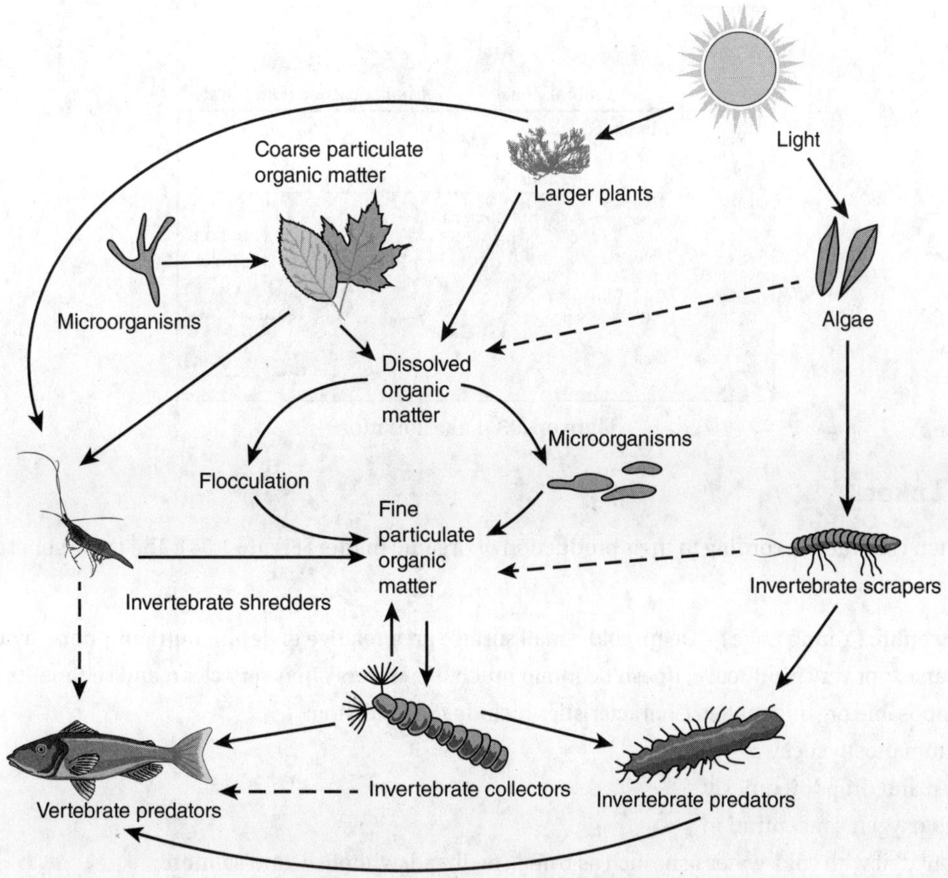

Figure 1.32 Typical lake food web
Source: U.S. Department of Agriculture

Productivity in aquatic environments is determined by temperature, depth, and nutrient and dissolved oxygen content. Oxygen can enter the water through photosynthesis and from mixing with air through wave action and determines the type of organisms found in a particular area.

Biological oxygen demand (BOD) is the amount of oxygen used by decomposers (e.g., bacteria and fungi) to break down a specific amount of organic matter. Larger amounts of organic matter increase the BOD and decrease the amount of oxygen available in the water.

Lake Zones

- **Benthic**—bottom of lake, organisms can tolerate cool temperatures and low oxygen levels (Figure 1.33)
- **Euphotic**—the uppermost layer of water in a water body, such as an ocean or lake, where light can penetrate, and photosynthesis occurs. There are enough nutrients for phytoplankton and other microscopic plants to grow.
- **Limnetic**—well-lit, open surface water, farther from shore, extends to depth penetrated by light, occupied by phytoplankton, zooplankton, and higher animals; produces food and oxygen that supports most of a lake's consumers
- **Littoral**—shallow, close to shore, extends to depth penetrated by light; rooted and floating plants flourish
- **Profundal**—deep, no-light regions, too dark for photosynthesis; low oxygen levels; inhabited by fish adapted to cool, dark waters

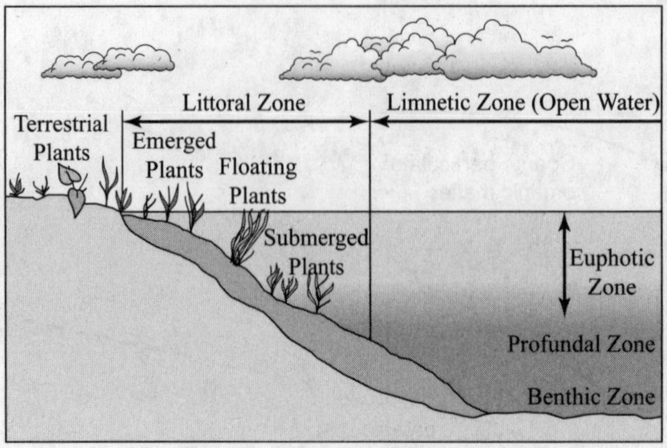

Figure 1.33 Lake zonation

Types of Lakes

Lakes are often classified according to their production of organic matter (Figure 1.34). The three general categories include the following:

1. **Oligotrophic (Young Lake)**—Deep, cold, small surface area relative to depth; nutrient-poor, phytoplankton are sparse; not very productive; doesn't contain much life; waters often very clear; and sediments are low in decomposable organic matter. Characteristics include the following:
   - Bottom mostly rocky
   - Deep and drops off quickly
   - High oxygen concentration
   - Populated with cold-water fish, such as trout, steelhead, whitefish, and salmon
   - Primarily conifer trees (pines) along the shore
   - Steep shorelines going down to the water's edge
   - Very few aquatic weeds
   - Water very clear
   - Example: Lake Superior
2. **Mesotrophic (Middle-Aged Lake)**—Moderate nutrient content and moderate amounts of phytoplankton; reasonably productive. Characteristics include the following:
   - A mixture of conifer and deciduous trees, such as oak, maple, and ash, along the shore
   - Bottom mostly sand, resulting from erosion and weathering of rocks
   - Few aquatic weeds and more plants growing on the shoreline
   - Less deep on average than oligotrophic lakes
   - Less steep shorelines
   - Supports both cold- and warmer-water fish, such as bass, perch, and bluegill
   - Water still very clear
   - Example: Lake Ontario
3. **Eutrophic (Old Lake)**—Shallow, warm, large surface area relative to depth; nutrient-rich, phytoplankton more plentiful and productive; waters often murky; high organic matter content in benthos (the community of organisms that live on, in, or near a seabed river, lake, or stream bottom, also known as "benthic zone"), which leads to high decomposition rates and potentially low oxygen. Eutrophication occurs over long periods of time as runoff brings in nutrients and silt. Pollution from fertilizers often causes algae populations to dramatically increase (algal bloom), causing a decrease in the oxygen content of the water, with detrimental consequences for life in the lake. Characteristics include the following:
   - Few, if any, cold-water fish; bass, pike, and carp thrive in old lakes
   - Gentle, mostly flat shorelines

- Heavy aquatic weed growth
- Mostly deciduous trees along the shore and an abundance of shoreline plants
- Usually quite shallow compared with sandy- or rocky-bottom lakes
- Very little oxygen in waters deeper than 30 feet
- Water
- Water murky from organic material and single-cell planktonic algae
- Example: Lake Erie

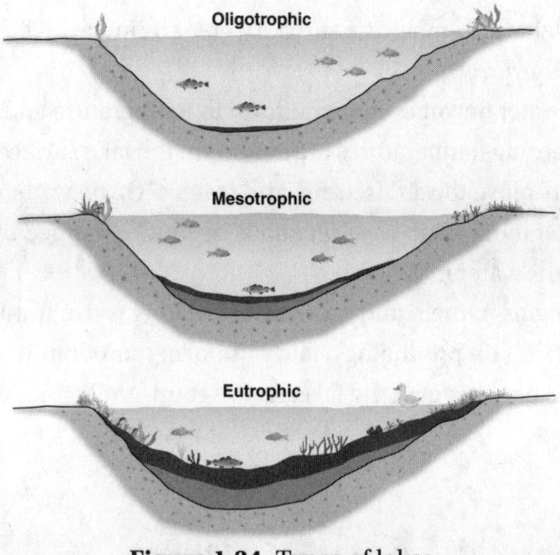

Figure 1.34 Types of lakes
Source: Vermont Agency of Natural Resources Department of Environmental Conservation

Lake Stratification

The stratification or layering of water in lakes is the result of density changes caused by shifts in temperature (Figure 1.35). The density of water increases as temperature decreases until it reaches its maximum density at about 39°F (4°C), causing thermal stratification—the tendency of deep lakes to form distinct layers in the summer months. Deep water is insulated from the sun and stays cool and denser, forming a lower layer called the hypolimnion. The surface and water near the shore are warmed by the sun, making them less dense, so that they form a surface layer called the epilimnion.

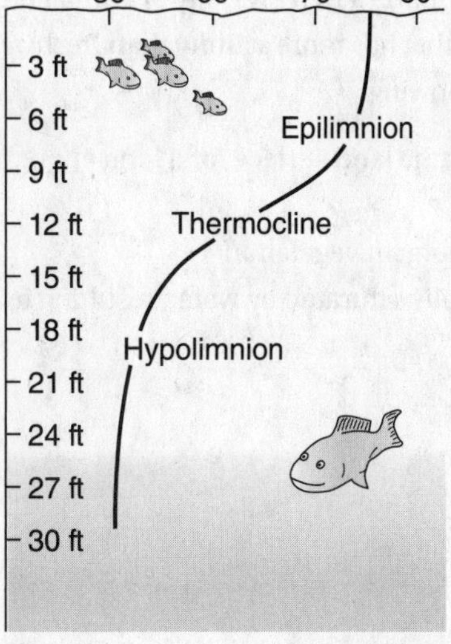

Figure 1.35 Thermal stratification

Seasonal Turnover

Seasonal turnover refers to the exchange of surface and bottom water in a lake or pond that happens twice a year (spring and fall) (Figure 1.36). During the summer, the sun heats water near the surface of lakes, which results in a well-defined warm layer of water occurring over a cooler one (stratification). As summer progresses, temperature differences increase between the layers, and a thin middle layer, or thermocline, develops, where a rapid transition in temperature occurs.

With the arrival of fall and cooler air temperatures, water at the surface of a lake begins to cool and becomes heavier. During this time, strong fall winds move the surface water around, which promotes mixing with deeper water—a condition known as *fall turnover*.

As the mixing continues, lake water becomes more uniform in temperature and oxygen level. As the winter approaches in areas where subfreezing temperatures are common, the lake surface temperatures approach the freezing mark. Thus, as lake waters move toward freezing and reach 4°C, the water sinks to the lake bottom (water is most dense at 4°C). Colder water remains above, potentially becoming capped by an ice layer, which further prevents the winds from stirring the water mass.

With spring, the surface ice begins to melt, and cold surface waters warm until they reach the temperature of the bottom waters, again producing a fairly uniform temperature distribution throughout the lake. When this occurs, winds blowing over the lake again set up a full circulation system known as *spring turnover*.

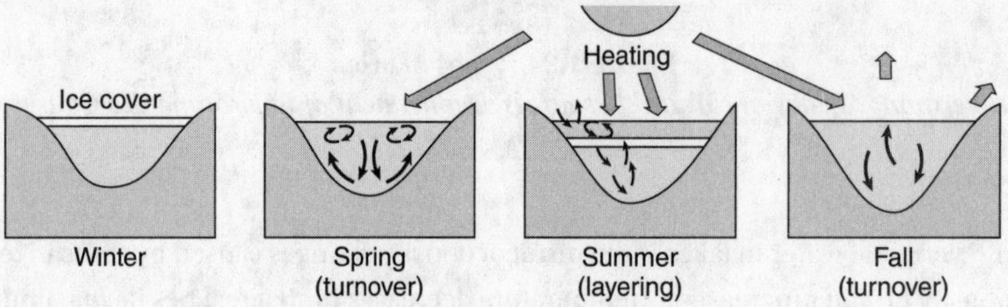

Figure 1.36 Seasonal lake turnover

Wetlands

Wetlands are areas that are covered with water at some point in the year and that support aquatic plants (Figure 1.37). High plant productivity supports a rich diversity of animal life. The water found in wetlands can be saltwater, freshwater, or brackish (water that has more salinity than freshwater but not as much as seawater).

Wetlands are characterized by the following:

- A water table that stands at or near the land surface for a long enough season each year to support aquatic plants
- Shallow or standing water with emergent vegetation
- Soil that is permanently or seasonally saturated by water, resulting in anaerobic conditions (also known as hydric soils)
- Vegetation that is water tolerant

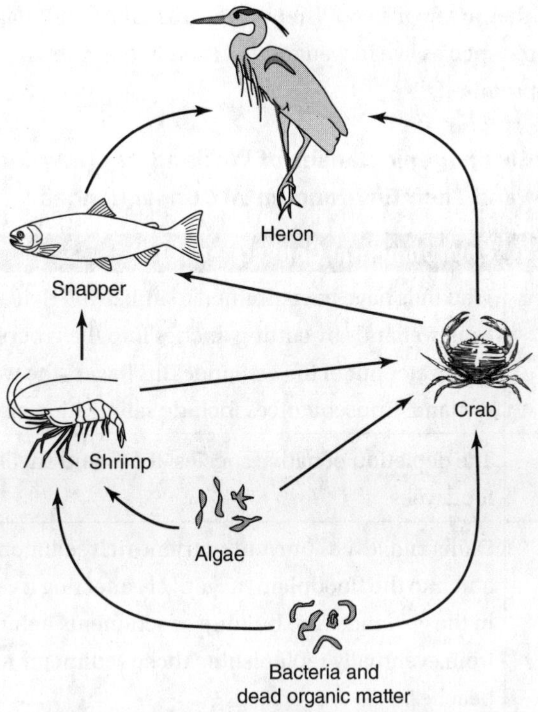

Figure 1.37 Wetland (estuarine) food web

Ecological Services of Wetlands

Wetlands provide many ecological benefits (ecological services), which include the following:

- Absorbing excess water from flooding or storm surges. The root systems of wetland plants stabilize soil at the water's edge and enhance soil accumulation at the shoreline. Wetland vegetation along shorelines dampens wave action and slows the speed of water currents, which provides erosion control and protects shorelines, thereby reducing the impact of flood damage to buildings, roads, infrastructure, and crops.
- Acting as carbon sinks. Wetlands can store large amounts of carbon dioxide or other forms of carbon to slow the atmospheric and marine accumulation of greenhouse gases, which are released by burning fossil fuels.
- As sediment flows through a wetland from the surrounding watershed, it becomes trapped, reducing the siltation into lakes, rivers, and streams. Ecological services can help with moving the siltation along. This reduces the amount of sedimentation buildup behind dams and in reservoirs and aids in coastline alteration.

  Ecological services can also help preserve important and/or sensitive aquatic habitats by reducing turbidity, which in turn helps to increase fishery resources, reduce changes in fish migration patterns, and reduce the loss of submerged vegetation.
- Providing areas for agriculture and timber (e.g., rice, various berries, and certain species of trees)
- Providing recreational areas
- Recharging groundwater, thereby reducing the need for water treatment facilities and replacing groundwater used for human and/or agricultural purposes
- Serving as nurseries for fish and shellfish

Wetland Degradation

Countries with the most wetlands are Canada (14% of land area), the Russian Federation, and Brazil. Wetlands were once about 10% of the land area of the United States but are currently about 5%, with Florida and Louisiana having the most wetlands. Ninety percent of the loss of wetlands is due to the conversion of land for agriculture, and 10% is the result of urbanization.

As important as wetlands are, they are fragile ecosystems that are being continually threatened by human activity (e.g., one-third of all endangered species live in wetlands). Listed in the following table are examples of how the world's wetlands have been compromised.

Anthropogenic Causes of Wetland Degradation and Their Environmental Consequences

Anthropogenic Causes	Environmental Consequence(s)
Agriculture	Wetlands have been drained to utilize the rich organic soil. Water is drained from wetlands by cutting ditches into the ground, which collect and transport water out of the wetland. This lowers the water table and dries out the wetland. Consequences include salinization and soil compaction.
Commercial fishing	The depletion of native species of fish and shellfish affects the wetland food webs.
Dams and levees	Dams and levees prevent nutrient-rich sediments from moving downstream and into the floodplain, negatively affecting a variety of food webs found in the wetlands. The buildup of sediments behind dams prevents them from eventually replenishing those sediments lost from barrier islands and beaches.
Development	Drained wetlands destroy natural habitats, which increases bank erosion and pollution. Dredging of streams lowers the surrounding water table and dries up adjacent wetlands. Water is diverted around wetlands, lowering the water table and increasing all forms of anthropogenic pollution (sediment, air, chemical, sewage, litter, etc.). Freshwater is depleted from wetlands for residential and commercial purposes.
Grazing	Soil compaction, removal of vegetation, and streambank destabilization are all consequences. Vegetation plays an important role in wetland ecology by removing water through evapotranspiration, altering water and soil chemistry, providing habitats for wildlife, and reducing erosion. Removal of vegetation can drastically and sometimes irreversibly alter wetland function.
Invasive species	Native species cannot always compete with introduced species. Common invasive species traits include fast growth, rapid reproduction, high dispersal ability, tolerance of a wide range of environmental conditions (i.e., ecological competence), ability to live off of a wide range of food types (i.e., generalist), association with humans, and prior successful invasions.
Logging	Logging decreases biodiversity in wetlands as natural habitats are destroyed, which may increase flooding.
Mining	Mine wastes (spoils) are often deposited in the floodplain.
Oil exploration and spills	Oil exploration and spills cause a disruption in wildlife both on land and in the sea, causing pollution and erosion as part of the drilling process. See Unit 6 for more on this topic.
Pumping groundwater	Pumping large quantities of water from springs lowers nearby groundwater and can result in the loss of wetland vegetation.

Anthropogenic Causes	Environmental Consequence(s)
Recreation	Boating, all-terrain vehicles, etc., disturb sediments, which affects breeding grounds for fish and other wildlife and also produces noise pollution, which affects wildlife behavior.
Roads and railroads	Roads and railroads narrow the floodplain, increase flooding, and create low-quality wetlands upslope of the roads by interrupting surface water and groundwater flows, which reduces sediment renewal, resulting in a loss of nutrients for native vegetation, which also affects higher trophic levels. Lack of sediment renewal also affects fish nurseries and bird breeding grounds. Dumping fill material buries hydric soil, which is permanently or seasonally saturated by water, resulting in anaerobic conditions, and lowers the water table so that water-loving plants (hydrophytic) cannot compete with upland plants.

Major Environmental Threats

- An invasion of exotic species can harm native animals and plants.
- Overexploitation and pollution threaten groundwater supplies.
- Runoff from agricultural and urban areas diminishes water quality.
- The creation of dams and water-diversion systems blocks migration routes for fish and wildlife and disrupts habitats.
- Water withdrawal for human use shrinks and degrades habitats.

Solutions

- Ban the use of inorganic fertilizers and pesticides in areas that drain into watersheds.
- Decrease the amount of impervious hardscaping—for example, asphalt roads, cement sidewalks, etc.—and substitute porous materials.
- Establish and maintain a vegetative buffer or greenbelt—a strip of upland surrounding the wetland that is maintained in a natural vegetated state.
- Establish programs to deal with invasive species once they are discovered.
- In areas where livestock grazing occurs in wetlands or where excessive human use is degrading wetlands, erect fencing to minimize erosion, water pollution, and sedimentation.
- Keep surface areas that wash into storm drains clean of pet feces, toxic chemicals, fertilizers, and motor oil, which eventually reach and impair wetlands.
- Minimize the amount of storm-water runoff that is generated beyond background levels while considering how and where to convey the storm water.
- Participate in local and government programs that help protect and restore wetlands.
- Plant only native species of trees, shrubs, and flowers to preserve the ecological balance of local wetlands.
- Reduce, reuse, and recycle household items and waste.
- Report illegal activity, such as filling, clearing, or dumping in wetlands, to government authorities, such as the U.S. Environmental Protection Agency or the Army Corps of Engineers.
- Use "living shoreline" techniques that make use of plant roots to stabilize soil.
- Use nontoxic products for household cleaning and lawn and garden care.
- Use phosphate-free laundry and dishwater detergents—phosphates encourage algae growth, which, when algae dies, depletes the oxygen levels in the water.

Rivers and Streams

The nutrient content of rivers and streams is largely determined by the terrain and vegetation of the area through which they flow and is also determined by adjacent and overhanging vegetation, the weathering of rocks in the area, and soil erosion.

Rivers and streams move continuously in a single downhill direction, and their inputs include

- groundwater recharge
- precipitation
- springs
- surface runoff;
- the release of stored water in ice and snowpack

The downward flow of surface water and groundwater from higher elevations to the sea can be separated into three zones (Figure 1.38):

- **Floodplain Zone**—As a result of large amounts of sediment and nutrients, the water is murky and warmer. Tributaries join to form rivers, which empty into oceans at estuaries.
- **Source Zone**—Contains headwaters or headwater streams and often begins as springs or snowmelt of cold, clear water with little sediment and relatively few nutrients. The rocky channels are usually narrow, creating swift currents. The water has relatively high oxygen levels and may include freshwater species such as trout.
- **Transition Zone**—Contains slower, warmer, wider, and lower-elevation moving streams, which eventually join to form tributaries. The water is less clear as it contains more sediment and nutrients, with the substrate beginning to accumulate silt. Species diversity is usually greater than in the source zone.

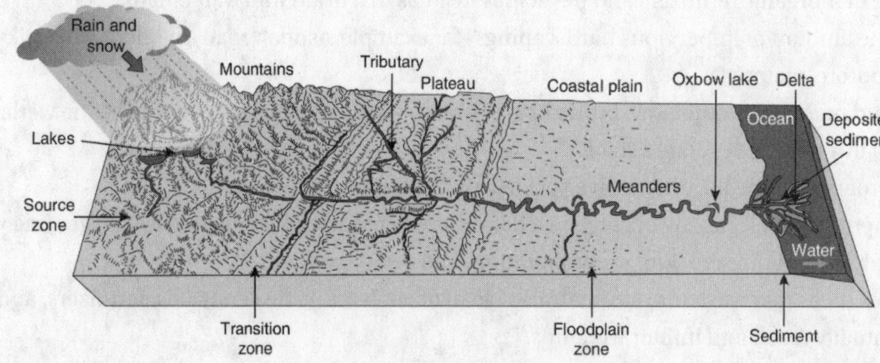

Figure 1.38 River zones

Riparian Areas

- Animals found in riparian zones include beavers, opossums, river otters, Canadian geese, cranes, ducks, frogs, salamanders, toads, and a wide variety of snakes and fish.
- Riparian areas are lands adjacent to creeks, lakes, rivers, and streams that support vegetation dependent upon free water in the soil (Figure 1.39).
- Vegetation consists of hydrophilic (water-loving) plants and trees (e.g., ferns, rushes, sedges, cottonwoods, and willows).

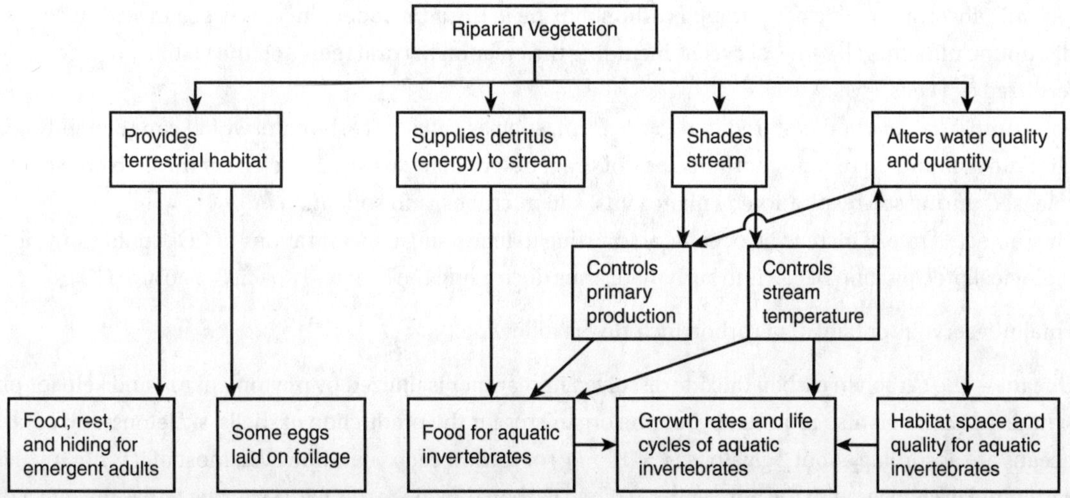

Figure 1.39 Riparian ecological services

1.4 The Carbon Cycle

Carbon is exchanged among the biosphere, geosphere, hydrosphere, and atmosphere and is the basic building block of life and the fundamental element found in carbohydrates, fats, proteins, and nucleic acids (DNA and RNA). Although carbon is found in rocks, it is a minor component when compared with the mass of either oxygen or silicon atoms in rocks. Carbon is also found in carbon dioxide (CO_2), which makes up less than 1% of the atmosphere (roughly 420 parts per million [ppm]) (Figure 1.40).

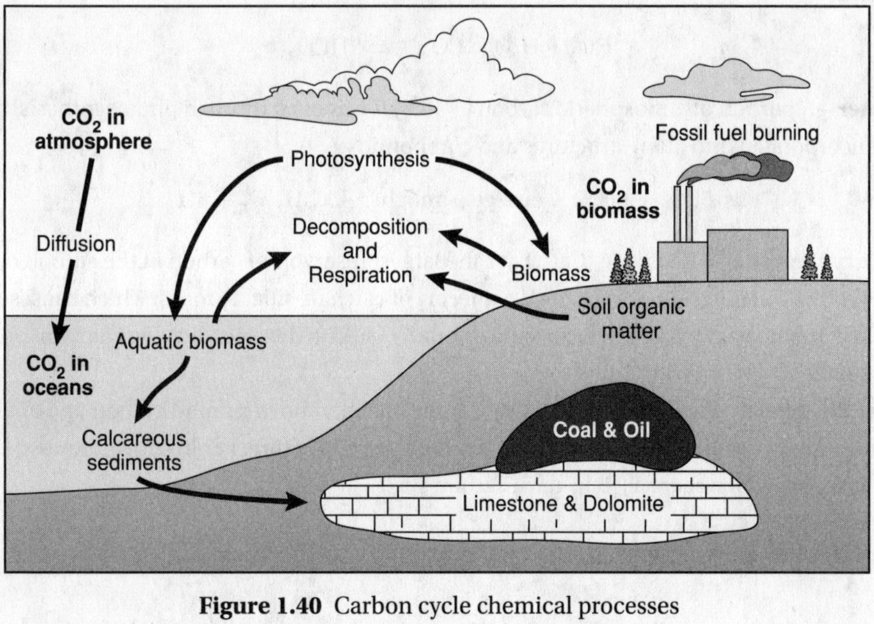

Figure 1.40 Carbon cycle chemical processes

Important facts to remember about the carbon cycle include the following:

- Carbon can be precipitated into the ocean's deeper, more carbon-rich layers as dead soft tissue or in shells as calcium carbonate ($CaCO_3$).
- Carbon enters the ocean primarily through the dissolution (dissolving) of atmospheric carbon dioxide (CO_2).

- Oceanic absorption of carbon through carbon dioxide in the atmosphere increases ocean acidity (lower pH), disrupting numerous biological cycles, including the creation of coral reefs and the viability of externally fertilized egg cells.
- One-third of soil carbon is stored in organic form, such as calcium carbonate, which can (depending on climatic conditions) remain for thousands of years before being washed into rivers through erosion or released during soil respiration by plant roots, soil microbes, and soil fauna.
- The projected rate of increasing oceanic acidity due to increasing concentrations of CO_2 could slow the biological precipitation of calcium carbonate, thus decreasing the ocean's capacity to absorb CO_2.

The major reservoirs or "sinks" of carbon include the following:

- **Oceans**—The carbon in carbon dioxide dissolved in seawater is utilized by phytoplankton and kelp for photosynthesis. Carbon is also required by marine organisms for the production of shells, skeletons, and coral. The oceans are absorbing about 2 gigatons (4×10^{12} kg) of carbon each year; however, most of it is not involved with rapid exchanges with the atmosphere. When carbon dioxide mixes with seawater, it has the effect of reducing the availability of carbonate (CO_3^{2-}) ions, which many organisms, such as corals, marine plankton, and shellfish, need to build their shells:

$$CO_2 + H_2O \rightarrow H_2CO_3 \rightarrow H^+ + HCO_3^-$$

  The increase in the hydrogen ion (H^+) concentration decreases the pH of seawater, making it more acidic. Simultaneously, carbonate ions (CO_3^{2-}) are consumed by the hydrogen ions to form even more bicarbonate ions (HCO_3^-). The net effect is that one unit of carbonate ion is consumed for each unit of carbon dioxide added to seawater. Because the forward and reverse reactions run simultaneously, both the pH and the availability of carbonate are reduced as the atmospheric concentration of carbon dioxide rises:

$$CO_2 + H_2O + CO_3^{2-} \rightleftarrows 2HCO_3^-$$

- **Plant Matter**—A portion of atmospheric carbon (~15%) is removed through photosynthesis, by which carbon is incorporated into plant structures and compounds:

$$6CO_2 + 6H_2O + \text{energy (sunlight)} \rightarrow C_6H_{12}O_6 + 6O_2$$

- **Sedimentary Deposits**—Limestone ($CaCO_3$) is the largest reservoir of carbon in the carbon cycle (Figure 1.41). The calcium comes from the weathering of calcium-silicate rocks, which causes the silicon in the rocks to combine with oxygen to form sand or quartz (silicon dioxide), leaving calcium ions available to form limestone.
- **Terrestrial Biosphere**—Forests store about 90% of the planet's above-ground carbon and about 75% of the planet's soil carbon. Carbon can be stored for very long periods of time in old-growth forests, limestone ($CaCO_3$), and peat, which serve as long-term carbon sinks.

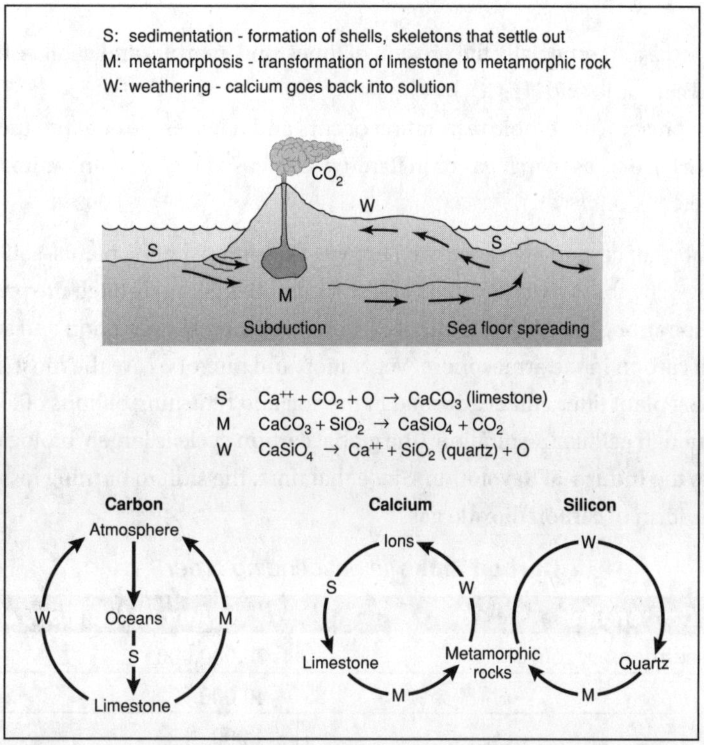

Figure 1.41 Inorganic carbon cycle

Carbon is released back into the atmosphere through the following:

- Burning fossil fuels (e.g., wood, coal, etc.)
- Cellular respiration of plants and animals that break down glucose into carbon dioxide and water:

$$C_6H_{12}O_6 + 6O_2 \rightarrow 6CO_2 + 6H_2O + \text{energy}$$

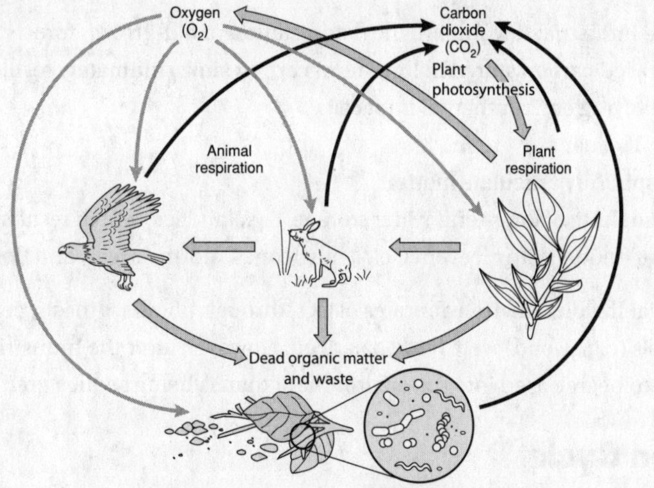

Figure 1.42 Carbon ecosystem cycle

- Decay of organic material by the action of decomposers; if oxygen is present, the carbon is released in the form of carbon dioxide; if oxygen is absent, it is released in the form of methane (CH_4)
- Incineration of wastes, which releases ash and soot (carbon) along with carbon dioxide and carbon monoxide
- Release of carbon dioxide by warmer ocean waters
- Strip-mining and deep plowing, which disturb the soil, releasing CO_2

- Volcanic eruptions
- Weatherization of rocks and especially the erosion of limestone, marble, and chalk, which break down to carbon dioxide and carbonic acid (H_2CO_3)
- When oxygen is not present, anaerobic respiration occurs and releases carbon into the atmosphere in the form of methane (CH_4); e.g., as marsh gas or in flatulence released by ruminant animals; e.g., cattle, dairy cows, deer, sheep, etc.

Every year, billions of tons of carbon are exchanged between the atmosphere, plants, soils, and the ocean. These exchanges are largely driven by the activity of plants and soil microbes going through the processes of photosynthesis and cellular respiration. The total amount of carbon contained in vegetation and soil sinks is more than three times the total carbon in the atmosphere. Vegetation and microbes are the most important biological components, decomposing plant litter and dead wood in the soil and returning billions of tons of carbon to the atmosphere each year through cellular respiration. The global carbon cycle is largely biological and was in a relatively steady state prior to the Industrial Revolution. Since that time, the shift to burning fossil fuels has released the trapped carbon in the form of carbon dioxide gas.

Carbon Sinks *(in descending order)*

Carbon Sink	Amount (Billions of Metric Tons)
Marine sediments and sedimentary rocks	~75,000,000
Oceans	~40,000
Fossil fuel deposits	~4,000
Soil organic matter	~1,500
Atmosphere	578 (in 1700 C.E.) to 766 (in 2000 C.E.)
Terrestrial plants	~580

Human Impact on the Carbon Cycle

Human activity since the Industrial Revolution has impacted the natural carbon cycle in the following ways:

- During and after the Industrial Revolution, the deforestation of old-growth forests and the combustion of fossil fuels have released carbon stored in long-term carbon sinks, ultimately resulting in climate change, with some of the following environmental impacts:
  - increased acidity of oceans
  - increase in atmospheric particulate matter
  - increased rate of melting of long-term water storage, e.g., land ice, which results in higher sea levels
  - stronger and more frequent storm events; e.g., hurricanes, storm surges, and floods
- Prior to the Industrial Revolution, transfer rates of CO_2 through photosynthesis, cellular respiration, and the burning of fossil fuels (e.g., wood) were fairly balanced. However, after the Industrial Revolution, more CO_2 was, and continues to be, released into the environment than is being sequestered (stored long term).

1.5 The Nitrogen Cycle

Nitrogen makes up 78% of the atmosphere. Nitrogen is also an essential element needed to make amino acids, proteins, and nucleic acids. Other nitrogen stores include organic matter in the soil and the oceans (one million times more nitrogen is found in the atmosphere than is contained in either land or ocean waters).

Important facts to remember about the nitrogen cycle include the following:

- Human activities such as fossil fuel combustion, use of inorganic fertilizers, and production of wastewater and sewage have dramatically altered the nitrogen cycle.

TIP

Know *all* steps involved in the nitrogen cycle. You will find more questions on the APES exam about the nitrogen cycle than about any other cycle.

- Increased concentrations of nitrogen in water can lead to water acidification, eutrophication, and toxicity.
- In plants, nitrogen is essential for the process of photosynthesis and is found in chlorophyll, which is essential for plant growth.
- Nitrogen availability can affect the rate of ecosystem processes, including primary production and decomposition.
- Nitrogen is a key component in nucleic acids (DNA and RNA) and proteins.
- Since nitrogen is often a limiting nutrient in terrestrial ecosystems, its presence often determines how much food can be grown on a piece of land. The natural cycling of nitrogen, in which atmospheric nitrogen is converted to nitrogen oxides by lightning and deposited in the soil by rain, where it is assimilated by plants and either eaten by animals (and returned as feces) or decomposed back to elemental nitrogen by bacteria, includes the following processes (Figure 1.43):
    - **Nitrogen Fixation**—Atmospheric nitrogen is converted into ammonia (NH_3) or nitrate ions (NO_3^-), which are biologically usable forms of nitrogen. The key participants in nitrogen fixation are legumes, such as alfalfa, clover, and soybeans, and nitrogen-fixing bacteria known as rhizobium.
    - **Nitrification**—Ammonia (NH_3) is converted to nitrite (NO_2^-) and nitrate (NO_3^-), which are the most useful forms of nitrogen to plants.
    - **Assimilation**—Plants absorb ammonia (NH_3), ammonium ions (NH_4^+), and nitrate ions (NO_3^-) through their roots.
    - **Ammonification**—Decomposing bacteria convert dead organisms and wastes, which include nitrates, uric acid, proteins, and nucleic acids, to ammonia (NH_3) and ammonium ions (NH_4^+)—biologically useful forms.
    - **Denitrification**—Anaerobic bacteria convert ammonia into nitrites (NO_2^-), nitrates (NO_3^-), nitrogen gas (N_2), and nitrous oxide (N_2O) to continue the cycle.
- Though atmospheric nitrogen (N_2) is abundant (composing 78% of Earth's atmosphere), it has limited use biologically, which leads to a scarcity of usable forms of nitrogen in terrestrial and aquatic ecosystems.

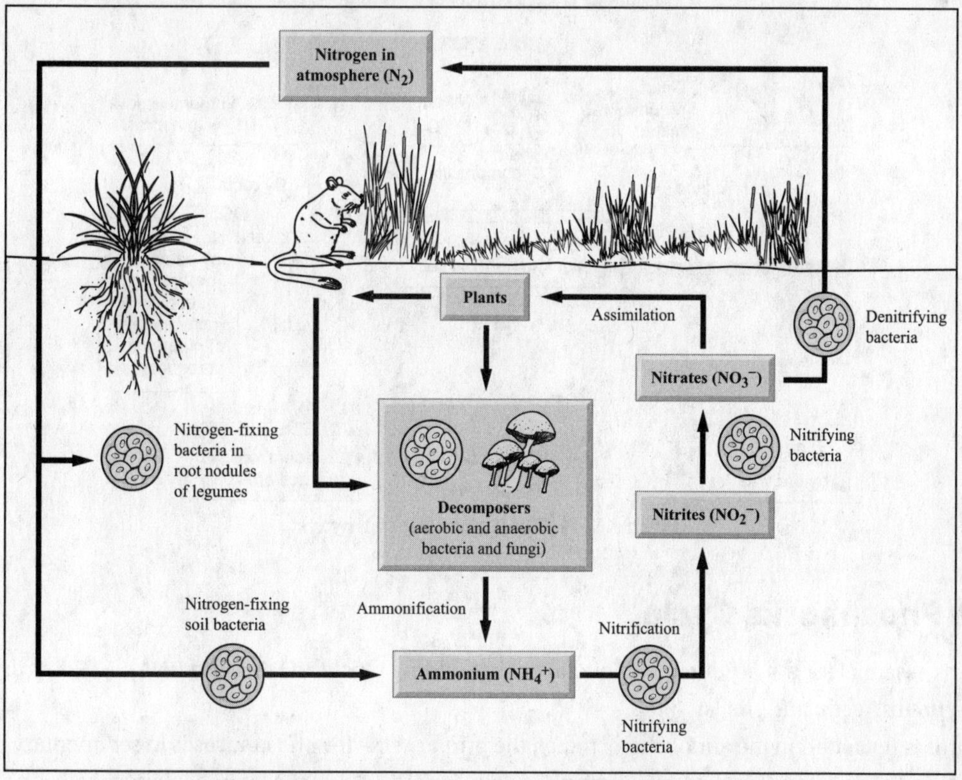

Figure 1.43 Nitrogen cycle

Effects of Excess Nitrogen

Fossil fuel combustion has contributed to a sevenfold increase in nitrogen oxides (NO_x) to the atmosphere, particularly nitrogen dioxide (NO_2). NO_x is a precursor of tropospheric (lower atmosphere) ozone production, contributes to smog and acid rain, and increases nitrogen inputs to ecosystems.

Ammonia (NH_3) in the atmosphere has tripled as a result of human activities since the Industrial Revolution. Ammonia acts as an aerosol and decreases air quality. Nitrous oxide (N_2O) is a significant greenhouse gas and has deleterious effects in the stratosphere, where it breaks down and acts as a catalyst in the destruction of atmospheric ozone. N_2O is in a large part emitted during nitrification (conversion of ammonium (NH_4^+) to nitrate (NO_3^-) and nitrite (NO_2^-)) and denitrification (converting oxides back to nitrogen gas or nitrous oxides for energy generation) processes that take place in the soil. The largest N_2O emissions are observed where nitrogen-containing fertilizer is applied in agriculture. Human activity has more than doubled the annual transfer of nitrogen into biological available forms through

- biomass burning;
- cattle and feedlots;
- extensive cultivation of legumes (particularly soy, alfalfa, and clover);
- industrial processes; and
- the extensive use of chemical fertilizers and pollution emitted by vehicles and industrial plants (NO_x)

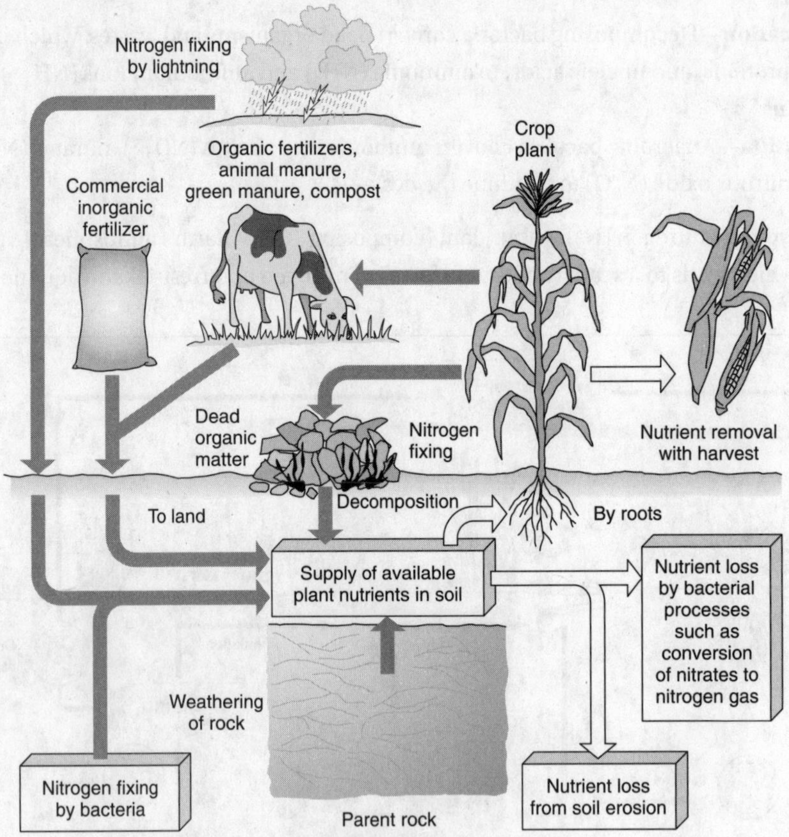

Figure 1.44 Plant nutrient pathways

1.6 The Phosphorus Cycle

Phosphorus is essential for the production of nucleotides (building blocks of DNA and RNA), ATP, fats in cell membranes, bones, teeth, and shells.

Phosphorus is not found in the atmosphere; rather, the primary sink for phosphorus is in sedimentary rocks. Generally, phosphorus is found in the form of the phosphate ion (PO_4^{3-}) or the hydrogen phosphate ion (HPO_4^{2-}).

Phosphorus is slowly released from terrestrial rocks by weathering and the action of acid rain and then dissolves into the soil and is taken up by plants. It is often a limiting factor for soils due to its low concentration and solubility, and it is a key element in fertilizer. A fertilizer labeled 6-24-26 contains 6% nitrogen (N), 24% phosphorus (P), and 26% potassium (K). Unlike the carbon, nitrogen, sulfur, or water cycles, the phosphorus cycle does not involve a gaseous or atmospheric phase (Figure 1.45). Since phosphorus is trapped in rocks and minerals, it must undergo a weathering process to be released before it can be utilized.

Humans have impacted the phosphorus cycle in several ways, as follows:

- Allowing runoff from feedlots, fertilizers, and the discharge of municipal sewage plants, which collects in lakes, streams, and ponds, causing an increase in the growth of cyanobacteria (blue-green algae), green algae, and aquatic plants. In turn, this growth results in decreased oxygen content in the water, which then kills other aquatic organisms in the food web.
- Applying phosphorus-rich guano and other phosphate-containing fertilizers to fields
- Clear-cutting tropical habitats for agriculture, which decreases the amount of available phosphorus, as it is contained in the vegetation
- Mining large quantities of rocks containing phosphorus for inorganic fertilizers and detergents

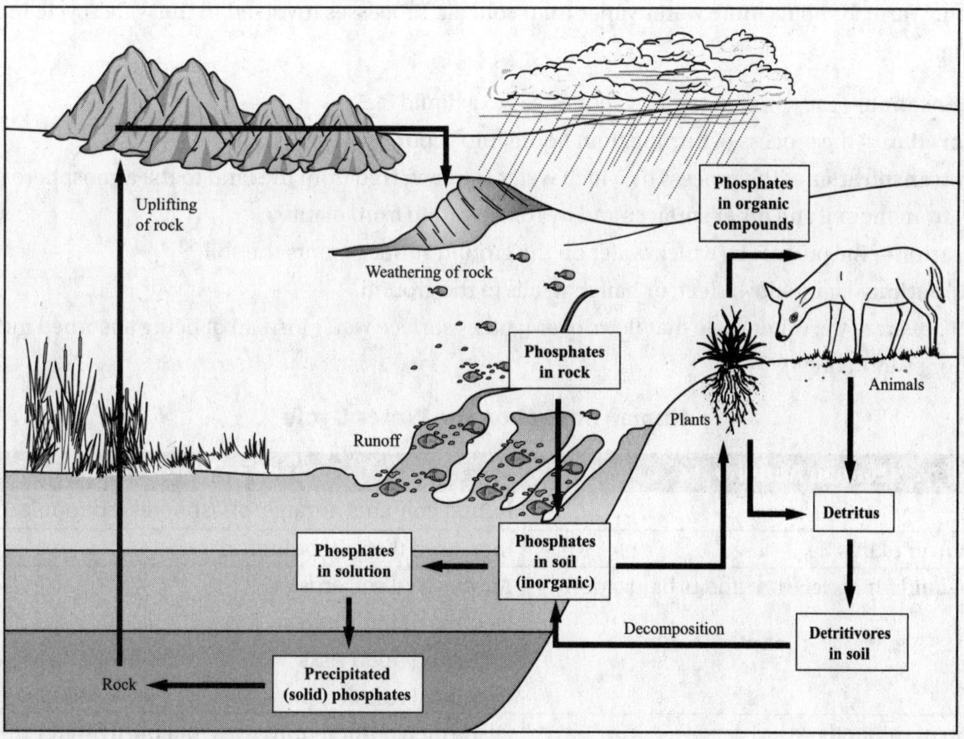

Figure 1.45 Phosphorus cycle

1.7 The Hydrologic (Water) Cycle

The water cycle is powered by energy from the sun, which evaporates water from oceans, lakes, rivers, streams, soil, and vegetation (Figure 1.46). The oceans hold 97% of all water on the planet and are the source of 78% of all global precipitation. Oceans are also the source of 86% of all global evaporation, with evaporation from the sea surface keeping Earth from overheating. If there were no oceans, surface temperatures on land would rise to an average of 153°F (67°C).

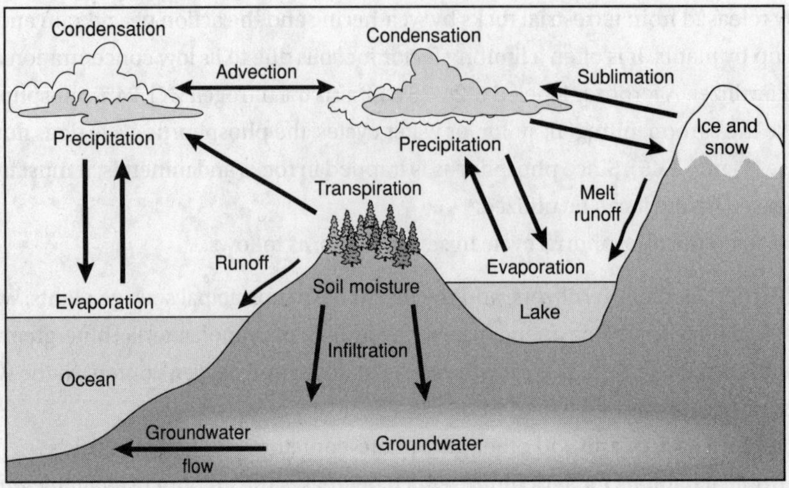

Figure 1.46 Water cycle

The water cycle is in a state of dynamic equilibrium by which the rate of evaporation equals the rate of precipitation. Warm air holds more water vapor than cold air. Processes involved in the water cycle include the following:

- **Condensation**—The conversion of a vapor or gas to a liquid
- **Evaporation**—The process of turning from liquid into vapor
- **Evapotranspiration**—The process by which water is transferred from the land to the atmosphere by evaporation from the soil and other surfaces and by transpiration from plants
- **Infiltration**—The process by which water on the ground surface enters the soil
- **Precipitation**—Rain, snow, sleet, or hail that falls to the ground
- **Runoff**—Part of the water cycle that flows over land as surface water instead of being absorbed into groundwater or evaporating

Human Impact on the Water Cycle

Human Activity	Impact on Water Cycle
Agriculture	Runoff contains nitrates, phosphates, ammonia, etc.
Building power plants	Increased thermal pollution
Clearing of land for agriculture and urbanization	Accelerated soil erosion Decreased infiltration Increased flood risks Increased runoff
Destruction of wetlands	Disturbing natural processes that purify water
Pollution of water sources	Increased occurrences of infectious agents such as cholera, dysentery, etc.
Sewage runoff, feedlot runoff	Cultural eutrophication
Withdrawing water from lakes, aquifers, and rivers	Groundwater depletion and saltwater intrusion

Water Distribution and Properties

Over 70% of Earth's surface is covered by water, with oceans holding about 97% of all water on Earth, and freshwater making up only about 3%. Of the freshwater that is available, most of it is trapped in glaciers and ice caps, with the rest found in (descending order) groundwater, lakes, soil moisture, atmospheric moisture, rivers, and streams (Figure 1.47).

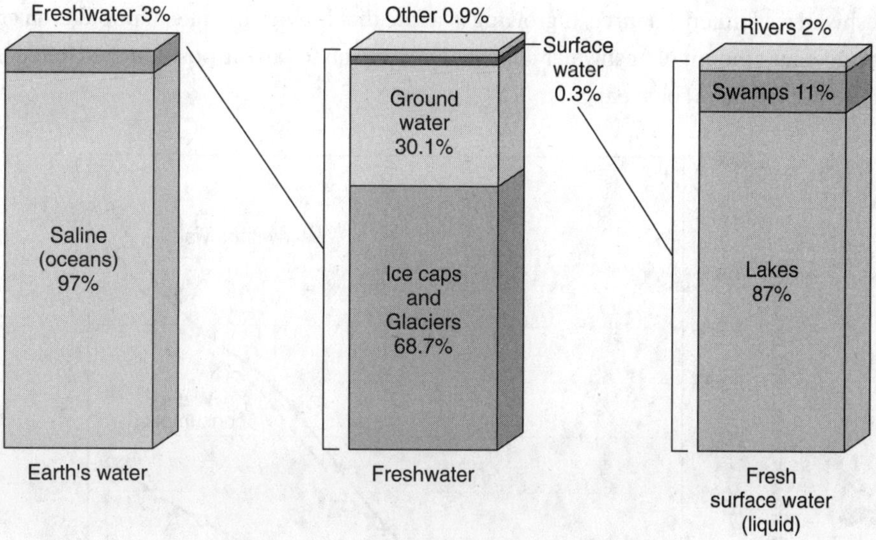

Figure 1.47 Distribution of Earth's water

Water has many unique properties, as follows:

- A lot of energy is needed to evaporate water.
- Strong hydrogen bonds hold water molecules together.
- The temperature of water changes slowly due to its high specific heat capacity.
- Water expands when it freezes.
- Water filters out harmful UV radiation in aquatic ecosystems.
- Water has a high boiling point.
- Water is a polar molecule, which means the following:
  - Capillary action, a result of hydrogen bonding, helps tree roots take up water, allowing trees to grow as large as they do.
  - Floating ice, essential to life near the poles, results from the different ways water molecules arrange themselves at different temperatures.
  - The polarity of water helps to dissolve many compounds.
  - Water's polarity allows interaction with non-polar molecules, e.g., fatty acids, and allows for the compartmentalizing of membranes, essential for living organisms.

Freshwater

The renewal of Earth's freshwater supply depends on the regular movement of water from Earth's surface into the atmosphere and back again. Listed in the following table are average water renewal rates.

Freshwater Renewal Rates

Source of H_2O	Average Renewal Rate
Groundwater (deep)	~10,000 years
Groundwater (near surface)	~200 years
Lakes	~100 years
Glaciers	~40 years
Water in the soil	~70 days
Rivers	16 days
Atmosphere	8 days

The use of freshwater, a limited resource, is growing at a rate that is twice the rate of population growth (Figure 1.48). The average amount of freshwater allocated per person for all purposes in the United States is about 500,000 gallons (~2 million liters) per year.

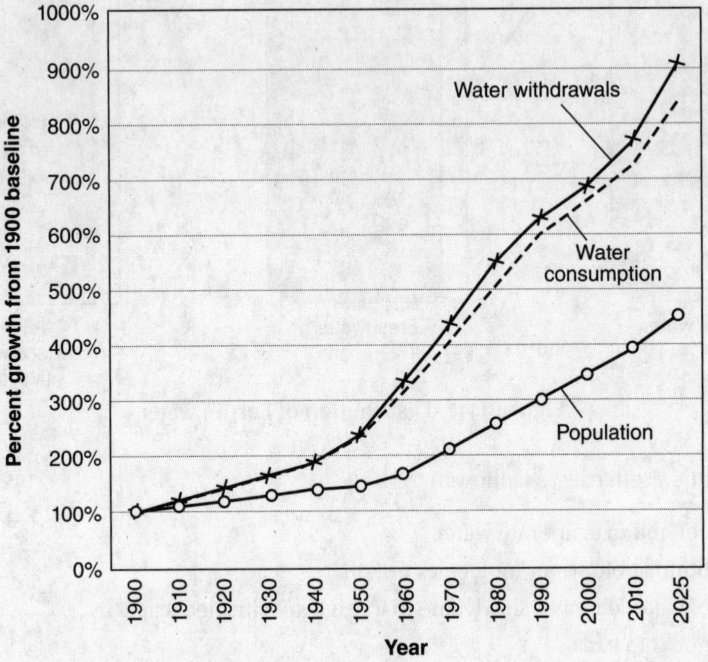

Figure 1.48 Global water use and global population, 1900–2025 (projected)
Source: United Nations Environmental Programme

Aquifers

An aquifer is a geologic formation that contains water in quantities sufficient to support a well or spring (Figure 1.49). Aquifers in the United States hold 30 times more water than all U.S. lakes and rivers combined, with groundwater supplying approximately 40% of all freshwater in the United States.

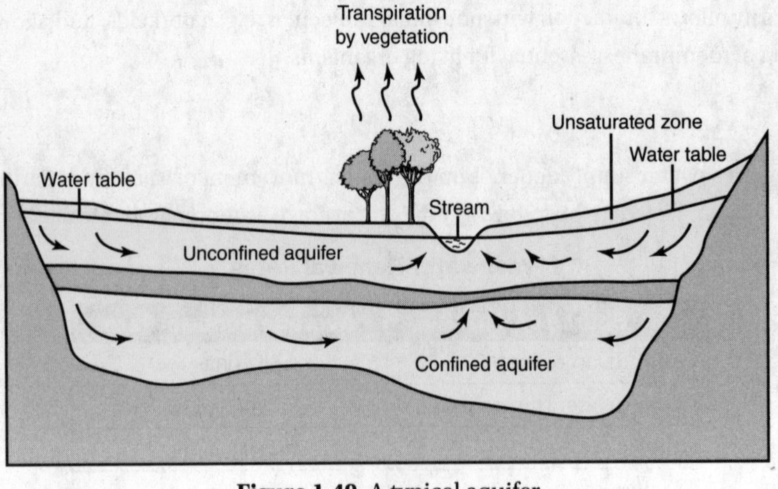

Figure 1.49 A typical aquifer

Important terms regarding aquifers include the following:

- **Confined "artesian well" aquifer**—An aquifer below the land surface that is saturated with water. As a result of impermeable material above and below the aquifer, the water is under pressure.
- **Recharge zone**—The surface area above an aquifer that supplies water to the aquifer

- **Unsaturated zone**—The zone immediately below the land surface where the open spaces (pores) in the soil contain both water and air, but are not totally saturated with water
- **Water table**—The level below which the ground is saturated with water

Factors That Threaten Aquifers

Aquifer (or groundwater) depletion is primarily caused by sustained groundwater pumping. When the rate of groundwater extraction is greater than the rate of aquifer recharge, the net effect is a drop in the water table. Though agriculture is the largest sector responsible for aquifer depletion, domestic and municipal withdrawals also affect groundwater levels in many areas around the world (Figure 1.50). As human populations increase, the rates of groundwater extraction likewise increase.

Changes in global weather patterns also reduce aquifer inputs, jeopardizing groundwater levels. This issue is an example of the "Tragedy of the Commons"* combined with high economic externalities (cost or benefit incurred or received by a third party; however, the third party has no control over the creation of that cost or benefit).

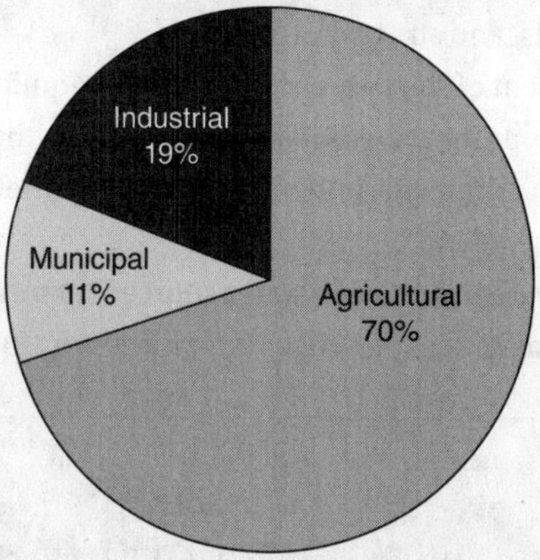

Figure 1.50 Global sources of water withdrawal
Source: Food and Agricultural Organization of the United Nations

Effects of Groundwater Depletion

- Increased costs as more energy is required for pumping
- Land subsidence—The sinking of land that results from groundwater extraction. Land subsidence is a major problem in developing countries, as large cities continue to increase in population without there being many alternatives for supplying freshwater to the growing population (Figure 1.51).

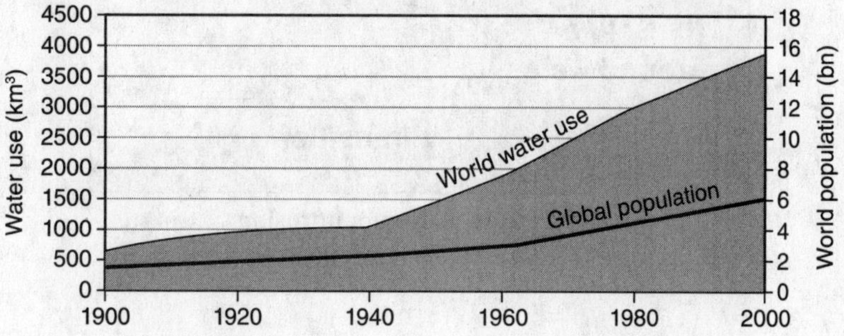

Figure 1.51 Global water use vs. population growth
Source: Food and Agricultural Organization—United Nations

*A situation in which individual users, who have open access to a resource, act independently according to their own self-interest and, contrary to the common good of all users, cause depletion of the resource through their uncoordinated action(s).

- Lowering of the water table (Figure 1.52)

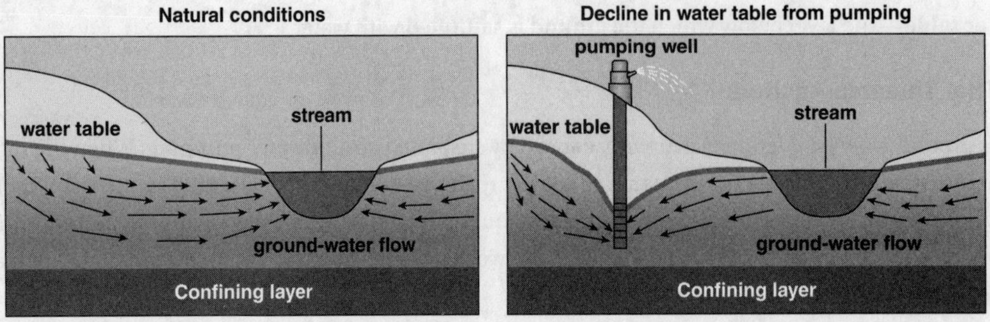

Figure 1.52 Decline in water table from groundwater pumping
Source: University of Georgia

- Overgrazing and the resulting erosion
- Reduction of water in lakes, ponds, and streams
- Saltwater intrusion—The movement of saltwater into freshwater aquifers, which can lead to contamination (Figure 1.53). Water extraction drops the water table, which reduces the water pressure and allows saltwater to flow farther inland. Agricultural, drainage, and navigational channels, hurricanes, and storm surges also provide ways for saltwater to move inland.
- Water shortages—Since groundwater is the main water source for many populations (especially in arid environments), residents of these areas may experience water insecurity for domestic and agricultural needs.

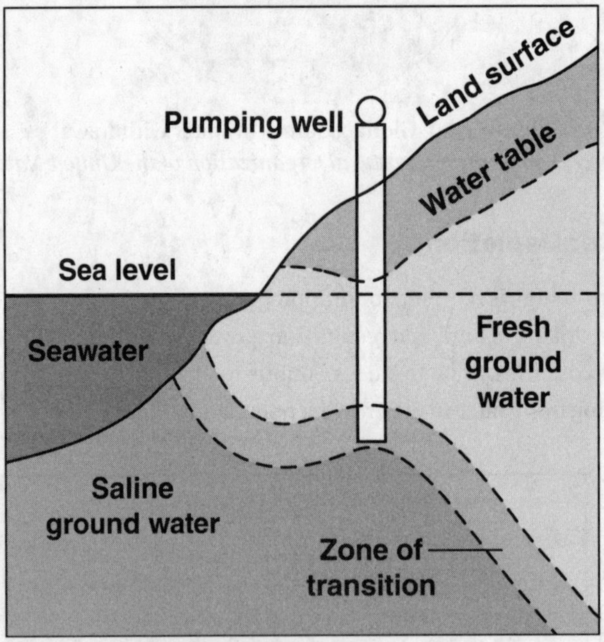

Figure 1.53 Saltwater intrusion
Source: U.S. Environmental Protection Agency

1.8 Primary Productivity

The ultimate source of energy is the sun. Plants (autotrophs) are able to use this light energy to create food through the process of photosynthesis, in which plants remove carbon dioxide from the atmosphere and use light energy to produce carbohydrates and other organic compounds:

$$6CO_2 + 6H_2O + \textit{sunlight} \rightarrow C_6H_{12}O_6 + 6O_2$$

Plants capture light primarily through the green pigment chlorophyll, which is contained in organelles called chloroplasts. The energy derived from the oxidation of glucose during cellular respiration is then used to form other organic compounds such as cellulose (for support), lipids (waxes and oils), and amino acids and eventually proteins. Oxygen gas is released into the atmosphere during photosynthesis, and plants emit carbon dioxide during respiration. Since plants produce less carbon dioxide than they absorb, they therefore become net sinks of carbon.

Factors that affect the rate of photosynthesis include

- carbon dioxide concentration;
- the amount of light and its wavelength;
- the availability of water; and
- temperature.

1.9 Trophic Levels

A trophic level is the position an organism occupies in a food chain and is the number of steps it is from the start of the chain. The food chain in Figure 1.54 starts at trophic level 1 with primary producers (grass—producer), which are eaten by herbivores (grasshopper—primary consumer), which are eaten by bluebirds (bluebird—secondary consumer), which are eaten by snakes (snake—tertiary consumer), which are finally eaten by the owl, which is known as the apex or quaternary predator.

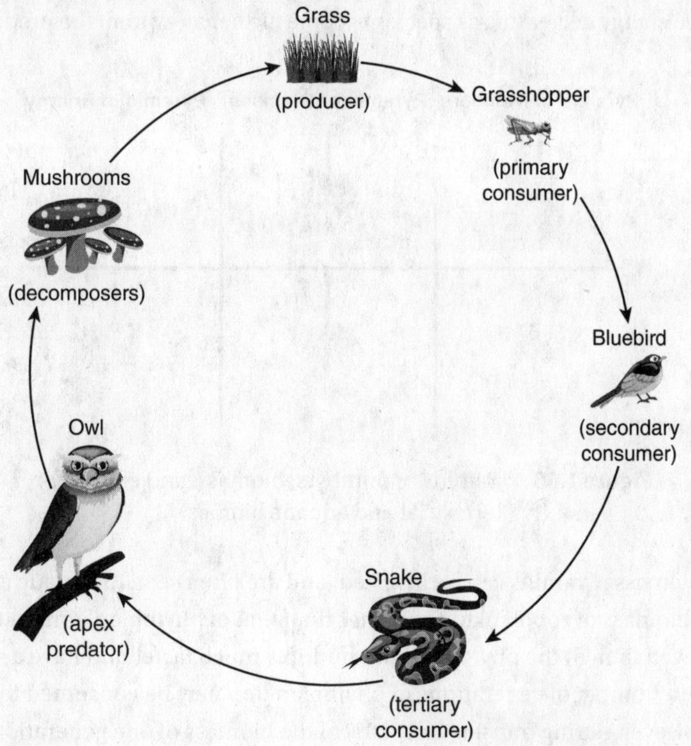

Figure 1.54 Food chain

A food web (Figure 1.55) is the natural interconnection of food chains.

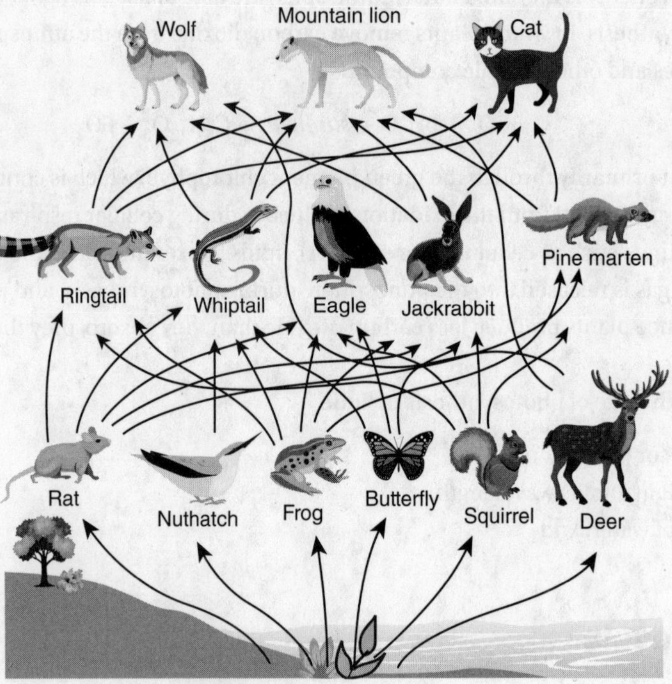

Figure 1.55 Food web

Ecological Pyramids

Ecological pyramids place the primary producers at the base of the pyramid and are able to show various properties of ecosystems, such as the numbers of individuals per unit of area, biomass (g/m^2), or energy (kcal · m^{-2} · yr^{-1}), with amounts of available energy decreasing as species become further away from the producers.

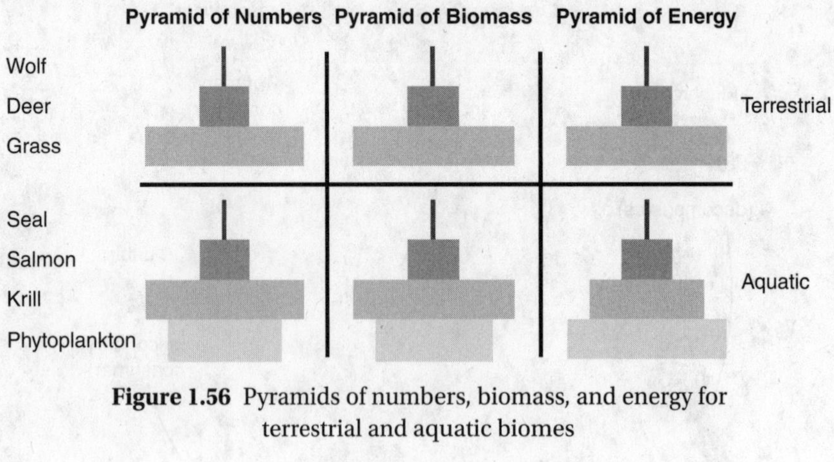

Figure 1.56 Pyramids of numbers, biomass, and energy for terrestrial and aquatic biomes

In some instances, biomass pyramids can be inverted, and are often seen in aquatic and coral reef ecosystems. For example, the biomass of zooplankton is higher than that of phytoplankton, as the life span of zooplankton is longer, even though the phytoplankton multiply much faster and have a shorter life span than zooplankton. Therefore, a number of generations of phytoplankton may be consumed by a single generation of zooplankton. However, during transfer, only 10% of the biomass of one generation is passed on to next trophic level.

Furthermore, primary consumers have longer life spans and slower growth rates, and accumulate more biomass than the producers they consume. Phytoplankton live just a few days, whereas the zooplankton eating the phytoplankton live for several weeks and the fish eating the zooplankton live for several years.

Aquatic predators also tend to have a lower death rate than the smaller consumers, which further contributes to the inverted pyramid pattern.

However, energy pyramids will always have an upright pyramid shape if all sources of food energy are included, according to the Second Law of Thermodynamics, which states that as energy is transferred or transformed, more and more of it is wasted (usually as heat), and that there is a natural tendency of any isolated system to degenerate from an ordered state into a more disordered state—also known as entropy.

1.10 Energy Flow and the 10% Rule

Cellular Respiration

$$C_6H_{12}O_6 + 6O_2 \rightarrow 6CO_2 + 6H_2O + \textit{energy}$$

Organisms dependent on photosynthetic organisms are called heterotrophs. In general, cellular respiration is the opposite of photosynthesis. In respiration, glucose is oxidized by the cells to produce carbon dioxide, water, and chemical energy. This energy is then stored in the molecule adenosine triphosphate (ATP).

Ecological Pyramids and the 10% Rule

During the transfer of energy from one trophic level to the next, only about 10% of the energy is used to convert organic matter into tissue. The remaining energy is generally lost in the form of heat during respiration, metabolic processes, temperature regulation, incomplete digestion (e.g., elephants only digest 40% of the material they consume), and decay of waste products.

Sunlight is the ultimate source of energy required for most biological processes. Of the sun's energy that reaches Earth, 35% is spent to heat water and land areas and to evaporate water. Only 8% of energy released from the sun is available to plants with only 1% being used for photosynthesis; about 15% is reflected back into space, and 80% is absorbed by Earth. The amount of available energy decreases approximately 10% from one stage to the next (10% Rule), with the majority of energy being lost in the form of heat.

Productivity refers to the rate of generation of biomass in an ecosystem and is expressed in units of mass per unit surface area (or volume) per unit time, e.g., grams per square meter per day ($g/m^2/d$), with mass referring to dry matter or to the mass of carbon generated. Productivity of autotrophs (plants) is called primary productivity, while that of heterotrophs (animals) is called secondary productivity.

Secondary production is the generation of biomass by heterotrophic consumers in a system, is driven by the transfer of organic material between trophic levels, and represents the quantity of new tissue created through the use of assimilated food by organisms responsible for secondary production, including animals, protists, fungi, and most bacteria.

Biomass Pyramids

A biomass pyramid shows how much organic mass is within each trophic level (Figure 1.57). For example, the vast majority of all mass within a rainforest is held within the trees.

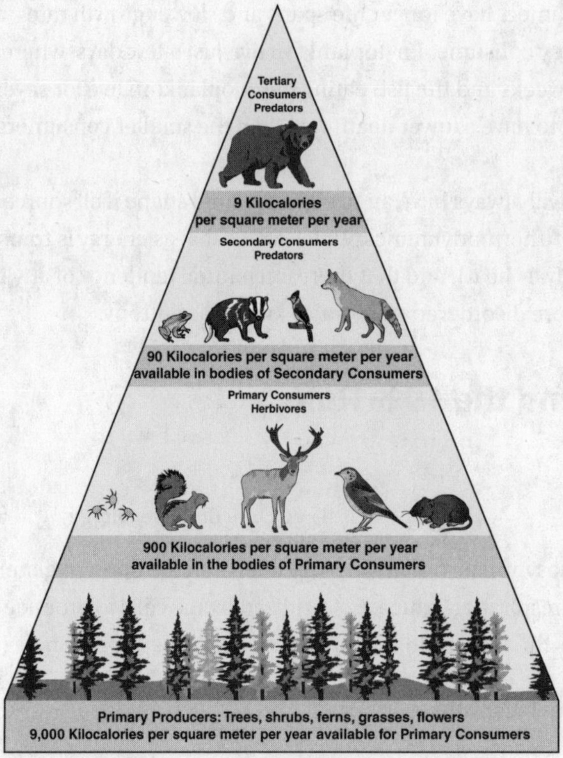

Figure 1.57 Pyramid of biomass

A marine pyramid of biomass is inverted (Figure 1.58) because

- the biomass of the trophic levels is dependent on the longevity of the members within that trophic level;
- the biomass of zooplankton is greater than that of phytoplankton (the producers) because of their small size and low weight, and predatory fish are much larger than zooplankton; and
- the producers in ocean or aquatic ecosystems are phytoplankton, which have short life spans, turn over (grow and reproduce) rapidly, and have lower mass than zooplankton (primary consumers).

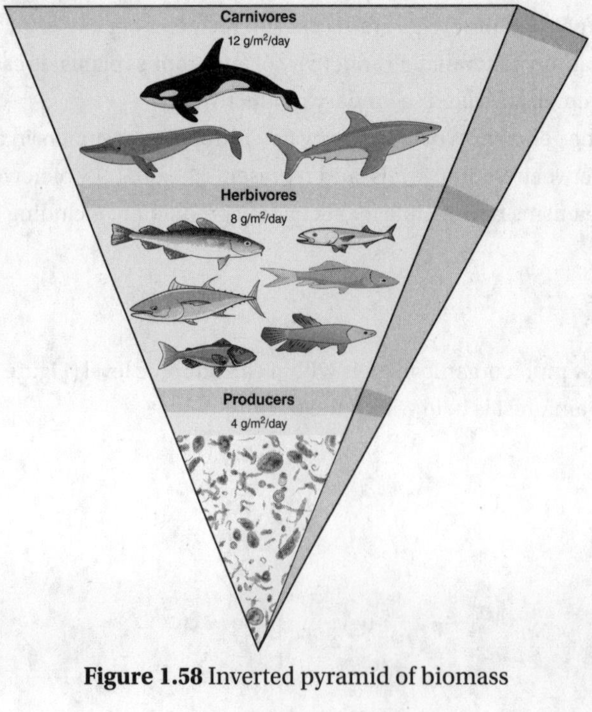

Figure 1.58 Inverted pyramid of biomass

Energy Pyramids

Energy pyramids show the proportion of energy passed from one trophic level to the next-level consumers in an ecosystem (Figure 1.59). The energy temporarily "trapped" within the mass of the trophic level is not counted.

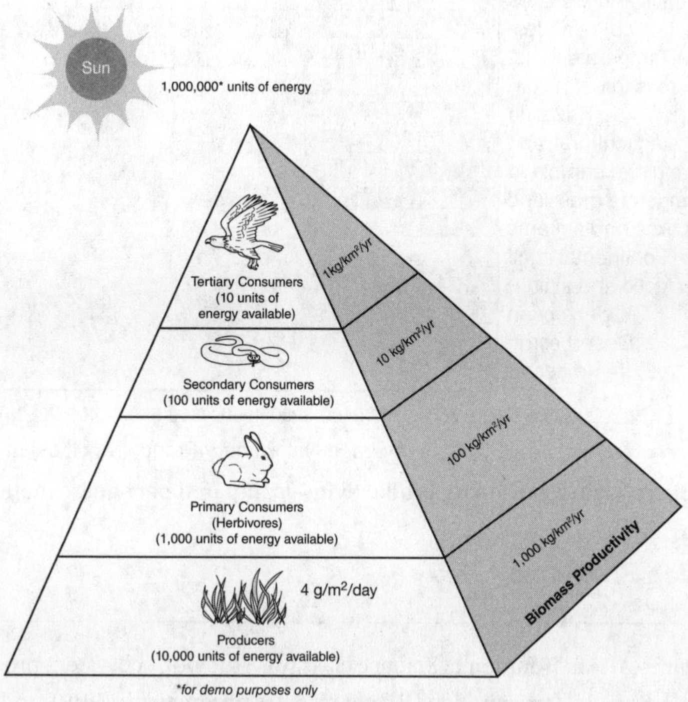

Figure 1.59 Energy pyramid

Ecosystem Productivity

Of all available sunlight that reaches Earth, less than 3% for land plants and less than 1% for aquatic plants is used for photosynthesis. This relatively low efficiency of the conversion of solar energy into energy stored in carbon compounds sets the overall amount of energy available to heterotrophs at all other trophic levels.

Gross Primary Production (GPP)

Gross primary production (GPP) is the rate at which plants capture and fix (store) a given amount of chemical energy as biomass in a given length of time. Some fraction of this fixed energy is used by primary producers for cellular respiration and the maintenance of existing tissues.

Net Primary Production (NPP)

The remaining fixed energy is referred to as net primary production (NPP) and is the rate at which all the plants in an ecosystem produce net useful chemical energy. NPP is equal to the difference between the rate at which the plants produce useful chemical energy known as gross primary productivity (GPP) and the rate (R) at which they use some of that energy during respiration.

$$NPP = GPP - R_{\textit{plant respiration}}$$

Some net primary production goes toward the growth and reproduction of primary producers, while some is consumed by herbivores.

Open oceans, due to their making up a large proportion of the Earth's surface, collectively have the highest net primary productivity. However, when compared on a one-to-one per square meter basis, estuaries have the highest net primary productivity (Figure 1.60).

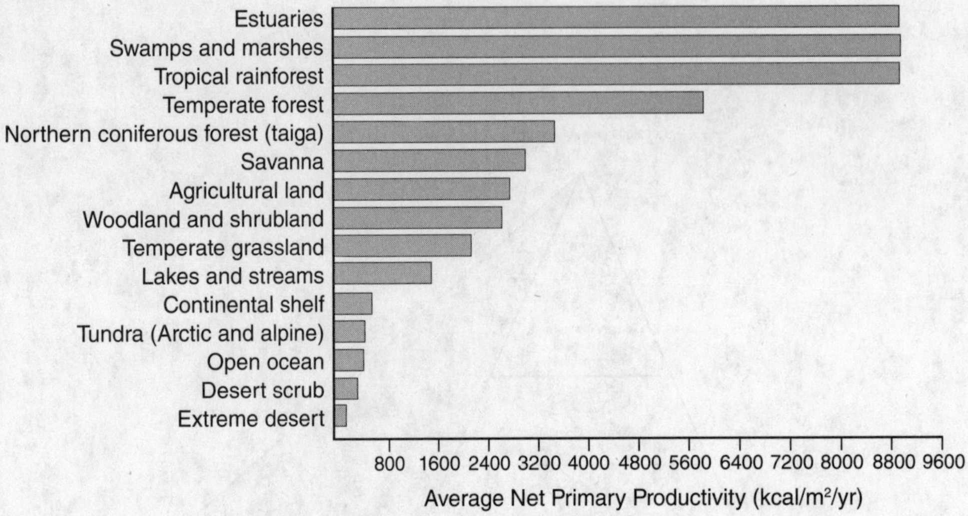

Figure 1.60 Net primary productivity—compared per square meter

Productivity Practice Problems

#1 Dr. Smith's AP Environmental Science class was studying ecosystem productivity, focusing on cellular respiration (R), net primary productivity (NPP), and gross primary productivity (GPP).

To help reinforce the concept, the class went on a field trip to a local pond which contained relatively large amounts of planktonic algae and microscopic organisms (e.g., amoebas, diatoms, euglena, paramecium, planaria rotifers, and stentors).

The relative rates of respiration and photosynthesis within the pond determine whether there is a net addition or decrease of carbon dioxide in the water and therefore whether the pH of the water rises (becoming more basic) or falls (becoming more acidic). Respiration rates are affected by water temperature and the biomass of plants, animals and microorganisms in the water and bottom sediment.

Upon arrival to the pond, the students used a portable dissolved oxygen (DO) meter and determined that the pond water had a dissolved oxygen (DO) concentration of 10 mg of O_2 per liter of pond water.

The students then took two identical sample bottles and filled them with the pond water and then tightly capped the bottles with one bottle being totally wrapped in aluminum foil and labelled "DARK" while the other bottle was left clear and labelled "LIGHT". The bottles were then carefully taken back to school and allowed to sit near a window for one week.

After one week, the students then used the same dissolved oxygen (DO) meter and determined that the DARK bottle had a DO concentration of 5 mg of O_2 per liter of pond water while the LIGHT bottle had a DO concentration of 11 mg of O_2 per liter of pond water. Calculate the respiration rate (R) of the organisms living in the pond water.

#2 Using the information above, calculate the net primary productivity.

#3 Using the information above, calculate the gross primary productivity of the algae.

1.11 Food Chains and Food Webs

You will find specific food chains and food webs in sections 1.2, "Terrestrial Biomes," and 1.3, "Aquatic Biomes."

Answers to Productivity Practice Problems

1. $Respiration_{(plant)}$ (R) = (Initial − Dark) = (10 mg/L − 5 mg/L) = **5 mg/L/week**
2. Net Primary Productivity (NPP) = (Light − Initial) = (11 mg/L − 10 mg/L) = **1 mg/L/week**
3. Gross Primary Productivity (GPP) = (Light − Dark) = (11 mg/L − 5 mg/L) = **6 mg/L/week**

Practice Questions

For Questions 1–3, choose from the following items:

(A) Savanna
(B) Taiga
(C) Temperate deciduous forest
(D) Tundra

1. Coniferous forests of cold climates of high latitudes and high altitudes.

2. Warm year-round; prolonged dry seasons; scattered trees.

3. Tree growth is hindered by low temperatures and short growing seasons. Vegetation is composed of dwarf shrubs, grasses, mosses, and lichens.

4. The annual productivity of any ecosystem is greater than the annual increase in biomass of the herbivores in the ecosystem because

(A) animals convert energy input into biomass more efficiently than plants do
(B) during each energy transformation, some energy is lost
(C) plants convert energy input into biomass more efficiently than animals do
(D) plants have a greater longevity than animals

5. Net primary productivity per square meter is highest in which biome listed below?

(A) Boreal forests
(B) Estuaries
(C) Grasslands
(D) Open oceans

Examine the nitrogen cycle diagram below to answer Question 6.

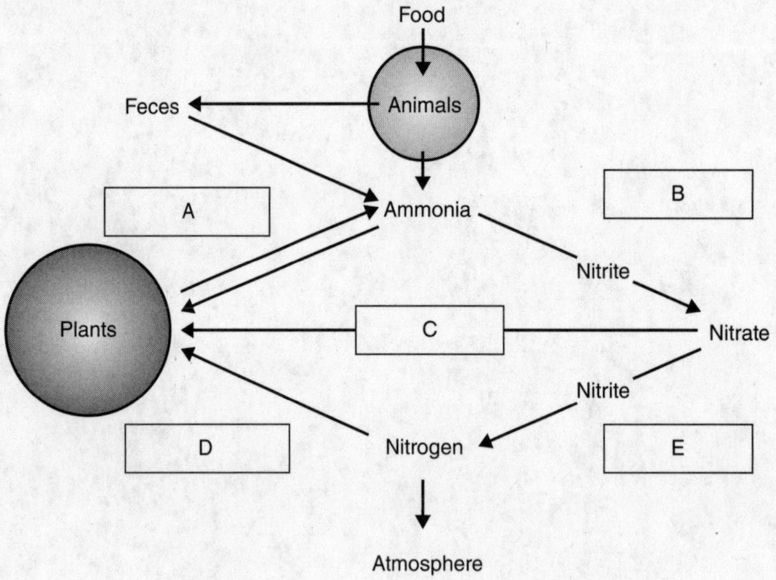

6. Which of the following choices is TRUE?

(A) A = Ammonification, B = Nitrification, C = Assimilation, D = Denitrification
(B) A = Nitrification, B = Ammonification, C = Denitrification, D = Nitrogen Fixation
(C) A = Assimilation, B = Nitrogen Fixation , C = Ammonification, D = Denitrification
(D) A = Ammonification, B = Nitrogen Fixation, C = Assimilation, E = Denitrification

Examine the following food web to answer Question 7.

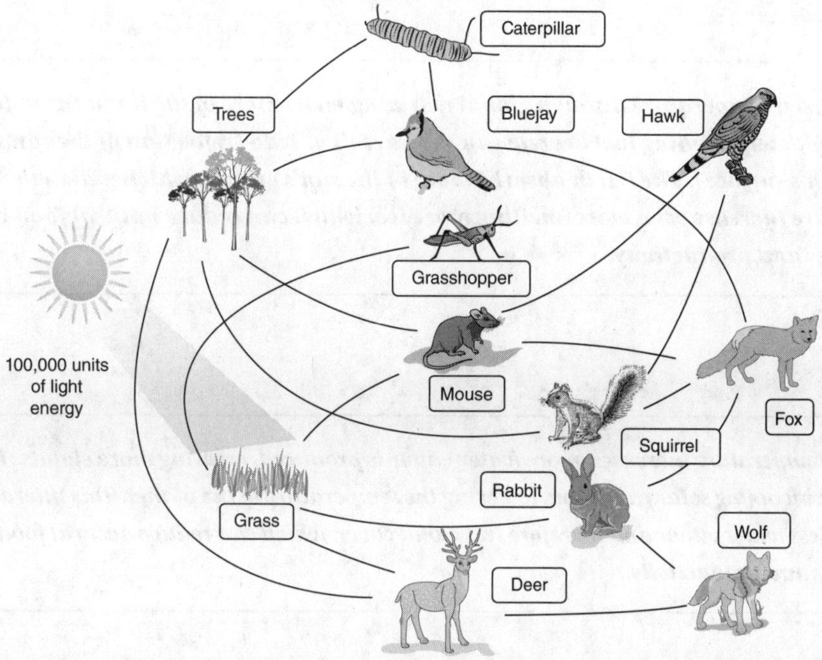

7. Assuming that the energy from the sun reaching a plant was measured at 100,000 units of energy, how many units of energy would be available for the hawk?

   (A) 10 units
   (B) 100 units
   (C) 1,000 units
   (D) 10,000 units

8. Whether a land area supports a deciduous forest or grassland depends primarily on

   (A) changes in length of the growing season
   (B) changes in temperature
   (C) consistency of rainfall from year to year and the effect that it has on fires
   (D) latitude north or south of the equator

9. Mrs. Lopez's AP Environmental Science class performed an experiment on chlorophyte (green) algae. The class placed equal amounts of algae into a light bottle and a dark (covered) bottle. They then immediately measured the dissolved oxygen (DO) in both bottles and found it was 10 mg/L. They then let both bottles sit for one week. After one week, the light bottle had a DO value of 11 mg/L and the dark bottle had a DO value of 5 mg/L. What were the gross primary productivity (GPP), net primary productivity (NPP), and respiration (R) values?

   (A) GPP = 1 mg/L/week; NPP = 5 mg/L/week; R = 6 mg/L/week
   (B) GPP = 6 mg/L/week; NPP = 1 mg/L/week; R = 5 mg/L/week
   (C) GPP = 5 mg/L/week; NPP = 6 mg/L/week; R = 1 mg/L/week
   (D) GPP = 6mg/L/week; NPP = 5 mg/L/week; R = 1 mg/L/week

Read the following two (true) statements to answer Question 10.

Statement A

The increased temperature through global warming melts more of the ice in the polar ice caps and glaciers, leading to a decrease in the Earth's albedo (reflection of the sun's energy from Earth's surface). The Earth absorbs more of the sun's energy, which makes the Earth's temperature increase even more, melting more ice, which can reduce natural food chain efficiencies and productivity.

Statement B

As planet temperature increases, more water vapor is produced, creating more clouds. The clouds then block incoming solar radiation, lowering the temperature of the planet. This interaction produces less water vapor and therefore less cloud cover, which can reduce natural food chain efficiencies and productivity.

10. Based upon the two statements above, which of the following choices listed below are FALSE?

    (A) Statement A describes a positive feedback loop, while Statement B represents a negative feedback loop.
    (B) Statement A describes a negative feedback loop, while Statement B represents a positive feedback loop.
    (C) Both Statements A and B represent positive feedback loops.
    (D) Both Statements A and B represent negative feedback loops.

11. Jan van Helmont began his research of the process of photosynthesis in the mid-17th century. In his own words

"I took an earthen pot and in it placed 200 pounds of earth which had been dried out in an oven. This I moistened with rain water, and in it planted a shoot of willow which weighed five pounds. When five years had passed, the tree which grew from it weighed 169 pounds and about three ounces. The earthen pot was wetted whenever it was necessary with rain or distilled water only. It was very large, and was sunk in the ground, and had a tin-plated iron lid with many holes punched in it, which covered the edge of the pot to keep air-borne dust from mixing with the earth. I did not keep track of the weight of the leaves which fell in each of the four autumns. Finally, I dried out the earth in the pot once more, and found the same 200 pounds, less about 2 ounces."

A diagram of his experiment is presented here:

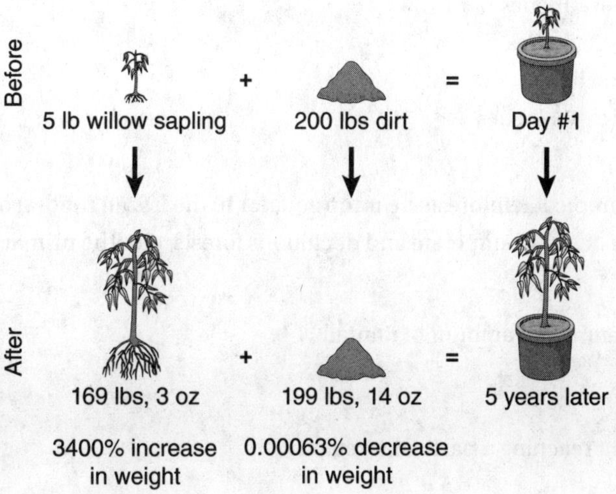

Which of the following statements about his soil experiment would be correct?

(A) The tree gained weight by absorbing nutrients (including water) from the soil.
(B) The tree gained weight by producing oxygen, which then combined with nutrients in the soil and air and eventually was converted to organic matter.
(C) The tree gained weight by producing and storing carbon dioxide (dark reaction) and its compounds in the branches, leaves, and roots.
(D) The tree gained weight by using carbon dioxide to produce organic compounds (e.g., glucose, lignin), which were then stored in the tissues of the tree.

For Questions 12 and 13, refer to the following climatographs.

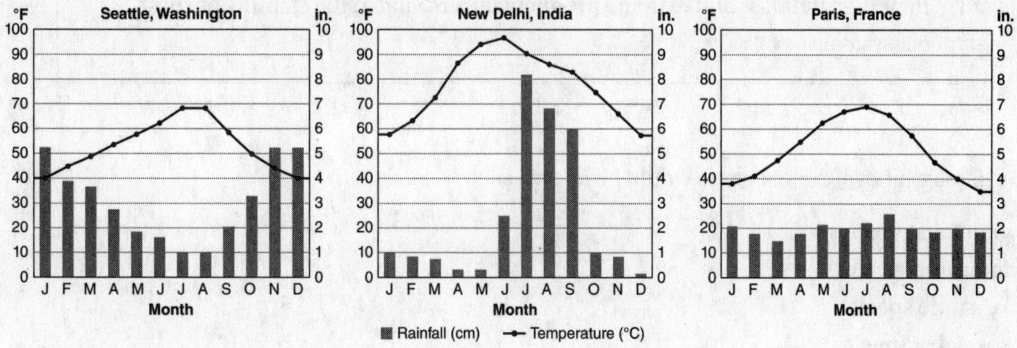

12. Which climatograph above shows the greatest annual range of temperatures and in what city did it occur?

(A) New Delhi, India, 1°F – 81°F
(B) New Delhi, India, 58°F – 96°F
(C) Paris, France, 37°F – 70°F
(D) Seattle, Washington, 10°F – 52°F

13. Which climatograph above shows the highest total precipitation during one month, and what was the amount?

(A) New Delhi, India, ~8.1 inches
(B) New Delhi, India, ~96 inches
(C) Paris, France, ~70 inches
(D) Seattle, Washington, ~70 inches

14. As one travels from the tropical rainforests near the equator to the frozen tundra, one passes through a variety of biomes, including deserts, temperate and deciduous forests, etc. The primary reason for the change in vegetation is due to

(A) changes in temperature and amount of rainfall
(B) evolution
(C) succession
(D) the hours of sunlight reaching a particular region

Read the following passage to answer Question 15.

> Sea otters consume sea urchins and crabs. Sea otter populations dropped significantly during the 1800s due to demand for their pelts, which were used for coats and hats. Around islands that became devoid of sea otters, the population of sea urchins and crabs had increased in size and in numbers and the forests of kelp that once grew in the area had disappeared. In their place, large sea urchins littered the barren sea floor, having consumed every kelp plant in sight.
>
> However, near islands where sea otters survived, or had been reintroduced, the kelp flourished. The discovery was important given the nourishment kelp's underwater forests provide for fish and other sea animals. Kelp forests, with their high biomass and extreme productivity, are vital in the stability of coast ecosystems.

15. In the passage above, the sea otters would be known as

(A) generalist species
(B) indicator species
(C) keystone species
(D) specialist species

Examine the following four pyramids of biomass to answer Question 16.

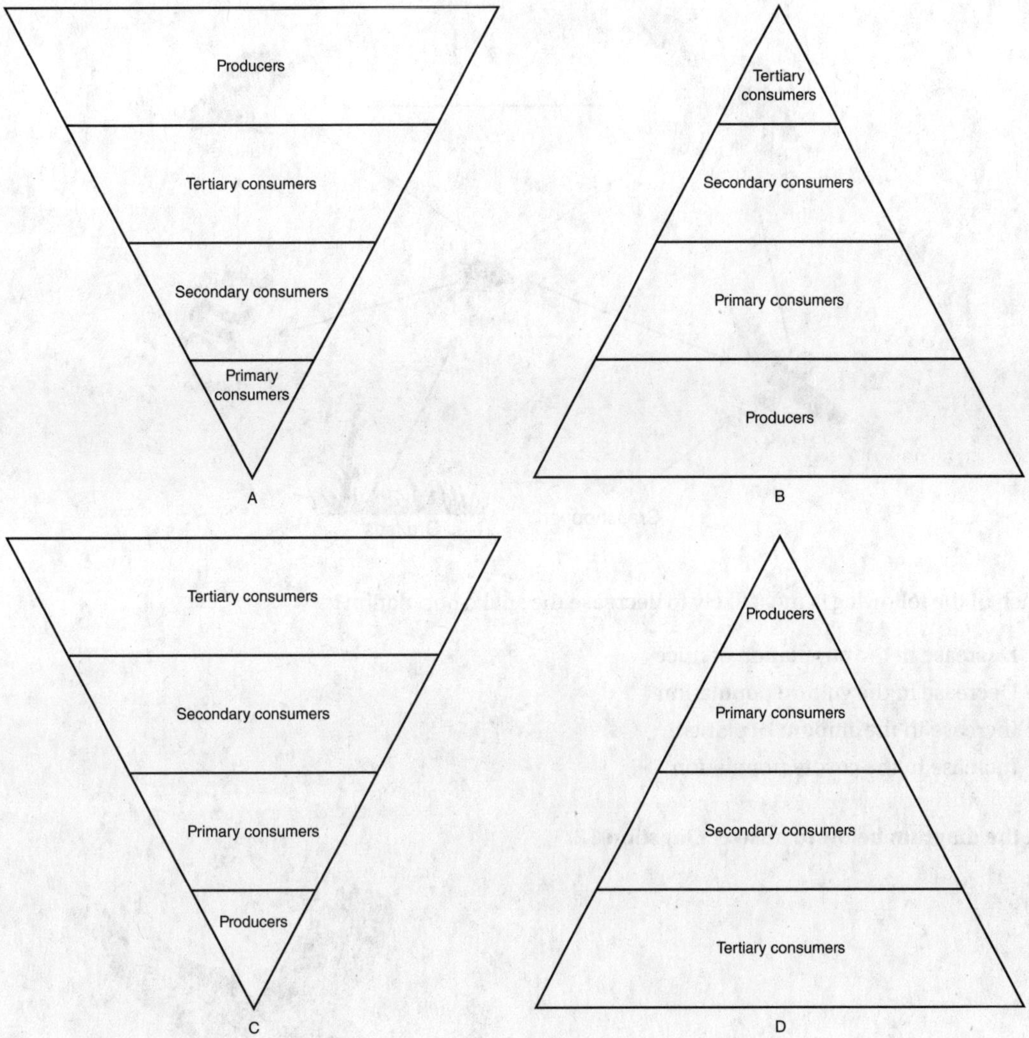

16. Which of the four pyramids of biomass shown above are consistent with a marine environment?

    (A) A
    (B) B
    (C) C
    (D) D

Refer to the following diagram to answer Question 17.

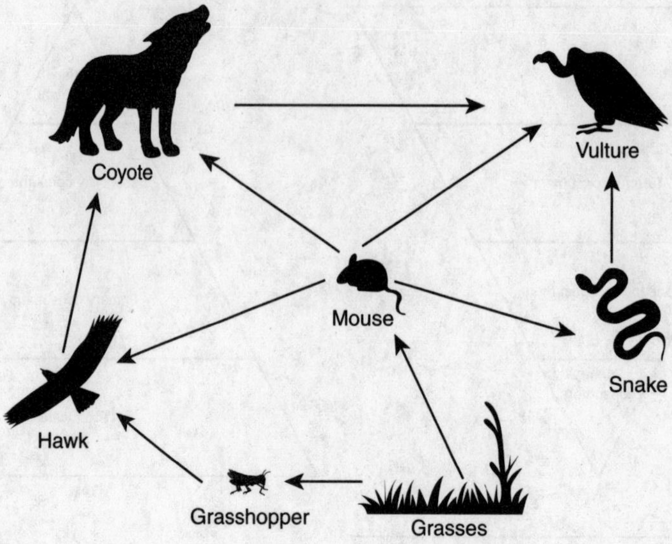

17. Which of the following is most likely to decrease the snake population?

(A) Decrease in the population of mice
(B) Decrease in the vulture population
(C) Increase in the amount of plants
(D) Increase in the coyote population

Refer to the diagram below to answer Question 18.

18. How would you calculate the percentage efficiency of energy transfer from the second to the third trophic level?

(A) Divide 8,350 kJ by 100,000 kJ and multiply by 100%
(B) Divide 719 kJ by 8,350 kJ and multiply by 100%
(C) (8,350 kJ − 719 kJ) divided by (100,000 kJ + 8,350 kJ + 719 kJ + 83 kJ) times 100
(D) (719 kJ − 8,350 kJ) divided by 8,350 kJ times 100

For Question 19, refer to the shaded locations in the world map below.

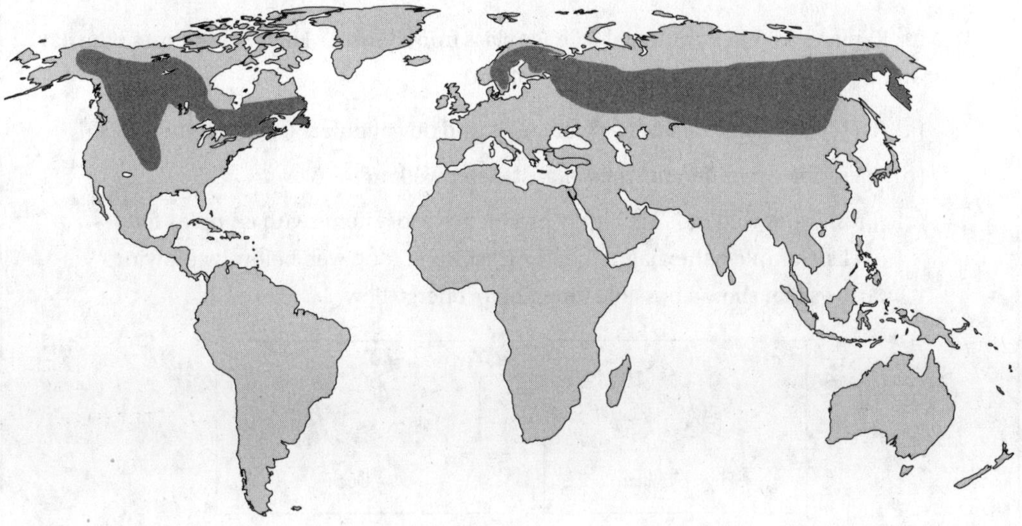

19. The darker shaded areas in the map above are known as

    (A) chaparral
    (B) grasslands
    (C) taiga (boreal forest)
    (D) tundra

20. How does global warming affect the hydrologic cycle?

    (A) It causes random variations in the hydrologic cycle.
    (B) It causes the hydrologic cycle to accelerate.
    (C) It causes the hydrologic cycle to decelerate.
    (D) It doesn't affect the hydrologic cycle at all.

Free-Response Questions

1. Mrs. Kaplan's AP Environmental Science class from Barron's High School spent the day traveling to and studying a freshwater wetland.

   (a) Describe what a freshwater wetland is and how it differs from an estuary.

   (b) Describe two different types of freshwater wetlands.

   (c) Freshwater wetlands are highly productive ecosystems with complex food webs. Complete the diagram of the freshwater food web below by drawing arrows that show a possible direction of energy flow.

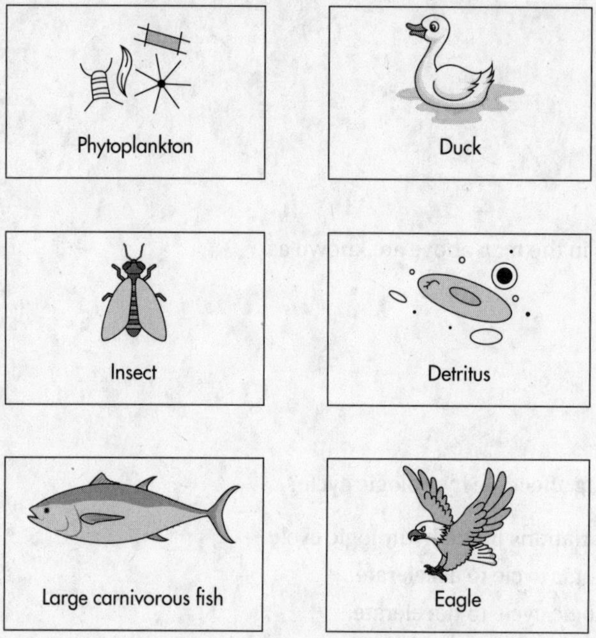

   (d) One of the activities that Mrs. Kaplan's APES students performed while on the field trip was a water-quality test known as a biological oxygen demand (BOD) test. Describe what a BOD test is and what the results can indicate.

   (e) The results of the BOD test that the students performed indicated the presence of organic wastes in the water. Describe ONE remediation technique that could be used to reduce the environmental impact to the wetland.

   (f) Describe what indicator species are and how they are used to assess environmental quality in a freshwater ecosystem.

   (g) Provide a specific example of an indicator species that students may have found in the wetland and describe how it functions in its role as an indicator species.

By Brian Palm, Catholic Memorial School, West Roxbury, MA
A.B., Dartmouth College, Hanover, NH
MSc, Oxford University, UK

2. Phosphorus mining provides humans with an important product called ammonium phosphate. This key ingredient is used as fertilizer that helps to increase crop yields in our agricultural system. Ammonium phosphate is derived from calcium phosphate rock, which is strip-mined and pulverized in several U.S. states. Producers extract the calcium phosphate and then add sulfuric acid to form phosphoric acid, which is later modified to produce the desired product, ammonium phosphate.

   As an unfortunate by-product of this process, phosphogypsum is produced. In fact, five tons of phosphogypsum are produced for every ton of ammonium phosphate. The runoff from the phosphogypsum has a pH between 1 and 2.

   Florida generates 75% of the ammonium phosphate used by U.S. farmers each year and currently has one billion tons of phosphogypsum stored at 25 sites around the state, and miners are adding 30 million tons of phosphogypsum to those piles each year.

   (a) Using the information above, what is the annual demand for ammonium phosphate by U.S. farmers?
   (b) If the United States is responsible for 20% of the world's production, how much ammonium phosphate is produced globally?
   (c) Based on this world total, how much of the toxin phosphogypsum is produced on an annual basis?
   (d) Identify and describe TWO environmental consequences that result from the extraction/processing of phosphorus.
   (e) Describe ONE strategy to mitigate the environmental damage caused by extraction or processing.
   (f) Describe ONE alternative strategy that might reduce the need for phosphorus mining.

IMPORTANT NOTE TO STUDENTS FOR ALL FRQS

Please note that your answers to the Free-Response Questions (FRQs) will most likely be much shorter than what you see in this book, and that is perfectly fine and expected. The answers presented in this book are very comprehensive and try to cover a wide range of possible responses. When comparing your responses with the responses presented in this book, try to see if there are any major topics that you may have left out or if you have made any factual errors in what you did write about. Remember, you only have on average ~23 minutes to write each response.

Answer Explanations

1. **(B)** Refer to the biome descriptions (*pages 16–17*).
2. **(A)** Refer to the biome descriptions (*pages 18–20*).
3. **(D)** Refer to the biome descriptions (*pages 20–22*).
4. **(B)** Less energy is available at each trophic level because energy is lost by organisms through cellular respiration and incomplete digestion of food sources.
5. **(B)** Estuaries form a transition zone between river environments and ocean environments and are subject both to marine influences, such as tides, waves, and the influx of saline water, and to riverine influences, such as flows of freshwater and sediment. The inflow of both seawater and freshwater provide high levels of nutrients in both the water column and sediment, making estuaries the most productive natural habitats in the world and also the most threatened.
6. **(A)** Refer to pages 40–41 if you missed this question to review the steps involved in the nitrogen cycle.
7. **(A)** The sun is the original source of energy. Plants capture ~1% of the available light energy from the sun for biomass production by way of photosynthesis ($100{,}000 \times 0.01 = 1{,}000$ units of energy captured by plants). Herbivores (exclusive plant eaters) are only able to utilize about 10% of this captured energy ($1{,}000 \times 0.10 = 100$ units of energy). The hawk that eats an herbivore is likewise only able to utilize 10% of the energy within the herbivore ($100 \times 0.10 = 10$ units of energy). The loss of energy at each step is known as the 10% Rule.
8. **(C)** The question of determining whether it is a deciduous forest or grassland is dependent on yearly patterns of rainfall since both biomes can exist over similar temperature ranges. Frequent fires are an important factor in determining grasslands.
9. **(B)** $GPP = (NPP + R) = (\text{Light} - \text{Dark}) = (11 - 5) = 6$ mg/L/week

   $NPP = ((GPP - R) = \text{Light} - \text{Initial}) = (11 - 10) = 1$ mg/L/week

   $R = (\text{Initial} - \text{Dark}) = (10 - 5) = 5$ mg/L/week
10. **(D)** To determine whether a scenario represents either a positive or negative feedback loop, simply look for the words pertaining to "increase" and "decrease" at the very beginning of the scenario and at the very end of the scenario. In Statement A, it begins with "The increased temperature" and ends with "can reduce," which is consistent with a negative feedback loop. In Statement B, it begins with "temperature increases" and ends with "reduce natural food chain productivity," which is also consistent with a negative feedback loop.
11. **(D)** Photosynthesis is a process used by plants to convert light energy into chemical energy that can later be released to fuel an organism's activities. This chemical energy is stored in carbohydrate molecules (e.g., sugars [glucose, fructose, sucrose, etc.], polysaccharides [e.g., starches], and fiber [e.g., pectin]), which are synthesized from carbon dioxide and water. Oxygen is also released as a waste product. Photosynthesis is largely responsible for producing and maintaining the oxygen content of Earth's atmosphere and supplies all of the organic compounds and most of the energy necessary for life on Earth.
12. **(B)** Lines on a climatograph are used for showing temperature. New Delhi, India, had a minimum temperature of ~58°F during the months of December and January and a maximum temperature of ~97°F in June (a difference of ~39°F); Seattle had a difference of ~29°F (~69°F – ~40°F); and Paris, France, had a difference of ~34°F (~69°F – ~35°F).
13. **(A)** Bars on a climatograph are used for showing monthly precipitation amounts. New Delhi, India, had the tallest bar (amount of monthly rain) of any of the three graphs: ~8.1 inches during the month of June.
14. **(A)** The distribution of biomes is primarily determined by global climatic zones as measured by average annual temperature and rainfall. An area's climate will determine what sort of biome can be sustained in that area. Other determining factors include sun angle, heat import, and the angle of Earth's rotation. Regional climates, which also

determine the distribution of biomes, are affected by rain showers and cold or warm ocean currents.

15. **(B)** An indicator species is a species whose presence, absence, or relative well-being in a given environment is indicative of the health of the ecosystem as a whole.

16. **(C)** Marine biomass pyramids show the relative level of biomass at each of the trophic levels for ocean ecosystems and tend to be inverted due to the dynamics of the producers and consumers.

    Many marine ecosystems rely on phytoplankton as their primary producer. Phytoplankton are very small (even microscopic) and reproduce and die very quickly. Therefore, at any given moment their biomass is relatively small, even though they supply energy for the entire ecosystem.

    The next level of primary consumers consists of zooplankton, another microscopic organism, and other small organisms like small fish, sea anemones, krill, or crustaceans. These animals are larger in size than the phytoplankton and thus have more biomass. The secondary consumers (e.g., larger fish) are much larger than the primary consumers and producers. These organisms tend to live for longer periods of time as well, so at any given moment their biomass is quite significant (refer to Figure 1.58).

17. **(A)** Mice are known as primary consumers—they consume plant matter (grasses). The snake in this food web is known as a secondary consumer, as it feeds on the primary consumer (mice). A decrease in the food supply for the snakes (in this case, mice) would decrease their population. A decrease in the vulture population would result in an increase in the number of snakes, as the vultures consume snakes. An increase in the amount of plants would result in an increase in mice for the snakes to eat and therefore increase the snake population. An increase in the coyote population would mean a decrease in the mice and vulture population, resulting in an increase in snakes.

18. **(B)** The amount of energy at each trophic level decreases as it moves through an ecosystem. As little as 10% of the energy at any trophic level is transferred to the next level; the rest is lost largely through metabolic processes as heat (10% Rule). Example—if a grassland ecosystem has 10,000 kilocalories (kcal) of energy concentrated in vegetation, only about 1,000 kcal will be transferred to primary consumers. Efficiency is defined as the measure of how much work or energy is conserved in a process. The formula to calculate the percentage efficiency of energy transfer between trophic levels, which is known as the trophic level transfer efficiency, or TLTE, is: TLTE = Total energy transferred to next higher level = Energy received during transfer $\times$ 100% = (719 kJ $\div$ 8,350 kJ) $\times$ 100% = ~8.6% which is close to 10% and is in accordance with the 10% Rule.

19. **(C)** The taiga, also known as boreal forest, is the world's largest land biome, and makes up almost a third of the world's forest cover, with the largest areas located in Russia and Canada. The taiga has a subarctic climate with very large temperature ranges between seasons, but the long and cold winter is the dominant feature.

20. **(B)** Global warming causes the hydrologic cycle to accelerate due to warmer weather conditions, which result in evaporation, and more evaporation results in more precipitation—all leading to frequent and extreme weather conditions.

Free-Response

Answer #1

10 Total Points Possible

(a) *Maximum 4 points.*

Describe what a freshwater wetland is (1 point) and how it differs from an estuary (1 point). (2 point maximum)

Wetlands are areas where standing water covers the soil or an area where the ground is very wet, and can be found along the boundaries of streams, lakes, ponds, or even in large shallow holes that fill up with rainwater. Freshwater wetlands may stay wet all year long, or the water may evaporate during the dry season. Unlike estuaries, freshwater wetlands are not connected to the ocean.

(b) *Maximum 2 points (1 point for each correct description).*

Describe two different types of freshwater wetlands.

A marsh is a wetland that is dominated by vascular plants containing xylem for conducting water and minerals throughout the plant and phloem to conduct the products of photosynthesis (e.g., sugars) that have no persistent woody stems above ground. Marshes can often be found at the edges of lakes and streams, where they form a transition between the aquatic and terrestrial ecosystems. Marshes are often dominated by grasses, rushes, or reeds. If woody plants are present, they tend to be low-growing shrubs. This form of vegetation is what differentiates marshes from other types of wetland such as swamps, which are dominated by trees, and mires or bogs, which are wetlands that have accumulated deposits of acidic peat.

(c) *Maximum 1 point. ALL boxes must have at least one connecting arrow, and no points are earned if ANY arrows are incorrect.*

Freshwater wetlands are highly productive ecosystems with complex food webs. Complete the diagram of the freshwater food web below by drawing arrows that show a possible direction of energy flow.

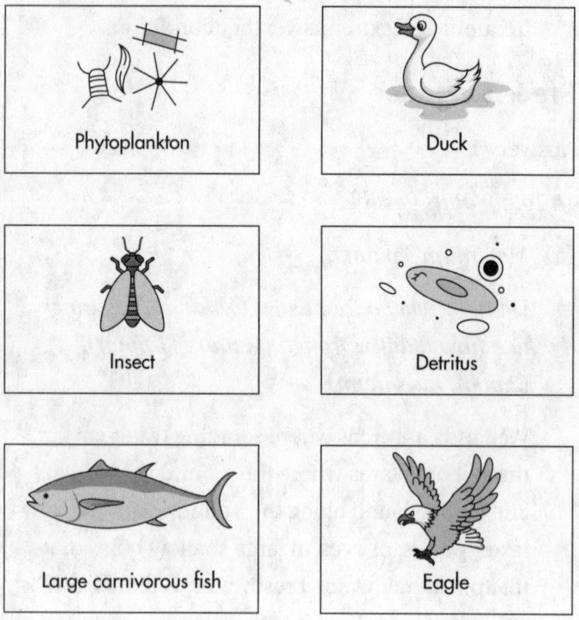

(d) *Maximum 2 points (1 point for describing what a BOD test is and 1 point for describing what the results can indicate).*

One of the activities that Mrs. Kaplan's APES students performed while on the field trip was a water-quality test known as a biological oxygen demand (BOD) test. Describe what a BOD test is and what the results can indicate.

Biological oxygen demand (BOD) is a measure of the oxygen used by microorganisms to decompose organic waste. If there is a large quantity of organic waste in the water supply, bacterial counts will be high. Therefore, the demand for oxygen will be high, resulting in a high BOD level. As the waste is consumed or dispersed through the water, BOD levels will begin to decline. Higher-than-normal nitrate and phosphate concentrations in a body of water (known as cultural eutrophication) also contribute to high BOD levels. Nitrates and phosphates are plant nutrients and can cause plant life and algae to grow quickly.

When plants grow quickly, they also die quickly which contributes to the organic waste in the water, which is then decomposed by bacteria and results in a high BOD level. When BOD levels are high, dissolved oxygen (DO) levels decrease because the oxygen that is available in the water is being consumed by the bacteria, and since less dissolved oxygen is available in the water, fish and other aquatic organisms may not survive.

(e) *Maximum 1 point.*

The results of the BOD test that the students performed indicated the presence of organic wastes in the water. Describe ONE remediation technique that could be used to reduce the environmental impact to the wetland.

The first step in reducing the environmental impact of low dissolved oxygen content in the water is to identify the source of the waste. By carefully testing various sites along a stream or within the wetland, it may be possible to identify exactly the source of the organic pollution, e.g., leaking sewer line, leaking septic tank, discharge from a factory or cattle feedlot, or other point sources of pollution. Once the source has been identified, some options are available: (1) contact the polluter and let them know the results; and/or (2) contact local, state, or national authorities.

Remediation techniques could involve changes in the type of fertilizer being used and its application, erosion- and sediment-control techniques, changes in animal feeding operations or cattle grazing management, and/or changes in irrigation water management. However, non-point pollution sources, such as agricultural runoff, may be more difficult to locate and identify.

(f) *Maximum 1 point.*

Describe what indicator species are and how they are used to assess environmental quality in a freshwater ecosystem.

Indicator species are species whose presence, absence, or relative well-being in an environment is a sign of the overall health of the ecosystem. By monitoring the condition and behavior of an indicator species, scientists can determine how changes in the environment are likely to affect other species that are more difficult to study.

(g) *Provide a specific example of an indicator species that students may have found in the wetland and describe how it functions in its role as an indicator species.*

Examples of a wetland indicator species are mayflies or salmon whose presence (or absence) are indicators of freshwater quality. Other wetland indicator species include lichen which can be used to assess air pollution and spotted owls whose absence or decline can be an indicator of loss of habitat.

Free-Response

Answer #2

10 Total Points Possible

(a) *Maximum 1 point total: 1 point for the correct answer, which must show the correct setup.*

Using the information above, what is the annual demand for ammonium phosphate by U.S. farmers?

The annual demand is calculated by using the annual production values from Florida to determine the total U.S. production. Since 30 million tons are produced in Florida AND because 75% of the U.S. total comes from Florida, 40 million tons are produced annually. This is because 30 represents 75% of 40. A range of calculation setups are possible. The correct answer is 40 million tons.

$$\frac{30\ \cancel{\text{million tons (FL)}}}{1} \times \frac{100\ \text{million tons (U.S.)}}{75\ \cancel{\text{million tons (FL)}}} = \mathbf{40\ million\ tons}$$

(b) *Maximum 1 point total: 1 point for the correct answer, which must show the correct setup.*

If the U.S. is responsible for 20% of the world's production, how much ammonium phosphate is produced globally?

Since the United States produces 20% of the world total, the correct answer is derived by multiplying 40 million by 5, yielding a correct answer of 200 million tons.

$$\frac{40\ \cancel{\text{million tons (U.S.)}}}{1} \times \frac{100\ \text{million tons (world total)}}{20\ \cancel{\text{million tons (U.S.)}}} = \mathbf{200\ million\ tons\ produced}$$

(c) *Maximum 2 points total: 1 point for the correct setup and 1 point for the correct answer.*

Based on this world total, how much of the toxin phosphogypsum is produced on an annual basis?

The following setup is suggested, though inclusion of units is not required. The correct answer can be displayed in the following formats: 1 billion tons, 1,000 million tons, 1×10^9 tons.

$$\frac{2 \times 10^8\ \cancel{\text{tons ammonium phosphate produced}}}{1} \times \frac{5\ \text{tons phosphogypsum}}{1\ \cancel{\text{ton ammonium phosphate produced}}} =$$

$$\mathbf{1 \times 10^9\ tons\ phosphogypsum}$$

(d) *Maximum 4 points total: 1 point for identifying each environmental consequence (2 points) and 1 point for describing each environmental consequence (2 points).*

Identify and describe TWO environmental consequences that result from the extraction/processing of phosphorus.

The low pH runoff that results from phosphogypsum was identified in the question's background information. Responses may include environmental consequences that result from acid mine drainage, such as low pH or the acidification of aquatic ecosystems adjacent to the mines/processing facilities. Habitat destruction or the direct consequences of acidic environments will result in species loss or reduction in biodiversity.

Strip-mining is one of the more impactful types of mining in terms of habitat destruction, as vegetation and soils are removed entirely to access the resources below. The use of large mining vehicles produces a range of air pollutants that can also be connected to environmental impacts. CO_2 additions will cause climate change, which would consequently result in species decline and habitat destruction. Particulates from earth-moving activities and/or the particulates released from fossil fuel consumption by mining equipment may reduce the photosynthetic capabilities of adjacent plant life.

(e) *Maximum 1 point total: 1 point for describing the mitigation strategy.*

Describe ONE strategy to mitigate the environmental damage caused by extraction or processing.

A range of mitigation strategies are acceptable.

Bioremediation, which uses toxin-tolerant plants to absorb and take up toxins from the soil or surrounding water bodies, is often used in this type of situation. Certain bacteria may also be used to convert some of the harmful toxins, turning them into more benign products. Photoremediation is a process that uses light to break down toxins. Topsoil from other areas might also be brought in to recreate a soil system that can support plant life, helping to stabilize the disturbed soil.

(f) *Maximum 1 point total: 1 point for describing the alternative strategy.*

Describe ONE alternative strategy that might reduce the need for phosphorus mining.

Credit can be earned for describing any strategies that reduce the demand/need for phosphorus fertilizers. This can include suggesting alternatives such as organic fertilizer substitutes; the use of sustainable farming practices, such as crop rotation that utilizes other crops; natural replenishment (through composting or something similar); or using crops that demand less phosphorus (generally modified crops are acceptable).

2

The Living World: Biodiversity

Learning Objectives

In this unit, you will learn:

- ➔ 2.1 Introduction to Biodiversity
- ➔ 2.2 Ecosystem Services
- ➔ 2.3 Island Biogeography
- ➔ 2.4 Ecological Tolerance
- ➔ 2.5 Natural Disruptions to Ecosystems
- ➔ 2.6 Adaptations
- ➔ 2.7 Ecological Succession

For added practice with this unit and topics, log onto Online Labs 2A–2B at online.barronsbooks.com

2.1 Introduction to Biodiversity

Biodiversity is defined as the variability among species, between species, and of ecosystems. It can be described and defined at the genetic, species, and ecosystem levels.

- **Ecosystem diversity**—describes the range of habitats that can be found in a specific area. Ecosystems that have high biodiversity are characterized by the following:
  - Abundant natural resources
  - Large genetic diversity
  - Complex food webs involving a variety of ecological niches
  - Large numbers of organisms of different species
  - Large numbers of different species
- **Genetic diversity**—describes the range of all genetic traits, both expressed and recessive, that makes up the gene pool for a particular species
- **Species diversity**—is the number of different species that inhabit a specific area (e.g., tropical rainforests have higher species diversity than deserts)

Diversity Increasers	Diversity Decreasers
Disturbance in the habitat (fires, storms, etc.)	Environmental stress
Diverse habitats	Extreme amounts of disturbance
Environmental conditions with low variation	Extreme environments
Evolution	Extreme limitations in the supply of a fundamental resource
Middle states of succession	Geographic isolation
Trophic levels with high diversity	Introduction of species from other areas

Biodiversity is important because it helps keep the environment in a natural balance. An ecosystem that is species-rich is more resilient and adaptable to external stress than one in which the range of species is limited. In a system where species are limited, the loss or temporary reduction of any one species could disrupt the food chain, with serious effects on other species in that same system. When biodiversity is sufficient, and if one nutrient-cycling path is affected, another pathway can function and the ecosystem can survive.

Anthropogenic Activities That Can Reduce Biodiversity

Anthropogenic Activity	How It Reduces Biodiversity	How the Activity Can Be Remediated
Burning fossil fuels	The acid rain produced from burning fossil fuels changes the pH of water habitats to the extent that many species cannot survive.	– Enact carbon taxes – Require scrubbers for all industries that burn coal – Have tax incentives for products that do not require burning fossil fuels – Use renewable energy (wind, solar)
Deforestation	Deforestation reduces the quality and quantity of suitable habitats for many species of flora and fauna	– Requires replanting – Use selective cutting
Modern industrial agriculture	Modern industrial agriculture involves the use of one or two crops that cover massive areas of land.	– Crop rotation – Intercropping – Interplanting – Organic farming – Polyculture – Polyvarietal cultivation
Overfishing	Overfishing negatively impacts keystone species and threatens endangered species. Overfishing of lower-trophic-level species affects all organisms in the niche.	– Enforce international treaties that monitor and penalize countries that overfish – Establish quotas for all species fished – Have tighter enforcement of the Endangered Species Act
Use of pesticides	Pesticides indiscriminately kill both pests and beneficial organisms.	– Require integrated pest management techniques
Using genetically modified organisms (GMOs)	GMOs decrease the genetic variation necessary to cope successfully with changes in the environment.	– Require package labeling for all products that use GMOs – Require that GMO crops be sterile – Require vigorous testing and research before allowing GMOs to be used
Water pollution	High nutrient levels or low dissolved oxygen levels resulting from water pollution can be lethal to some species.	– Require secondary and tertiary treatment methods for all water treatment plants – Enact stricter penalties and tougher regulations for companies that pollute – Use recycled water

Population Bottleneck

A population bottleneck is a large reduction in the size of a single population (not at a community or larger ecosystem level) due to a catastrophic environmental event (e.g., disease, drought, fire, or flood). As a result of the smaller population, there is less genetic diversity in the gene pool for future generations. Genetic diversity within the population only increases when there is gene flow from another population or over time, as random mutations occur that are better suited for the changed environment.

The effects of a population bottleneck often depend on the number of individuals remaining after the bottleneck and how that compares to the smallest possible size at which a population can exist without facing extinction from a natural disaster, also known as the minimum viable population size.

POPULATION BOTTLENECK EXAMPLE

The population of the northern elephant seal fell to about 30 in the 1890s due to overhunting for ivory. However, currently the population numbers are in the hundreds of thousands. Dominant male elephant seals (bulls) mate with the largest number of females, often over 100. With so much of a colony's offspring descended from just one dominant male, genetic diversity is limited, making the species more vulnerable to diseases and genetic mutations.

Figure 2.1 illustrates how maintaining or increasing diversity can stabilize how ecosystems function. Each box represents a community of two species of annual plants: ○ and ●. ○ plants do better in warm years, whereas ● plants do better in colder years. In the diverse community containing both ○ and ●, over time and with fluctuating temperatures, genetic variation remains, whereas in the communities dominated by either ○ or ● plants, genetic diversity is diminished, resulting in a decrease in productivity (total circles).

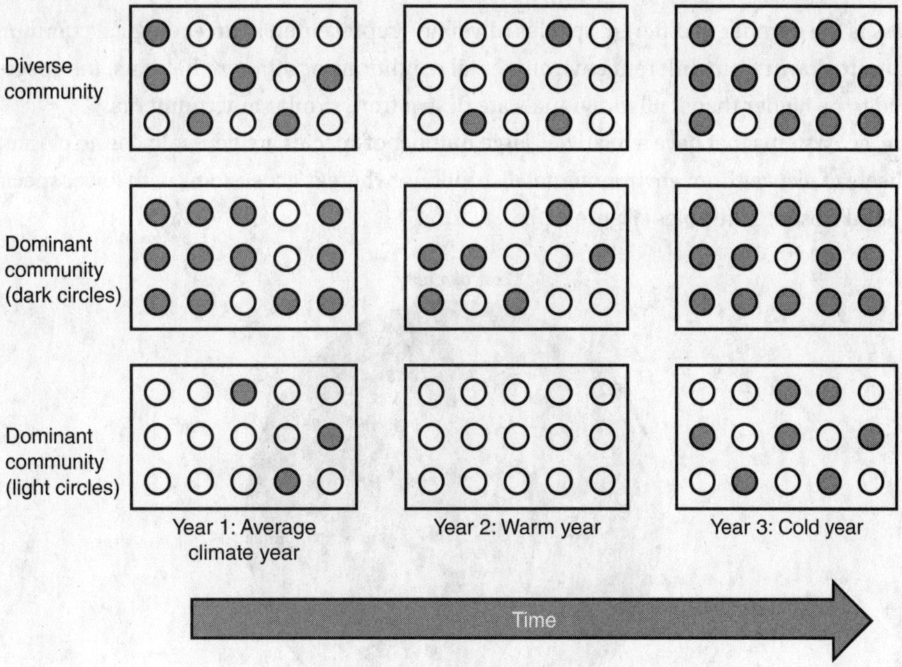

Figure 2.1 Population bottleneck

Loss of Habitat = Loss of Specialist Species

Species that live in different types of environments and have varied diets are known as generalist species. An example of a generalist species is the raccoon, which lives throughout much of North and Central America. Raccoons are classified as omnivores as they are able to survive on a large variety of food types. It is this ability to survive in a wide range of environments that has enabled raccoons to maintain large and stable populations.

Specialist species, however, require unique resources and often have a very limited diet; they often need a specific habitat in which to survive. An example of a specialist species is the giant panda bear, which survives almost entirely on bamboo and lives in remote bamboo forests in China. Studies have shown that the giant panda has the digestive system of a carnivore and, as a result, does not receive many calories or proteins from exclusively eating bamboo. This results in a lower metabolic rate and is demonstrated by their sedentary (very little physical activity) lifestyle. Because a panda's diet consists almost entirely of bamboo, it must eat between 25 pounds and 50 pounds of it every day to survive.

Poaching and habitat destruction in China over the past 3,000 years have brought the total panda population in the wild down below 2,000 individuals, which exist in an area that is less than 1% of their historical range.

Current conservation programs to save pandas have begun to increase the panda population; however, this success is being offset by a faster rate of global warming, which will make the current bamboo habitat unsuitable for panda survival, as it is estimated that the habitat may be all but gone by the end of this century, with half of it vanishing by 2070.

So far, average global temperatures have risen by 1.5°F over the last 100 years, with predictions that by the year 2100, average global temperatures could rise another 1°F to 6°F. Research suggests that half of the bamboo forests will disappear with an additional 2°F increase in temperature.

Conservation methods that can save bamboo forests and the pandas that depend on them include the following:

- Begin planting bamboo in new areas.
- Ensure connectivity between current and future panda nature reserves to allow for migration and gene flow.
- Protect habitats that will soon become climatically suitable for bamboo growth.

Species Richness

Species richness is the number of different species (diversity) represented in an ecological community or region.

If individuals are drawn from different environmental conditions or different habitats, the species richness can be expected to be higher than if all individuals are drawn from similar environments.

Furthermore, ecosystems that have a relatively large number of species are generally found closer to the equator and are more likely to recover from environmental disruptions, whereas ecosystems with lower species richness are generally found closer to the poles (Figure 2.2).

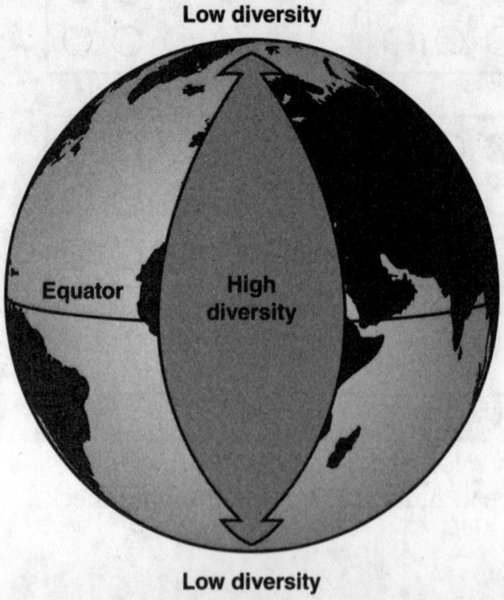

Figure 2.2 Species diversity as a function of latitude

2.2 Ecosystem Services

Humans benefit from ecosystems in a multitude of ways. Collectively, these benefits are known as ecosystem services. Ecosystem services can be grouped into four broad categories:

1. Cultural (e.g., spiritual and recreational benefits)
2. Provisioning (e.g., production of food and water)
3. Regulating (e.g., control of climate and disease)
4. Supporting (e.g., nutrient cycles and crop pollination)

Ecosystem Service Examples	Benefit
Cultural: Sustainable fisheries and aquaculture can directly support recreational services.	- Recreational fishing is linked to healthy aquatic ecosystems.
Provisioning: Ecosystems provide diversity of materials and products.	- Livestock provide different types of raw material such as fiber (wool), meat, milk
Regulating: Keep pest populations in balance through natural predators.	- Keeps food prices lower - Reduces the need for pesticides - Achieved in ecosystems through the actions of predators and parasites as well as by the defense mechanisms of their prey (e.g., birds and bats that eat insects)
Supporting: Form new soil and renew soil fertility.	- Allows for greater crop yields, which can feed more people - Reduces the need for fertilizer

2.3 Island Biogeography

The term *island* refers to a suitable habitat for a specific ecosystem that is surrounded by a large area of unsuitable habitat and may refer, for example, to actual islands, isolated lakes, mountaintops, or natural habitats surrounded by human-altered landscapes; e.g., expanses of grassland surrounded by highways or dense urban developments.

Island biogeography examines the factors that affect the richness and diversity of species living in these isolated natural communities (islands).

The theory of island biogeography proposes that the number of species found on an "island" is determined by immigration and extinction of isolated populations. The species may follow different evolutionary routes (e.g., finches on different islands in the Galapagos Islands). Islands that are more isolated are less likely to receive immigrants than islands that are less isolated.

Diagram A in Figure 2.3 shows the effect of an island's *distance* from the mainland on the amount of species richness. The sizes of the two islands are approximately the same. Island 1 receives more random dispersion of organisms, while island 2, since it is farther away, receives less random dispersion of organisms. Therefore, island 1 would have a higher level of diversity or species richness than island 2.

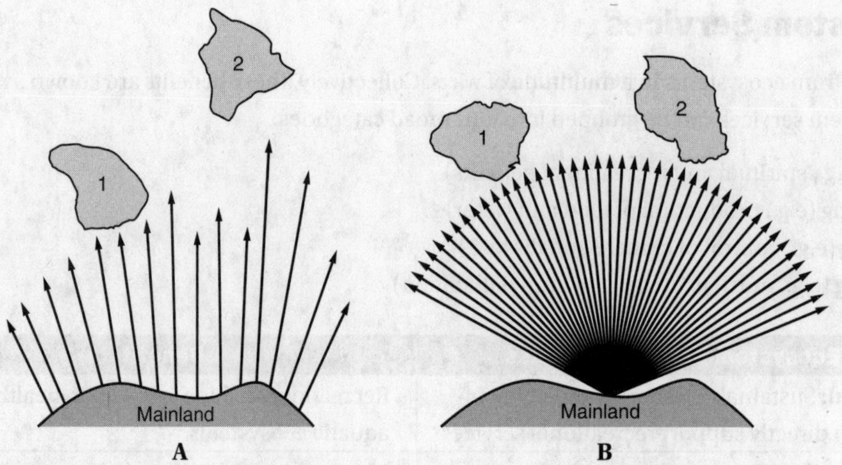

Figure 2.3 How island distance and size from the mainland affect diversity

Diagram B shows the effect of an island's *size* on the amount of species diversity when the two islands are the same distance from the mainland. Larger islands generally have larger and more diverse habitat areas and opportunities for more diverse habitats. This larger habitat size reduces the probability of extinction due to chance events, while differences in habitat increase the number of species that would be successful after immigration. In this case, island 1 would receive less random dispersion of organisms while island 2 would have more random dispersion of organisms; therefore, island 2 would have a higher level of species diversity than island 1.

In addition to island size and distance from the mainland, island biogeography as it relates to species diversity is also influenced by the following:

- Degree of isolation (distance to nearest island and/or mainland)
- Habitat fragmentation occurs when a habitat is broken into pieces by development, industry, logging, roads, etc., and can cause an edge effect (changes in population or community structures that occur at the boundary of two or more habitats). Generally, the edges of a habitat have fewer species because species that thrive in the interior of a habitat can fail to find food or reproduce near the edge; in addition, some animals may face more predators near the edges of a habitat.
- Habitat suitability, which includes
  - climate (e.g., tropical versus arctic, arid versus humid, etc.);
  - initial plant and animal composition, if previously attached to a larger land mass; and
  - the current species composition.
- Human activity and subsequent level of disruption
- Location relative to ocean currents (influences bird, fish, and nutrient patterns and pathways) (Figure 2.4)

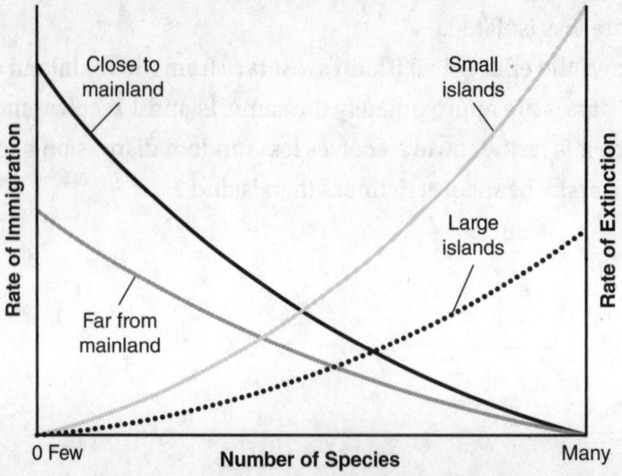

Figure 2.4 Size of islands and distance to mainland as a function of the number of species

CASE STUDY—DARWIN'S FINCHES

In 1835, while visiting the Galapagos Islands, Charles Darwin discovered species (e.g., finches) that were not found anywhere else on Earth (and still aren't). For example, in addition to the Galapagos Islands' being isolated, each island had its own unique species* of finch. Darwin concluded that the islands had broken off separately from the mainland many thousands of years ago. Over a great amount of time, the once single species had become reproductively and geographically isolated from the other finches and eventually developed into new species.

**the largest group of organisms in which two individuals of the opposite sex can produce fertile offspring*

Points to Remember

- Closer islands are also easier to find for migrating species.
- Habitat fragmentation is currently the main threat to terrestrial biodiversity.
- Islands closer to the mainland have more biodiversity.
- Island biogeography is used to predict biodiversity and extinction rates in habitat fragmentation on the continents.
- Larger islands are bigger targets, so migrating species can find them more easily.
- Larger islands have more biodiversity.
- Larger islands have higher populations of species and therefore lower extinction rates because there is less of a chance for population numbers to drop to zero with an ecosystem disturbance.

2.4 Ecological Tolerance

Earth's ecosystems are affected by both biotic (living) and abiotic (nonliving) factors, and are regulated by the Law of Tolerance, which states that the existence, abundance, and distribution of species depend on the tolerance level of each species to both physical and chemical factors within its environment (Figure 2.5). In other words, an organism's success is based on a complex set of conditions, and each organism has certain minimum, maximum, and optimum environmental factors or combination of factors that determine its success.

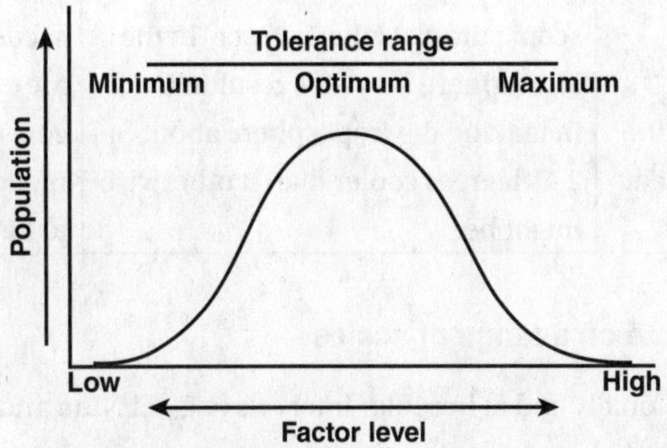

Figure 2.5 Law of Tolerance

The abundance or distribution of an organism can be controlled by certain factors, such as its biological, climatic, and topographic requirements. In situations where levels of these exceed the maximum or minimum limits of tolerance of that organism, species numbers will decline.

2.5 Natural Disruptions to Ecosystems

An ecosystem is a community of organisms that interact with each other and their environment and that can change over time. Natural and sudden disruptions such as flooding, volcanic eruptions, or wildfires can dramatically affect which species will thrive in an environment and which species will not and will possibly become extinct. Extinction not only affects the species that becomes extinct, but also can dramatically affect or disrupt all other organisms in that food web and dramatically influence local biodiversity.

Natural environmental disruptions also have the potential to affect ecosystems as much as or more than disruptions caused by humans. The following are three types of natural disruption and their potential effects on an ecosystem.

Flooding	Volcanic Eruptions	Wildfires
Kills wildlife and their food source (vegetation and/or other animals)	Kills wildlife and their food source (vegetation and/or other animals)	Kills wildlife and their food source (vegetation and/or other animals)
Soil erosion—Soil is no longer held in place by roots.	Soil erosion—Soil is no longer held in place by roots.	Soil erosion—Soil is no longer held in place by roots.
Flooding can result in water-saturated soils. Plant roots require oxygen, so saturated soils can kill plants by drowning the plant roots.	Volcanic materials ultimately break down and weather to form some of the most fertile soils on Earth, cultivation of which has produced abundant food and fostered civilizations.	Helps the ecosystem by clearing out much of the dead and dying vegetation, allowing surviving plants to benefit from increased light
Flooding may also cause water and nutrients to run off across land surfaces. Rushing water can cause soil to wash away, particularly bare soil. Burrows, dens, and nests can be destroyed, forcing surviving animals to relocate.	Over 4.5 billion years, the condensed steam from volcanoes and cooling magma produced all of the water found on the Earth today. Volcanoes also contributed to a large portion of Earth's early atmosphere.	Ash and charcoal left from burnt vegetation can help add nutrients to depleted soil. These nutrients provide a rich environment for surviving vegetation and sprouting seeds.
Flood plains, which are flat areas along rivers that flood when the river rises above its banks, are inhabited by species that have adapted to occasional flooding. The flooding deposits nutrient-rich sediment along stream banks.	Sulfur gas combines with water in the atmosphere, creating microscopic droplets that can stay in the atmosphere for years, resulting in making the troposphere about 2–3 degrees cooler than it otherwise might be.	Several plants actually require fire in their life cycles. For example, seeds from many pine tree species are enclosed in pine cones that are covered in pitch, which must be melted by fire in order for the seeds to be released.

Earth system processes operate on a range of scales

- **Episodic**—occurring occasionally and at irregular intervals (e.g., El Niño and La Niña [~ every 2–7 years])
- **Periodic**—occurring at repeated intervals; e.g., tides
- **Random**—lacking a regular pattern; e.g., meteorite impacts

Earth's climate has changed over geologic time for many reasons

- Changes in ocean and atmosphere circulation patterns
- Plate tectonics—continents "drifting" to different latitudes
- Varying concentrations of atmospheric carbon dioxide
- Volcanic eruptions

Sea level has varied significantly as a result of changes in the amount of glacial ice on Earth over geological time

Global sea level has changed significantly over Earth's history, with sea level being affected by the amount and volume of available water and the shape and volume of the ocean basins.

The temperature of ocean water (which affects density) and the amount of water retained in aquifers, glaciers, lakes, polar ice caps, rivers, and sea ice, along with the changing shape over time of the ocean basins, combined with tectonic uplift and land subsidence, have a direct influence on sea level. However, the primary reason for changes in sea level today are glacier and sea ice melts caused by global warming (current sea levels are about 425 feet [130 m] higher than the historical minimum, which occurred about 20,000 years ago—a period known as the Last Glacial Maximum—while the last time sea level was higher than today occurred about 130,000 years ago (Figure 2.6).

Here are some things to keep in mind:

- ~30% of sea-level change is due to the melting of glaciers and ice sheets on land.
- ~30% of sea-level change is due to thermal expansion—as the oceans warm (climate change), water expands.
- ~40% of sea-level change is due to coastal land subsidence (sinking).

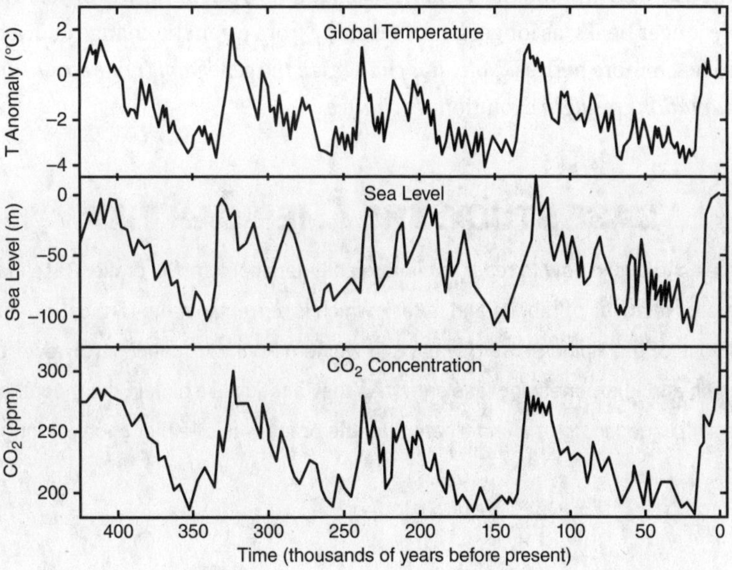

Figure 2.6 Changes in global temperatures, CO_2 concentrations, and sea level over the last 400,000 years

Wildlife engages in both short- and long-term migration for a variety of reasons, including natural disruptions

- Escaping harsh weather; e.g., seeking warmer water for breeding and raising young but returning to colder water for feeding as there is more food available
- Escaping natural disasters and their environmental aftermath (e.g., wildfires, floods, and storm events)
- Finding natural resources of food

2.6 Adaptations

Adaptation is the biological mechanism by which organisms adjust to new environments or to changes in their current environment. It is an evolutionary process that works through the mechanism of natural selection and consists of three types:

1. Behavioral—such as instincts (e.g., hunting for food), mating behavior, or vocalizations (e.g., bird calls)
2. Physiological—such as methods of temperature control or how food is digested
3. Structural—involve physical features such as body coverings (e.g., feathers, fur, scales, or skin)

Adaptation, whether it be behavioral, physiological, or structural, is a characteristic of all living organisms and can be classified as either short term or long term. A short-term adaptation

- develops in response to temporary changes in the environment
- involves temporary changes
- is not inherited, nor does DNA change
- plays no role in evolutionary processes.

Example 1: A particular bird species on a remote island eats a specific type of berry found on that remote island. A specific disease wipes out all of those berries, but all remaining birds of that species are capable of surviving by eating another type of berry not affected by the plant disease. This example is a short-term adaptation not involving evolutionary processes, as all birds were capable of eating the surviving berries and nothing changed in the gene pool (e.g., beak length, behavioral preference, digestive enzymes).

Long-term adaptations may involve DNA changing over long time periods in response to natural selection involving evolutionary processes.

Example 2: Imagine the same environmental scenario as Example 1, but hundreds of years later the birds on the island are found to have longer beaks, as longer beaks resulting from a genetic change allowed more of the birds to survive by being able to reach more berries. *Note: It is changes in the gene pool (not necessarily the time span) that determines whether adaptations result in* evolutionary *change.*

CASE STUDY: THE AMERICAN PIKA

Animals only have three choices in how to react to climate change: adapt, move, or die. Case in point, the American pika (herbivorous, smaller relatives of rabbits and hares), which is currently at the farthest extent of their range in the cool, moist conditions of the alpine Sierra Nevada and western Rocky Mountains. However, their habitat has become hotter and drier, and is experiencing less snow. As they already live high in the mountains, there is no place left to go; global warming is occurring at a faster rate than the processes of either short- or long-term adaptation and natural selection.

2.7 Ecological Succession

Ecological succession is the gradual and orderly process of ecosystem development brought about by changes in community composition and the production of a climax community and describes the changes in an ecosystem through time and disturbance.

Here are some helpful terms to know:

- **Facilitation** is when one species modifies an environment to the extent that it meets the needs of another species.
- **Inhibition** is when one species modifies the environment to an extent that is not suitable for another species.
- **Tolerance** is when species are not affected by the presence of other species.

Earlier successional species, frequently called pioneer species, are generalists. Pioneer plants have short reproductive times (annuals), and pioneer animals have low biomass and fast reproductive rates. Later successional species include larger perennial plants and animals with greater biomass, longer generational times, and higher parental care.

The following table lists the characteristics of succession within plant communities.

Characteristics of Succession Within Plant Communities

Characteristic	Early Successional Stage	Late Successional Stage
Biomass	Limited	High in tropics and wetlands; limited in deserts.
Consumption of soil nutrients	Nutrients are quickly absorbed by simpler plants.	Since biomass is greater and more nutrients are contained within plant structures, nutrient cycling between the plant and soil tends to be slower.
Impact of macro-environment	Early plants depend primarily on conditions created by macro-environmental changes (fires, floods, etc.).	These plant species appear only after macro-environmental changes (fires, floods, etc.), and after pioneer plant communities have adequately prepared the soil.
Life span of seed	Long. Seeds may become dormant and able to withstand wide environmental fluctuations.	Short. Not able to withstand wide environmental fluctuations.
Life strategy	*r*-strategists: mature rapidly; short-lived species; number of organisms within a species is high; low biodiversity; niche generalists.	*K*-strategists: mature slowly; long-lived; number of organisms within a species is lower; greater biodiversity; niche specialists.
Location of nutrients	In the soil and in leaf litter.	Within the plant and top layers of soil.
Net primary productivity	High	Low
Nutrient cycling by decomposers	Limited	Complex
Nutrient cycling through biogeochemical cycles	Because nutrient sinks have not fully developed, the nutrients are available to cycle through established biogeochemical cycles fairly rapidly.	Because of nutrient sinks (carbon trapped in vegetation), nutrients may not be readily available to flow through cycles.
Photosynthesis efficiency	Low	High
Plant structure complexity	Simple	More complex
Recovery rate of plants from environmental stress	Plants quickly and easily come back.	Recovery is slow.
Seed dispersal	Widespread	Limited in range
Species diversity	Limited	High
Stability of ecosystem	Since diversity is limited, ecosystem is subject to instability.	Due to high diversity, ecosystem can withstand stress.

Primary vs. Secondary Succession

Ecological succession is the process of change in the species structure of an ecological community over time, which can be millions of years in the case of primary succession (e.g., mass extinction) or decades in the case of secondary succession (e.g., wildfire).

Primary succession—The evolution of a biological community's ecological structure in which plants and animals first colonize a barren, lifeless habitat. Species that arrive first in the newly created environment are called pioneer species, and through their interactions they build a simple initial biological community that becomes more complex as new species arrive.

Secondary succession—Type of ecological succession in which plants and animals recolonize a habitat after a major disturbance; e.g., a devastating flood, lava flow, or wildfire significantly alters an area but has not rendered it completely lifeless (Figure 2.7).

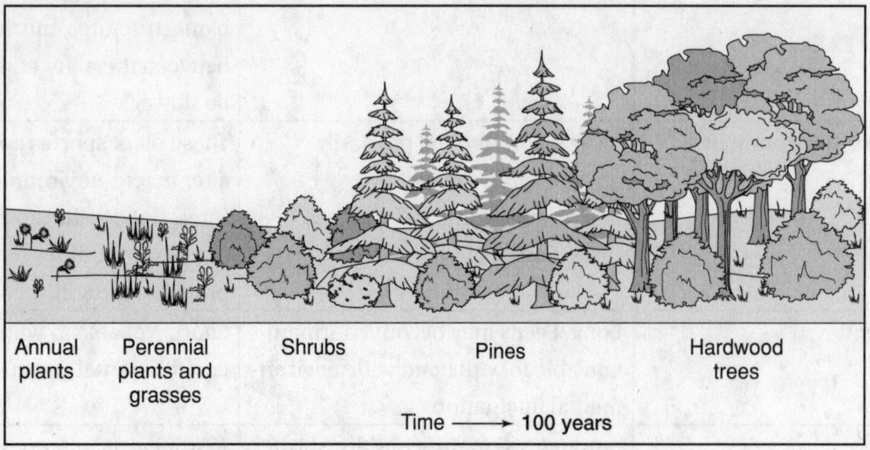

Figure 2.7 Stages of secondary succession

Ecological succession in a disturbed ecosystem and how it affects total biomass, species richness, and net productivity

An ecological disturbance is an event or force that can result in mortality to organisms and changes in the spatial patterns in their ecosystem and plays a significant role in shaping the structure of individual populations within the ecosystem. Examples of ecological disturbances include disease outbreaks, droughts, floods, forest fires, and volcanic eruptions. The impact that a disturbance has on an ecosystem depends upon

- intensity and frequency
- season
- size and spatial pattern
- topography (shape and features of land surfaces)

Succession is a directional, non-seasonal, cumulative change in the types of plant species that occupy a given area through time and involves the processes of (in order) colonization, establishment, and extinction, and demonstrates the progressive changes that occur within an ecosystem after an ecological disturbance (e.g., fire).

Species richness (number of different species represented in an ecological community) generally increases as succession proceeds and generally peaks when it reaches the climax community, but the diversity growth rate gradually slows down as succession advances to the climax community.

During succession, the patterns of energy flow, gross and net productivity, diversity, and mineral cycling change over time

In the early stages of succession, gross productivity is low due to the initial environmental conditions and low numbers of producers. The proportion of energy lost through community respiration is also relatively low, so net productivity is high; that is, the system is growing and biomass is accumulating.

In later stages of succession nearer to the climax community, gross productivity (GP) may be high; however, this is balanced by increased respiration (R), so net productivity approaches zero and the gross production respiration (GP:R) ratio approaches 1:1.

Changes that occur during succession include the following:

- Biodiversity increases and then falls as the climax community is reached.
- The biomass production-respiration ratio falls.
- Early stages of succession have few species.
- Energy flow becomes more complex.
- Net primary productivity (NPP) and gross primary productivity (GPP) rise and then fall.
- Soil depth, humus, water-holding capacity, mineral content and cycling increase.
- Species-diversity increase continues until a balance is reached between (1) existing species to expand their range; (2) possibilities for new species to establish; and (3) local extinction.
- Species diversity increases with succession.
- The size of organisms increases.

	Primary Succession	Secondary Succession
Definition	Occurs on barren, uninhabited areas; e.g., rock, sand dunes, etc.	Life has existed there before but disappeared due to natural disturbances; e.g., fires, floods, etc.
Environment	In the beginning, unfavorable	Favorable from the beginning
Occurs on	Barren or lifeless areas	Where life once existed
Pioneer community	Arrives from outside the area	Develops from previous occupants and from migrating species
Soil	No soil at the beginning	Soil and some organisms are present
Time to complete	1,000+ years	50–200 years
Stages (Example)	1. Bare rocks 2. Pioneer microorganisms 3. Plants (lichens and mosses) 4. Grassy stage 5. Smaller shrubs, and trees 6. Animals begin to return when there is food there for them to eat. 7. When it is a fully functioning ecosystem, it has reached the climax community stage.	1. A stable deciduous forest community exists. 2. A disturbance (e.g., wildfire) destroys the forest and burns it to the ground. 3. The fire leaves behind soil. 4. Grasses and other herbaceous plants grow back first. 5. Small bushes and trees begin to colonize the area. 6. Fast-growing evergreen trees develop while shade-tolerant trees develop in the understory. 7. The short-lived and shade-intolerant evergreen trees die as the larger deciduous trees survive. The ecosystem is now back to a state similar to how it began.

Keystone Species

A keystone species is a species whose very presence contributes to a diversity of life and whose extinction would lead to the extinction of other forms of life. Through various interactions, a small number of individuals from a keystone species have a very large and disproportionate impact on how ecosystems function. An ecosystem may experience a dramatic shift if a keystone species is removed, even if that keystone species was a small part of the ecosystem as measured by biomass or productivity. Examples include the following:

- Certain bat species pollinate critical trees in the rainforest and help disperse their seeds.
- Grizzly bears transfer nutrients from oceanic to forest ecosystems.

- Prairie dog burrowing aerates the soil and improves soil structure, while other animals use prairie dog burrows for shelter and hibernation.
- Sea stars (starfish) prey on sea urchins, mussels, and other shellfish that have no other natural predators, keeping their populations in check.

Indicator Species

Indicator species are organisms whose presence, absence, or abundance reflects a specific environmental condition and can indicate the health of an ecosystem. Examples include the following:

- Caddisflies, mayflies, and stoneflies—require high levels of dissolved oxygen in the water
- Lichens—some species indicate air pollution
- Mollusks—indicate water pollution
- Mosses—indicate acidic soil
- Sludge (Tubifex) worms—indicate stagnant, oxygen-poor water

Practice Questions

Examine the following diagram to answer Question 1.

Community 1

Community 2

1. Species richness refers to the number of different species represented in an ecological community, landscape, or region. Which of the following statements is/are TRUE regarding species evenness?

   I. In Community 1, the abundance of the light-colored mushrooms is high relative to the other species, resulting in a high species richness.

   II. In Community 2, the abundance of the dark-colored mushrooms is low relative to the other species, resulting in low species richness.

   III. Both communities have the same species richness.

   (A) I
   (B) II
   (C) III
   (D) All statements are false.

2. Which of the following relationships between species richness (diversity) and recovery from natural disturbances is TRUE?

   (A) Communities with less diversity find it easier to recover from natural disturbances than communities with more diversity.
   (B) Communities with more diversity recover less rapidly from natural disturbances.
   (C) Communities with more species richness recover more rapidly from natural disturbances.
   (D) There is no correlation between community diversity and recovery from natural disturbances.

Examine the food chain and the food web shown below to answer Question 3.

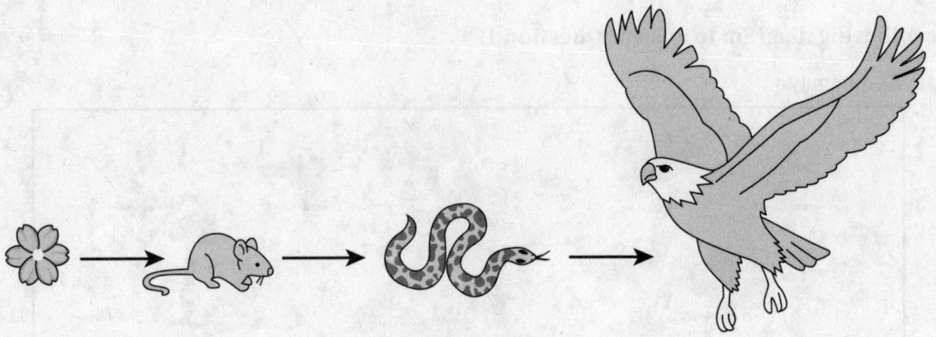

Food chain

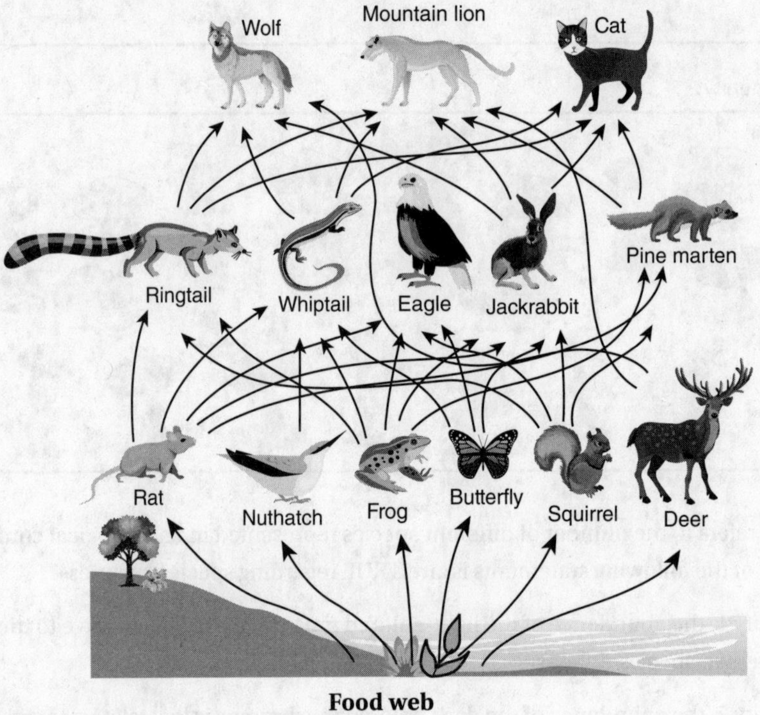

Food web

3. Which conclusion based on the diagrams shown above is TRUE?

(A) The food chain is more stable than the food web.
(B) The food chain is less stable than the food web.
(C) The food web is less stable than the food chain.
(D) The food chain and the food web are equal in complexity and therefore are both equally stable.

4. Relatively short-term ecological succession within a community can result from all of the following EXCEPT

(A) Short-term evolutionary change
(B) Variation in growth and maturation rates
(C) Variation in the adaptability potential of a species over time
(D) Variation in the interspecific differences in dispersal patterns of seeds

5. A national park consists of two forests. Refer to the chart below to answer Question 5.

Forest 1	Forest 2
10 different tree species	10 different tree species
60 total trees	60 total trees
40 trees of one species	6 trees of each species

Which forest would be the most diverse?

(A) Forest 1
(B) Forest 2
(C) Both Forest 1 and Forest 2 have equal number of species; therefore, they are equally diverse.
(D) Cannot be determined based upon the data provided.

6. High species richness on a large tropical island that is relatively close to the mainland is probably a result of all of the following EXCEPT ______________.

(A) a mild climate
(B) high annual rainfall
(C) seasonality
(D) year-round availability of resources

7. A habitat with ______________ disturbance generally has the greatest species richness.

(A) a large amount of
(B) an intermediate amount of
(C) sporadic
(D) very little

Select from the following answer choices below for Questions 8–11.

(A) Adaptation
(B) Artificial selection
(C) Natural selection
(D) Resistance

8. Farmers raise cattle with specific characteristics.

9. Farmers use insecticides to kill pests that destroy crops. However, no matter how much they spray, some of the insects survive and return again.

10. A large number of insects that ate and destroyed corn crops and that were killed by a certain insecticide are, for the most part, no longer affected by it.

11. The creosote bush is a desert-dwelling plant that produces toxins that prevent other plants from growing nearby, thus reducing competition for nutrients and water.

Read the following passage to answer Question 12.

Oh, Home on the Range—Where the Bison (Used to) Roam

It is estimated that there were more than 30 million bison* in North America when the first European settlers arrived in the 1600s. Not long after, a series of events nearly wiped the population out, as follows:

- A drought dried out the animals' grassland habitat, which was already overburdened by new populations of horses and cattle.
- Farmers and ranchers began killing bison to make room for their animals.
- Indigenous American tribes began acquiring horses and guns and were then able to kill bison in larger numbers than ever before.
- Railroads were laid through the bison's territory, dividing herds and accelerating the arrival of hunters, whose kills fed the high demand for bison hides back East and in Europe.
- Some soldiers killed bison to spite their American Indian enemies, who depended on the animals for food and clothing.
- Sport shooters traveled west to shoot the animals by the dozens, sometimes from the open windows of moving trains, and often left their bodies out on the plains to rot once the hunt was over.

By the beginning of the 20th century, there were only several hundred bison left in North America. However, thanks to the efforts of early 20th-century conservation organizations and former president Theodore Roosevelt, the American bison were saved from its impending extinction, with today's bison population being estimated at more than 500,000 and steadily growing.

Today, most bison are genetically different from their wild ancestors. At the low point of the bison population—what geneticists call the "bottleneck"—in the late 19th and early 20th centuries, the ranchers who owned a lot of the remaining bison population bred their bison with cattle in an attempt to create better meat animals and, as a result, only about 1.6 percent of today's bison population (8,000 animals) is not hybridized.

Currently, only about 20,000 bison, or about 4% of the overall population, make up the wild herds that graze national parks and private reserves, whereas the other 96% are livestock animals, raised commercially for meat and hides.

**The animals inhabiting the early American plains were bison, NOT buffalo. Some of the major differences are shown in the following chart.*

Bison	Buffalo
Bison have a shoulder hump.	Buffalos do not have a shoulder hump.
Found in Europe and North America	Found in South Asia and Africa (e.g., water buffalo)
Have a thick beard	No beard
Thick coat of hair	Thin coat of hair

12. Genetically speaking, the American bison herd became susceptible to (the)

(A) bottleneck effect
(B) exploitation
(C) extinction
(D) all of the above

Use the following choices for Questions 13–16 to determine which type of ecosystem service is being described.

(A) Cultural
(B) Provisioning
(C) Regulating
(D) Supporting

13. Carbon sequestration

14. Ecotourism, outdoor sports, and recreation

15. Sources of energy; e.g., biomass, coal, nuclear, solar

16. Photosynthesis

Refer to the following diagram to answer Question 17.

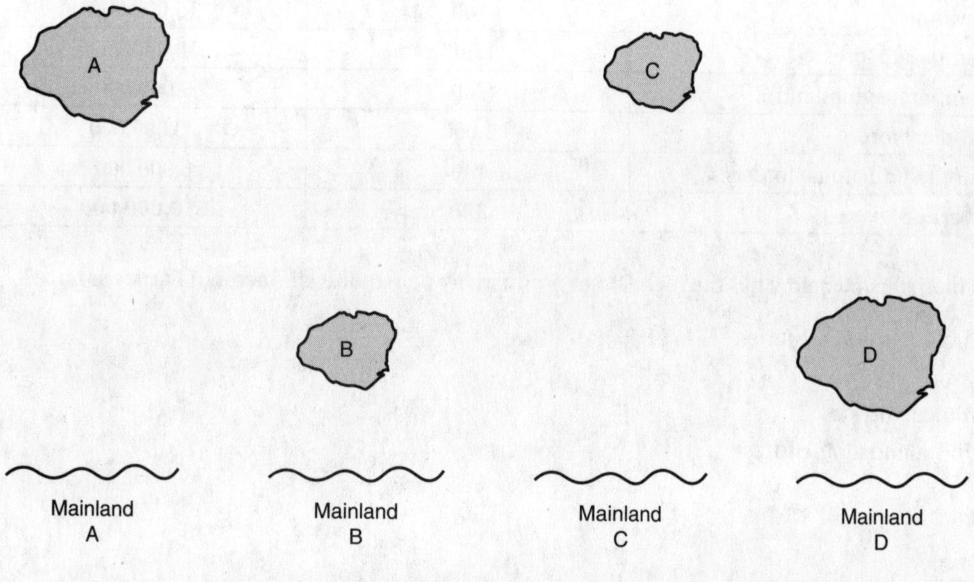

17. Based on the theory of island biogeography, which of the following islands would most likely have the highest number of species living on it?

(A) Island A is 50 hectares in size and is 50 kilometers off the coast of the mainland.
(B) Island B is 20 hectares in size and is 20 kilometers off the coast of the mainland.
(C) Island C is 20 hectares in size and is 50 kilometers off the coast of the mainland.
(D) Island D is 50 hectares in size and is 20 kilometers off the coast of the mainland.

18. According to the Law of Tolerance, organisms with large or wide tolerance limits generally exhibit

(A) narrow distribution with large populations
(B) narrow distribution with small populations
(C) wide distribution with large populations
(D) wide distribution with small populations

19. Biodiversity hotspots are defined by which two criteria?

(A) Ecological diversity and species diversity
(B) Species endemism and degree of threat
(C) Species richness and ecosystem integrity
(D) Species richness and size

Refer to the following chart to answer Question 20.

Ecosystem	Net Primary Productivity ($g/m^{-2}/year^{-1}$)	Area on Earth (km^2)
Agricultural land	650	14,000,000
Boreal forest	800	12,000,000
Continental shelf	350	27,000,000
Coral reefs and estuaries	2,000	2,000,000
Desert scrub	70	18,000,000
Extreme desert (ice, rock, sand)	3	24,000,000
Open ocean	125	332,000,000
Savanna	700	15,000,000
Temperate forest	1,300	18,000,000
Temperate grassland	500	9,000,000
Tropical forest	2,000	20,000,000
Tundra and alpine forests	140	8,000,000
Average	**320**	**510,000,000**

20. Which biome listed below has the highest net productivity per square kilometer of Earth's surface?

(A) Coral reefs and estuaries
(B) Deserts
(C) Tropical forests
(D) Tundra and alpine forests

Free-Response Question

The Gray Wolf Is Back!

In 1872, Yellowstone National Park was created. In 1907, under political pressure from the western cattle and livestock industries, a U.S. government program called Animal Damage Control was responsible for killing ~2,000 wolves and ~23,000 coyotes in 39 U.S. national forests, and by 1926, all gray wolves had been eliminated in Yellowstone.

Once the wolves were gone in Yellowstone, the elk populations began to multiply and destroy native plants and trees due to overgrazing, which also resulted in land erosion. In response, the park service began killing off the elk. However, in response to public pressure, (i.e., hunters) the park began to reintroduce elk. Due to the elimination of gray wolves, the coyote population had increased, which dramatically impacted the pronghorn antelope and fox populations. To stabilize and reduce the coyote population, wolves were released into the park again.

In 1973, the Endangered Species Act became law and elk populations rose; the quality of the rangeland decreased, which resulted in the elk's having to extend their range, which in turn reduced the pressure on willow, a plant that is required for beavers to survive through the winter. As more willow was now available, the beaver population increased, resulting in more beaver dams, which had a dramatic and positive impact on the quality of the watershed; e.g., creating more stable water sources, reducing erosion, and providing habitats for a variety of wildlife; e.g., amphibians, fish, moose, otters, wading birds, waterfowl, etc.

With increased predation of elk by the wolves, the native grizzly bear population also benefited due to increases in their food supply; e.g., berries.

(a) The gray wolf is classified as an apex predator. Define *apex predator*, define its role in an ecosystem, and provide an example of a benefit that apex predators provide to a healthy ecosystem.

(b) The gray wolf can also be classified as a keystone species. Define *keystone species* and provide another example of a current keystone species and how it functions in its ecosystem.

(c) The story of the gray wolf in Yellowstone Park illustrates trophic cascade. Define *trophic cascade* and explain how the story of the gray wolf in Yellowstone Park illustrates the concept of a trophic cascade. You must draw a correct sketch to help illustrate your explanation.

Answer Explanations

1. **(C)** The ecological impact of a disturbance (e.g., wildfire) depends upon the (1) intensity, (2) frequency, and (3) spatial distribution of the disturbance, which are themselves influenced by the season in which the disturbance occurs, the history of the disturbed site, and the site's topography or landscape. The more (genetic) diversity or variety there are in communities that are left available in a disturbed site, the greater the chances are that the site will recover sooner.

2. **(C)** The changes a terrestrial ecosystem goes through as it recovers from a disturbance depend on the intensity and magnitude of the disturbance. In secondary succession, which follows a disturbance in an area with *existing* communities of organisms, biological remnants (such as buried seeds) survive, and the recovery process begins sooner.

3. **(B)** The food web is more stable since organisms have more than a single food source to rely upon. Food webs illustrate the interaction of multiple food chains within a certain ecosystem, showing the mutual dependency of species and the natural balance of habitats that sustain animal and plant life and are helpful in explaining how disruptions in populations due to overhunting, poaching, global warming, and habitat destruction, for example, result in food scarcities that can lead to extinction.

4. **(A)** In ecological succession, the populations that make up a community change, but the inheritable characteristics of the individuals within the population do not change over a short time period, as ecological succession is the process of change in the species structure of an ecological community over time. The time scale can be decades (e.g., after a wildfire), or even millions of years, such as after a mass extinction. Bottom line: ecological succession is something that happens within communities, whereas evolution occurs within populations.

5. **(B)** Species diversity is defined as the number of species and abundance of each species that live in a particular location. The number of species that live in a certain location is called species richness. Think of it this way: if country A had 300 million people (299,999,995 Caucasians, 1 African American, 1 Native American, 1 Asian American, 1 Hispanic American, and 1 Pacific Islander) and country B also had 300 million people but with 500,000 people of each ethnicity mentioned—which country would be the most diverse?

6. **(C)** Tropical locations (near the equator) are characterized by their abundance of water and food resources, high vegetation diversity (therefore, more complex food chains), large and consistent amounts of solar radiation, numerous microhabitats, and stable temperatures (no seasons). While the number of species (e.g., birds, insects, mammals) increases closer to the equator, the number of genetically distinct groups within each species (subspecies) is greater in more extreme environments typical of higher latitudes, the reason being that animals in these more extreme environments are more likely to freeze during cold winters or die during unusually hot and/or dry summers. If catastrophic extreme weather events wipe out a population but don't wipe out an entire species, the population(s) that survive may be geographically separated from the other populations of that species and could start to diverge enough to become reproductively isolated, forming a new species.

7. **(B)** At low levels of environmental disturbance, more competitive organisms will push less competitive species to extinction and dominate the ecosystem, whereas at high levels of disturbance (frequent floods or wildfires fires), all species are at risk of going extinct, as shown in the graph on the following page. However, at intermediate levels of disturbance, diversity is maximized since species that thrive at both early and late successional stages are able to coexist.

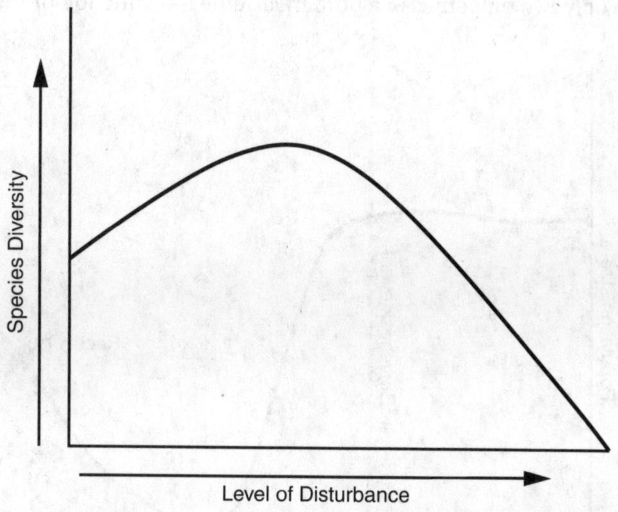

8. **(B)** Artificial selection is a process in the breeding of animals and in the cultivation of plants by which the breeder chooses to perpetuate only those forms having certain desirable inheritable characteristics.
9. **(C)** Natural selection is the process by which forms of life having traits that better enable them to adapt to specific environmental pressures, such as predators, changes in climate, or competition for food or mates, will tend to survive and reproduce in greater numbers than others of their kind, thus ensuring the perpetuation of those favorable traits in succeeding generations.
10. **(D)** Resistance is the natural or genetic ability of an organism to avoid or repel attack by biotic agents (e.g., pesticide) or pathogens (e.g., specific type of bacteria, pests, parasites, etc.).
11. **(A)** Adaptation means to undergo changes to become more suited or fit to its environment. For a trait to be considered an adaptation, it has to be inheritable, functional, and increase fitness. According to Charles Darwin's theory of evolution by natural selection, the organisms adapt to their environment so that they can persist and pass their genes on to the next generation. Adaptations are essential to the species' survival. The adaptive traits that the species will acquire through time may be structural, behavioral (e.g., vocalizations, courtship rituals, nesting, and mating), or physiological (e.g., developing resistance to diseases or to toxic chemicals).

12. **(A)** The key to getting this question correct is the word "genetically." A population bottleneck or genetic bottleneck is a sharp reduction in the size of a population due to environmental events such as famines, earthquakes, floods, fires, disease, and droughts, or human activities such as genocide and human population planning. Such events can reduce the variation in the gene pool of a population; thereafter, a smaller population, with smaller genetic diversity, remains to pass on genes to future generations of offspring through sexual reproduction. There are only two outcomes of a bottleneck effect—extinction or recovery.

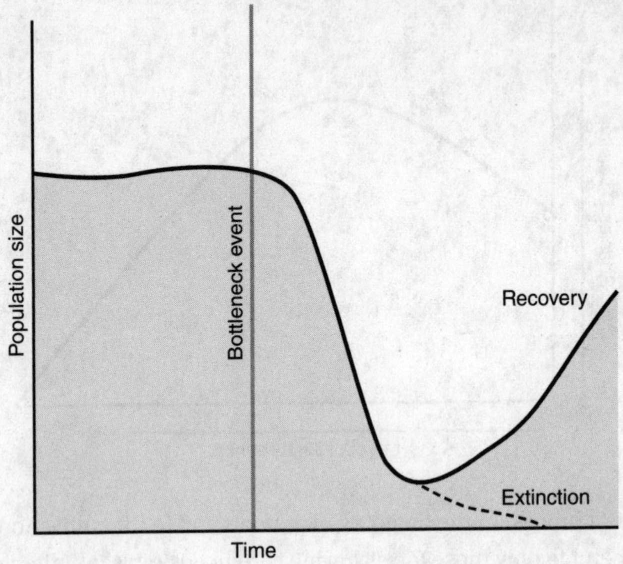

13. **(C)** A regulating service is the benefit provided by ecosystem processes that moderate natural phenomena. Regulating services include carbon storage, climate regulation, decomposition, erosion and flood control, pollination, and water purification.
14. **(A)** The non-material benefits people obtain from ecosystems are known as cultural services and include aesthetic and spiritual experiences related to the natural environment.
15. **(B)** Material benefits that people obtain from the ecosystems are called provisioning services. Examples include food, freshwater, and raw materials (e.g., lumber).
16. **(D)** Supporting services make it possible for the ecosystems to provide living spaces for plants or animals and to maintain a diversity of plants and animals. Supporting services include pollination, nutrient cycling, primary production, and soil formation.
17. **(B)** As the size of an island increases, the number of species living on it should (in theory) increase. Furthermore, as an island's distance from the mainland increases, the number of species living on it should theoretically decrease. This choice represents the largest island that is closest to the mainland.
18. **(C)** The Law of Tolerance states that the ability of an organism to survive in a particular habitat or the size of a population is determined by its ability to tolerate a range of physical and chemical factors present in a particular area. The highest concentration of the individuals of a population is found within the optimum range where the environmental conditions are most conducive to growth and reproduction of the individuals of the population. The upper and lower limits of this optimum range vary with geographical region, season, and stages of growth of the species.

19. **(B)** A biodiversity hotspot is a biogeographic area notable for sustaining significant levels of biological diversity and that is threatened by destructive activities. Biodiversity hotspots are home to unique flora and fauna, most of which are endemic to the particular environment. For a region to be recognized as a biodiversity hotspot, it must meet two criteria:

(1) At least 1,500 or more of the vascular plant species found in the hotspot must be endemic (native to a single defined geographic location). Vascular plants have specific tissues known as xylem for conducting water and minerals throughout the plant, and they must have specialized tissue known as phloem for conducting the products of photosynthesis (i.e., sugars and oxygen) through the plant. Examples of vascular plants include angiosperms (flowering plants), conifers (cone-bearing trees), and ferns. Vascular plants are critical to the sustainability of the ecosystem since they allow the circulation of water, photosynthetic products, and minerals.

(2) The area must have lost 70% or more of its original habitat. Currently, there are 36 biodiversity hotspots around the world, which are home to ~60% of the world's flora and fauna.

20. **(A)** The key to getting this question correct are the words "per square kilometer *of Earth's surface*." If a particular ecosystem has a NPP of 10,000 g/m^{-2}/year BUT only occupied 1 km^2 of Earth's surface, then it would not have the highest NPP per square kilometer of Earth's *total* surface. To solve this, let's create a third column that allows us to factor in the amount of area on Earth that this biome occupies:

Ecosystem	Net Primary Productivity (g/m^{-2}/year^{-1})	Area on Earth (km^2)
Agricultural land	650	14,000,000
Boreal forest	800	12,000,000
Continental shelf	350	27,000,000
Coral reefs and estuaries	2,000	2,000,000
Desert scrub	70	18,000,000
Extreme desert (ice, rock, sand)	3	24,000,000
Open ocean	125	332,000,000
Savanna	700	15,000,000
Temperate forest	1,300	18,000,000
Temperate grassland	500	9,000,000
Tropical forest	2,000	20,000,000
Tundra and alpine forests	140	8,000,000
Average	**320**	**510,000,000**

The key part of the question is the word "per," which means to divide. Therefore, you must divide the net primary productivity by the amount of area it occupies on Earth. For example, ecosystem A might have a net primary productivity of 5,000 g/m^{-2}/year^{-1} (very high), but if it only exists in an area of one square meter on Earth, then its NPP considering the total Earth's surface would be extremely small.

Tropical rainforests have high NPP and the highest biodiversity of any *terrestrial* ecosystem, as the physical environment favors vast photosynthetic output and high growth rate. Furthermore, plants have constant high levels of water and light (at the canopy level), and the nutrient supply is as high as possible due to rapid decomposition. Because the light is strong, it is able to sustain several layers of plant growth, and high levels of water allow plants to maintain a constant flow from their roots, bringing up more nutrients. The plants in the tropical rainforest have also evolved to be able to keep their stomata open all day to optimize CO_2 uptake without becoming short of water, and high and constant temperatures allow rapid enzyme activity without the seasonal dormancy typical of many other types of forests.

While coral reefs have the same net primary productivity per square meter as tropical rainforests, they occupy significantly less area on Earth than tropical rainforests; and remember, the question asked "per square kilometer of Earth's surface." However, both coral reefs (due to rising ocean temperatures) and tropical rainforests (due to turning the land into farms) are dramatically decreasing in size.

Deserts are areas of low available water. In an extreme desert, be it hot or cold, the absence of water will prevent most plants from surviving and, hence, productivity is negligible. Plants may show similar adaptations to conserve water in both environments (xerophytism).

Tundra and alpine environments do not permit plants to grow very high because of the desiccation caused by wind, which means there can be fewer layers of vegetation and less efficient light capture. Furthermore, the cold temperature limits net primary productivity by reducing the rate at which enzymes can catalyze reactions, and even when the air temperatures rise sufficiently to allow photosynthesis, the soil nutrients may be locked up permanently in frozen soil (permafrost).

Free-Response

10 Total Points Possible

(a) *Maximum 3 points.*

> *Define* apex predator, *define its role in an ecosystem, and provide another example of a current apex predator. (1 point for correct definition of an apex predator, 1 point for correctly defining an apex predator's role in an ecosystem, and 1 point for providing a correct example of a current apex predator.)*

An apex predator, also known as a top or alpha predator, is an animal at the top of a food chain that does not have natural predators of its own. An example of an apex predator would be an African lion or a polar bear.

Apex predators serve to keep prey numbers in check by "weeding out" the slower, weaker, and dying animals, which ultimately results in an increase in the health of the prey population as a whole. Without apex predators, a herd of herbivores, such as deer, would stay in one location and eat most or all of the plants they live on before they would move on, resulting in a cascading effect of starvation for animals in lower trophic levels. By keeping lower trophic levels in population balance, it leaves smaller plants and grasses for smaller herbivores, prevents erosion, and allows more saplings to mature, which reduces runoff into rivers and streams, which in turn reduces flood damage.

(b) *Maximum 3 points.*

> *Define* keystone species *and provide another example of a current keystone species and how it functions in its ecosystem. (1 point for correct definition of* keystone species, *1 point for correctly identifying another current keystone species, and 1 point for correctly explaining how it functions as a keystone species.)*

A keystone species is a species that has a disproportionately large effect on its natural environment relative to its abundance.

Keystone species maintain the local biodiversity of an ecosystem and have a powerful influence on the abundance and type of other species in a habitat. They are a critical component of food webs.

Keystone species fill a critical ecological role that no other species within the food web perform, and without keystone species, the entire ecosystem would radically change—or cease to exist.

An example of a keystone species mutualistic relationship would be the ecological relationship between hummingbirds and specific species of plants.

The ability of hummingbirds to hover is necessary for gathering food from flowers, and being small makes this possible. Furthermore, the flower's nectar, which is high in sugar, is necessary for the high metabolic rate needed to sustain the hummingbirds' feeding behavior. In turn, the plants that the hummingbirds feed upon have, over time, evolved traits that are beneficial to hummingbirds but not so for bees (also pollinators); the flowers are brightly colored, which is critical for hummingbirds, but lack the odor necessary for the bees.

Plants have been equally affected by this relationship. Species that rely on hummingbirds for pollination, instead of the insects their ancestors did, have acquired a number of "pro-bird" and "anti-bee" traits: nectar that is particularly sucrose rich, flowers that are brightly colored but unscented (smell is vital for insects to find flowers, but vision is the key for birds), and various adaptations to their flowers to allow easy access for hummingbirds. Currently, over 7,000 species of plants now depend for pollination on one or more of the 361 known species of hummingbird.

(c) *Maximum 4 points. 1 point for correctly defining* trophic cascade *and 2 points for correctly explaining how the story of the gray wolf in Yellowstone Park illustrates the concept of a trophic cascade. You must also include a sketch to help illustrate your answer and, if correct, 1 additional point will be added.*

Trophic cascades occur when predators limit the density and/or behavior of their prey and thereby enhance survival of *at least* the next lower trophic level; by definition, trophic cascades must occur across a minimum of three trophic levels. By consuming prey, predators influence the prey's abundance, density, and behavior, and in these situations, indirectly benefit and increase the abundance of their prey's prey.

In the sketch on the following page, the picture on the left shows that a gray wolf (an apex predator) keeps the moose population in check and by doing so allows willows to thrive, which in turn allows beavers to flourish. The benefits that beavers and their dams provide to the local environment include increasing the rate of groundwater recharge, as the dams slow the flow of water through a stream or another body of water. The wetland created behind the dam also provides great habitat for wildlife, including birds and amphibians. Water quality in the streams downstream improves as sediment and nutrients are filtered in these wetlands. Finally, the water stored behind the dam helps recharge groundwater levels by allowing the water to filter down into the ground rather than simply slip by in the stream.

However, in the sketch on the right, without the presence of the wolf, the moose population increases. In time, the increase of moose depletes the population of willow trees, which negatively impacts the beaver population and the building of dams.

3

Populations

Learning Objectives

In this unit, you will learn:

- ➔ 3.1 Generalist and Specialist Species
- ➔ 3.2 *K*-Selected and *r*-Selected Species
- ➔ 3.3 Survivorship Curves
- ➔ 3.4 Carrying Capacity
- ➔ 3.5 Population Growth and Resource Availability
- ➔ 3.6 Age Structure Diagrams
- ➔ 3.7 Total Fertility Rate
- ➔ 3.8 Human Population Dynamics
- ➔ 3.9 Demographic Transition

For added practice with this unit and topics, log onto Online Labs 3A–3C at online.barronsbooks.com

3.1 Generalist and Specialist Species

A generalist species is able to thrive in a wide variety of environmental conditions and can make use of a variety of different resources (for example, a heterotroph with a varied diet), whereas a specialist species can thrive only in a narrow range of environmental conditions and generally has a limited diet.

Generalist	Specialist
Able to use a variety of environmental resources	Use a specific set of resources
Adaptable to a wide range of environments	Less adaptable due to specialized needs
Have a high level (range) of tolerance	Generally, have a low level (range) of tolerance
Have an advantage when environmental conditions change	Easily affected when environmental conditions change
Less likely to become extinct	More likely to become extinct
Example: human	Example: panda

3.2 *K*-Selected and *r*-Selected Species

Organisms have adapted either to maximize growth rates in environments that lack limits or to maintain population size at close to the carrying capacity in stable environments. Species that have high reproductive rates are known as *r*-strategists, while species that reproduce later in life and with fewer offspring are known as *K*-strategists.

Reproductive Strategies

r-Strategists	*K*-Strategists
Generally, not endangered	Most endangered species are *K*-strategists
Have many offspring and tend to overproduce	Have few offspring
Low parental care	High parental care
Mature rapidly	Mature slowly
Population size limited by density-independent limiting factors, including climate, weather, natural disasters, and requirements for growth	Density-dependent limiting factors to population growth stem from intraspecific competition and include competition, predation, parasitism, and migration
Short lived	Long lived
Tend to be prey	Tend to be both predator and prey
Tend to be small	Tend to be larger
Type III survivorship curve	Type I or II survivorship curve
Wide fluctuations in population density (booms and busts)	Population size stabilizes near the carrying capacity
Examples: most insects, algae, bacteria, rodents, and annual plants	Examples: humans, elephants, cacti, and sharks

3.3 Survivorship Curves

Survivorship curves show age-distribution characteristics of species, reproductive strategies, and life history (Figure 3.1). Reproductive success is measured by how many organisms are able to mature and reproduce, with each survivorship curve representing a balance between natural resource limitations and interspecific and intraspecific competition. For example, humans could not survive in Type III survivorship mode, in which human females would produce hundreds or thousands of offspring. Likewise, ants could not survive in a Type I mode, whereby each queen ant would produce only a few eggs during her lifetime and would spend most of her time and energy raising offspring.

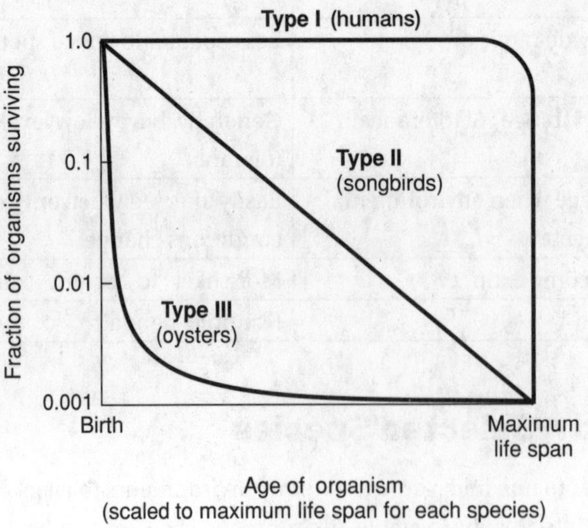

Figure 3.1 Survivorship curve

Survivorship Curves Table

Type	Description
Type I. Late Loss	Reproduction occurs fairly early in life, with most deaths occurring at the limit of biological life span. Low mortality at birth with a high probability of surviving to advanced age. Death rates decrease in younger years due to advances in prenatal care, nutrition, disease prevention, and cures, including immunization. Examples: humans, annual plants, sheep, and elephants.
Type II. Constant Loss	Individuals in all age categories have fairly uniform death rates, with predation being the primary cause of death. Typical of organisms that reach adult stages quickly. Examples: rodents, perennial plants, and songbirds.
Type III. Early Loss	Typical of species that have great numbers of offspring and reproduce for most of their lifetime. Death is prevalent for younger members of the species due to environmental loss and predation and declines with age. Examples: sea turtles, trees, internal parasites, fish, and oysters.

3.4 Carrying Capacity

Carrying capacity (represented by "*K*") refers to the number of individuals that can be supported sustainably in a given area. The carrying capacity varies from species to species and is subject to changes over time. As an environment degrades, the carrying capacity decreases. Factors that keep population sizes in balance with the carrying capacity are called regulating factors and include the following:

- Amount of sunlight available
- Food availability
- Nutrient levels in soil profiles
- Oxygen content in aquatic ecosystems
- Space

When populations are below the carrying capacity, they tend to increase in size. Population size cannot be sustained above the carrying capacity for extended periods of time; eventually the population will crash (Figure 3.2).

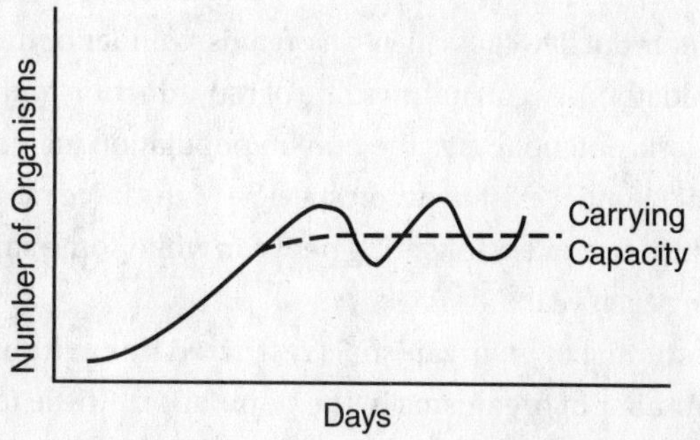

Figure 3.2 Fluctuations around the carrying capacity

3.5 Population Growth and Resource Availability

Population Dispersal Patterns

A population dispersal pattern is how individuals or species of animal become distributed in different spaces over certain periods of time. Key population dispersion patterns follow:

- **Clumped**—Some areas within a habitat are dense with organisms, while other areas contain few members. Found in environments with "patch" resources; e.g., water. Living in groups provides advantages (e.g., division of labor in bees—some build the hive while others protect it) and is common for animals. Examples include the following:
  - Animals living in social families; e.g., elephants
  - Animals that feel safer living in groups; e.g., zebras
  - Animals that serve as prey; e.g., ducks
  - Animals that work together to trap or corner prey; e.g., lions
  - Animals with the inability of their offspring to independently move from their habitat; e.g., chimpanzees
- **Random**—There is little interaction among members of the population. Occurs in habitats where environmental conditions and resources are consistent. Individuals are distributed randomly; occurs with dandelions and other plants that have wind-dispersed seeds.
- **Uniform**—Space is maximized between individuals to minimize competition; e.g., creosote bushes, which produce chemicals that influence the germination, growth, survival, and reproduction of other organisms (allelopathy). See Figure 3.3.

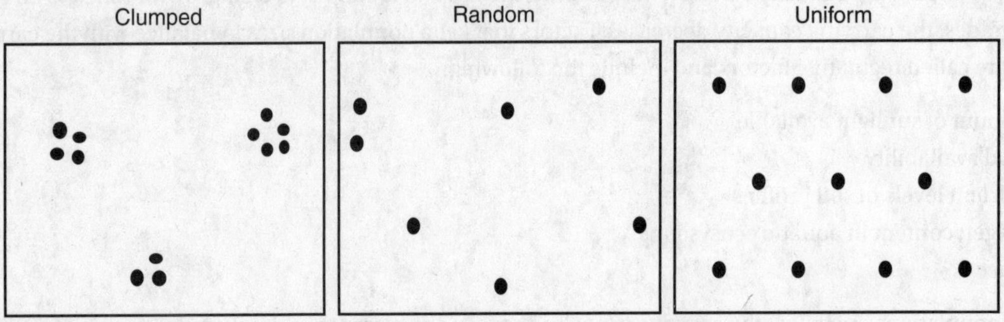

Figure 3.3 Clumped, random, and uniform distribution patterns

Biotic Potential

Biotic potential is the maximum reproductive capacity of an organism under optimum environmental conditions (e.g., sufficient food supply, no predators). An organism's rate of reproduction and the size of each litter are the primary determining factors for biotic potential; e.g., the human population increased by 1.2% last year.

Significant differences in biotic potential exist between species. Many large mammals, such as humans or elephants, will typically only produce one or two offspring per year, while some small organisms, such as insects, will produce thousands of offspring each year.

Full expression of the biotic potential of an organism is restricted by environmental resistance—any factor that inhibits an increase in the number of organisms in the population. These factors include unfavorable climatic conditions such as the lack of space or light, nutritional deficiencies, and the inhibiting effects of predators, parasites, or disease.

Factors That Influence Biotic Potential

Increase Biotic Potential	Decrease Biotic Potential
Able to adapt	Unable to adapt
Able to migrate	Unable to migrate
Adequate resistance to disease and parasites	Little or no suitable defense mechanisms against disease or parasites
Favorable environmental conditions; e.g., light, nutrients, and/or temperature	Unfavorable environmental conditions; e.g., insufficient light, poor supply of nutrients; and/or temperature extremes
Few competitors	Too many competitors
Generalized niche	Specialized niche
High birth rate	Low birth rate
Satisfactory habitat	Unsatisfactory habitat
Sufficient food supply	Inadequate food supply
Suitable predatory defense mechanism(s)	Unsuitable predatory defense mechanism(s)

J-Curves

A J-curve representing population growth occurs in a new environment when the population density of an organism increases rapidly in an exponential or logarithmic form, but then stops abruptly as environmental resistance or another factor (e.g., shortage of food) suddenly impacts the population growth. This type of population growth rate is known as "density dependent," as the regulation of the growth rate is not tied to the population density until the resources are exhausted and the population growth rate crashes. Examples include some insect cycles (e.g., locusts) and algal blooms (Figure 3.4).

S-Curves

An S-shaped growth curve (also known as logistical growth) occurs when, in a new environment, the population density of an organism initially increases slowly but then stabilizes due to the finite amount of resources available. This slowing of the growth rate reflects the increasing environmental resistance, which becomes proportionately more significant at higher population densities. This type of population growth is termed "density dependent" since the growth rate depends on the number of organisms in the population. This point of stabilization is known as the carrying capacity of the environment, and it denotes the point at which the upward growth curve begins to level out (Figure 3.5).

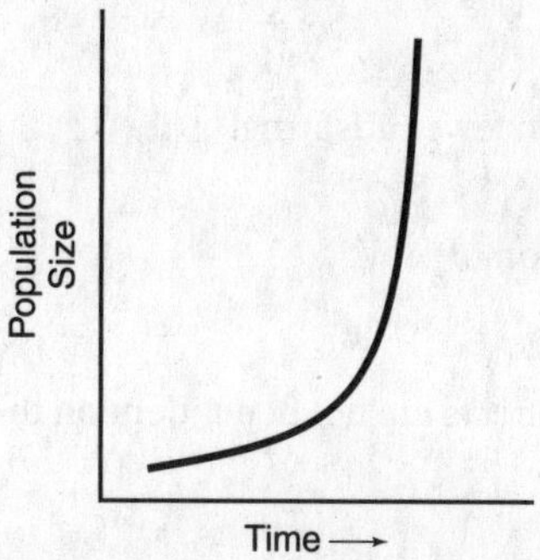

Figure 3.4 J-shaped population curve

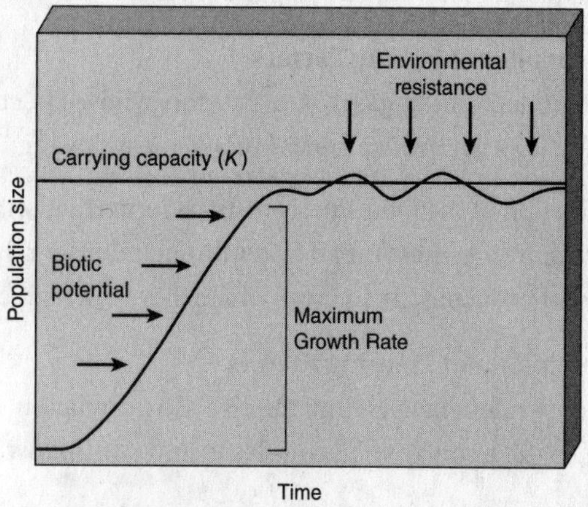

Figure 3.5 S-shaped population curve

Feedback Loops

Positive feedback loops stimulate change and are responsible for sudden or rapid changes within ecosystems. When part of the system increases, another part of the system also changes in a way that makes the first part increase even more. For example, when there is a surplus of resources, it allows a plant or animal population to grow without limit, and having a higher number of individuals in the population leads to even more births.

Negative feedback loops often provide stability; e.g., population regulation. Because the resources that sustain populations are limited and no population can exceed the carrying capacity of the ecosystem for long, limiting factors are good examples of inputs that contribute to a negative feedback loop. Predators and prey work to keep animal populations within the limits of the carrying capacity of their environment and thus help maintain some form of population stability. An increase in the amount of prey will contribute to more available energy for the predator population and results in an increase in the number of predators, which will then cause a decrease in the number of prey (Figure 3.6).

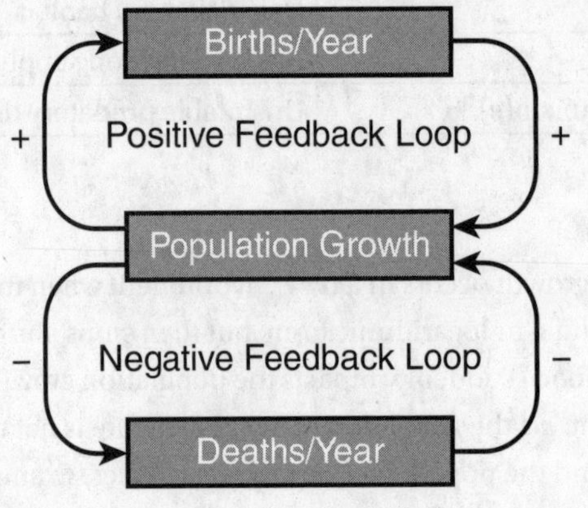

Figure 3.6 Positive and negative population feedback loops

Liebig's Law of the Minimum

Population growth is determined not by the total resources available, but by the scarcest resource, known as a limiting factor. Examples of limiting factors include food supply, sunlight, and water.

Limiting Factors

A limiting factor can be any resource or environmental condition that limits the abundance, distribution, and/or growth of a population. Based on Liebig's law of the minimum, even if all other factors are favorable, the one that is least favorable will dictate the growth, abundance, or distribution of the population of a species. There are two forms of limiting factors: density-dependent and density-independent.

Density-Dependent Limiting Factors

Density-dependent limiting factors are factors whose effects on the size or growth of the population vary with the density of the population. Examples include the following:

- A reduction in the food supply restricts reproduction, resulting in fewer offspring.
- Contagious diseases spread faster through denser populations.
- Predators concentrate in areas where there is a high concentration of prey.

Density-Independent Limiting Factors

Density-independent factors limit the size of a population, and their effects are not dependent on the number of individuals in the population. Examples include the following:

- Droughts, earthquakes, floods, tsunamis, and volcanic eruptions
- Extreme environmental pollution

Rule of 70

The Rule of 70 helps to explain the time periods involved in exponential population growth occurring at a constant rate. Doubling time is the amount of time it takes for a population to double in size. To find how long it takes for a population to double in size (assuming the rate is constant), we can use the following formula:

$$dt = \frac{70}{r}$$

where *dt* represents the change in time and *r* represents the growth rate, which must be entered as a whole number; e.g., 2% = 2. A population with a 2% annual growth rate would have a doubling time of 35 years:

$$\frac{70}{2} = 35$$

Key points to remember about population doubling times are as follows:

- Populations cannot double forever. Resistance factors, such as availability of food, contribute to a leveling off in population size over time once the carrying capacity is reached.
- The growth rate varies considerably among organisms (e.g., most small-bodied organisms grow faster and have larger rates of population increase than larger organisms; for example, bacteria increase in number faster than elephants do.
- The larger the growth rate (*r*), the faster the doubling time.

Important Population Formulas

BIRTH RATE (%) = [(total births/total population)] × 100

CRUDE BIRTH RATE (CBR) refers to the number of live births (*b*) in a year divided by the total midyear population (*p*), with the ratio multiplied by 1,000 to arrive at the number of births per 1,000 people. For example, in 2022, there were 5,000 births in a city with population of 225,000. Therefore: CBR = $[(b \div p) \times 1{,}000]$ = $[(5{,}000 \div 225{,}000) \times 1{,}000] = {\sim}22$. Therefore, there were ~22 births for every 1,000 people in the city.

DEATH RATE (%) = [(total deaths/total population)] × 100

CRUDE DEATH RATE (CDR) refers to the number of deaths (*d*) in a year divided by the total midyear population (*p*), with the ratio multiplied by 1,000 to arrive at the number of deaths per 1,000 people. For example, in 2022, there were 2,500 deaths (*d*) in a city with population (*p*) of 225,000. Therefore: CDR = $[(d \div p) \times 1{,}000]$ = $(2{,}500 \div 225{,}000) \times 1{,}000 = {\sim}11$. Therefore, there were ~11 deaths for every 1,000 people in the city.

DOUBLING TIME = 70/% growth rate

EMIGRATION = number leaving a population

GLOBAL POPULATION GROWTH RATE (%) = [(CBR − CDR)]/10

IMMIGRATION = number entering a population

NATIONAL POPULATION GROWTH RATE = [(CBR + IMMIGRATION) − (CDR + EMIGRATION)]/10

PERCENT RATE OF CHANGE = [(new # − old #)/old #] × 100

POPULATION DENSITY = total population size/total area

POPULATION GROWTH RATE (%) = $\frac{[(\text{births} + \text{immigration}) - (\text{deaths} + \text{emigration})]}{\text{total population}} \times 100$

Population Practice Problems

(1) The United States had a birth rate of 14.6 live births per 1,000 people in one year, while the death rate that year in the United States was 8.3 deaths per 1,000 people. Calculate the population growth rate for the United States that year.

(2) A population is growing by 2% per year. How long will it take for the population to double in size?

(3) A country is expected to double in population size in 35 years. What is the growth rate (in %)?

(4) On January 1, 1950, the population of a small suburb in Los Angeles, California, was 20,000. The birth rate at that time was 25 per 1,000 people, while the death rate was 7 per 1,000 people. Immigration at the time was 600 people per year, while emigration was 200 people per year. By how much did the population increase (or decrease) by January 1, 1951?

(5) A small city in Ohio had a population of 1,000 in 1920. Over the next 10 years, there were 100 births, 50 deaths, 200 people who left the city, and 600 people who arrived in the city. What was the population of the city in 1930?

(6) The city of Barronsville has a population of 240,000. Last year, there were 12,000 children born, while 2,400 people died. There was no immigration or emigration that year. What were the % birth and death rates?

(7) Given the information for Barronsville above, what were the crude birth and crude death rates?

(8) In 1980, the population of a city was 250,000. In 1990, the population of the city grew to 280,000. What was the % change in population growth?

(9) A city had a growth rate of 12% over 10 years. What was the annual percentage growth rate?

(10) If a city had an annual population growth rate of 1.4%, how many years would it take for the city to double in population?

(11) How would the population growth rate of a country change if it had a crude birth rate of 32 but was able to reduce its crude death rate through improved medical services from 28 to 12, and how would its doubling time be affected?

See next page for complete solutions to these problems.

Answers

1. $\frac{\text{birth rate} - \text{death rate}}{10} = \frac{14.6 - 8.6}{10} = \mathbf{0.6}$
2. Doubling time = 70/% growth = 70/2 = **35 years**
3. % growth rate = 70/doubling time = 70/35 years = **2%**
4. $\frac{20{,}000 \ \text{~~population~~}}{1} \times \frac{25 \text{ births}}{1{,}000 \ \text{~~population~~}}$ = 500 births + 600 immigrants = 1,100 new people

   $\frac{20{,}000 \ \text{~~population~~}}{1} \times \frac{7 \text{ deaths}}{1{,}000 \ \text{~~population~~}}$ = 140 deaths + 200 emigrants = 340 less people

   ∴ 1,100 new people − 340 less people = **+760 people**
5. $N_1 = (N_0 + B + I) - (D + E) = (1{,}000 + 100 + 600) - (50 + 200) = \mathbf{1{,}450}$
6. Birth rate (as a %) = (12,000 children born/240,000 population) × 100% = **5.0%**

   Death rate (as a %) = (2,400 children died/240,000 population) × 100% = **1.0%**
7. Crude birth rate is the number of live births occurring among the population of a given geographical area during a given year per 1,000 people.

   (12,000 children/240,000 total population) × 1,000 = **50 births per 1,000 people**

   (2,400 died/240,000 total population) × 1,000 = **10 deaths per 1,000 people**
8. % change = [(new − old)/old] × 100% = [(280,000 − 250,000)/250,000] × 100% = **12%**
9. % annual growth rate = % growth rate/number of years = 12%/10 years = **1.2% per year**
10. 70/1.4 = **50 years**
11.

GROWTH RATE CHANGE

Original	Later
32 − 28/10 = **0.4%**	32 − 12/10 = **2.0%**

DOUBLING TIME CHANGE

Original	Later
70/0.4 = **175 years**	70/2 = **35 years**

Strategies for population growth and sustainability based upon resource availability include the following:

- Improve prenatal and infant health care. Women would not need to have more children if the ones they had survived.
- Increase economic development in developing countries through free trade and private investment coupled with tax incentives.
- Provide economic incentives for having fewer children. Possible drawbacks would be that rewards may be given to people who already have the number of children they want, and the cost of the program.
- Provide free education, housing subsidies, monthly subsidies, free healthcare, higher pension benefits, tax incentives, or other economic incentives to women who only have one or two children. Higher education for women usually results in having children later in life. Drawbacks include the fact that it may not work, as children may be needed for labor either on farms or in the city to provide extra income for the family, and the cost of the program.
- Provide government family-planning services. Examples include education, birth control, sterilization, abortion, and raising the age for marriage. Drawbacks may include cultural and social issues, confusion as to alternatives should a child die, and interference with religious teachings.

- Provide greater access to education and more job and empowerment opportunities for women. The low status of women is a major problem in overpopulation. Higher education levels for women as well as higher family incomes usually result in having fewer children. A drawback may be resentment by men.

CASE STUDIES

China

Between 1972 and 2000, China dramatically reduced its crude birth rate (the total fertility rate [TFR] dropped from 5.7 to 1.8) by adopting the "One-Child Rule." Methods used to achieve this goal included incentives such as extra food, larger pensions, better housing, free school tuition, free medical care, and salary bonuses. A mobile program that reached both urban and rural areas of China, offering family-planning services including sterilization, was also implemented. However, an unforeseen consequence was that with the advent of ultrasounds, many families chose to abort female fetuses because they preferred to have a male child (as a result, over 35 million females are currently "missing"). Over time, this has resulted in a very large discrepancy in the male-to-female ratio in China (there are 130 males for every 100 females).

In 2016, in response to an aging population and the large discrepancy in the sex ratio, China adopted a "Two-Child Rule."

India

In 1952, India (with a population of 400 million at the time) began its first family-planning program. In 2018, India's population was 1.4 billion, or 18% of the world's population. Each day there are over 50,000 live births in India. One-third of the population of India earns less than 40 cents per day, and cropland has decreased 50% per capita since 1960. In the 1970s, India instituted a mandatory sterilization program involving vasectomies. Some of the reasons for India's family-planning failures were poor planning, low status of women, preference for male children, and insensitivity to cultures and religion. Tubal ligation is the preferred method of family planning in India today. Condoms are free from the Indian government but have less than 10% use.

Impacts of Population Growth

- **Biodiversity:** The Earth's biological diversity is crucial to the continued vitality of agriculture and medicine. Yet, human activities are pushing many thousands of plant and animal species into extinction, with two-thirds of the world's species in decline.
- **Coastlines and oceans:** Half of all coastal ecosystems are pressured by high population densities and urban development. Ocean fisheries are being overexploited, estuaries (nurseries of the sea) are being drained and filled in because of population growth, and fish catches are down.
- **Food supply and malnutrition:** Today, one-quarter of the world's population is malnourished (Figure 3.7). In 64 of 105 developing countries, most notably in Africa, Asia, and parts of Latin America, the population has been growing faster than available food supplies. Population pressures have also degraded some two billion hectares of arable land—an area the size of Canada and the United States combined.

  The issue is not that the world does not produce enough food; rather, the issue is that too many people cannot afford food and that it is not distributed equally. If all food grown and raised on Earth were distributed equally, it would result in 4.3 pounds (2 kg) of food per person per day. Advances made during the first and second green revolutions, which focused on increasing food production, have not ended world famine and malnutrition. Even though poorer countries (e.g., India, the Philippines, and many African countries) have been able to increase their food production through technological improvements, the food produced (often by large, foreign, multinational corporations whose motivation is profit) is often sold to other wealthier countries that are able to pay higher prices.

- **Forests:** Nearly half of the world's original forest cover has been lost, and each year another 16 million hectares are cut, bulldozed, or burned. Forests provide over $400 billion to the world economy annually and are vital to maintaining healthy ecosystems. Yet, the current demand for forest products may exceed the limit of sustainable consumption by 25%.

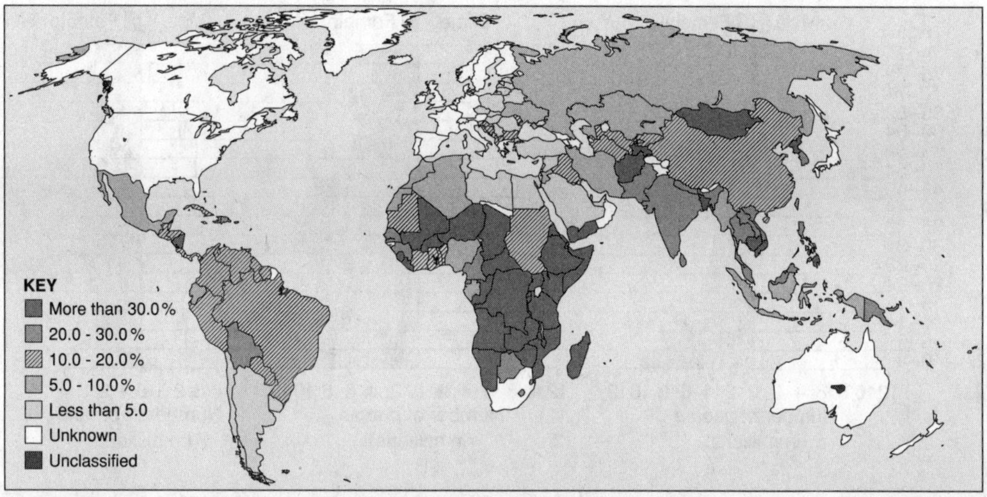

Figure 3.7 Worldwide malnutrition

- **Freshwater:** The supply of freshwater is finite, but demand is soaring as the population grows and per-capita use rises.
- **Global climate change:** Earth's surface is warming due to greenhouse gas emissions, largely from burning fossil fuels. If the global temperature rises as projected, sea levels would rise by several meters, causing widespread flooding, with global warming projected to cause droughts and disrupt agriculture.
- **Public health:** Unclean water, along with poor sanitation, kills over 12 million people each year, most in developing countries. Air pollution kills nearly three million more. Heavy metals and other contaminants also cause widespread health problems. Tobacco-related illnesses, including heart disease, cancer, and respiratory disorders, are the world's leading cause of death and are responsible for more deaths than AIDS, tuberculosis, road accidents, murder, and suicide combined. In 2017, the economic cost of smoking tobacco was more than $300 billion, which included $170 billion in direct medical care and more than $156 billion in lost productivity due to premature death and exposure to secondhand smoke.
- **Unequal distribution of wealth and governmental priorities:** Rapid population growth rates can make it politically difficult for countries to raise standards of living and, at the same time, protect the environment because of government priorities, financial restraints, and special interest groups (e.g., spending money on food, healthcare, education, housing, jobs, energy, etc., vs. spending that same money on wetland restoration). Adding more people to a country's population means that the wealth must be redistributed among more people, causing the gross domestic product (GDP) per capita to decrease.

3.6 Age-Structure Diagrams

A good indicator of future trends in population growth and changes in demographic transition are age-structure diagrams. Age-structure diagrams are determined by birth rate, generation time, death rate, and sex ratios. There are three major age groups in a population: pre-reproductive, reproductive, and post-reproductive (Figure 3.8).

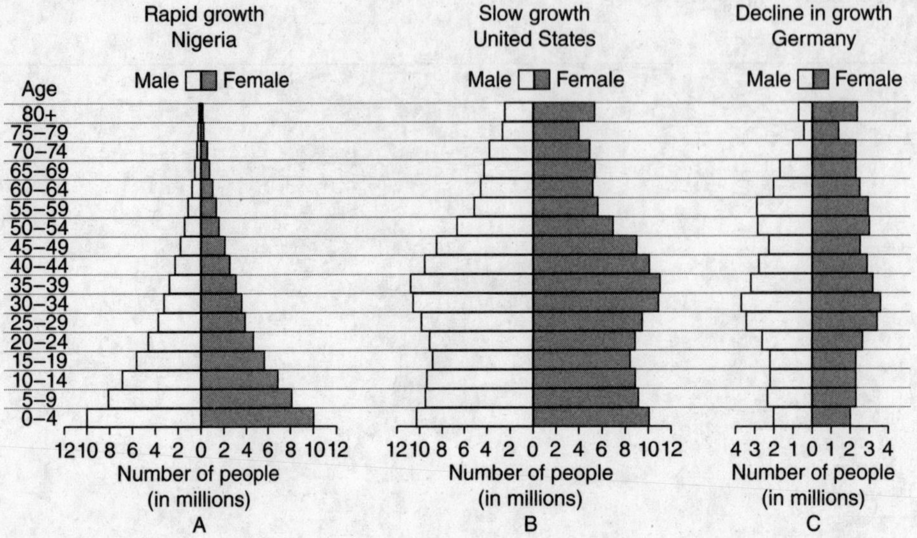

Figure 3.8 Age-structure diagrams for countries with rapid, slow, and declining birth rates

A pyramid-shaped age-structure diagram (e.g., Nigeria) indicates that the population has high birth rates and the majority of the population is in the reproductive age group (generally late teens to mid-forties).

A bell shape indicates that pre-reproductive and reproductive age groups are more nearly equal, with the post-reproductive group being smallest due to mortality. This is characteristic of stable populations (for example, the United States).

An urn-shaped diagram indicates that the post-reproductive group is largest and the pre-reproductive group is smallest, a result of the birth rate's falling below the death rate, and is characteristic of declining populations (e.g., Germany).

The age-structure diagrams in Figure 3.9 show age distributions of developing countries compared with more-developed countries. When the base is large (greater number of younger individuals in the population), there is a potential for an increase in the population as these younger individuals mature and have children of their own (population momentum). However, when the top of the pyramid is larger, it indicates a large segment of the population is past their reproductive years (post-reproductive) and indicates a future slowdown in population growth.

In Mexico, the large family sizes are the result of the necessity for farm labor, the need to support parents when they no longer work, the need to increase family income, and cultural and religious beliefs. The death rate has declined due to advances in social and medical programs.

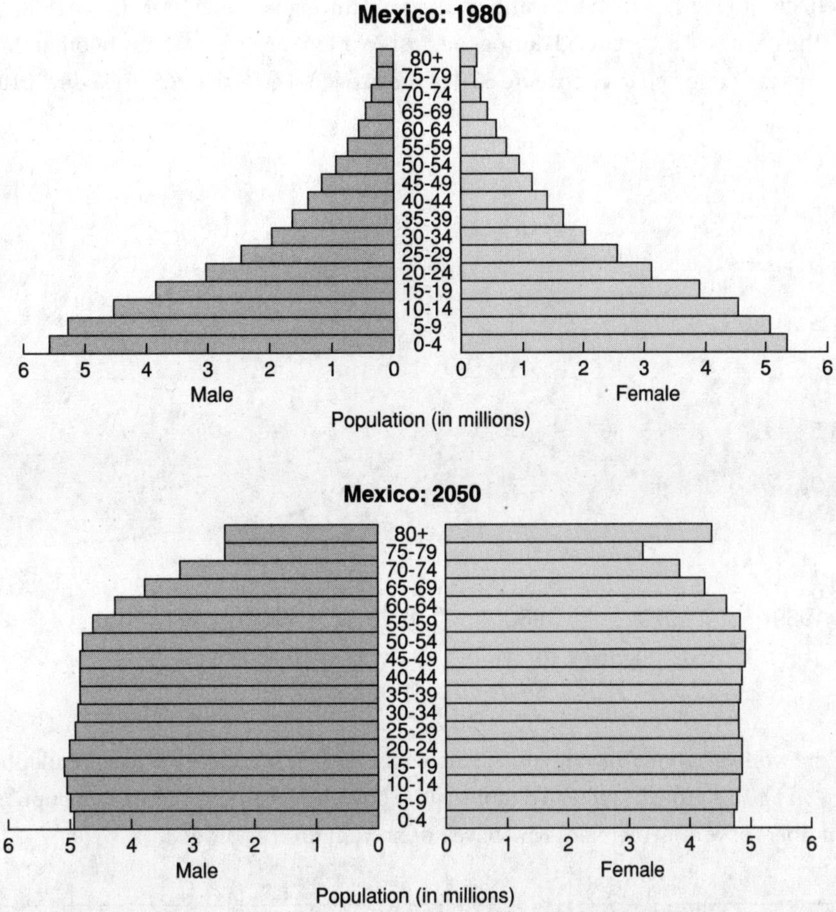

Figure 3.9 Age-structure diagrams for Mexico, 1980 and 2050 (estimated)
Source: U.S. Census Bureau, International Database

3.7 Total Fertility Rate

The total fertility rate (TFR) is the average number of children that each woman will have during her lifetime. The country of Niger in Africa leads the world's TFR at 7.63.

Worldwide Total Fertility Rate

Country	TFR
Niger	7.63
India	2.43
Mexico	2.24
United States	1.87
Russia	1.61
China	1.60
Japan	1.41
World average	**2.59**

Source: U.S. Central Intelligence Agency

The two main effects of TFRs less than 2.1 without additions through immigration are population decline and population aging. The greatest TFR occurred during the post–World War II years (baby boomers). New immigrants and their descendants are projected to contribute 66% of the expected growth by 2050 (Figure 3.10).

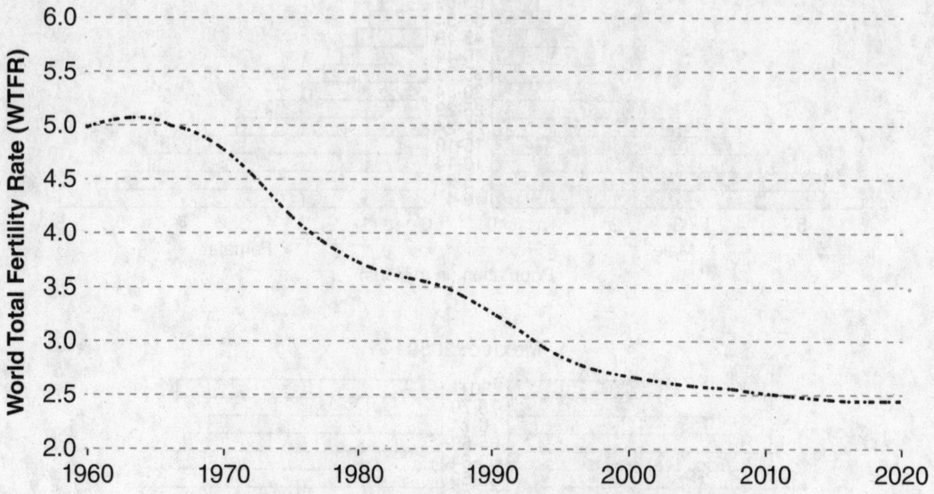

Figure 3.10 World TFRs by year (1960–2020)
Source: The World Bank

Despite half of the world's nations' having sub-replacement fertility rates, the world's population is still growing quickly (Figure 3.11). This growth is due to nations with above-replacement TFRs and a population momentum caused by large numbers of younger females who have not had children as of yet.

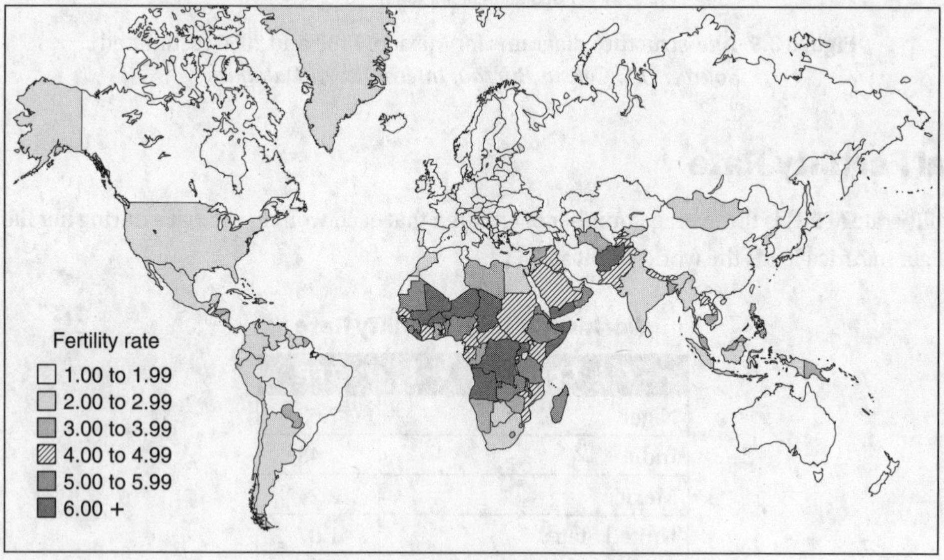

Figure 3.11 Worldwide total fertility rates

Declines in fertility rates can be attributed to several factors, as follows:

- As developing countries transition to developed countries, there is greater access to primary healthcare and family-planning services.
- Female educational opportunities are increasing.
- Many "millennials" are postponing marriage until their careers are established.
- More individuals desire to increase their standard of living by having fewer children.
- The number of females in the workforce has increased.
- There is greater personal acceptance and government encouragement of contraception.
- Urbanization results in a higher cost of living and reduces the need for extra children to work on farms.

3.8 Human Population Dynamics

Many factors affect the human population, such as historical population sizes, population distribution, fertility rates, growth rates, and doubling times. Age-structure diagrams act as indicators of future population trends.

Historical Population Sizes

The recent rapid growth of the world's human population has been due primarily to a decrease in death rates. In 1900, the overall death rate in the United States was 1.7%. By 2018, the death rate had dropped to 0.8% (almost half). Children in 1900 were ten times more likely to die than children in 2000. Several factors have reduced human death rates, as follows:

- Increased food and more efficient distribution, resulting in better nutrition
- Improvements in medical and public health programs, resulting in better access to anesthetics, antibiotics, and vaccinations
- Improvements in sanitation and personal hygiene
- Improvements in the safety of water supplies

The human population has had four surges in growth as a result of the following:

- The use of tools (3.5 million years ago)
- The discovery of fire (1.5 million years ago)
- The first agricultural revolution, which allowed the change from hunting and gathering to crop growing (~ 10,000 B.C.E.)
- The industrial and medical revolutions (within the last ~ 200 years).

Human Population Growth

Time Period	Description	Practicing Worldview
Before Agricultural Revolution	~1 million to 3 million humans. Hunter-gatherer lifestyle.	
8000 B.C.E. to 5000 B.C.E.	~50 million humans. Increases due to advances in agriculture, domestication of animals, and the end of a nomadic lifestyle.	Earth Wisdom—Natural cycles that can serve as a model for human behavior.
5000 B.C.E. to 1 B.C.E.	~200 million humans. Rate of population growth during this period was about 0.03 to 0.05%, compared with today's growth rate of 1.3%.	
0 C.E. to 1300 C.E.	~500 million humans. Population rate increased during the Middle Ages because new habitats were discovered. Factors that reduced population growth rate during this time were famines, wars, and disease (density-dependent factors).	Frontier Worldview—Viewed undeveloped land as a hostile wilderness to be cleared and planted, then exploited for its resources as quickly as possible.
1300 C.E. to 1650 C.E.	~600 million humans. Plagues reduced population growth rate. Up to 25% mortality rates are attributed to the plagues that reached their peak in the mid-1600s.	
1650 C.E. to present	Currently ~7.5 billion humans. In 1650 C.E., the growth rate was ~0.1%. Today it is ~1.2%. Healthcare, health insurance, vaccines, medical cures, preventative care, advanced drugs and antibiotics, improvements in hygiene and sanitation, advances in agriculture and distribution, and education are factors that have increased the growth rate.	Planetary Management—Beliefs that as the planet's most important species, we are in charge of Earth; we will not run out of resources because of our ability to develop and find new ones; the potential for economic growth is essentially unlimited; and our success depends on how well we manage Earth's life-support systems mostly for our own benefit.
Present to 2050 C.E.	Estimates are as high as 9.8 billion.	Earth Wisdom—Beliefs that nature exists for all Earth's species and we are not in charge of Earth; resources are limited and should not be wasted. We should encourage Earth-sustaining forms of economic growth, and our success depends on learning how Earth sustains itself and integrating such lessons from nature into the ways we think and act.

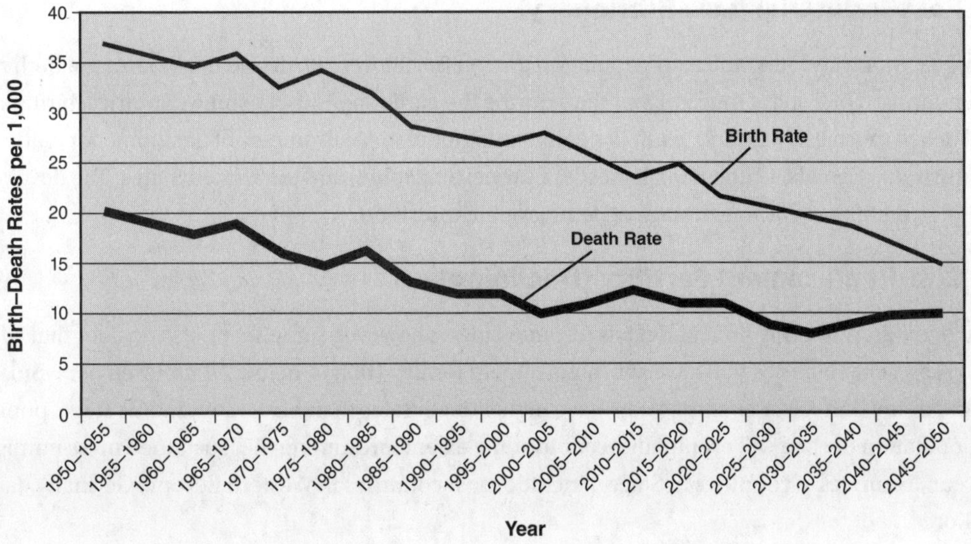

Figure 3.12 Global birth and death rates, 1950–2050 (projected)

3.9 Demographic Transition

Demographic transition is the transition from high birth and death rates to lower birth and death rates as a country or region develops from a pre-industrial to an industrialized economic system. It leads to a stabilization of population growth in the more highly developed countries and is generally characterized as having four separate stages: pre-industrial, transitional, industrial, and post-industrial (Figure 3.13).

Stage 1: Pre-Industrial (High Stationary)

Living conditions are severe, medical care is poor or nonexistent, and the food supply is limited due to poor agricultural techniques and pestilence. Birth rates are high to replace individuals lost through high mortality rates, with a net result of little population growth. Sub-Saharan Africa is the region hardest hit by AIDS-HIV; the region is home to ~6% of the world's population but 54% of all AIDS-HIV cases. However, overall, new infections in the region have declined by 28% and death rates have fallen by 44% since 2010 as a result of advances in drug therapy.

Stage 2: Transitional (Early Expanding)

This stage occurs after the start of industrialization. Standards of hygiene, advances in medical care, improved sanitation, cleaner water supplies, vaccinations, and higher levels of education begin to drive down the death rate, leading to a significant upward trend in population size. The net result is a rapid increase in population. Examples include Afghanistan, Pakistan, Guatemala, and sub-Saharan Africa.

Stage 3: Industrial (Late Expanding)

Urbanization decreases the economic incentives for large families. The cost of supporting an urban family grows, and parents are more actively discouraged from having large families. Educational and work opportunities for women decrease birth rates. Obtaining food is not a major focus of the day, and leisure time is available. Retirement safety nets are in place, reducing the need for extra children to support parents. In response to these economic pressures, the birth rate starts to drop, ultimately coming close to the death rate. Countries currently in Stage 3 include India, Mexico, and South Africa.

Stage 4: Post-Industrial (Low Stationary)

Birth rates equal mortality rates, and zero population growth is achieved. Birth and death rates are both relatively low, and the standard of living is much higher than during the earlier periods. In some countries, birth rates may actually fall below mortality rates and result in net losses in population. Examples of declining populations include Argentina, Australia, Canada, China, Brazil, most of Europe, Singapore, and the United States. The developed world currently remains in this fourth stage of demographic transition.

Stage 5: Sub-Replacement Fertility (Declining)

The original demographic transition model has just four stages; however, some theorists consider that a fifth stage is needed to represent countries that have sub-replacement fertility (that is, below 2.1 children per woman). Most European and many East Asian countries now have higher death rates than birth rates. In this stage, population aging and population decline will eventually occur to some extent, presuming that sustained mass immigration does not occur. Examples of countries in Stage 5 include most countries in Western Europe, Germany, Japan, and South Korea.

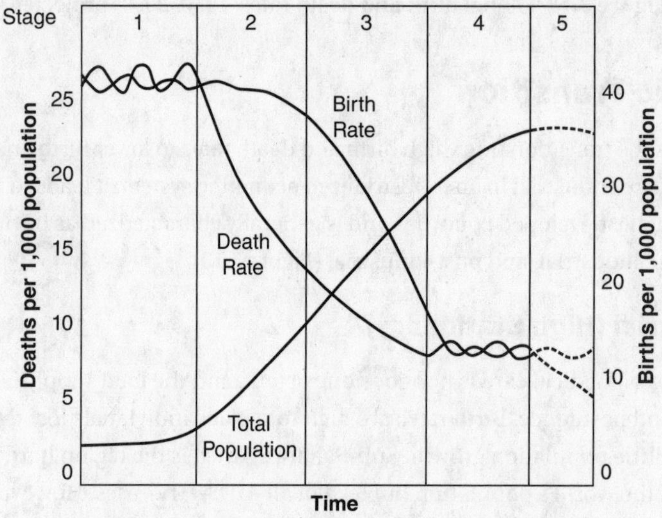

Figure 3.13 Demographic transition model

Practice Questions

1. Which would be least likely to be affected by a density-dependent limiting factor?

   (A) A large, dense population
   (B) A population with a high birth rate
   (C) A population with a high immigration rate
   (D) A small, scattered population

2. A population showing a growth rate of 20, 40, 60, 80, ... would be characteristic of

   (A) arithmetic (linear) growth
   (B) exponential growth
   (C) logarithmic growth
   (D) static growth

3. If a population doubles in about 70 years, it is showing a ________ % growth rate.

   (A) 1
   (B) 5
   (C) 35
   (D) 140

4. An island off Costa Rica includes 500 birds of a particular species. Population biologists determined that this bird population was isolated, with no immigration or emigration. After one year, the scientists were able to count 60 births and 10 deaths. The net growth for this population was

   (A) 0.5
   (B) 1.0
   (C) 1.1
   (D) 1.5

5. Afghanistan has a current growth rate of 4.8%, representing a doubling time of approximately

   (A) 9.6 years
   (B) 14.5 years
   (C) 35 years
   (D) 70 years

6. Biotic potential refers to

   (A) a factor that influences population growth and that increases in magnitude with an increase in the size or density of the population
   (B) an estimate of the maximum capacity of living things to survive and reproduce under optimal environmental conditions
   (C) events and phenomena of nature that act to keep population sizes stable
   (D) the ratio of total live births to total population

7. For a given population, the average number of children that would be born to a woman over her lifetime if she was to (1) experience the exact current age-specific fertility rates throughout her lifetime and (2) survive from birth to the end of her reproductive life span is known as the

   (A) Crude Birth Rate (CBR)
   (B) Crude Fertility Rate (CFR)
   (C) Replacement Level Fertility Rate (RLFR)
   (D) Total Fertility Rate (TFR)

8. An APES class performed an experiment in which they released and recaptured light and dark moths in the city and in a rural setting. The results are presented below:

Location		Light Moths	Dark Moths
City	Released	400	400
	Captured	200	300
Rural	Released	300	300
	Captured	200	100

   Which of the following statements is TRUE?

   (A) A higher percentage of light moths were recaptured in the country than were recaptured in the city.
   (B) A higher percentage of light moths were recaptured in the city than dark moths recaptured in the city.
   (C) A lower percentage of dark moths were recaptured in the city than were recaptured in the country.
   (D) A lower percentage of light moths were recaptured in the country than dark moths recaptured in the country.

9. All of the following are characteristics of developing countries EXCEPT

(A) high energy production and consumption
(B) high rates of illiteracy
(C) large percentage of the population is under the age of 15
(D) low incomes and dependence on subsistence farming

Refer to the following diagram to answer Question 10.

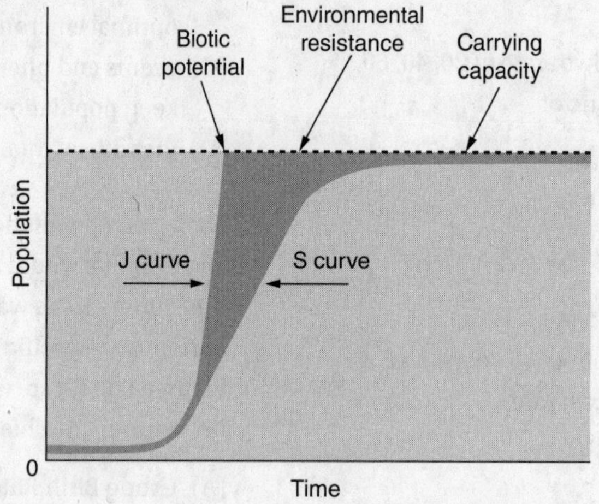

10. Which statement below best describes the difference between exponential (J curve) and logistic (S curve) growth?

(A) Exponential growth depends on birth and death rates; logistic growth does not.
(B) Exponential growth depends on density; logistic growth depends on the carrying capacity.
(C) Exponential growth follows an S (sigmoidal) curve; logistic growth follows a linear (J) function.
(D) Logistic growth reflects density-dependent effects of birth and death rates; exponential growth is independent of density.

The graph below shows the shift of people in the United States moving to the suburbs for the years 1900–2000.

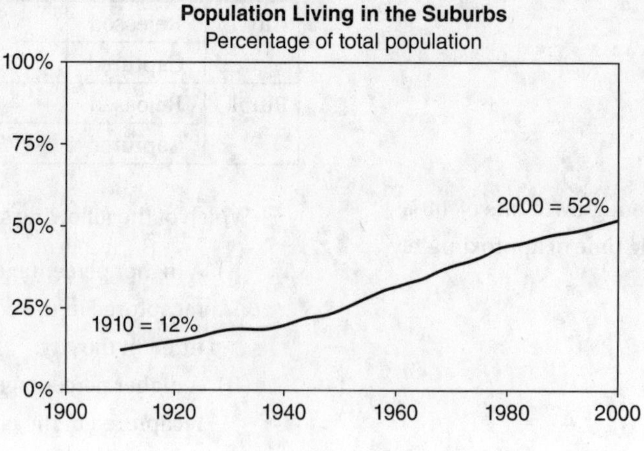

11. What was the primary reason that allowed people to begin moving to the suburbs?

(A) Better transportation between the city and their homes in the suburbs
(B) Less crime in the suburbs
(C) More housing opportunities in the suburbs
(D) More job opportunities in the suburbs

12. The following age-structure diagram would be typical of

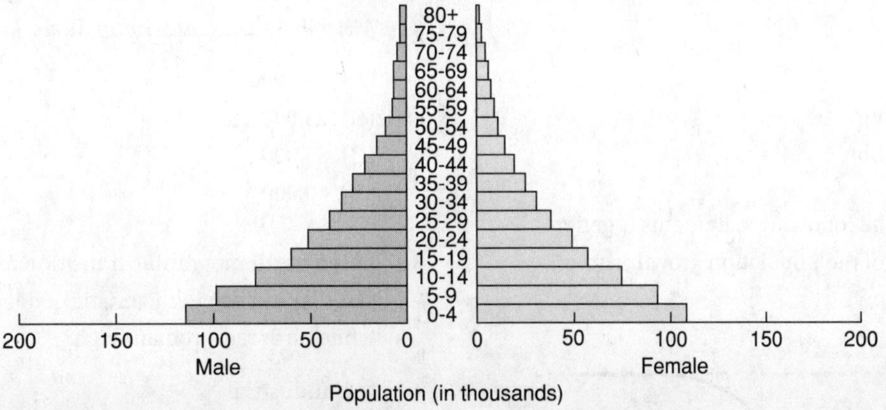

Source: U.S. Census Bureau, International Database

(A) China
(B) India
(C) Russia
(D) the United States

13. Examine the following age-structure diagram.

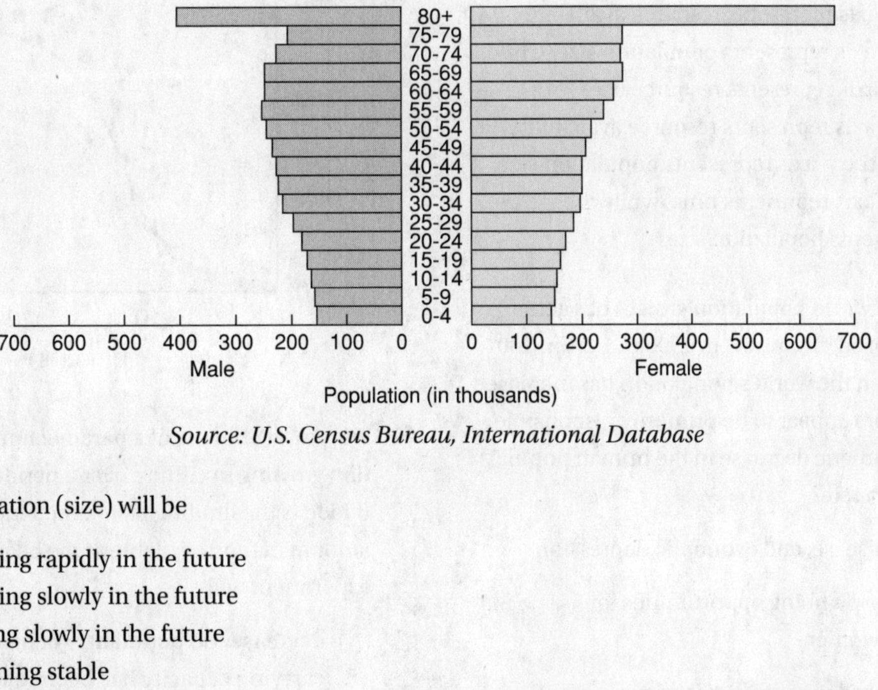

Source: U.S. Census Bureau, International Database

This population (size) will be

(A) declining rapidly in the future
(B) declining slowly in the future
(C) growing slowly in the future
(D) remaining stable

14. Which of the following examples does NOT demonstrate a density-dependent factor affecting the population size?

(A) An outbreak of influenza in a hospital
(B) The "black plagues" that occurred in Europe during the mid-14th century and which are estimated to have killed up to one-third of the European population
(C) The destruction of a rainforest in Brazil due to drought
(D) The number of lions inhabiting a grassland in Africa

15. Density-independent factors would include all of the following EXCEPT

(A) drought
(B) fires
(C) flooding
(D) predation

16. Which of the following statements is generally TRUE of the population growth curve shown below?

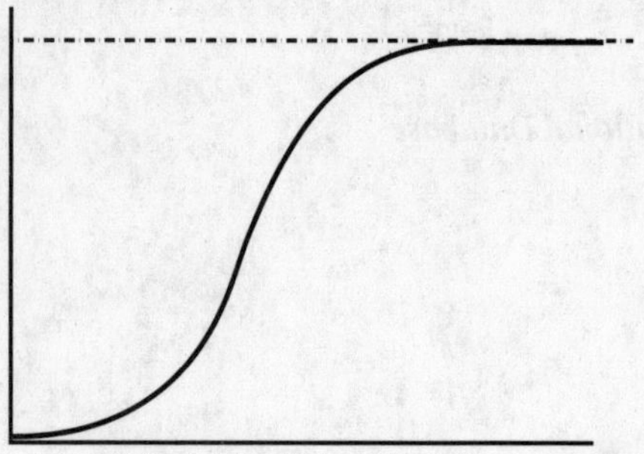

(A) The x-axis represents population size, while the y-axis represents resource availability.
(B) The x-axis represents population size, while the y-axis represents resource use.
(C) The x-axis represents resource availability, while the y-axis represents population size.
(D) The x-axis represents time, while the y-axis represents population size.

17. The rate of world population growth has actually decreased, from 1.8% in 1990 to 1.2% in 2007, even though the world's population has increased. What factors appear to be primarily responsible for this dramatic decrease in the human population growth rate?

I. War, disease, and economic depression

II. Family-planning opportunities for more women

III. Increased urbanization and more opportunities for women

(A) I
(B) II
(C) III
(D) II and III

18. After 35 years, the size of an initial population of 10,000 individuals, whose annual growth rate is 4%, will be ________ individuals.

(A) 15,000
(B) 25,000
(C) 35,000
(D) 40,000

19. Using the demographic transition model, what stage would be characteristic of death rates falling while birth rates remain high?

(A) Industrial
(B) Pre-industrial
(C) Post-industrial
(D) Transitional

Refer to the following graph to answer Question 20.

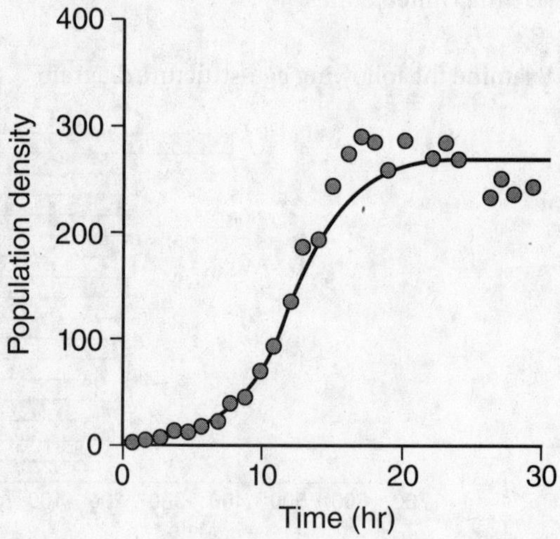

20. The graph above is of a paramecium population growing in culture over a period of 30 hours. If food is the limiting factor, then increasing the amount of food available at the beginning of the experiment will

(A) decrease the population density after the carrying capacity has been reached
(B) decrease the time for the population to reach its carrying capacity
(C) have no effect on the population density between 0 and 15 hours
(D) increase the size of the population density at all times during the experiment

Free-Response Question

Gold was discovered in 1848 in the foothills of the Sierra Nevada mountain range in California. The fictional town of Barronsville soon sprang up, serving the needs of the miners and others who flocked to the area. Below is a chart of several population parameters of Barronsville from 1850 through 1890.

Year	Total Number of Immigrants During Last 10 Years	Total Number of Emigrants During Last 10 Years	Total Number of Births During Last 10 Years	Total Number of Deaths During Last 10 Years	Population
1850	500	0	0	0	500
1860	2,480	0	50	30	3,000
1870	500	?	30	40	1,450
1880	10	1,000	?	7	620
1890	0	200	0	2	?

(a) Calculate the following. Show all work.
  (i) The number of emigrants between 1860 and 1870.
  (ii) The number of births between 1870 and 1880.
  (iii) The population in 1890.

(b) The following chart shows the number of births and deaths in Barronsville from 1850 through 1859.

Year	Births/1,000	Deaths/1,000	Population
1850	0	0	500
1851	0	0	600
1852	0	0	700
1853	0	1.25	800
1854	1.11	1.11	900
1855	2.00	1.00	1,000
1856	3.84	1.54	1,300
1857	4.00	1.71	1,750
1858	6.36	3.18	2,200
1859	7.90	5.66	2,650

On the graph below, plot the crude birth rate and the crude death rate for Barronsville between 1850 and 1859. Clearly label the axes and the curves.

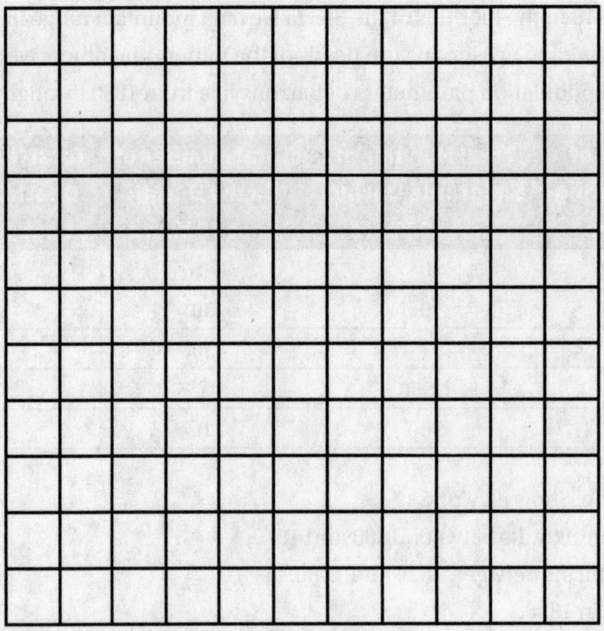

(c) Indicate TWO factors that may have accounted for the rapid decline in the population of Barronsville between 1850 and 1890.

(d) Cyanide and mercury are toxic pollutants that are still used in gold mining today. Choose ONE of these chemicals and explain how it is used in gold mining and its effect on the environment and human health.

Answer Explanations

1. **(D)** An increasing population size reduces available resources, thus limiting population growth. In restricting population growth, a density-dependent factor intensifies as the population size increases, affecting each individual more strongly. Population growth declines because of death rate increase, birth rate decrease, or both. There is a reduction in the food supply, which restricts reproduction, resulting in fewer offspring. The competition for space to establish territories is a behavioral mechanism that may restrict population growth.

   Predators concentrate in areas where there is a high concentration of prey. As long as the natural resources are available in sufficient quantity, the population will remain constant. As the prey population decreases, so does that of predators. Intrinsic factors may play a role in limiting a population size; e.g., the accumulation of toxic wastes. High densities may also cause stress syndromes, resulting in hormonal changes that may delay the onset of reproduction. It has also been reported that immune disorders are related to stress in densely populated areas.

   Density-independent factors include weather, climate, and natural disasters such as freezes, seasonal changes, hurricanes, and fires. These factors affect all individuals in the population, regardless of population size.

2. **(A)** Arithmetic or linear growth is characterized by a constant increase per unit of time. In this case, the constant is an increase of 20.

3. **(A)** A 1% growth rate would cause a population to double in 70 years. Hint: divide 70 by the annual percentage growth rate to get the doubling time in years.

4. **(C)** Population size = original size (500) + births (60) − deaths (10) + immigration (0) − emigration (0) = 550. Net growth rate = 550/500 = 1.1.

5. **(B)** Using the Rule of 70, 70/4.8 = 14.5 years.

6. **(B)** The maximum reproductive rate is called the biotic potential.

7. **(D)** The total fertility rate (TFR) is a better index of fertility than the crude birth rate (CBR), which is the annual number of births per thousand population, because it is independent of the age structure of the population. Niger currently has the highest TFR of 7.63, the United States comes in at 1.87, and the lowest TFR is 1.22 for Taiwan.

8. **(A)** To find the percentage of moths recaptured, divide the number of moths recaptured by the number of moths released and multiply by 100%. Sixty-six percent [(200/300) × 100%] of the light moths were recaptured in the rural setting, but only 50% [(200/400) × 100%] of them were recaptured in the city.

9. **(A)** Developing countries are countries that, according to the United Nations, exhibit the lowest indicators of socioeconomic development in the world; e.g., Angola, Ethiopia, and Uganda. A country is classified as developing if it meets the following three criteria:

   - Economic vulnerability (based on instability of agricultural production, instability of exports and services, percentage of population displaced by natural disasters, and overreliance on nontraditional economic activities)
   - Human resource weaknesses (based on poor levels of nutrition, health, education, and adult literacy)
   - Low income (Gross National Income per capita of less than $750)

10. **(D)** Logistic growth (S curve) is defined as density-dependent population growth in which the growth rate decreases, with increasing numbers of individuals, until it becomes zero when the population reaches a maximum known as the carrying capacity. The model assumes that when populations increase in size, the per capita birth rate decreases as a result of competition for resources. Thus, there is a population size at which the per capita birth rate equals the per capita death rate; it is known as the carrying capacity and is the point at which the population growth rate is zero. Exponential growth (J curve) describes population growth without initial limits since there are no limiting resources; i.e., there's plenty of food, space, and other resources for all. However, over time, as the population growth begins to outpace the resources, the resources begin to run out and the population levels off.

11. **(A)** The key word in the question is "allowed." The other choices may have been why people *wished* to move to the suburbs. However, if there were not mechanisms in place that allowed people to live away from the cities (rail systems, freeways, subways, etc.), such large shifts in population toward the suburbs would not have been possible.

12. **(B)** Age-structure diagrams with a wide base are populations that have a high proportion of young people, which results in a powerful, built-in momentum to increase population size, assuming death rates do not unexpectedly increase.

13. **(A)** This graph is the reverse of the diagram in Question 12. In this case, the majority of the population is beyond reproductive years, and the death rate exceeds the birth rate. This projected age-structure diagram is for Hong Kong in 2050.

14. **(C)** Choices (A), (B), and (D) are examples of density-dependent factors, wherein large, dense populations are more strongly affected than small, less-crowded ones. Drought is a naturally occurring event and does not depend on the density of any organism(s) occurring in the area. However, had the destruction of the rainforest been due to ranching, timber harvesting, and/or agriculture (as is the most common cause), then an increase in the number of humans in the rainforest would increase the amount of rainforest destruction, in which case the destruction would be due to density-dependent elements.

15. **(D)** Density-independent factors influence population growth and do not depend on the size or density of the population. Predation rates are affected by population size.

16. **(D)** Stable populations occupying a fixed geographic space demonstrate an S-shaped population growth curve and typically follow three key stages: (1) an exponential growth phase; (2) a transitional phase; and (3) a plateau phase. Initially, population growth is slow, as there are few reproductive individuals and they are generally widely dispersed; however, over time, there is a rapid increase in population size due to abundant environmental resources and minimal environmental resistance. As the population continues to grow, resources eventually become limited, which leads to competition for survival, resulting in birth rates starting to fall and mortality rates beginning to rise, leading to a slowing of population growth. Eventually the increasing death rate equals the decreasing birth rate and population growth becomes static, reaching a point of equilibrium as the population has reached the carrying capacity (K) of the environment, with limiting factors keeping the population stable.

17. **(D)** Factors influencing fertility rates include economic development, urbanization, family-planning practices, and sex education awareness. Developed countries where urbanization and economic development are the norms show the lowest fertility rates, while developing countries have higher rates. Economic and urbanization factors may work to influence priorities within developed countries, along with more accessible educational and planning resources. As world populations become more developed, an overall decline in fertility rates is expected. Although war, disease, and economic depression do decrease the human population count, they do not have a significant effect on the overall fertility rate.

18. **(D)** $70 \div 4.0 = 17.5$ years doubling time. 17.5 is two doubling times ($17.5 \times 2 = 35$ years). The first doubling would result in 20,000 individuals. The second doubling would result in 40,000 individuals.

19. **(D)** In the pre-industrial stage, living conditions are harsh, birth and death rates are high, and there is little increase in population size. In the transitional stage, living conditions improve, the death rate drops, and birth rates remain high. In the industrial stage, growth slows. In the post-industrial stage, zero population growth is reached, and the birth rate falls below the death rate.

20. **(C)** Having extra food in the water between 0 and 15 hours would generally not affect the population density in this experiment. Assuming ideal conditions with more than enough food available during this time period, the paramecia are already reproducing at their maximum capacity; however, if there is extra food available at all times during the 30 hours of the experiment, then the time period of exponential growth is extended and the population density increases until a point is reached when another limiting factor (e.g., dissolved oxygen content of the water) begins to affect the number of paramecia that can survive, at which point a new carrying capacity is established.

Free-Response

10 Total Points Possible

(a) *Maximum 3 points total: 1 point maximum for each correct answer. Must show work.*

(i) *The number of emigrants between 1860 and 1870. (1 point maximum)*

$$N_{1870} = (N_{1860} + B + I) - (D + E) \text{ where}$$
$$1{,}450 = (3{,}000 + 30 + 500) - (40 + E)$$
$$E = 2{,}040$$

(ii) *The number of births between 1870 and 1880. (1 point maximum)*

$$N_{1880} = (N_{1870} + B + I) - (D + E)$$
$$620 = (1{,}450 + B + 10) - (7 + 1{,}000)$$
$$B = 167$$

(iii) *The population in 1890. (1 point maximum)*

$$N_{1890} = (N_{1880} + B + I) - (D + E)$$
$$N_{1890} = (620 + 0 + 0) - (2 + 200)$$
$$N_{1890} = 418$$

(b) *Maximum 3 points total.*

Plot the crude birth rate and the crude death rate for Barronsville between 1850 and 1859. Clearly label the axes and the curves. (1 point maximum for correctly scaling and labeling the axes. Each axis must include all data between 1850 and 1859. 2 points for correctly plotting both rates.)

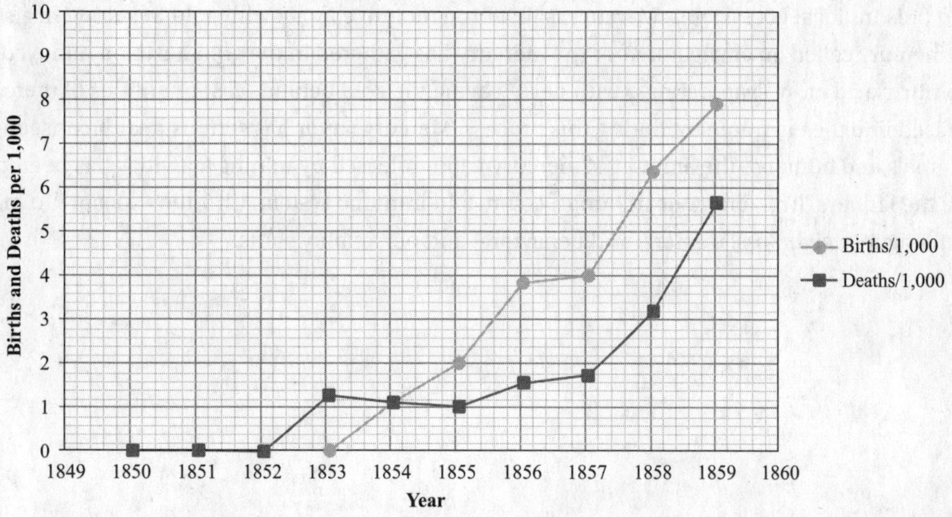

(c) *Maximum 2 points total: 1 point for each factor described.*

Identify TWO factors that may have accounted for the rapid decline in the population in Barronsville between 1850 and 1890. (1 point maximum for each possible cause for the population to decline. Only the first two factors provided are considered.)

Possible factors might include:

- What gold existed in the area was extracted to the point where it was not economical to mine anymore.
- Lack of proper sanitation and/or healthcare, which could have created conditions for disease or an epidemic.
- News of a richer strike in another location.
- Pollution and health effects caused by chemicals used during the gold-extraction process.

(d) *Maximum 2 points total.*

Cyanide and mercury are toxic pollutants that are still used in gold mining today. Choose ONE of these chemicals and explain how it is used in gold mining and its effect on the environment and human health. (1 point for a correct explanation of the process, and 1 point for a correct explanation of the health and/or environmental effect[s].)

After being brought to the surface, the ore must be processed to extract the mineral, which generates huge quantities of waste. The amount of recoverable metal in even high-grade ores is generally just a small fraction of the total mass. Every ounce of gold produced results in 30 tons of mine waste. Gold is commonly extracted from the ore through a technique called "heap leaching." The ore containing the gold is crushed, piled into heaps, and sprayed with a solution containing cyanide, which trickles down through the ore and bonds with the gold. The resulting gold–cyanide solution is collected at the base of the heap and pumped to a mill, where the gold and cyanide are chemically separated. The cyanide is then stored in artificial ponds for reuse. Each bout of leaching takes a few months, after which the heaps receive a layer of fresh ore. Given the scale and duration of these operations, contamination of the surrounding environment with cyanide is almost inevitable. To dispose of the leftover ore (tailings) contaminated with cyanide and other toxins, a dam is constructed, which allows percolation of these toxins into the groundwater. These dams are also often structurally unsound.

OR

Amalgamation is a commonly used gold-extraction process that unleashes widespread mercury contamination and poisons local ecosystems. Mercury is first brought into contact with gold, resulting in a solution of gold in mercury called an amalgam. After the mercury has gathered in the gold it can be removed by dissolving it in nitric acid or by evaporating it with heat, leaving the gold behind. Vast quantities of mercury vapor are released into the environment during this process. Mercury vapor has serious health consequences for both animals and humans. The amount of mercury vapor released by mining activities has been proven to damage the kidneys, liver, brain, heart, lungs, colon, and immune system. Chronic exposure to mercury may result in fatigue, weight loss, tremors, and behavioral and personality shifts.

4

Earth Systems and Resources

Learning Objectives

In this unit, you will learn:

- ➔ 4.1 Plate Tectonics
- ➔ 4.2 Soil Formation and Erosion
- ➔ 4.3 Soil Composition and Properties
- ➔ 4.4 Earth's Atmosphere
- ➔ 4.5 Global Wind Patterns
- ➔ 4.6 Watersheds
- ➔ 4.7 Solar Radiation and Earth's Seasons
- ➔ 4.8 Earth's Geography and Climate
- ➔ 4.9 El Niño and La Niña

For added practice with this unit and topics, log onto Online Labs 4A–4B at online.barronsbooks.com

4.1 Plate Tectonics

Plate tectonic theory states that Earth's lithosphere is divided into a small number of plates that float on and travel independently over the mantle, with much of Earth's seismic activity occurring at the boundaries of these plates. Plate tectonic theory arose from two separate geological observations: continental drift and seafloor spreading.

Continental Drift Theory

In 1915, Alfred Wegener proposed that all present-day continents originally formed one landmass he called Pangaea (Figure 4.1).

Figure 4.1 Pangaea

Wegener believed that this supercontinent began to break up into smaller continents around 200 million years ago. He based his theory on the following six factors:

1. Fossils of extinct land animals were found on separated landmasses.
2. Fossilized tropical plants were discovered beneath Greenland's ice caps.
3. Glaciated landscapes occurred in the tropics of Africa and South America.
4. Similarities existed in rocks between the east coasts of North and South America and the west coasts of Africa and Europe.
5. The continents fit together like pieces of a puzzle.
6. Tropical regions on some continents had polar climates in the past, based on paleo-climatic data.

Seafloor Spreading Theory

During the 1960s, alternating patterns of magnetic properties were discovered in rocks found on the seafloor. Similar patterns were discovered on either side of mid-oceanic ridges found near the center of the oceanic basins. Dating of the rocks indicated that as one moved away from the ridge, the rocks became older, and suggested that new crust was being created at volcanic rift zones (Figure 4.2).

The lithosphere is the solid, outer part of the Earth and is broken into huge sections called plates, which are slowly moving. When one plate moves beneath another (subduction) or when two plates converge, it can result in earthquakes and volcanoes.

Subduction zones are areas on Earth where two tectonic plates meet and move toward each other, with one sliding underneath the other and moving down into the mantle.

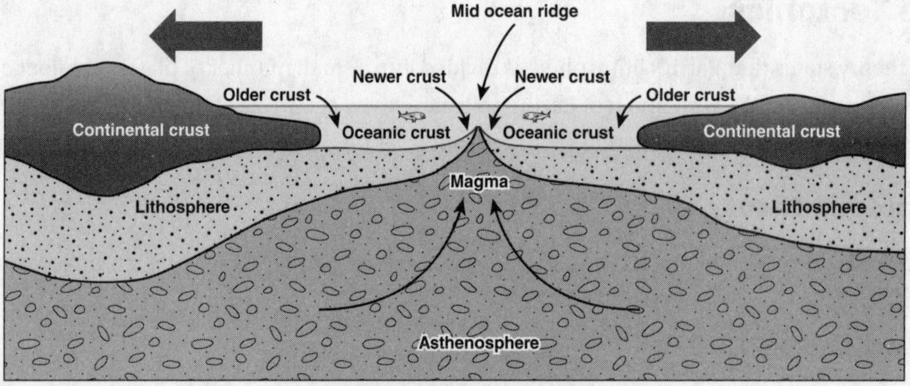

Figure 4.2 Seafloor spreading

Types of Boundaries

Tectonic plates move slowly over time, forming volcanic island chains, folded mountain belts, deep-sea trenches, ridges, and rift valleys. These plates move in relation to one another at one of three types of plate boundaries: convergent, divergent, and transform.

Convergent Boundaries

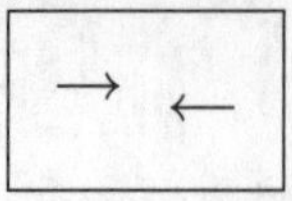

Convergent boundaries occur where two plates slide toward each other, commonly forming either (a) a subduction zone, where one plate moves underneath the other; or (b) an orogenic belt, if the two plates collide and compress. When a denser oceanic plate moves underneath (subducts) a less dense continental plate, an oceanic trench may be produced on the ocean side and a mountain range may be produced on the continental side. Example: Cascade Mountain Range.

Divergent Boundaries

← →

Divergent boundaries occur where two plates slide apart from each other, and the space that is created is filled with molten magma from below. Divergent boundaries can create massive fault zones in the oceanic ridge system and are areas of frequent oceanic earthquakes. Examples: (1) oceanic divergent boundary—Mid-Atlantic Ridge and the East Pacific Rise; and (2) continental divergent boundary—East African Great Rift Valley.

When two oceanic plates converge, they create an island arc—a curved chain of volcanic islands rising from the deep seafloor and near a continent (Figure 4.3). They are created by subduction processes and occur on the continental side of the subduction zone. Their curve is generally convex toward the open ocean. A deep undersea trench is located in front of such arcs where the descending plate dips downward. Examples include Japan and the Aleutian Islands in Alaska.

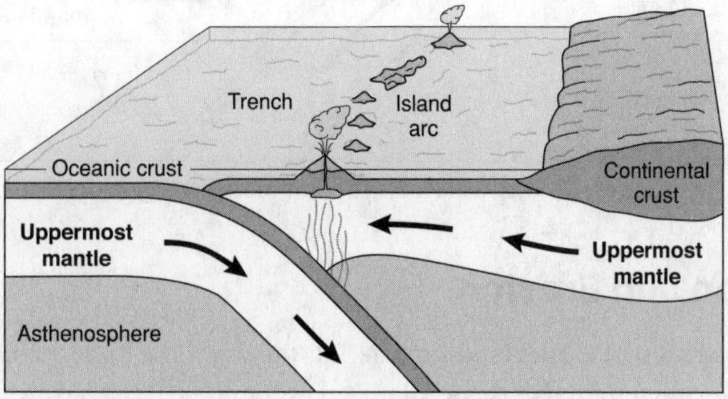

Figure 4.3 Oceanic-oceanic plate convergence

When two continental plates collide, mountain ranges are created as the colliding crust is compressed and pushed upward (Figure 4.4). An example is the northern margins of the Indian subcontinental plate, which is being thrust under a portion of the Eurasian plate, lifting it and creating the Himalayas.

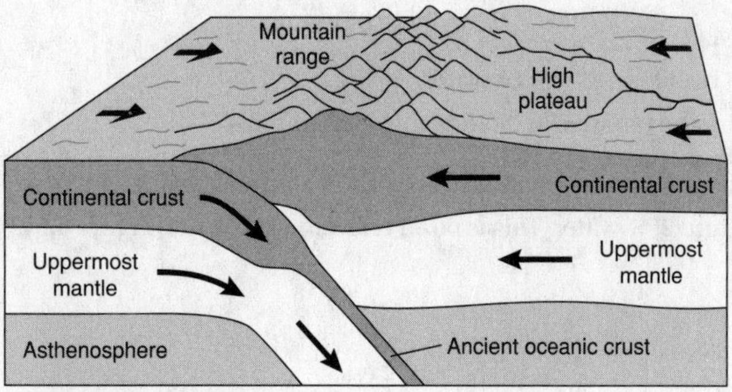

Figure 4.4 Continental-continental plate convergence

Transform Boundaries

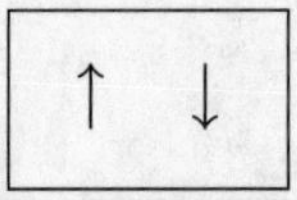

Transform boundaries occur where plates slide past each other in opposite directions. The friction and stress buildup from the sliding plates frequently causes earthquakes, a common feature along transform boundaries. Example: The San Andreas fault, which is near the western coast of North America, is where the Pacific plate and the North American plate meet. In 50 million years, western California will be a separate island near Alaska.

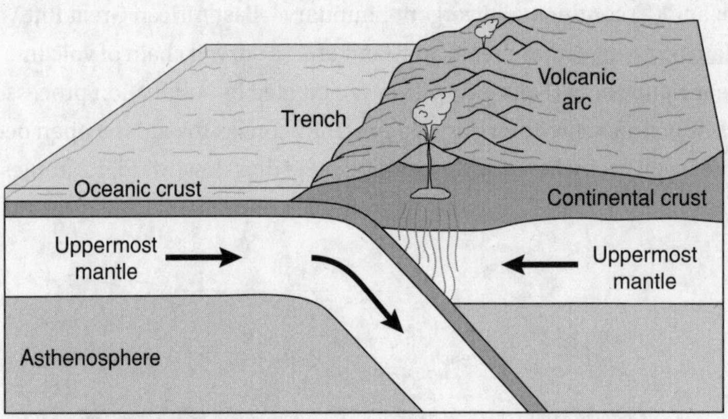

Figure 4.5 Oceanic–continental plate convergence

4.2 Soil Formation and Erosion

Soils are a thin layer on top of most of Earth's land surface. This thin layer is a basic, natural resource, and its characteristics deeply affect every other part of the ecosystem. For example:

- Soils affect the chemistry of ground and surface water as well as the amount of water that returns to the atmosphere, via evaporation, to form rain.
- Soils hold nutrients and water for plants and animals.
- Water is filtered and cleansed as it flows through soils.

Soils are composed of three main ingredients:

1. Minerals of different sizes
2. Open spaces that can be filled with air or water
3. Organic materials from the remains of dead plants and animals

A good soil for growing most plants should have about 45% minerals (with a mixture of sand, silt, and clay), 5% organic matter, 25% air, and 25% water. This is often referred to as loam (Figure 4.6).

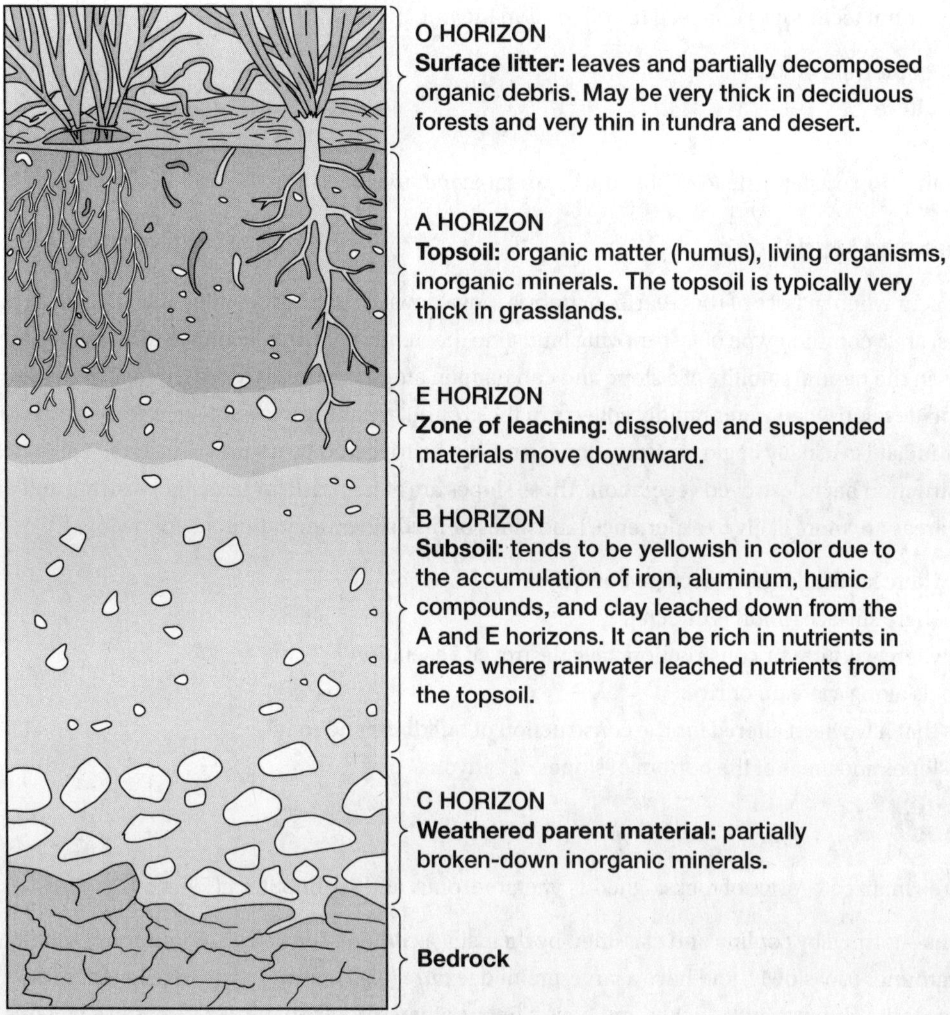

Figure 4.6 Soil profile

Soils develop in response to the following factors:

- **Climate**—measured by precipitation and temperature, which results in partial weathering of the parent material, which forms the substrate for soil.
- **Living organisms**—include the nitrogen-fixing bacteria *Rhizobium*, fungi, insects, worms, snails, etc., that help to decompose litter and recycle nutrients.
- **Parent material**—refers to the rock and minerals from which the soil derives. The nature of the parent rock, which can be either native to the area or transported to the area by wind, water, or glacier, has a direct effect on the ultimate soil profile.
- **Topography**—refers to the physical characteristics of the location where the soil is formed. Topographic factors that affect a soil's profile include drainage, slope direction, elevation, and wind exposure.

Soil Erosion

Soil erosion is the movement of weathered rock and/or soil components from one place to another caused by flowing water, wind, and human activity (burning of native vegetation, cultivating inappropriate land, and/or deforestation). Soil erosion decreases the soil's water-holding capacity, destroys the soil profile, and increases soil compaction. Because not all precipitation is able to percolate through the soil, it can run off the land, taking more soil with it (positive feedback loop), and crops grown in areas of soil erosion frequently suffer from water shortages and/or droughts.

Poor agricultural techniques that lead to soil erosion include the following:

- Improper plowing of the soil
- Monoculture
- Overgrazing
- Removing crop wastes instead of plowing the organic material back into the soil

Landslides and Mudslides

Landslides occur when masses of rock, earth, or debris move down a slope. Mudslides, also known as debris flows or mud flows, are a common type of fast-moving landslide that tends to flow in channels. Landslides are caused by disturbances in the natural stability of a slope and can happen after heavy rains, droughts, earthquakes, or volcanic eruptions and develop when water rapidly collects in the ground, resulting in a surge of water-soaked rock, earth, and debris. Mudslides usually begin on steep slopes and can be triggered by natural disasters in areas where wildfires or construction have destroyed vegetation. These slopes are at high risk for landslides during and after heavy rains. Some areas are more likely to experience landslides or mudslides, including the following:

- Areas where landslides have occurred before
- Areas where surface runoff is directed
- Areas where wildfires or construction have destroyed vegetation
- Channels along a stream or river
- Slopes that have been altered for the construction of buildings and roads
- Steep slopes and areas at the bottom of slopes or canyons

Rock Types

There are three main categories of rocks: igneous, metamorphic, and sedimentary (Figure 4.7).

- **Igneous**—formed by cooling and classified by their silica content. Intrusive igneous rocks solidify deep underground, cool slowly, and have a large-grained texture (e.g., granite). Extrusive igneous rocks solidify on or near the surface, cool quickly, and have a fine-grained smooth texture (e.g., basalt). Igneous rocks are broken down by weathering and water transport. Most soils come from igneous rocks.
- **Metamorphic**—formed by intense heat and pressure, high quartz content (e.g., gneiss). Common examples: diamond, marble, asbestos, slate, and anthracite coal.
- **Sedimentary**—formed by the piling and cementing of various materials (diatoms, weathered chemical precipitates, fragments of older rocks) over time in low-lying areas. Fossils form only in sedimentary rock.

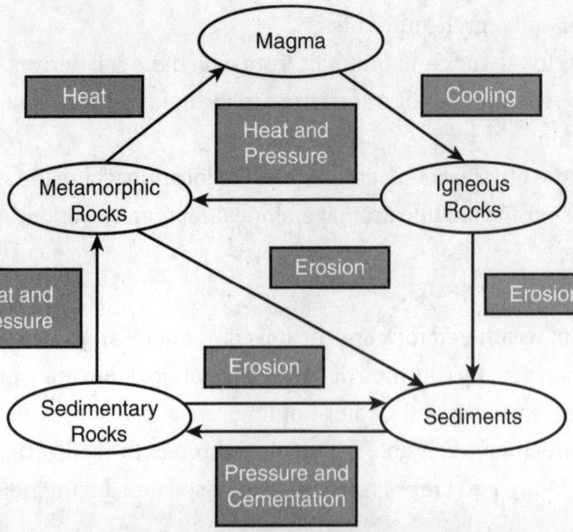

Figure 4.7 Rock formation

4.3 Soil Composition and Properties

With sufficient time, a mature soil profile reaches a state of equilibrium whereby feedback mechanisms involving both abiotic and biotic factors work to preserve the mature soil profile. The relative abundance of sand, silt, and clay is called the soil texture.

Soil Components/Soil Properties

Component*	Description
Gravel	Coarse particles. Consists of rock fragments.
Sand	Sedimentary material coarser than silt. Water flows through too quickly for most crops. Good for crops and plants requiring low amounts of water.
Loam	About equal mixtures of clay, sand, silt, and humus. Rich in nutrients. Holds water but does not become waterlogged. Particle size can vary.
Silt	Sedimentary material consisting of very fine particles between the sizes of sand and clay. Easily transported by water.
Clay	Very fine particles. Compacts easily. Forms large, dense clumps when wet. Low permeability to water; therefore, upper layers become waterlogged.

**Particle size listed in order from largest to smallest.*

Humus

Humus is the dark organic material that forms in soil when plant (e.g., leaf litter) and animal matter decays. As this material decays, it breaks down into its most basic chemical elements and compounds, which are important nutrients for plants and animals that depend upon soil for life. The thick brown or black substance that remains after most of the organic litter has decomposed is called humus. Earthworms often help mix humus with minerals in the soil. Soil containing humus will crumble, allowing air and water to move easily through the loose soil, making root growth easier, reducing erosion, and stabilizing the pH.

Soil Quality

Soil quality reflects biodiversity and productivity. Components of soil quality are listed in the table below.

Components of Soil Quality

Component	Role in Determining Soil Quality
Aeration	Refers to how well a soil is able to absorb oxygen, water, and nutrients. Aeration, which reduces soil compaction, involves perforating the soil with small holes to allow air (especially oxygen), water, and nutrients to penetrate to the roots. This helps the roots grow deeply and produce a stronger, more vigorous plant. When there's little or no light, plants require oxygen to break down the plant's sugar(s) to release CO_2, water, and energy.
Degree of soil compaction	Heavily compacted soils contain few large pores and have a reduced rate of both water infiltration and drainage from the compacted layer.

Component	Role in Determining Soil Quality
Nutrient-holding capacity	The ability of soil to absorb and retain nutrients so they will be available to the roots of plants. The process of weathering greatly influences the availability of plant nutrients. Initially, as soil particles begin to weather, primary minerals release nutrients into the soil. As these particles decrease in size, the soil is able to retain greater amounts of nutrients. The capacity to hold and retain nutrients is greatly reduced in highly weathered soils since most nutrients have been lost due to leaching. Primary plant nutrients are nitrogen (N), phosphorus (P), and potassium (K).
Permeability	The measure of the capacity of the soil to allow water and oxygen (needed by roots and soil organisms) to pass through it. Low permeability can lead to soil salinization.
pH	pH is the measure of how acidic or basic a soil is. Various plants have different soil pH requirements. Acidic soils can be caused by pollutants, such as acid rain and mine spoiling, and are most often found in areas of high rainfall. Alkaline (basic) soils have a high amount of potassium (K^+), calcium (Ca^{2+}), magnesium (Mg^{2+}), and/or sodium (Na^+) ions.
Pore size	Describes the space between soil particles. Pore size determines how much water, air, and nutrients are available for plant roots.
Size of soil particles	Soil particle size (gravel > sand > silt > clay) determines the amount of moisture, nutrients, and oxygen that the soil can hold along with the capacity for water to infiltrate.
Water-holding capacity	Water-holding capacity is controlled primarily by the soil texture and the soil organic matter content. Soil texture is a reflection of the particle size distribution of a soil. After a soil is saturated with water, all of the excess water and some of the nutrients and pesticides that are in the soil solution are leached downward in the soil profile.

Water-Holding Capacity and Soil

Evaporation from the soil surface, deep percolation, and transpiration by plants all reduce soil moisture between water applications (precipitation, irrigation, etc.), but if the water content becomes too low, plants become stressed. The water-holding capacity of a soil provides a buffer that allows plants to withstand dry spells, and each type of soil will have different water-retention characteristics.

Much of the water in the soil is held in spaces (pores) between the soil particles. The amount of moisture that a soil can store and the amount it can supply to plants are dependent on the number and size of its pore spaces. When the soil is saturated with water, all of the pores are full of water, but later, due to gravity and evaporation, the soil begins to dry out.

The amount of soil water available to plants is governed by the depth of soil that roots can explore (the root zone) and the nature of the soil material, which is affected by soil porosity, the sizes of the soil particles (texture), and the arrangement of these particles (structure). In general, sandier (light-textured) soils need to be irrigated more frequently than soils with a greater clay content (heavier-textured).

Soil Food Web

The soil food web is the community of organisms living all or part of their lives in the soil, and it describes a complex living system in the soil and how it interacts with the environments, plants, and animals (Figure 4.8).

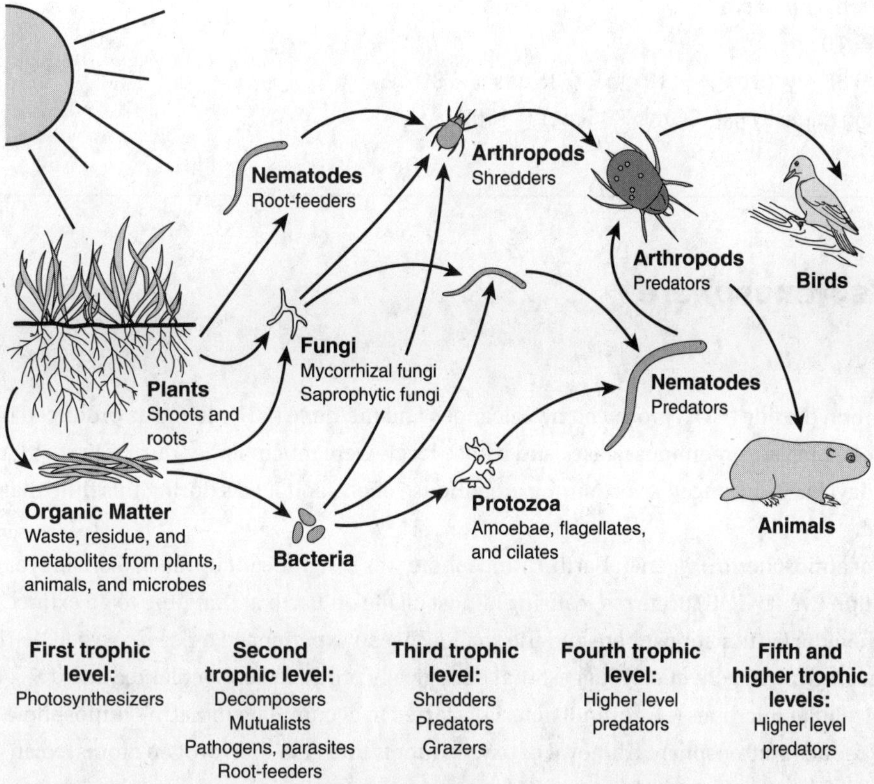

Figure 4.8 Soil food web

PRACTICE PROBLEM

Using the values provided, calculate the water-holding capacity of a sample of soil.

- Compute the holding capacity in % as shown here:

$$\textbf{\% Water-Holding Capacity} = (V_w/V_t) * 100\%$$

  where V_w is the volume of the water required to saturate the soil and V_t is the total volume of the saturated soil (1 cm^3 = 1 mL).
- Determine how much water was required to totally saturate the soil: 100 mL.
- Determine the volume of a cylindrical can ($\pi r^2 \times h$): diameter = 10 cm; height = 100 mm. Note: The can has several small holes in the bottom for water to drip out of.
- Dry a soil sample in an oven at a low temperature until it is completely dry.
- Fill the can completely with the dry soil.
- Fill a graduated cylinder with exactly 200 mL of water.
- Slowly add the water to the can until water drips out. Add the water that dripped out from the can back into graduated cylinder.

Answer to Water-Holding Capacity Problem

$V_w = 100\text{ mL} = 100\text{ cm}^3$
$d = 10\text{ cm} = \text{radius } (r)\ 5\text{ cm}$
$h = 100\text{ mm} = 10\text{ cm}$
$V_t = \pi r^2 \times h = 3.14 \times (5\text{ cm})^2 \times 10\text{ cm} = 785\text{ cm}^3 = 785\text{ mL}$
% water-holding capacity $= (100\text{ mL}/785\text{ mL}) \times 100\%$
$= \sim 13\%$

4.4 Earth's Atmosphere

Early History

Atmospheric carbon dioxide (CO_2) produced by volcanoes and methane (CH_4) that was produced by early microbes, both of which are greenhouse gases and whose levels were much higher during Earth's early history than they are today, likely produced a very strong greenhouse effect, and it was during this time that the earliest life forms developed.

The amount of atmospheric oxygen in Earth's atmosphere was insignificant until ~2.5 billion years ago when the Great Oxidation Event (GOE) occurred, causing almost all life on Earth at that time to go extinct. The GOE was a time period when the Earth's atmosphere and the shallow ocean experienced a rise in oxygen. Various forms of evidence (e.g., geological and chemical) suggest that biologically produced molecular oxygen (O_2), due to photosynthesizing unicellular microbes (i.e., cyanobacteria), started to accumulate in Earth's atmosphere and changed it from a weakly reducing atmosphere (removal of oxygen atoms and gain of hydrogen atoms) to an oxidizing atmosphere (gain of oxygen atoms and removal of hydrogen atoms), causing many existing life forms on Earth to die out over time.

As oxygen began to accumulate in the atmosphere, it is believed that there were two major consequences:

1. The free oxygen oxidized the atmospheric methane (greenhouse warming potential [GWP] of 25) to carbon dioxide (GWP of 1), weakening the greenhouse effect of Earth's atmosphere and causing planetary cooling, which triggered a series of ice ages.
2. The increased oxygen concentrations provided an opportunity for biological diversification, as well as major changes in the nature of chemical interactions between Earth's clay, rocks, and sand, and the Earth's atmosphere and oceans.

The result of the GOE is that the free energy available to living organisms greatly increased, allowing them to evolve greater complexity and diversity and to inhabit new environments.

Current Composition

Earth's atmosphere (today) is composed of four primary gases.

Nitrogen (N_2) 78%

- Deposits on Earth through nitrogen fixation (primarily by bacteria and fungi) and reactions involving lightning and subsequent precipitation
- Found in all organisms, primarily in amino acids (proteins) and nucleic acids (DNA and RNA)
- Fundamental nutrient for living organisms
- Makes up about 3% of the human body by weight
- Returns to the atmosphere through combustion of biomass and denitrification

Oxygen (O_2) 21%

- By mass, the third most abundant element in the universe, after hydrogen and helium
- Free elemental oxygen (O_2) began to accumulate in the atmosphere about 2.5 billion years ago (Earth is 4.5 billion years old)
- Highly reactive nonmetallic element that readily forms compounds (notably oxides)
- Product in photosynthesis and reactant in cellular respiration
- The most abundant element by mass in Earth's crust, making up almost half of the crust's mass as silicates

Water Vapor (H_2O) 0% to 4%

- Besides evaporation, other sources of atmospheric water include combustion, respiration, volcanic eruptions, and the transpiration of plants.
- Largest amounts are found near the equator, over oceans, and in tropical regions.
- Polar areas and deserts lack significant amounts of water vapor.

Carbon Dioxide (CO_2) $< 1\%$

- Average lifetime of a CO_2 molecule in the atmosphere is $\sim$100 years.
- Major greenhouse gas contributing to global warming
- Produced during cellular respiration ($C_6H_{12}O_6 + 6O_2 \rightarrow 6CO_2 + 6H_2O$), the combustion of fossil fuels, and the decay of organic matter
- Required for photosynthesis ($6CO_2 + 6H_2O \rightarrow C_6H_{12}O_6 + 6O_2$)

Structure

The atmosphere consists of several different layers (Figure 4.9). The two most important layers are the troposphere and the stratosphere (the layer that contains stratospheric ozone, which serves to absorb harmful UV radiation).

Troposphere

The troposphere is the lowest portion of Earth's atmosphere, 0–6 miles (0–10 km) above Earth's surface. Seventy-five percent of the atmosphere's mass and almost all of the water vapor on the planet is contained within the troposphere, with weather also occurring in this layer. The atmospheric pressure within the troposphere is highest at the surface and decreases with height, whereas the temperature of the troposphere decreases with height.

Stratosphere

The stratosphere is located 6–30 miles (10–50 km) above Earth's surface. In the stratosphere, ozone (O_3) absorbs high-energy ultraviolet radiation (UVB and UVC) from the sun and is broken down into atomic oxygen (O) and diatomic oxygen (O_2): $O_3 \rightarrow O + O_2$. Temperature increases with altitude in the stratosphere.

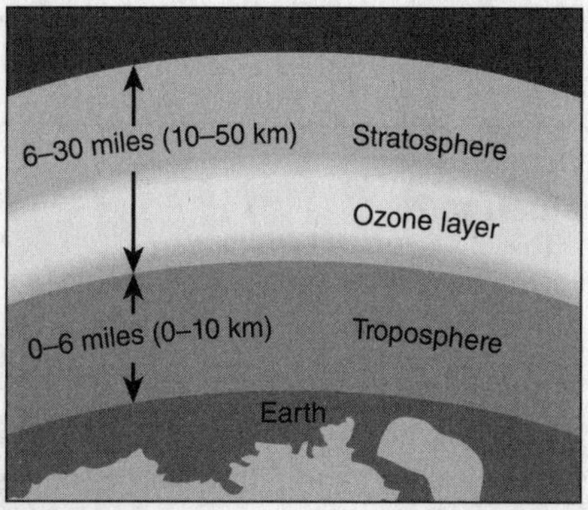

Figure 4.9 Important layers of the atmosphere

Weather and Climate

> Weather describes whatever is *currently* happening outdoors, whereas climate describes weather patterns in a place *over a period of years*.

Weather is caused by the movement or transfer of heat energy, which results from the unequal heating of Earth's surface by the sun and influences the following physical properties:

- Air pressure
- Air temperature
- Humidity
- Precipitation
- Sunlight reaching Earth affected by cloud cover
- Wind direction and speed

Convection is the primary way energy is transferred from hotter to colder regions in Earth's atmosphere and is the primary determinant of weather patterns. Convection involves the movement of the warmer and therefore more energetic molecules in air and takes place both vertically and horizontally. When air near the ground becomes warmer and therefore less dense than the air above it, the air rises, and the pressure differences that develop because of temperature differences result in wind.

Regions nearer to the equator receive much more solar energy than do regions nearer to the poles and are consequently much warmer. These latitudinal differences in surface temperature create global-scale flows of energy within the atmosphere, giving rise to the major weather patterns of the world. Without convection and the transfer of energy, the equator would be about 27°F (15°C) warmer and the Arctic would be about 45°F (25°C) colder than they actually are.

> **TIP**
>
> **Many different factors influence climate:**
> - **Air mass**
> - **Albedo**
> - **Altitude**
> - **Angle of sunlight**
> - **Carbon cycle**
> - **Clouds**
> - **Distance to oceans**
> - **Fronts**
> - **Greenhouse effect**
> - **Heat**
> - **Human activity**
> - **Land changes**
> - **Landmass distribution**
> - **Latitude**
> - **Location**
> - **Moisture content of air**
> - **Mountain ranges**
> - **Pollution**
> - **Precession**
> - **Rotation**
> - **Solar output**
> - **Volcanoes**
> - **Wind patterns**

Heat Index

Heat is the #1 weather-related killer in the United States. Data from the U.S. National Weather Service show that heat results in more deaths in the United States than floods, lightning, tornadoes, or hurricanes. Case in point: For the years 2000–2009, excessive heat resulted in 162 deaths in the United States, whereas hurricanes resulted in 117; floods, 65; tornadoes, 62; and lightning, 48.

The heat index (HI) is a measure of how warm it feels when factoring in relative humidity.

> **QUICK QUIZ**
>
> Using the formula shown below, where temperature is measured in °F and RH represents the relative humidity, what would be the heat index if the outdoor temperature were 82°F and the relative humidity were 62%?
>
> $$\text{Heat Index} = 0.5\{T + 61.0 + [(T - 68.0) \times 1.2] + (RH \times 0.094)\}$$
>
> Solution on page 168

Climate and Factors That Influence It

Climate is the average weather in a geographic region over a period of time (a minimum of 20–30 years is often suggested).

Evidence for changes in the climate can come from tree rings, fossilized plants, insect and pollen samples, gas bubbles trapped in glaciers, deep-ice core samples, lake sediments, stalactites and stalagmites, marine fossils including coral analysis, sediments including rafted debris, dust analysis, and isotope ratios in fossilized remains. This evidence clearly shows Earth's climate has gone through many cycles of warming and cooling trends.

Air Mass

An air mass is a large body of air that has similar temperature and moisture content. Air masses can be categorized as equatorial, tropical, polar, Arctic, continental, or maritime.

Albedo

Albedo is reflectivity. Materials like ocean water have low albedo, whereas landmasses have moderate albedo. Snow and ice have the highest albedo. Hence, periods when polar ice is highly extended will promote further cooling. This is a positive feedback mechanism. Dust in the atmosphere from periods of dry climates, volcanic eruptions, or meteor impacts has the same effect, as it forms a high albedo "cloud" around Earth so that a significant amount of solar radiation is reflected before it reaches the surface.

Altitude

For every 1,000 feet (300 m) of rise in elevation, there is a 3°F (1.5°C) drop in temperature, and every 300 feet (90 m) of rise in elevation is equivalent to a shift of 62 miles (100 km) north in latitude and biome similarity (Figure 4.10).

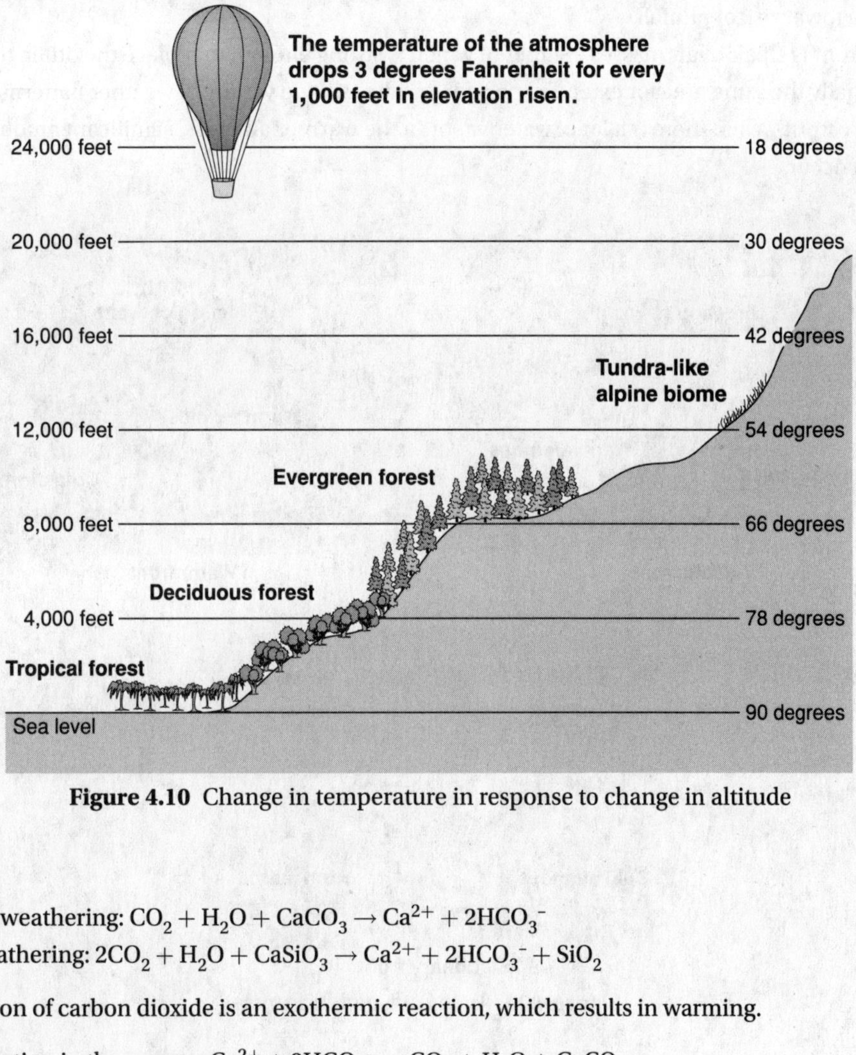

Figure 4.10 Change in temperature in response to change in altitude

Carbon Cycle

Carbonate rock weathering: $CO_2 + H_2O + CaCO_3 \rightarrow Ca^{2+} + 2HCO_3^-$
Silicate rock weathering: $2CO_2 + H_2O + CaSiO_3 \rightarrow Ca^{2+} + 2HCO_3^- + SiO_2$

The production of carbon dioxide is an exothermic reaction, which results in warming.

Carbonate formation in the oceans: $Ca^{2+} + 2HCO_3^- \rightarrow CO_2 + H_2O + CaCO_3$
Breakdown of carbonate: $SiO_2 + CaCO_3 \rightarrow CO_2 + CaSiO_3$

Both carbonate formation in the oceans and the breakdown of carbonate yield carbon dioxide.

Distance to Oceans

Oceans are thermally more stable than landmasses; the specific heat (heat-holding capacity) of water is five times greater than that of air. Because of this, changes in temperature are more extreme in the middle of the continents than on the coasts.

Fronts

When two different air masses meet, the boundary between them forms a "front." The air masses can vary in temperature, dew point (the temperature below which water droplets begin to condense), and wind direction (Figure 4.11).

A cold front is the leading edge of an advancing mass of cold air and is associated with thunderhead clouds, high surface winds, and thunderstorms. Clouds form along a cold front as dense, cold air wedges under less dense, warmer air. The resulting uplift causes cloud formation and precipitation. After a cold front passes, the weather is usually cool with clear skies.

A warm front is the boundary between an advancing warm air mass and the cooler one it is replacing. Since warm air is less dense, it moves over the dense, cool air. Because there is less chance of uplift, warm fronts often are associated with a lower risk of rainfall.

A stationary front is a pair of air masses, neither of which is strong enough to replace the other, that tend to remain in essentially the same area for extended periods of time. A wide variety of weather patterns can be found along a stationary front. When there is a lot of water vapor in the warmer air mass, significant amounts of rain or freezing rain can occur.

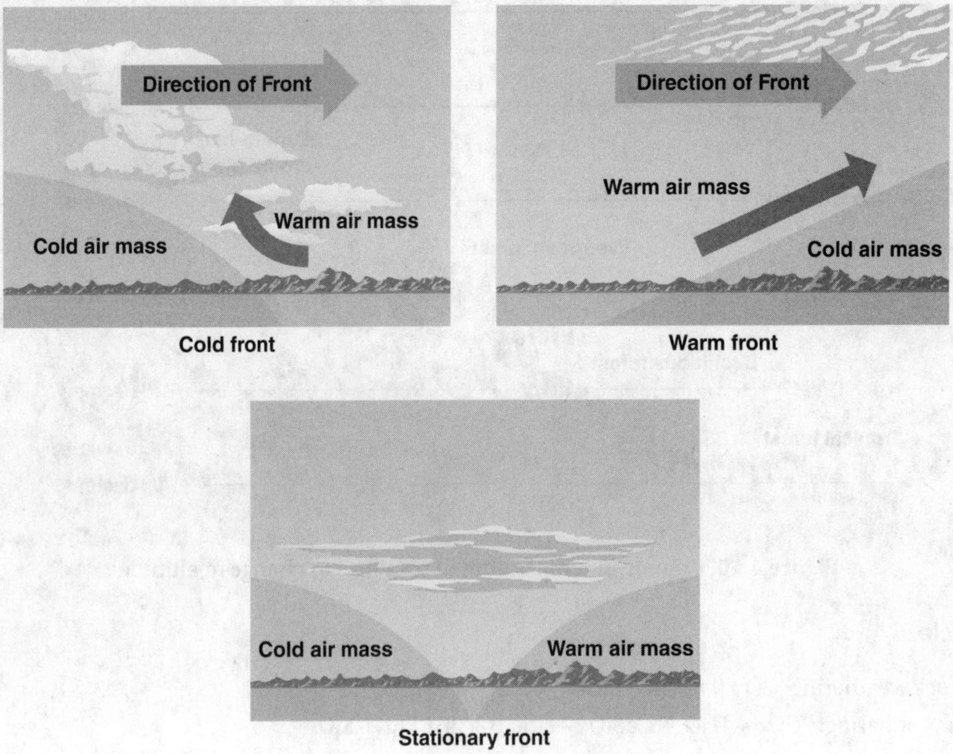

Figure 4.11 Types of weather fronts

Greenhouse Effect

The most important greenhouse gases are water vapor (H_2O), carbon dioxide (CO_2), methane (CH_4), and nitrous oxide (N_2O). Without this effect, Earth would be cold and inhospitable. If taken too far, however, Earth could evolve into a hothouse.

Heat (Convection)

Climate is influenced by how heat energy is exchanged between air over the oceans and the air over land.

Human Activity and Climate

Climate can also be influenced by human activity. Deforestation; urbanization; heat island effects; release of pollutants, including greenhouse gases from the burning of fossil fuels; and the production of acid rain are examples of how humans have altered climatic patterns. Increased pollution alone, combined with an increase in convectional uplift in urban areas, tends to increase the amount of rainfall in urban areas as much as 10% when compared with undeveloped areas. Climate is influenced by urbanization and deforestation.

Materials absorb and reflect solar radiation differently; e.g., ocean water absorbs more solar energy than do landmasses. Earth receives more solar radiation at low latitudes (near the equator) than near the poles.

Latitude and Location

Latitude is the measurement of the distance of a location on Earth from the equator. The farther away from the equator, the less sunlight is available.

At the poles, the sun's rays strike Earth at an acute angle, which spreads the heat over a larger area. More heat is lost to the atmosphere, as the rays travel a longer distance through the atmosphere (Figure 4.12).

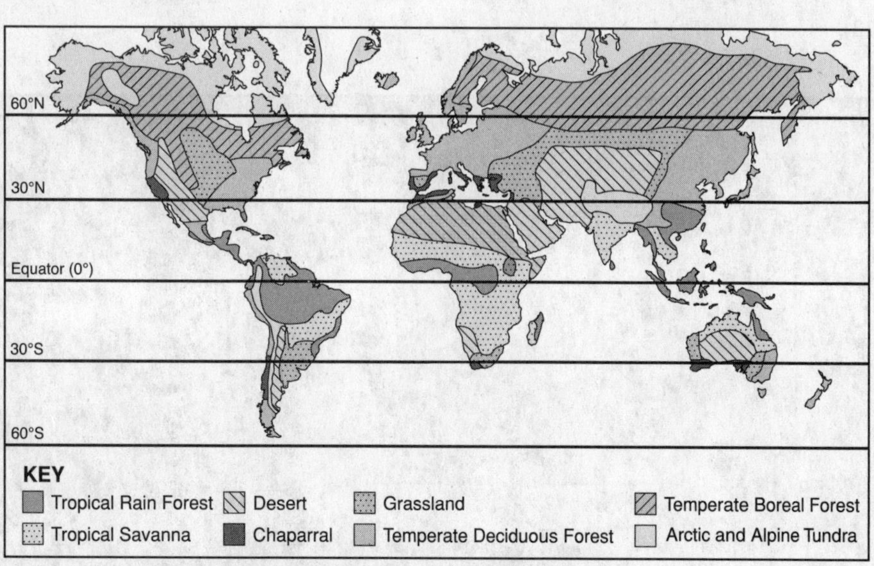

Figure 4.12 Latitude vs. temperature

Climate is influenced by the location of high and low air pressure zones and where landmasses are distributed.

Moisture Content of Air (Humidity)

The moisture content of air is a primary determinant of plant growth and distribution and is a major determinant of biome type (e.g., desert vs. tropical forest). Atmospheric water vapor supplies moisture for clouds and rainfall, and plays a significant role in energy exchanges within the atmosphere.

Pollution

Greenhouse gases are emitted from both natural sources (e.g., volcanoes) and anthropogenic (human) sources (e.g., industry and transportation).

Rotation

Daily temperature cycles are primarily influenced by Earth's rotation on its axis (once every 24 hours). At night, heat escapes from Earth's surface, and daily minimum temperatures occur just before sunrise.

Volcanoes

Sulfur-rich volcanic eruptions can eject material into the stratosphere, potentially causing tropospheric cooling and stratospheric warming. Volcanic aerosols exist in the atmosphere for an average of one to three years. Volcanic aerosols injected into the stratosphere can also provide surfaces for ozone-destroying reactions. Over the course of millions of years, large volumes of volcanic ash deposited in the oceans can increase the iron content in seawater and promote biotic activity, which can lower the CO_2 concentration of seawater, and hence atmospheric CO_2 levels, resulting in global cooling. Over the course of weeks to years, ongoing production of ash from volcanoes may locally change the climate by modifying the local atmosphere.

4.5 Global Wind Patterns

Land and Sea Breezes

A land breeze occurs during relatively calm, clear nights when the land cools down faster than the sea, resulting in the air above the land becoming denser than the air over the sea. As a result, air moves from the land toward the coast (Figure 4.13).

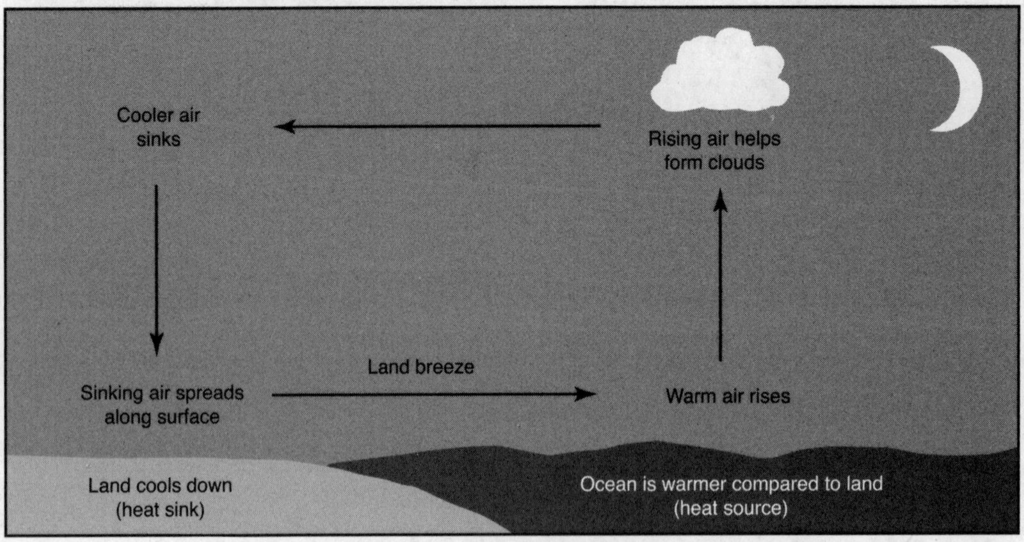

Figure 4.13 Land breeze

During relatively calm, sunny days, the land warms up faster than the sea, causing the air above it to become less dense, resulting in a sea breeze (Figure 4.14).

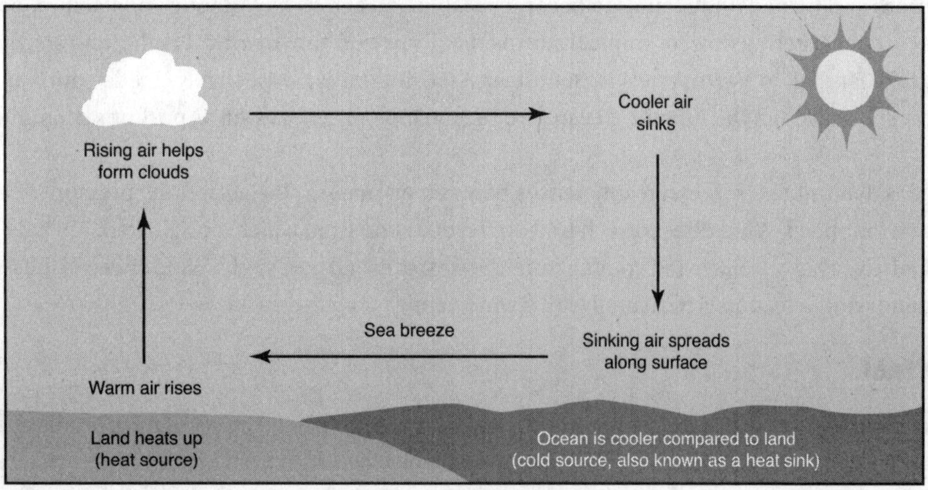

Figure 4.14 Sea breeze

Atmospheric Circulation-Pressure

Due to the rotation of Earth on its axis, its revolution around the sun, and the tilt of its axis, the sun heats the atmosphere unevenly, as air closer to Earth's surface is the warmest and rises. Air at higher elevations is cooler and, as such, denser, and sinks, which sets up convection processes and is the primary cause of winds. Global air circulation is also affected by uneven heating of Earth's surface, seasons, the Coriolis effect, the amount of solar radiation reaching Earth over long periods of time, convection cells created by warm ocean waters that commonly lead to hurricanes, and ocean currents. A low-pressure weather system has lower pressure at its center than in the areas around it. Winds blow toward the low pressure, and the air rises in the atmosphere where they meet. As the air rises, the water vapor within it condenses, forming clouds and precipitation.

A high-pressure weather system has higher pressure at its center than in the areas around it, which results in winds blowing away from the high pressure. Winds of a high-pressure system swirl in the opposite direction as those of a low-pressure system—clockwise north of the equator and counterclockwise south of the equator—with air from higher in the atmosphere sinking down to fill the spaces left as air blows outward. High-pressure masses contain cool, dense air that descends toward Earth's surface and becomes warmer, and are usually associated with fair weather (Figure 4.15).

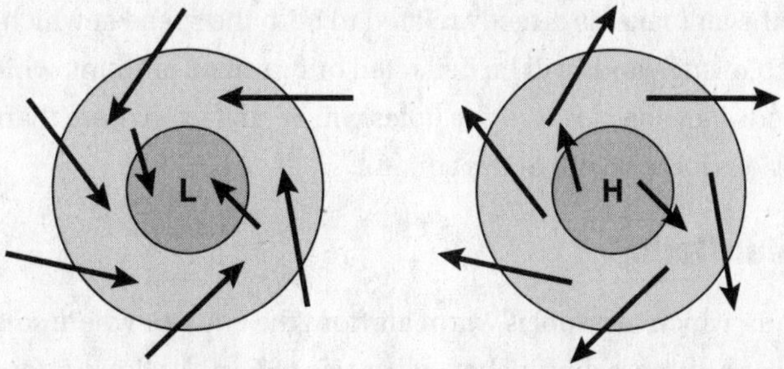

Figure 4.15 Circulation wind patterns of high and low pressure in the Northern Hemisphere. The pattern reverses in the Southern Hemisphere.

The trade winds are the prevailing pattern of easterly surface winds found in the tropics near Earth's equator, within the troposphere or lower portion of Earth's atmosphere. The trade winds blow predominantly from the northeast in the Northern Hemisphere (northeast trade winds) and from the southeast in the Southern

Hemisphere (southeast trade winds), strengthening during the winter. For centuries, the trade winds have been used by captains of sailing ships to cross the world's oceans; they also enabled European empire expansion into the Americas and helped trade routes to become established across the Atlantic and Pacific Oceans. The trade winds act as the steering flow for tropical storms that form over the Atlantic, Pacific, and southern Indian Oceans and make landfall in North America, Southeast Asia, and India, respectively. Trade winds also steer African dust westward across the Atlantic Ocean into the Caribbean Sea, as well as portions of southeast North America.

Wind speed is determined by pressure differences between air masses. The greater the pressure difference is, the greater the wind speed. Wind direction is based on the direction from which it originated. A wind coming from the east is called an easterly, while wind coming from the west is called a westerly. Wind speed is measured with an anemometer, and wind direction is measured with a wind vane.

Coriolis Effect

Earth's rotation on its axis causes winds to not travel straight, a phenomenon known as the Coriolis effect, which causes prevailing winds in the Northern Hemisphere to spiral clockwise out from high-pressure areas and spiral counterclockwise toward low-pressure areas (Figure 4.16).

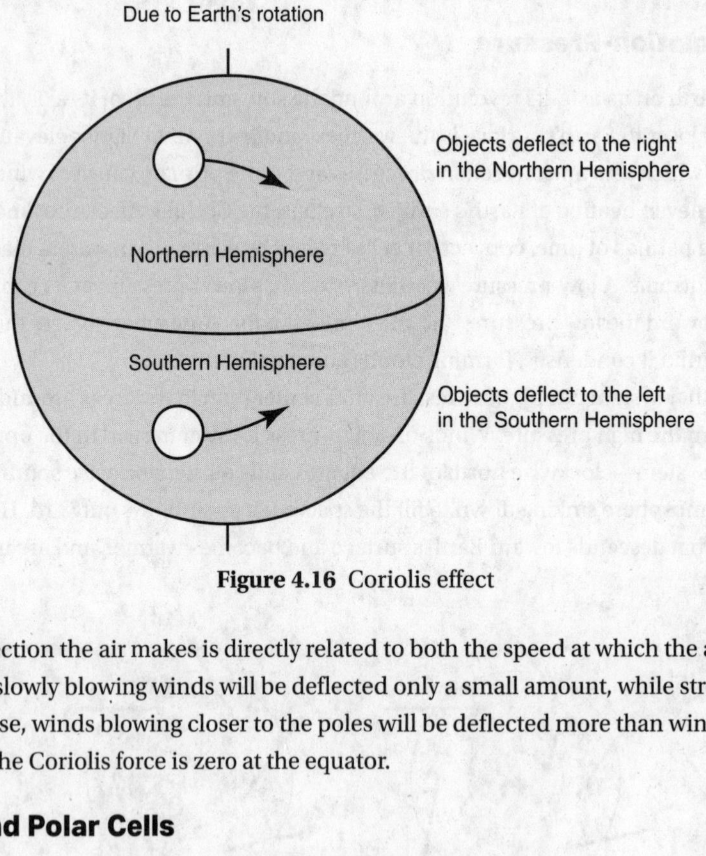

Figure 4.16 Coriolis effect

The amount of deflection the air makes is directly related to both the speed at which the air is moving and its latitude. Therefore, slowly blowing winds will be deflected only a small amount, while stronger winds will be deflected more. Likewise, winds blowing closer to the poles will be deflected more than winds at the same speed closer to the equator. The Coriolis force is zero at the equator.

Hadley, Ferrel, and Polar Cells

The worldwide system of winds, which transports warm air from the equator where solar heating is greatest toward the higher latitudes where solar heating is diminished, gives rise to Earth's climatic zones. Three types of air circulation cells associated with latitude exist—Hadley, Ferrel, and Polar (Figure 4.17).

Hadley Air Circulation Cells

Air heated near the equator rises and spreads out north and south. After cooling in the upper atmosphere, the air sinks back to Earth's surface within the subtropical climate zone (between 25° and 49° north and south latitudes).

Surface air from subtropical regions returns toward the equator to replace the rising air. The equatorial regions of the Hadley cells are characterized by high humidity, high clouds, and heavy rains. The monthly average temperatures are around 90°F (32°C) at sea level, and there is no winter. The vegetation is tropical rainforest. Temperature variation from day to night (diurnal) is greater than from season to season.

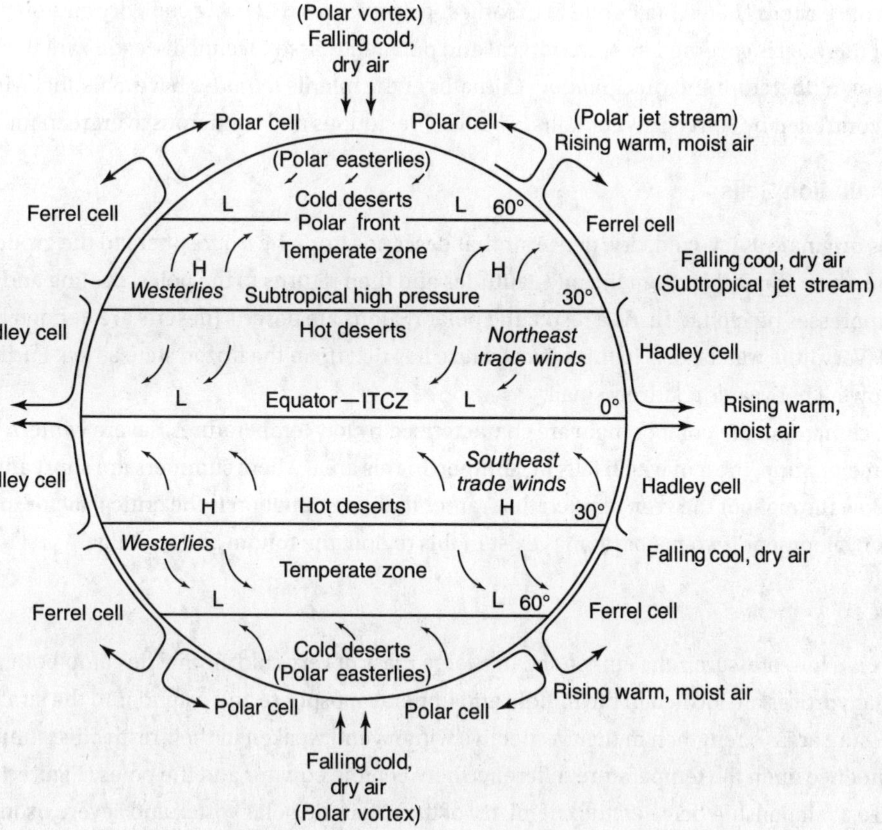

Figure 4.17 Hadley, Ferrel, and Polar cells

Subtropical regions of the Hadley cell are characterized by low relative humidity, little cloud formation, high ocean evaporation due to the low humidity, and many of the world's deserts. The climate is characterized by warm to hot summers and mild winters. The tropical wet and dry (or savanna) climate has a dry season more than two months long. Annual losses of water through evaporation in this region exceed annual water gains from precipitation (Figure 4.18).

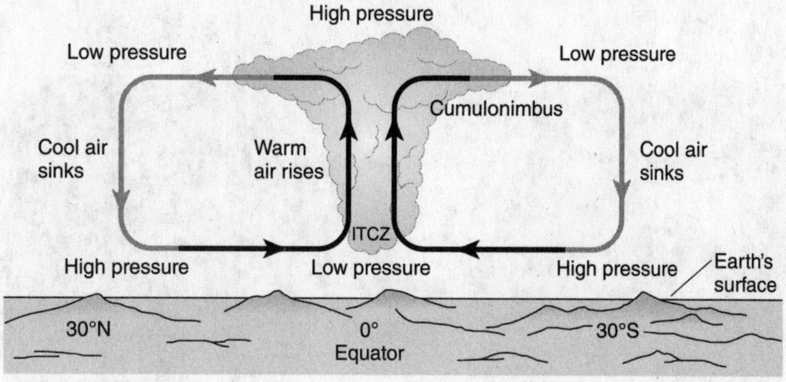

Figure 4.18 Hadley cell

Ferrel Air Circulation Cells

Ferrel cells develop between 30° and 60° north and south latitudes. The descending winds of the Hadley cells diverge as moist tropical air moves toward the poles in winds known as the westerlies. Mid-latitude climates can have severe winters and cool summers due to mid-latitude cyclone patterns. The western United States is drier in summer than the eastern United States due to oceanic high pressures that bring cool, dry air down from the north. The climate of this area is governed by both tropical and polar air masses. Defined seasons are the rule, with strong annual cycles of temperature and precipitation. Climates of the middle latitudes have a distinct winter season. The area of Earth controlled by Ferrel cells contains broadleaf deciduous and coniferous evergreen forests.

Polar Air Circulation Cells

The Polar cells originate as icy-cold, dry, dense air that descends from the troposphere to the ground. This air meets with the warm tropical air from the mid-latitudes and then returns to the poles, cooling and then sinking. Sinking air suppresses precipitation. As a result, the polar regions are deserts (deserts are defined by moisture, not temperature). Very little water exists in this area because it is tied up in the frozen state as ice. Furthermore, the amount of snowfall per year is relatively small.

In general, climates in the polar domain are characterized by low temperatures, severe winters, and small amounts of precipitation, most of which falls in summer. In this area, where summers are short and temperatures are generally low throughout the year, temperature rather than precipitation is the critical factor in plant distribution and soil development. Two major biomes exist in this region, the tundra and the taiga.

Polar Vortex

A polar vortex is a low-pressure zone embedded in a large mass of very cold air that lies atop both poles. The bases of the two polar vortices are located in the middle and upper troposphere and extend into the stratosphere. These cold, low-pressure areas strengthen in their respective winters and weaken in their respective summers as a result of their dependence upon the temperature difference between the equator and the poles (Figure 4.19).

There is also a relationship between the chemistry of the Antarctic polar vortex and severe ozone depletion (i.e., the nitric acid in polar stratospheric clouds reacts with chlorofluorocarbons to form chlorine, which catalyzes the photochemical destruction of ozone). Since these clouds can only form at temperatures below –112°F (–80°C), chlorine concentrations build up during the polar winter, and the consequent ozone destruction is greatest when the sunlight returns in the spring. Since there is greater air exchange between the Arctic and the mid-latitudes, ozone depletion is more pronounced at the South Pole, where air exchange is less.

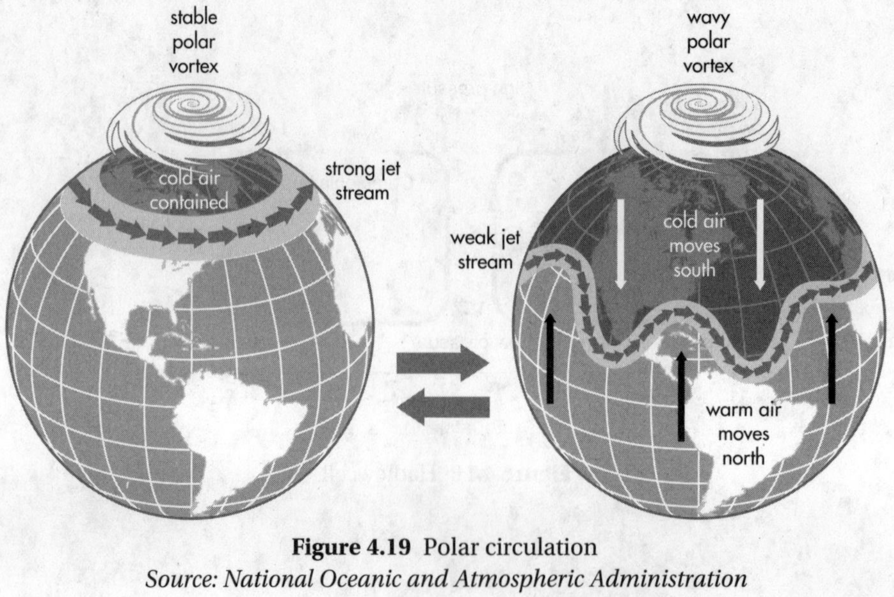

Figure 4.19 Polar circulation
Source: National Oceanic and Atmospheric Administration

Hurricanes, Tornadoes, and Monsoons

Hurricanes

Hurricanes, cyclones, and typhoons are all the same weather phenomenon. In the Atlantic and Northeast Pacific, the term *hurricane* is used. The same type of disturbance in the Northwest Pacific is called a *typhoon*, and *cyclones* occur in the South Pacific and Indian Oceans. The ingredients for these storms include a pre-existing weather disturbance, warm tropical oceans, atmospheric moisture, and relatively light winds in the upper troposphere. If the right conditions persist long enough, they can combine to produce violent winds, very large waves, torrential rains, and floods.

Hurricanes are the most severe weather phenomenon on the planet. Hurricanes begin over warm oceans in areas where the trade winds converge. A subtropical high-pressure zone creates hot daytime temperatures with low humidity that allow for large amounts of evaporation, with the Coriolis effect initiating the cyclonic flow.

The elements of hurricane development include the presence of separate thunderstorms that have developed over tropical oceans, and cyclonic circulation that begins to cause these thunderstorms to move in a circular motion. This cyclonic circulation allows them to pick up moisture and latent heat energy from the ocean. In the center of the hurricane is the eye, an area of descending air and low pressure. The energy of a hurricane dissipates as it travels over land or moves over cooler bodies of water. Rainfall can be as much as 24 inches (0.6 m) in 24 hours. A storm surge, which results from the increase in the height of the ocean near the eye of a hurricane, can cause extensive flooding.

Hurricane Storm Surges

A storm surge is a rise in sea level that occurs during tropical cyclones, typhoons, or hurricanes (Figure 4.20). These storms produce strong winds that push the seawater toward the shore, which often leads to flooding. The severity of flooding is affected by the shallowness and orientation of the water body relative to the storm path, as well as the timing of the tides. This makes storm surges very dangerous for coastal regions.

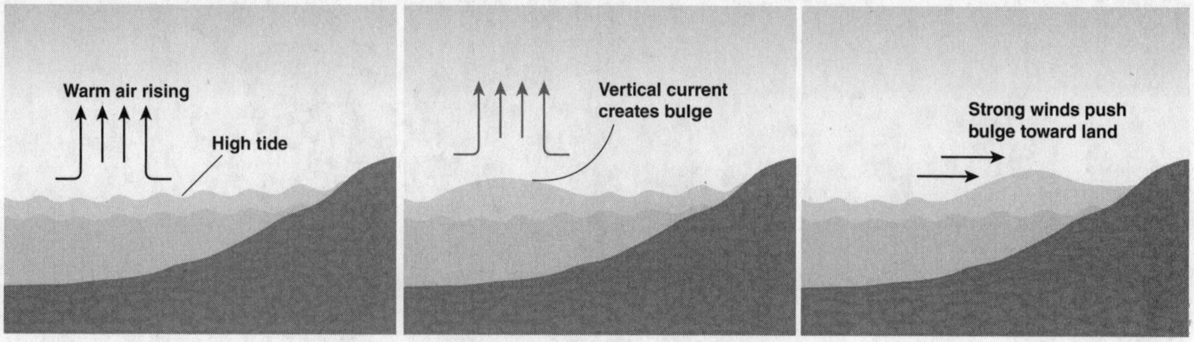

Figure 4.20 How storm surges form

Tornadoes

Tornadoes are swirling masses of air with wind speeds close to 300 miles per hour (485 kph). Like hurricanes, the center of the tornado is an area of low pressure (Figure 4.21). In the United States, tornadoes are frequent from April through July and occur in the center of the United States in an area known as Tornado Alley (Figure 4.22). As a result of advances in weather forecasting, modeling, and warning systems, the death rate due to tornadoes has decreased significantly.

Figure 4.21 Formation of a tornado

1. The thunderstorm or hailstorm creates strong winds.
2. The strong winds begin to rotate (due to updrafts and downdrafts) and form a column of spinning air called a mesocyclone.
3. The mesocyclone meets warm air moving up and cold air moving down and creates a funnel.
4. The funnel, made up of dust, air, and debris, reaches the ground, and a tornado is formed.

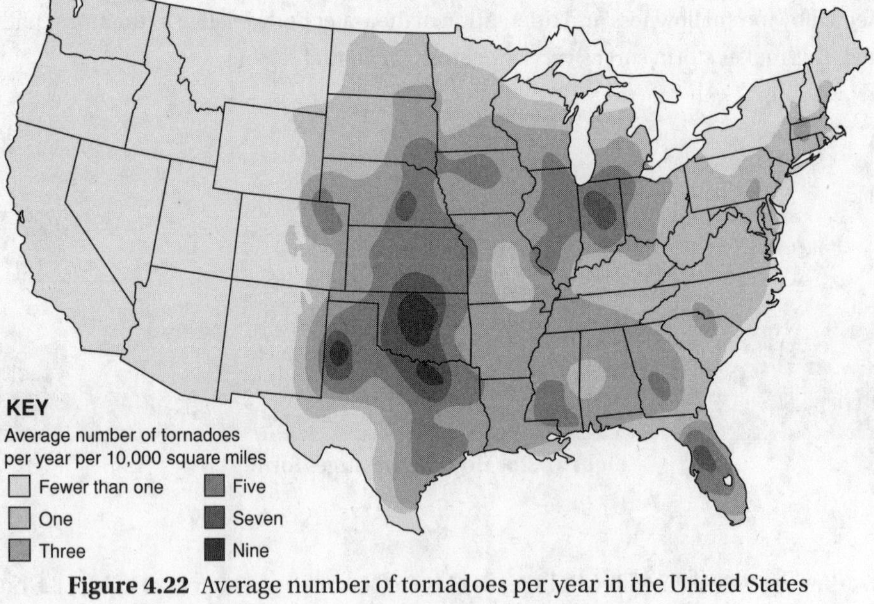

Figure 4.22 Average number of tornadoes per year in the United States

Tornado and hurricane mitigation can include the following actions:

- Building community shelters and tornado-safe rooms to help protect people
- Creating zoning ordinances that steer development away from areas subject to flooding, storm surges, and/or coastal erosion
- Enforcing stringent building codes, flood-proofing requirements, and wind-bracing requirements
- Having better weather prediction models and ways to communicate such information to the public
- Reducing development in flood-prone areas
- Retrofitting buildings to withstand hurricanes and tornadoes

Tornadoes vs. Hurricanes

Tornadoes	Hurricanes
Diameters of hundreds of meters	Diameters of hundreds of kilometers
Produced from a single convective storm, e.g., a thunderstorm	Composed of many convective storms
Occur primarily over land	Occur primarily over oceans
Require substantial vertical shear of the horizontal winds (change of wind speed and/or direction with height)	Require very low values of vertical shear in order to form and grow
Typically last less than an hour	Last for days

Monsoons

Monsoons are strong, often violent winds that change direction with the season. Monsoon winds blow from cold to warm regions because cold air takes up more space than warm air. Monsoons blow from the land toward the sea in winter and from the sea toward land in the summer.

India's climate is dominated by monsoons. During the Indian winter, which is hot and dry, the monsoon winds blow from the northeast and carry little moisture. The temperature is high because the Himalayas form a barrier that prevents cold air from passing onto the subcontinent. Furthermore, most of India lies between the Tropic of Cancer and the equator, so the sun's rays shine directly on the land. During the summer, the monsoons move onto the subcontinent from the southwest. The winds carry moisture from the Indian Ocean and bring heavy rains from June to September. Farmers in India rely on these torrential summer rainstorms to irrigate their land. Additionally, a large amount of India's electricity is generated by waterpower provided by the monsoon rains (Figure 4.23).

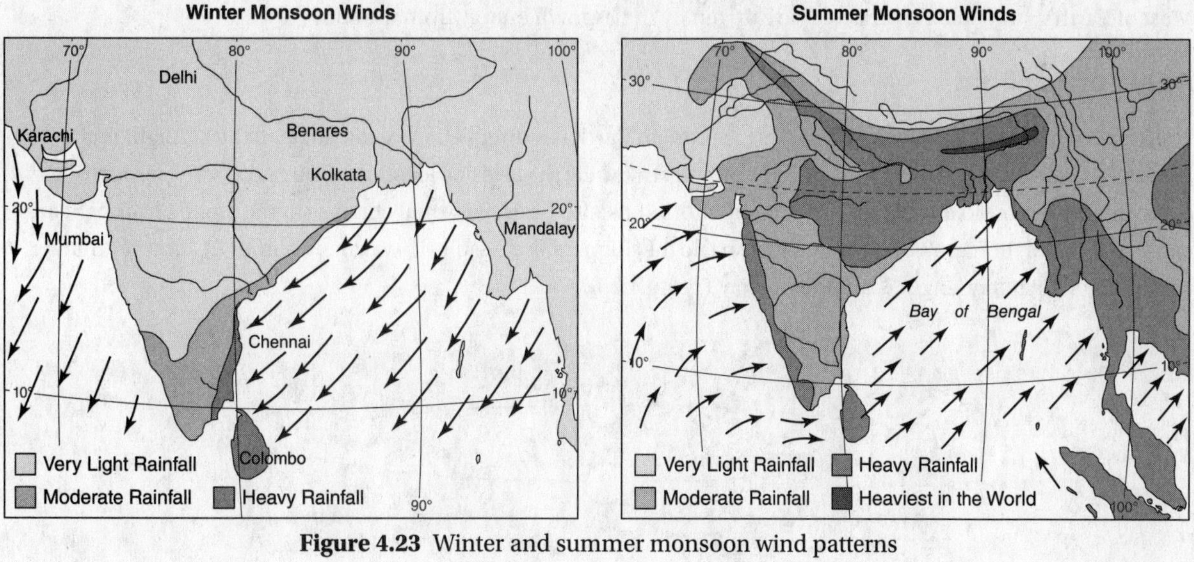

Figure 4.23 Winter and summer monsoon wind patterns

4.6 Watersheds

A watershed is an area of land that ultimately drains water from rainfall and snowmelt into a specific body of water such as a lake or the ocean or allows it to seep into the ground and collect in underground reservoirs known as aquifers (Figure 4.24).

The largest watershed in the United States is the Mississippi River watershed, which drains more than one million square miles (three million square kilometers) of land.

Watershed management involves the use of land, forest, and water resources in ways that do not harm the plants and animals living there, such as reducing the amount of pesticides and fertilizers that wash off farm fields and into nearby waterbodies.

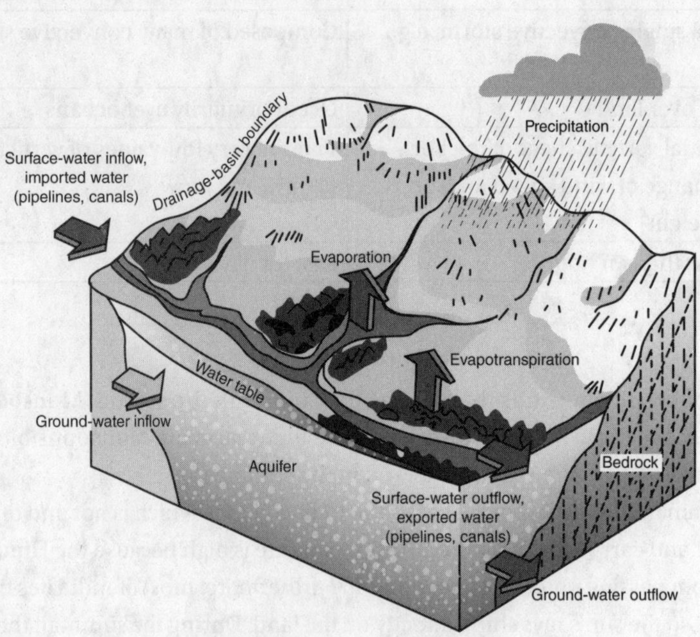

Figure 4.24 Watershed

4.7 Solar Radiation and Earth's Seasons

Most of Earth's atmospheric processes are driven by the input of energy from the sun.

Angle of Sunlight

The amount of heat energy received at any location on Earth is a direct effect of the angle of the sunlight reaching the Earth's surface. The angle at which sunlight strikes Earth varies by location, time of day, and season due to Earth's orbit around the sun and its rotation around its tilted axis. Seasonal changes in the angle of sunlight are caused by the tilt of Earth's axis, which is the basic mechanism that results in warmer weather in summer than in winter. Change in day length is another factor (Figure 4.25).

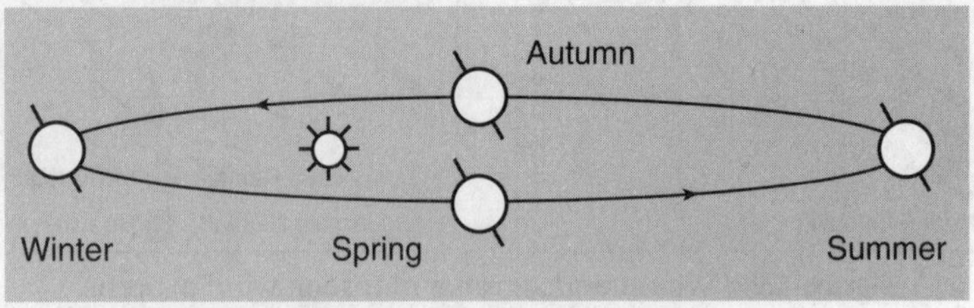

Figure 4.25 Seasons in the Northern Hemisphere

When sunlight shines on Earth at a lower angle (sun is closer to the horizon), the energy of the sunlight spreads out over a larger area and is therefore weaker (winter) than if the sun were higher overhead, when its energy is concentrated on a smaller area (summer). At this lower angle, the sunlight must also pass through more atmosphere, which causes diffraction and a reduction in the amount of sunlight that is able to reach Earth's surface.

The following figure (Figure 4.26) depicts a sunbeam one mile (1.6 km) wide striking the ground from directly overhead and another striking the ground at a 30° angle. Notice that the radiant energy striking the ground at a 30° angle diffused the same amount of radiant energy over twice as much distance (2 miles) as the radiant energy directly overhead. Therefore, the radiant energy coming from directly overhead is more intense (i.e., brighter, warmer) than the radiant energy striking at the 30° angle over a larger area.

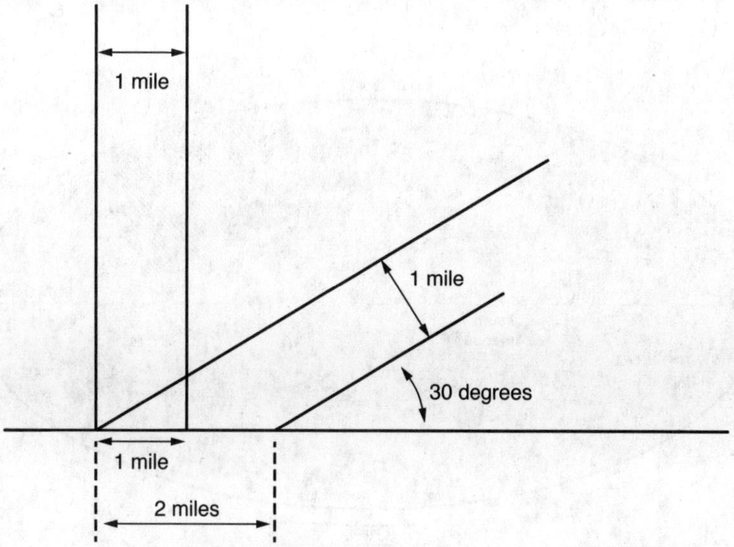

Figure 4.26 Angle of sunlight

Solar Intensity

Factors that affect the amount of solar energy at the surface of Earth (which directly affects plant productivity) include the following:

- Atmospheric conditions
- Earth's rotation around the sun (once per year)
- Earth's rotation on its axis (once every 24 hours)
- The tilt of Earth's axis (23.5°)

4.8 Earth's Geography and Climate

Climate is defined as the prevailing patterns of temperature and precipitation in a defined region on Earth, whereas weather is the condition of the atmosphere in one area at a particular time.

Bodies of Water Moderate Climate and Regulate Precipitation

Over 70% of the Earth's surface is covered in water. Oceans and lakes store solar radiation (heat), and as the water heats up it adds moisture to the air above it, beginning a process that drives the major air currents around the world. Large water bodies also tend to stabilize the climate of adjacent land masses by absorbing extra heat during warm periods and releasing it during cooler periods. Warm, moist ocean air is a driving force for precipitation patterns around the world as it is carried over cooler land masses.

Higher Elevations Have Cooler Climates

Climates become cooler and the cold season lasts longer as elevation increases; e.g., every 1 mile (1.61 kilometers) in elevation is equivalent to moving 800 miles (1,290 kilometers) farther from the equator. Furthermore, higher elevations have lower air pressure due in part to there being fewer atoms and molecules per unit of air and, thus, cooler temperatures. Higher elevations frequently receive more precipitation than the surrounding lower altitudes, but many high-altitude plains are technically deserts due to their location on the downwind (leeward) side of a mountain range or continental mass.

Latitude is a measure of distance either north or south from the equator. Locations between the Tropic of Cancer (~23° north of the equator), which is the northernmost latitude reached by the overhead sun, and the Tropic of Capricorn (~23° south of the equator), which is the southernmost latitude reached by the overhead sun, are considered tropical (Figure 4.27).

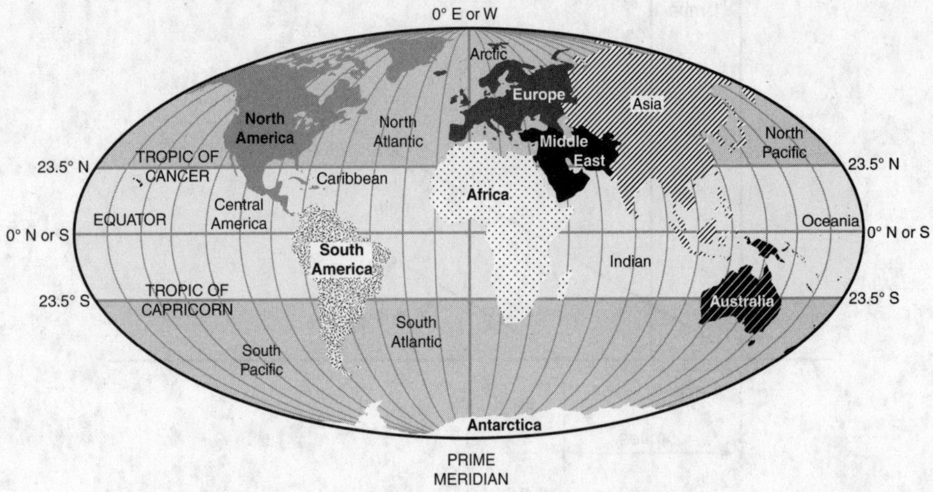

Figure 4.27 Tropic of Cancer and Tropic of Capricorn

As one gets farther away from the equator, climates shift through subtropical, temperate, subarctic, and, finally, Arctic at the poles. The tilt of the Earth on its axis results in the fact that the farther you get from the equator, the longer the area spends tilted away from the sun each year, and the cooler and more seasonal the climate.

Mountains Affect Air Flow—Rain Shadow Effect

Mountain ranges are barriers to the smooth movement of air currents across continents. When an air mass encounters mountains, it is slowed and cooled because the air is forced up into cooler parts of the atmosphere in order to move over the mountains. The cooled air can no longer hold as much moisture and releases it as precipitation on the windward side of the mountain range. Once the air is over the mountain and is on the leeward side of the mountain, it no longer has as much moisture, resulting in the leeward (back) side of the mountain range's being drier than the windward side. This drier condition, known as a rain shadow effect, is directly responsible for the type of flora found there, which in turn directly affects the type of fauna that live there (Figure 4.28).

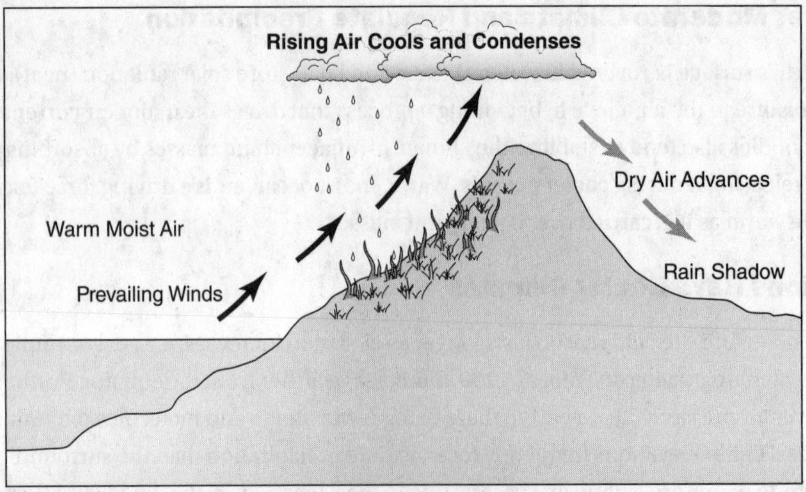

Figure 4.28 Rain shadow effect

4.9 El Niño and La Niña

El Niño–Southern Oscillation (ENSO)

La Nada (Normal Conditions)

During normal conditions, easterly trade winds move water and air toward the west. The ocean is generally around 24 inches (60 cm) higher in the western Pacific, and the water there is about 14°F warmer.

The trade winds, in piling up water in the western Pacific, make a deep warm layer in the west that pushes the thermocline down while it rises in the east. This shallow (90 feet or 30 m) eastern thermocline allows the winds to pull up nutrient-rich water from below, a phenomenon known as upwelling, which increases fishing stocks.

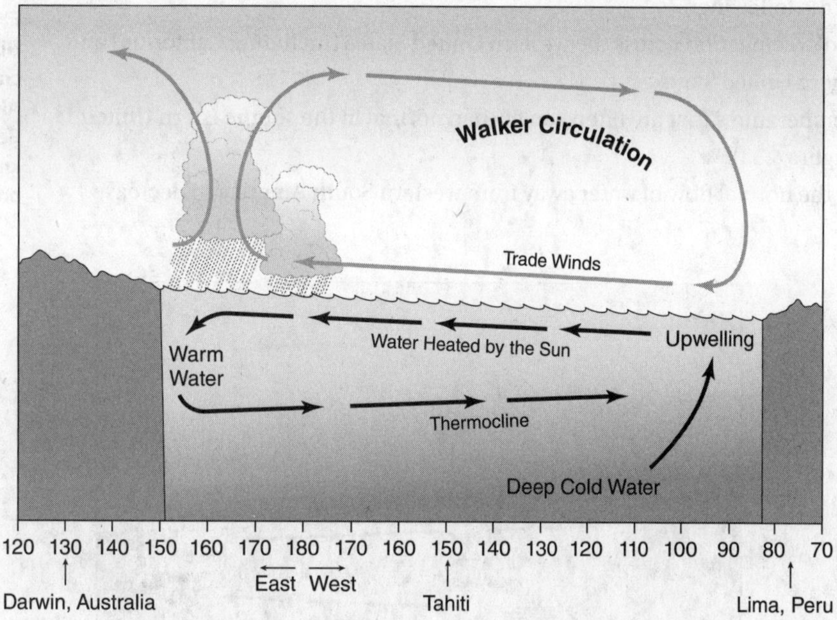

Figure 4.29 Normal conditions

Upwellings

Upwellings occur when prevailing winds, produced through the Coriolis effect and moving clockwise in the Northern Hemisphere, push warmer, nutrient-poor surface waters away from the coastline (Figure 4.30). This surface water is then replaced by cooler, nutrient-rich deeper waters. The deeper waters contain high levels of nitrates and phosphates, which result from the decomposition and sinking of surface water plankton. When these nutrients are brought to the surface through upwelling, they supply necessary nutrients for phytoplankton, which form the base of the oceanic food chain.

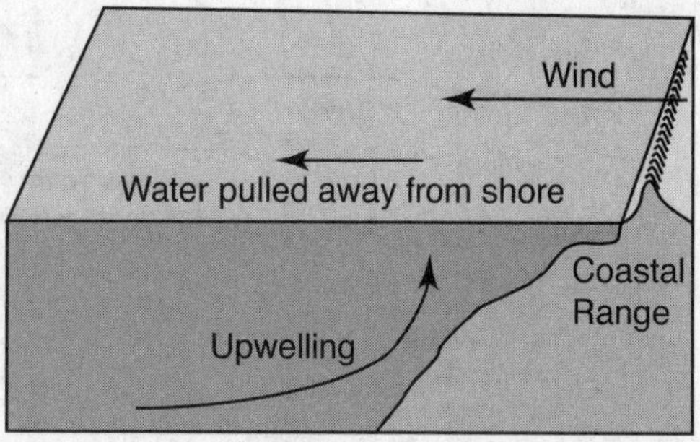

Figure 4.30 Upwelling

El Niño (Warm Phase)

The steps of forming an El Niño are as follows (Figure 4.31):

- A band of warmer-than-average ocean water temperatures develops off the Pacific coast of South America.
- Air pressure patterns reverse direction, causing trade winds to decrease in strength.
- As a result, the thermocline off western South America becomes deeper and there is a decrease in the upwelling of nutrients, which causes extensive fish kills.
- Effects are strongest during the Northern Hemisphere winter because ocean temperatures worldwide are at their warmest.
- Increased ocean warmth enhances convection, which then alters the jet stream, resulting in the following:
  - Enhanced precipitation across the western United States (including California) and the southern United States.
  - Winter temperatures that are often cooler than normal in the southeastern United States (Figure 4.32).
- This causes the normal flow of water away from western South America to decrease "pile up."

TIP

Questions about El Niño and La Niña are very common on the APES exam. Be sure you know these two processes!

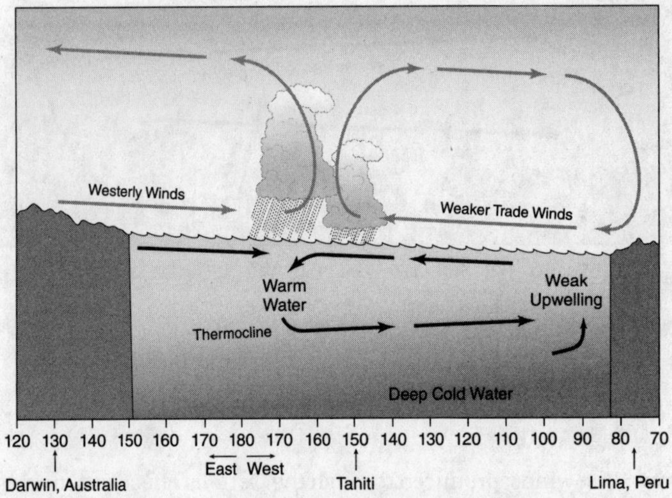

Figure 4.31 The development of El Niño

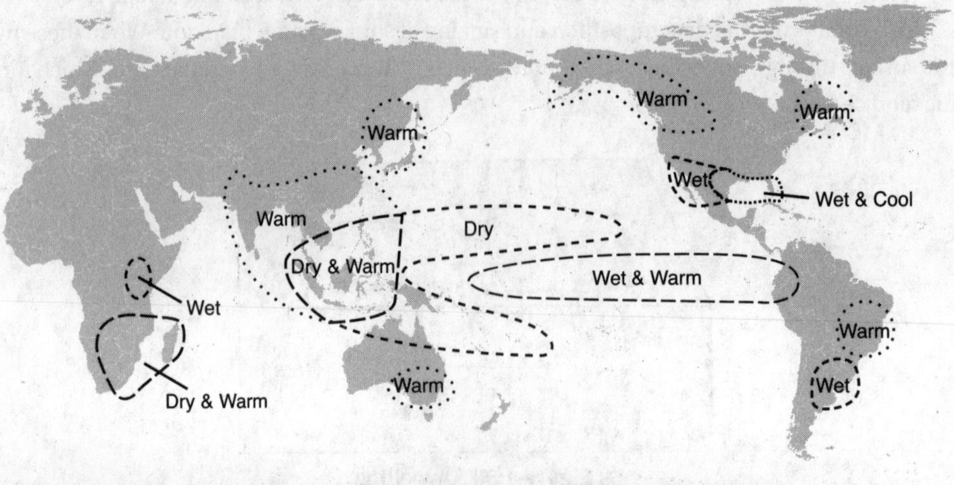

Figure 4.32 El Niño effects (December–February)
Source: National Weather Service

La Niña (Cool Phase)

The steps of forming a La Niña are as follows (Figure 4.33):

- Trade winds that blow west across the tropical Pacific are stronger than normal.
- This then results in an increase in the upwelling off of South America.
- This then results in cooler-than-normal sea surface temperatures off of South America.
- This then results in wetter-than-normal conditions across the Pacific Northwest, and both drier- and warmer-than-normal conditions in the southern United States.
- This then results in an increase in the number of hurricanes.
- Winter temperatures are warmer than normal in the southeastern United States and cooler than normal in the northwest, whereas there are heavier-than-normal monsoons in India and southeast Asia (Figure 4.34).

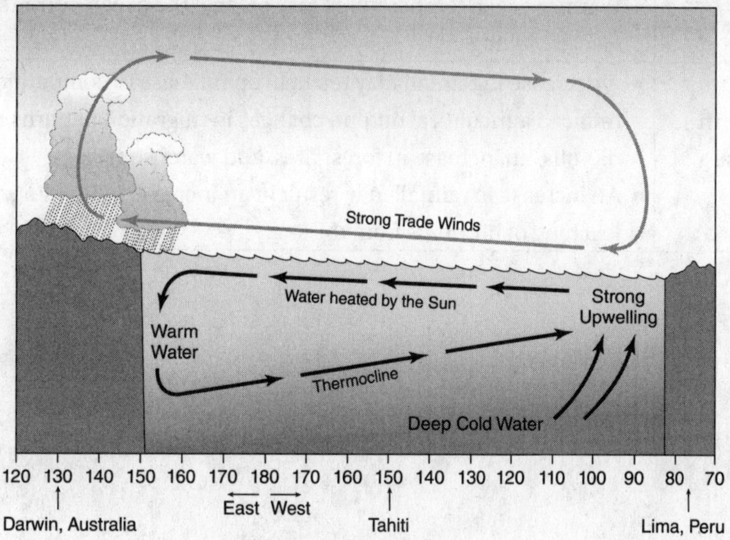

Figure 4.33 La Niña

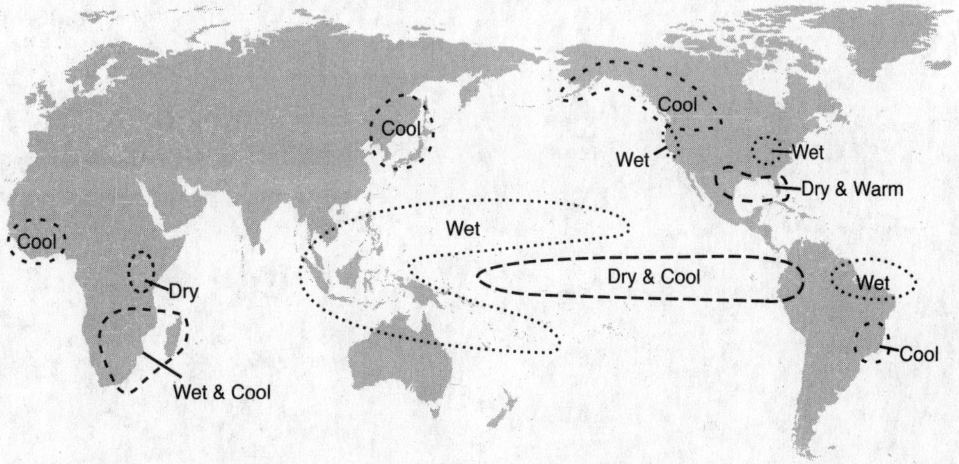

Figure 4.34 La Niña effects (December–February)
Source: National Weather Service

Environmental Effects of ENSO Weather Patterns

Cause	Environmental Effects
Warmer or cooler ocean temperatures	▪ A decrease in upwelling, resulting in die-offs. ▪ A negative impact on coral reefs (e.g., bleaching). ▪ Animal migration patterns may become disrupted. ▪ Shifts in traditional weather patterns, which may result in an increase in insect-borne diseases. ▪ Some marine species may not be able to tolerate warmer or cooler water temperatures, resulting in a disruption of marine food webs and biodiversity. ▪ The amount of CO_2 that warmer ocean water can hold decreases, which directly affects global warming. ▪ The strength and frequency of hurricanes and/or tornadoes may increase. ▪ Warmer ocean temperatures may change the ocean current flows and affect glacial melting.
Increase or decrease in the amount of normal rainfall	▪ A decrease in rainfall may result in an increase in competition for food resources, reduced agricultural output, changes in migration patterns, starvation, species die-offs, an increase in forest fires, and water shortages. ▪ An increase in rainfall may result in an increase in flooding, soil erosion, and leaching of nutrients from the soil.

Practice Questions

1. The zone of the atmosphere in which weather occurs is known as the

(A) ionosphere
(B) stratosphere
(C) thermosphere
(D) troposphere

2. Examine the following chart that shows the size of soil particles.

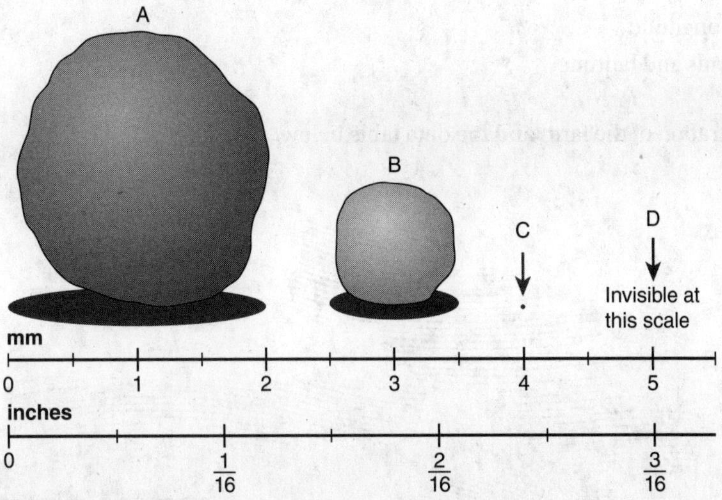

Which of the soil particles shown above match the following description?

> *"It can be tightly packed, not allowing water to pass through easily. The particles are very heavy and thus roots of plants may break while penetrating. It is low in organic matter and thus cultivation is possible only through the addition of soil amendments. When rubbed between thumb and forefinger it feels sticky."*

(A) Gravel
(B) Sand
(C) Silt
(D) Clay

3. Ninety-nine percent of the volume of gases in the lower atmosphere, listed in descending order of volume, are

(A) H_2O, N_2, O_2, CO_2
(B) N_2, O_2, H_2O, CO_2
(C) O_2, CO_2, N_2, H_2O
(D) O_2, N_2, CO_2, H_2O

4. Which of the following environmental factors does NOT lower the pH of soil?

   (A) Acid rain
   (B) Use of fertilizers
   (C) Weathering of minerals
   (D) All of these choices can possibly lower the pH of soil.

5. Regional climates are most influenced by

   (A) altitude and longitude
   (B) latitude and altitude
   (C) latitude and longitude
   (D) prevailing winds and latitude

6. Examine the illustration of the farm and the data table below.

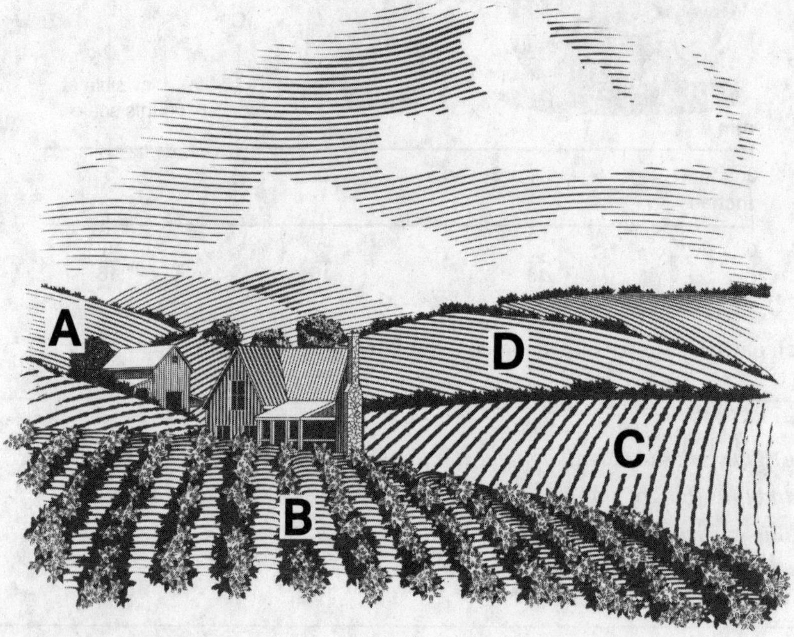

Which field would probably be the most productive based on the highest percentage of available soil nutrients?

Field	% Clay	% Gravel	% Loam	% Sand	% Silt
A	5	20	20	30	25
B	10	15	25	25	25
C	15	10	30	25	30
D	15	15	40	15	15

(A) A
(B) B
(C) C
(D) D

7. Examine the weather map below. The areas of the map indicating low pressure (L) would generally be associated with

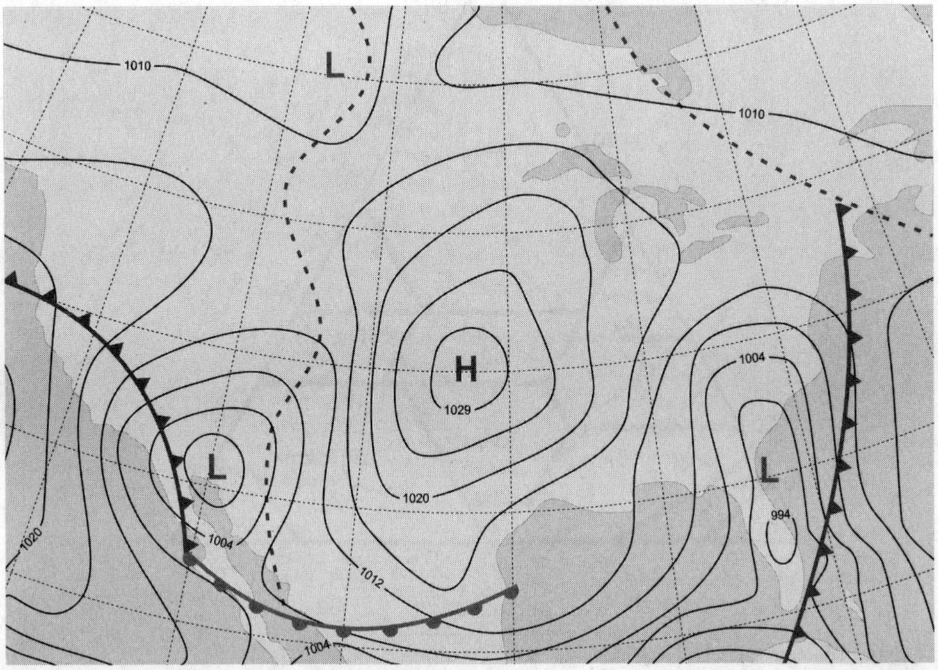

(A) cloudy or stormy weather
(B) fair weather
(C) hot, humid weather
(D) tornadoes

8. La Niña would produce all of the following effects EXCEPT

(A) more Atlantic hurricanes
(B) warmer and drier winters in the southwestern and southeastern United States
(C) warmer winters in Canada and the northeastern United States
(D) wetter winters in the Pacific Northwest region of the United States

Refer to the soil chart below to answer Question 9.

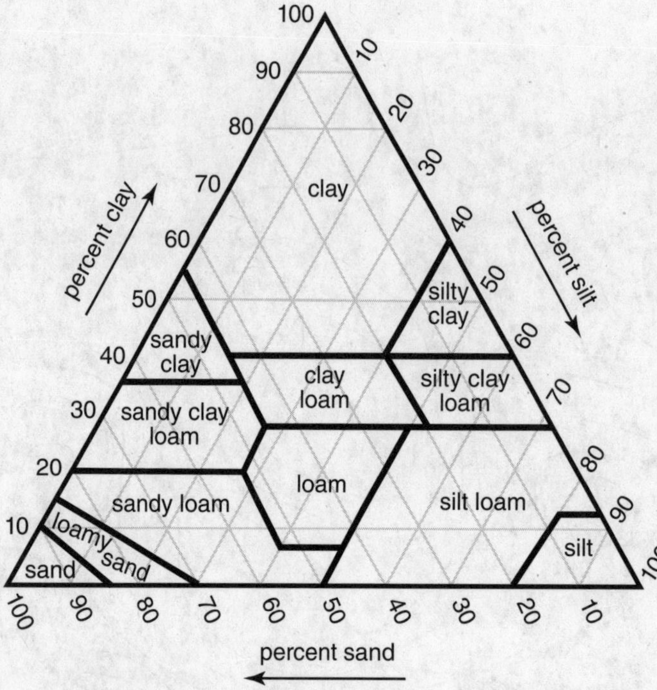

9. An APES class is doing soil studies and determined that a sample of soil contained 30% clay, 20% silt, and 50% sand. Using the soil texture chart above, the type of soil would be

(A) loamy sand
(B) sandy clay loam
(C) sandy loam
(D) silt clay loam

10. On the leeward side of a mountain range, one would expect

(A) fewer clouds and less rain than on the windward side
(B) more clouds and rain than on the windward side
(C) more clouds but less rain than on the windward side
(D) no significant difference in climate compared with the windward side

11. Seasons on Earth are primarily determined by

(A) Earth's longitude
(B) ocean temperatures
(C) the distance between Earth and the sun during a particular season
(D) the tilt of Earth's axis as it revolves around the sun

12. Poor nutrient-holding capacity, good water-infiltration capacity, and good aeration properties are examples of what type of particle found in soil?

(A) Clay
(B) Silt
(C) Sand
(D) Loam

Refer to the following diagram for Question 13.

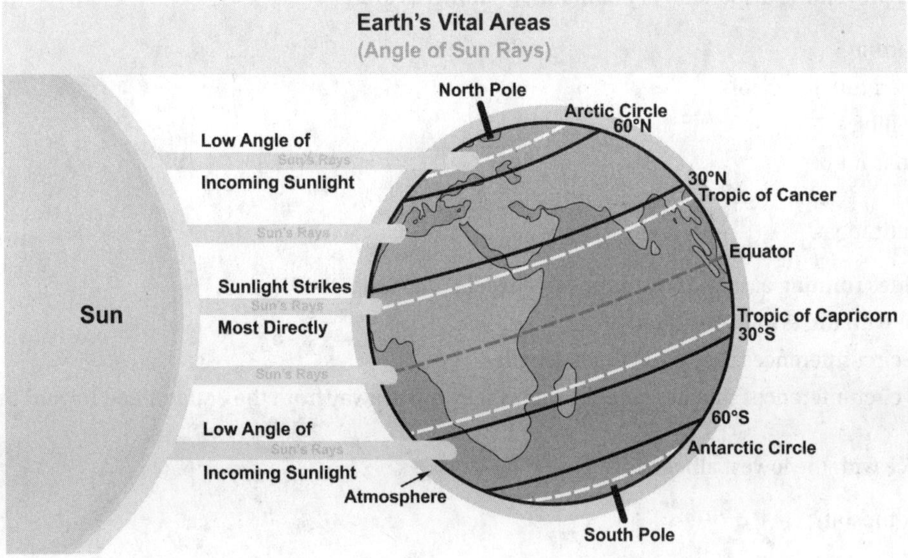

13. At the equator,

    (A) cool, dry air ascends
    (B) cool, moist air descends
    (C) warm, moist air ascends
    (D) warm, moist air descends

14. Most of Earth's deserts are at approximately 30° north and south latitudes because these latitudes are characterized by

    (A) a slow-moving jet stream
    (B) descending dry air currents
    (C) enhanced solar radiation
    (D) predominately low atmospheric pressure

15. Characteristics or requirements of a monsoon include all of the following EXCEPT

    (A) a seasonal reversal of wind patterns
    (B) different heating and cooling rates between the ocean and the continent
    (C) heating and cooling rates between the oceans and the continents that are equal
    (D) large land areas cut off from continental air masses by mountain ranges and surrounded by large bodies of water

16. Which factor is the most important in determining the characteristics of soil?

    (A) Topography
    (B) Climate
    (C) The type of bedrock
    (D) The type of vegetation that is growing

17. An atmospheric condition in which the air temperature rises with increasing altitude, holding surface air down and preventing dispersion of pollutants, is known as a/an

(A) cold front
(B) temperature inversion
(C) upwelling
(D) warm front

18. Lines of latitude

(A) are lines running east and west from Greenwich, England
(B) begin with the Prime Meridian
(C) have circumferences that are of equal length
(D) have circumferences that become shorter as they move away from the equator and toward the poles

19. The surface with the lowest albedo is

(A) black topsoil
(B) desert
(C) ocean water
(D) snow

20. Jet streams over the United States travel primarily

(A) east to west
(B) north to south
(C) south to north
(D) west to east

21. Which of the following statements about El Niño is TRUE?

(A) Conditions are more favorable for hurricanes in the Caribbean and central Atlantic area.
(B) Tornadoes occur more frequently in those areas of the United States already vulnerable to them.
(C) Mild winter temperatures occur over western Canada and the northwestern United States.
(D) Temperatures are above average in the southeastern United States and below average in the northwestern United States.

TIP

Be careful that, when answering questions that involve placing items in order, you have the order going in the proper direction. A common trick is to have the correct choices in the opposite direction.

22. Areas of low pressure are typically characterized by ________ (rising, sinking) air and move toward regions where the pressure is ________ (falling, rising) with time.

(A) rising, falling
(B) rising, rising
(C) sinking, falling
(D) sinking, rising

23. The horizon of soil, also known as the topsoil layer, that contains humus, minerals, and roots, and that is rich in living organisms, is known as the

(A) A horizon
(B) B horizon
(C) C horizon
(D) O horizon

24. The global circulation pattern that dominates the tropics is called the

(A) Ferrel cell
(B) Hadley cell
(C) Polar cell
(D) Tropical cell

25. Which of the following statements about sea and land breezes is FALSE?

(A) Land breezes form mostly during the daytime hours.
(B) Sea and land breezes are examples of thermal circulations.
(C) Sea breezes can sometimes result in rain showers and thunderstorms near the shore.
(D) Smaller versions of sea/land breezes often form in the vicinity of large lakes.

26. Examine the diagram below:

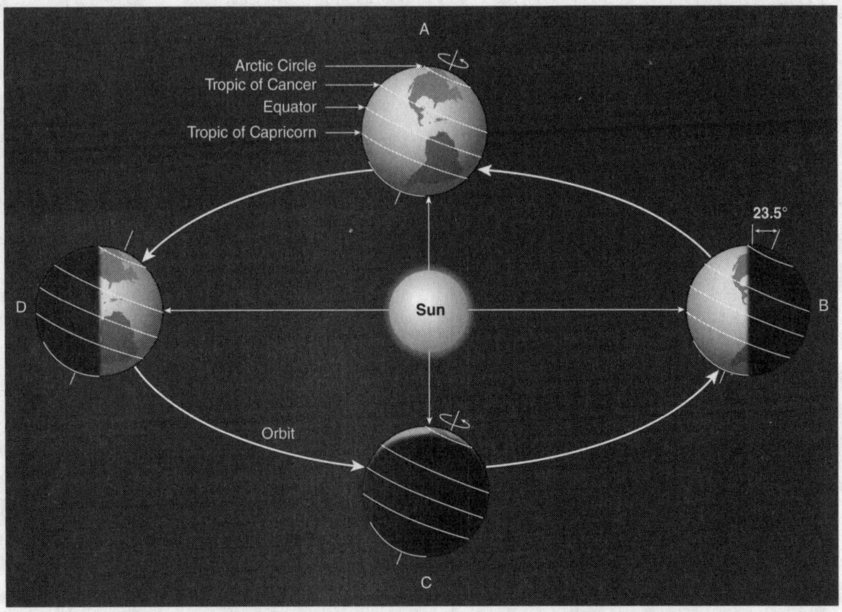

Which of the statements listed below is true?

(A) A = spring in the Northern Hemisphere, B = winter in the Northern Hemisphere, C = fall in the Northern Hemisphere, and D = summer in the Northern Hemisphere
(B) A = fall in the Northern Hemisphere, B = winter in the Northern Hemisphere, C = spring in the Northern Hemisphere, and D = summer in the Northern Hemisphere
(C) A = spring in the Southern Hemisphere, B = winter in the Southern Hemisphere, C = fall in the Southern Hemisphere, and D = summer in the Southern Hemisphere
(D) A = fall in the Southern Hemisphere, B = summer in the Northern Hemisphere, C = spring in the Southern Hemisphere, and D = winter in the Northern Hemisphere

27. Jet streams follow the sun in that as the sun's elevation __________ (increases, decreases) each day in the spring, the jet stream shifts by moving __________ (north, south) during the Northern Hemisphere spring.

    (A) decreases, north
    (B) decreases, south
    (C) increases, north
    (D) increases, south

28. Water in areas of upwelling is generally

    (A) cold and high in nutrients
    (B) cold and lacking nutrients
    (C) warm and high in nutrients
    (D) warm and lacking nutrients

29. The three necessary ingredients for thunderstorm formation are

    (A) lifting mechanism, fronts, moisture
    (B) lifting mechanism, mountains, oceans
    (C) moisture, lifting mechanism, instability
    (D) stability, moisture, heat

30. Which of the following descriptions best defines a watershed?

    (A) A large body of water
    (B) An area of land where all water under or on it drains to one place
    (C) A place where water is stored
    (D) The water underground

Free-Response Question

Using the Ferrel cell for this question:

(a) Describe a Ferrel cell by:
    (i) Drawing a sketch and showing where it is in relation to the Earth's equator and to the other two types of cells—i.e., Hadley and Polar.
    (ii) Explaining how air circulates within a Ferrel cell.
(b) Describe ONE prevailing North American continental wind pattern that occurs within a Ferrel cell in the Northern Hemisphere.
(c) Describe the reason for the climatic conditions found on Earth in the area where Ferrel cells are found and describe ONE biome that would exist (near sea level) within the latitudes where Ferrel cells are found.
(d) For the biome that you chose in (c), identify and briefly describe ONE environmental threat affecting that biome today.
(e) For the environmental threat that you chose in (d), describe a possible solution to the threat.

TIP

Before writing your responses, be sure to map out or brainstorm what you are going to write about. A few minutes planning and organizing your responses will get you a much higher score.

Answer Explanations

1. **(D)** The troposphere is the atmospheric layer closest to Earth and extends for about 11 miles (18 km) above Earth at the equator and about 5 miles (8 km) above Earth at the poles. Temperature declines as altitude increases.

2. **(D)** The description of clay particles is given in the question. The following are the descriptions of the other choices available: **Sand** (SiO_2) has no water-holding capacity and the infiltration rate is high. **Gravel** is a loose rock produced by weathering or crushing rocks. **Silt** particles are intermediate in size between clay and sand. Silt particles have limited ability to retain plant nutrients or release them to the soil solution for plant uptake. Silt tends to have a spherical shape, giving high-silt soils a soapy or slippery feeling when rubbed between the fingers when wet. Because of its spherical shape, silt also retains a large amount of water, but it releases the water readily to plants. Soils rich in silt are generally considered very fertile for the growth of plants, largely due to their water characteristics and ease of cultivation.

3. **(B)** Nitrogen (78%), oxygen (21%), water vapor (about 0– 4%), and carbon dioxide (below 1%).

4. **(D)** When atmospheric water reacts with sulfur and nitrogen compounds that result from industrial processes, the result can be the formation of sulfuric (H_2SO_4) and nitric (HNO_3) acid in rainwater. However, the amount of acidity that is deposited in rainwater is much less, on average, than that created through agricultural activities. Acid soils are most often found in areas of high rainfall. Additionally, rainwater has a slightly acidic pH of 5.7 due to a reaction with CO_2 in the atmosphere that forms carbonic acid. Ammonium (NH_4^+) fertilizers react in the soil in a process called nitrification to form nitrate (NO_3^-), and, in the process, they release H^+ ions. Highly weathered soils are often characterized by high concentrations of Fe and Al oxides. Severely acidic conditions can form in soils near mine spoils due to the oxidation of pyrite, also known as iron sulfide (FeS_2). In addition to the choices listed, potential acid sulfate soils naturally formed in waterlogged coastal and estuarine environments can become highly acidic when drained or excavated, and decomposition of organic matter by microorganisms can release carbon dioxide (CO_2), which can form carbonic acid (H_2CO_3) when mixed with soil water.

5. **(B)** Latitude expresses how far north or south of the equator a location is. The equator is 0° latitude, and the poles are at 90°. For every 1,000 feet (300 m) in altitude, there is a 3°F (1.5°C) drop in temperature.

6. **(D)** Loam is soil composed of sand, silt, and clay in relatively even concentrations. Loam soils generally contain more nutrients and humus than sandy soils, have better infiltration and drainage than silty soils, and are easier to till than clay soils.

7. **(A)** A low-pressure air mass (low) occurs when warm air, which is less dense than cooler air, spirals inward toward the center of a low-pressure area. Since the center of the low-pressure area is of even less density and pressure, the air in this section rises and the warm air cools as it expands. The temperature begins to fall and may go below the dew point—the point at which air condenses into water droplets. These water droplets make up clouds. If the droplets begin to coalesce on condensation nuclei, rain follows.

8. **(C)** During La Niña, large portions of central North America experience increased storminess, increased precipitation, and an increased frequency of significant cold-air outbreaks, while the southern states experience less storminess and precipitation. Also, there tend to be considerable month-to-month variations in temperature, rainfall, and storminess across central North America during the winter and spring seasons.

9. **(B)** The **clay** percentages are listed on the left side of the triangle. Lines corresponding to clay percentages extend from the percentages reading left to right. Draw a horizontal line at 30% clay. The **silt** percentages are on the right side, with lines extending downwardly, diagonally right to left. Draw a diagonal line at 20% silt. The sand percentages are at the bottom, with lines extending upwardly, diagonally right to left. Draw a diagonal line at 50% sand. Now find the spot on the triangle where all three lines intersect. *Note: you only need two of the soil type percentages to identify the soil type.*

10. **(A)** As air descends down the leeward slope of a mountain, it has already lost much of its moisture on the windward side. The air on the leeward side also warms as it descends, lowering humidity even more. Rain shadows on the leeward side are also more prevalent when the windward side of a mountain is steep, and thus warm air cools more rapidly over a shorter distance, creating more windward-side precipitation. Thus, the leeward-side air is drier since the saturated air on the windward side has lost even more if its moisture.

11. **(D)** During the year, the seasons (e.g., winter, summer) change depending upon the amount of sunlight that reaches Earth's surface as it revolves around the sun. As Earth travels in a loop around the sun, the hemisphere that is tilted toward the sun receives the most solar radiation (summer), and the hemisphere tilted away from the sun receives the least amount of sunlight (winter). At the equator there are no seasons because the sun strikes this area at about the same angle throughout the year. At the north and south poles the temperature is much cooler throughout the year because they are never tilted in a direct path of sunlight.

12. **(C)** Water flows through sandy soils too fast for many crops and requires frequent irrigation.

13. **(C)** Hadley cells occur between 0° and 25° north and south latitudes (equatorial region). In this area, there is upward air motion, cooling of the air due to uplift, high humidity, high clouds, and heavy rains.

14. **(B)** Deserts are arid regions defined by the amount of rainfall an area receives in a year (less than 10 inches [approximately 25 cm]), not by temperature. Deserts are often hot, but dry; cold places, such as Antarctica, qualify as deserts too. Since there is more direct sunlight at the equator, warm air rises and begins to cool. Since the rising cooler air holds less water than warmer air, precipitation is common at the equator (e.g., tropical rainforests). This drier air mass then moves both north and south. At around ±30 degrees north and south latitudes, the dry air begins to sink and warm. The warmer air can now hold more water, resulting in the evaporation of water and the formation of deserts.

15. **(C)** During monsoon season, winds blow from cooler ocean areas (higher pressure) to warmer landmasses (lower pressure). As the air rises over the landmasses, it cools and is unable to retain water, producing great amounts of precipitation. In winter, the ocean is now warmer and the cycle reverses. Drier air travels from the land out to the ocean. Monsoons occur in Africa, Australia, South Asia (e.g., India), and the Pacific coast of Central America.

16. **(B)** Climate is the major factor in determining the kind of plant and animal life on and in the soil. It determines the amount of water available for weathering minerals, transporting the minerals, and releasing elements. Climate, through its influence on soil temperature, determines the rate of chemical weathering. Warm, moist climates encourage rapid plant growth and thus high organic matter production. The opposite is true for cold, dry climates. Organic matter decomposition is also accelerated in warm, moist climates. Climate controls freezing, thawing, wetting, and drying, which break parent material apart.

    Rainfall causes leaching. Rain dissolves some minerals, such as carbonates, and transports them deeper into the soil. Some acidic soils have developed from parent materials that originally contained limestone. Rainfall can also be acidic, especially downwind from industrial processes.

17. **(B)** Temperature inversions are atmospheric conditions in which the air temperature rises with increasing altitude, holding surface air down and preventing the dispersion of pollutants.

18. **(D)** Latitude gives the location of a place on Earth north or south of the equator. Lines of latitude are the horizontal lines shown running east to west on maps. Technically, latitude is an angular measurement in degrees ranging from 0° at the equator to 90° at the poles (90°N for the North Pole and 90°S for the South Pole).

19. **(A)** Albedo is a measure of reflection of sunlight from a surface. Of the choices, dark topsoil absorbs the most energy and therefore reflects the least amount of energy, resulting in the lowest albedo.

20. **(D)** Jet streams are large-scale upper air flows that travel from west to east and are produced by differences in temperature. They can travel as fast as 250 miles per hour (400 kph) and travel between 3 and 8 miles (5–13 km) above Earth's surface.

21. **(C)** El Niños occur about once every five years, and what usually happens is warming in the oceans caused by the winds leads to diffusion of this warmth all over Earth, which changes atmospheric pressures and has consequences for rainfall, wind patterns, and sea surface temperatures. Climatic changes that occur in North America include the following:
    - Above average precipitation in the Gulf Coast, including Florida
    - A drier than average period in the Pacific Northwest
    - Milder winter temperatures over western Canada and the northwestern United States.

22. **(A)** In a high-pressure system, air pressure is greater than the surrounding areas. This difference in air pressure results in wind. In a high-pressure area, air is denser than in areas of lower pressure. The result is that air will move from the high-pressure area to an area of lower pressure. Clear skies and fair weather usually occur in these regions. On the other hand, winds tend to blow into a low-pressure system because air moves from areas of higher pressure into areas of lower pressure. As winds blow into a low-pressure system, the air moves up. This upward flow of air can cause clouds, strong winds, and precipitation to form.

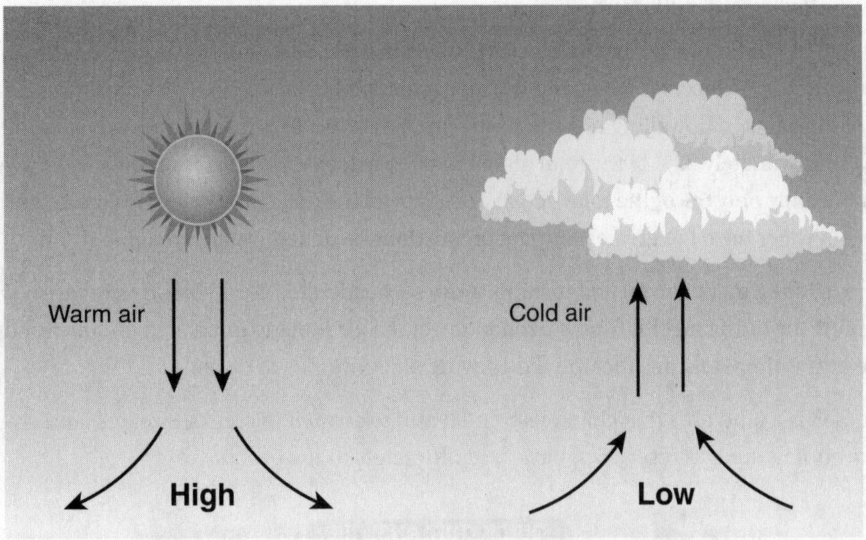

High Pressure	Low Pressure
Air leaves (diverges)	Air enters (converges)
Cold	Warm
Dry	Moist, humid
Air sinks	Air rises
Generally fair weather	Generally rainy weather

23. **(A)** Topsoil is usually the top 5–10 inches (13–25 cm) of a soil profile and has the highest concentration of organic matter and microorganisms. Topsoil is composed of mineral particles, organic matter, water, and air. There is generally a high concentration of roots in this layer as plants obtain much of their vital nutrients here.

24. **(B)** Hadley cells dominate the tropics. *See pages 144–145 for a complete description of a Hadley cell.*

25. **(A)** A sea breeze, or onshore breeze, is a gentle wind that develops over bodies of water near land due to differences in air pressure created by their differences in heat capacity and associated rates of heating and cooling. It is a common occurrence along coasts during the morning as solar radiation heats the land more quickly than it does the water. A land breeze, or offshore breeze, is the reverse effect, caused by land cooling more quickly than water in the evening. The sea breeze dissipates and the wind flows from the land toward the sea. Both are important factors in coastal regions' prevailing winds and more moderate temperature profile.

26. **(A)** During the year, the seasons change depending on the amount of sunlight reaching Earth as it revolves around the sun. The seasons are caused by Earth, tilted on its axis, traveling in a loop around the sun each year. Summer happens in the hemisphere tilted toward the sun, and winter happens in the hemisphere tilted away from the sun. As Earth travels around the sun, the hemisphere that is tilted toward or away from the sun changes. The hemisphere that is tilted toward the sun is warmer because sunlight travels more directly to Earth's surface, so less gets scattered in the atmosphere. When it is summer in the Northern Hemisphere, it is winter in the Southern Hemisphere. The hemisphere tilted toward the sun has longer days and shorter nights.

27. **(C)** The position of the jet stream also determines where the storm track is. As the jet stream moves north during spring, the storm track moves north, leaving the southern plains of Texas and Oklahoma and moving into the northern plains near the Dakotas.

28. **(A)** Upwelling is a phenomenon that involves the wind-driven motion of dense, cooler, and usually nutrient-rich water toward the surface, replacing the warmer, usually nutrient-depleted surface water. The increased availability of nutrients in upwelling regions results in high levels of primary productivity and thus fishery production. Approximately 25% of the total global marine fish catch comes from only five upwelling regions on Earth that occupy only 5% of the total ocean area. Upwellings that are driven by coastal currents or diverging open-ocean water have the greatest impact on nutrient-enriched waters and global fishery yields.

29. **(C)** Moisture, a lifting mechanism, and instability are all needed for thunderstorms to form. The moisture is needed for rain. The lifting mechanism is needed to get the air moving initially in an upward direction, and the unstable atmosphere ensures the upward-moving air continues to do so.

30. **(B)** A watershed is a land area that channels rainfall and snowmelt to creeks, streams, and rivers and eventually to outflow points such as reservoirs, bays, and ultimately to the ocean.

QUICK QUIZ ANSWER

$$HI = 0.5\{82 + 61.0 + [(82 - 68.0) \times 1.2] + (62 \times 0.094)\} = \mathbf{84.3°F}$$

Free-Response Question and Answer Exercise

TIP

When writing your FRQ responses, a picture is worth a thousand words! If you can sketch out a labeled diagram of what you are describing, it will go a long way in improving your score.

Let's do this free-response question (FRQ) together and make it an exercise that will help you write FRQs faster and more efficiently—and we won't worry about awarding points for this FRQ. Remember, you have about 23 minutes for each of the three FRQs you will need to answer on the actual APES exam in May, so practicing the most efficient method now will definitely help you earn that 5.

STEP 1 Read through the entire question first—before you write. Doing this will give you the overall tasks that you have ahead and help you start the process of organizing your answers.

STEP 2 Brainstorm the first question, writing down key points that you want to cover. Let's do that now:

(a) Describe a Ferrel cell by:

(i) Drawing a sketch and showing where it is in relation to Earth's equator and to the other two types of cells, i.e—Hadley and Polar.

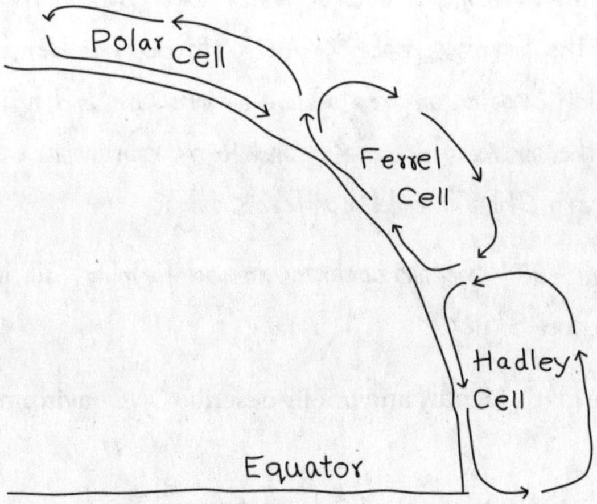

(ii) Describe how air circulates within a Ferrel cell.

Sinking air ~30° which ascends farther toward the poles Surface air flowing poleward is deflected east - Coriolis and ascends ~60°

Now, let's put these two ideas into paragraph form.

The Ferrel cell is characterized by sinking air near 30 deg N and S latitudes and rising air farther poleward. At the surface, air flowing poleward is deflected to the east by the Coriolis force, resulting in westerly surface winds.

Now, let's brainstorm the rest of the questions in this same way and, once we've done that, go back and write out our short answers.

(b) Describe ONE prevailing continental wind pattern that occurs within a Ferrel cell in the Northern Hemisphere.

Westerlies - W → E

Strongest winters - air pressure low over poles

Role - carry warm H_2O, equatorial waters-winds → western coast of U.S.

The prevailing winds known as westerlies (winds blowing from west to east) are strongest in the winter, when air pressure is lower over the poles. The winds are strongest in the winter when the pressure is lower over the poles and weakest when pressure is higher over the poles.

The westerlies play an important role in carrying warm, equatorial waters and winds to the western coasts of continents, especially in the Southern Hemisphere, because of its much larger oceanic area.

(c) Describe the reason for the climatic conditions found on Earth in the area where Ferrel cells are found and describe ONE biome that would exist (near sea level) within the latitudes where Ferrel cells are found.

~30° – dry air descends → high air pressure

Moves – (poles = get warmer → absorbs water) + equator ~60° – density ↓, → low pressure; air rises = temps ↓ = rain

At around 30° latitude, a warm, dry air mass descends earthward generating high air pressure. As this air reaches the Earth's surface, it travels towards the equator and the poles. As the air mass moves toward the poles from 30°, it begins to warm due to the warmer temperatures that occur at lower latitudes. As the air mass warms, it absorbs water vapor due to the increased water-holding capacity of warmer air. As the air mass reaches ~60° latitude, the density of the air mass decreases, causing the air to rise, which creates an area of low air pressure and with temperatures declining, the water-holding capacity of the air mass decreases and forms rain in places like northern Europe, Russia, southern Canada, and northern China in cold, coniferous taigas.

As the air moves away from ~60° the air continues to cool forming rain in places like cool, temperate forests of the northern United States.

(d) For the biome that you chose in (c), identify and briefly describe ONE environmental threat affecting that biome today.

clear cutting = removes trees, soil impact, CO_2, O_2, "homes" Environmental threats facing coniferous forests today include:

- clear-cutting – removes trees which stabilize soil, absorb and sequester carbon dioxide, serve as 'homes' for wildlife, and produce oxygen.

(e) For the environmental threat that you chose in (d), describe a possible solution to the threat.

zero deforestation – recycle – permits = Destroy 1 – Replace With 2

- Companies and federal and local governments can adopt "zero deforestation" policies.
- Look for the term "Recycled Wood Product" on furniture other wood products or provide tax incentives to stores that sell only these labelled products.
- Restrict construction permits with a "Destroy One, Replace With Two" clause.

Free-Response Question #2

Mount Pinatubo, a volcano located in the Philippines, erupted in 1991 for the first time in 600 years and caused widespread environmental devastation. The eruptions produced columns of ash and smoke that rose more than 19 miles (30 km) high, with rock debris falling about the same distance from the volcano. The resulting heavy ashfalls left about 100,000 people homeless, forced many thousands of people to flee the area, caused 300 human deaths, and resulted in devastating environmental damage. The eruption of Mt. Pinatubo was the result of a lithospheric plate's sinking (subducting) below another plate at a convergent plate boundary, largely due to the differences in densities (the denser plate sinking below the less dense plate).

(a) Create a sketch that illustrates the geological features that were involved in the eruption of Mt. Pinatubo. Your sketch must label and include the following:
(A) Asthenosphere; (B) Continental crust; (C) Lithosphere;
(D) Oceanic crust; (E) Trench; and (F) Volcanic arc.

(b) Below is a graph of how the eruption of Mt. Pinatubo affected average global temperatures.

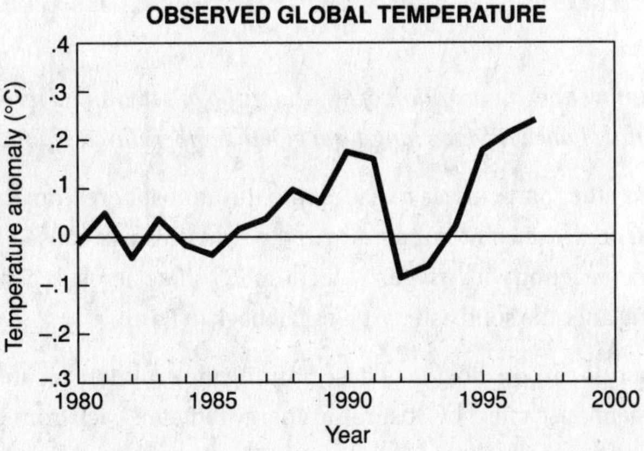

Explain how the eruption of Mount Pinatubo affected both short-term and long-term climate change.

(c) One of the events that occurred during the eruption of Mt. Pinatubo was the release of up to 17 million tons of sulfur dioxide into the atmosphere. Describe one health effect when this amount of sulfur dioxide is released into the troposphere.

(d) The release of massive amounts of sulfur dioxide during the eruption of Mt. Pinatubo also increased the global amount of wet and dry acid deposition. Explain the difference between wet and dry acid deposition and give one environmental effect of this type of precipitation.

(e) Volcanic eruptions are part of Earth's evolutionary history and have had, and continue to have, some positive environmental consequences. Describe one positive environmental consequence of volcanic activity.

(f) Charles Darwin was the botanist, geologist, and zoologist on the research vessel *H.M.S. Beagle* from 1831 to 1836. In 1859, he published *On the Origin of Species*, in which the theory of evolution through the process of natural selection was first introduced. A century later, scientists developed the theory of plate tectonics, which states that Earth's lithosphere is divided into a small number of plates that float on and travel independently over the mantle, with much of Earth's seismic and volcanic activity occurring at the boundaries of these plates. Describe one example in which evolution and/or speciation may have occurred as a consequence of volcanoes.

Free-Response Answer #2

10 Total Points Possible

(a) *1 point for correctly showing and correctly identifying any four of the five required features listed in the question. 4 points maximum.*

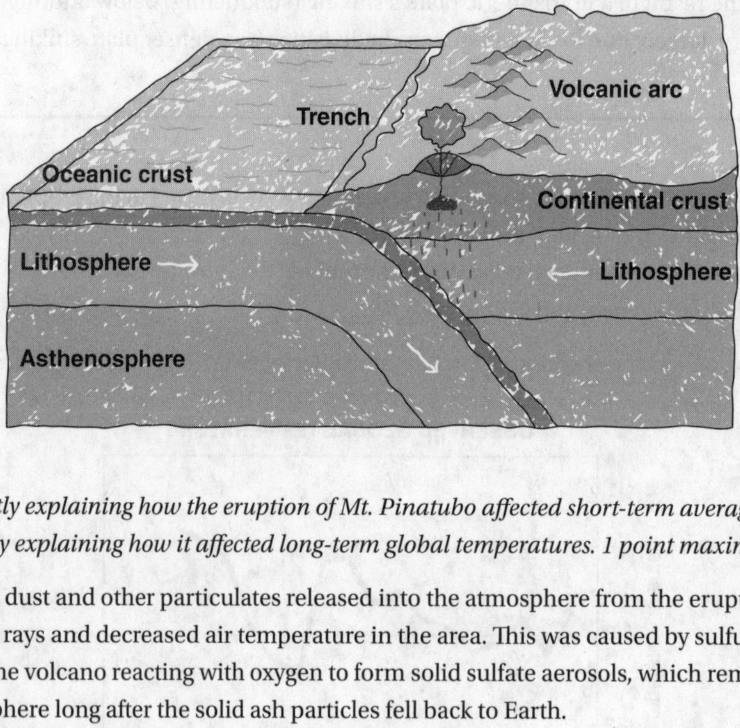

(b) *1 point for correctly explaining how the eruption of Mt. Pinatubo affected short-term average global temperatures and correctly explaining how it affected long-term global temperatures. 1 point maximum.*

In the short term, dust and other particulates released into the atmosphere from the eruption of Mt. Pinatubo blocked the sun's rays and decreased air temperature in the area. This was caused by sulfur dioxide gas that was released by the volcano reacting with oxygen to form solid sulfate aerosols, which remained suspended in the upper atmosphere long after the solid ash particles fell back to Earth.

In the long term, gases such as carbon dioxide released during the eruption accumulated in the stratosphere and contributed to the greenhouse effect by absorbing energy radiated back from Earth, which over time contributed to an increase in average global temperatures. The degree of temperature change was much less than that caused by dust blocking the sun's radiant energy, but the effects will last much longer.

(c) *1 point for correctly describing one human health effect of large amounts of sulfur dioxide in the troposphere. 1 point maximum.*

Short-term exposures to sulfur dioxide (SO_2) can harm the human respiratory system and make breathing difficult. People with asthma and other breathing problems, particularly children, are sensitive to these effects of SO_2. SO_2 emissions in the air can also lead to the formation of other sulfur oxides (SO_x). SO_x can react with other compounds in the atmosphere to form small particles, which contribute to particulate matter (PM) pollution. These small particles are able to penetrate into the lungs of humans and in sufficient quantity can contribute to health problems.

(d) *1 point for correctly explaining the difference between wet and dry acid deposition and 1 point for providing one environmental effect of either of these two forms of precipitation. 2 points maximum.*

Acid rain (wet deposition) can be caused by emissions of sulfur dioxide (SO_2) and nitrogen oxides (NO_x), which react with the water molecules in the atmosphere to produce rain, snow, fog, or hail with elevated levels of hydrogen ions (H^+ or H_3O^+), which by definition have a pH less than 7. Acid rain can have direct, harmful effects on plants and aquatic animals.

Acidic particles and gases can also deposit from the atmosphere in the absence of moisture. This is referred to as dry deposition. The acidic particles and gases land directly on surfaces (e.g., lakes, vegetation, buildings). When the acids are washed off a surface by rain, this acidic water flows over and through the ground, and can harm plants and wildlife.

The amount of acidity in the atmosphere that deposits to Earth through dry deposition depends on the amount of rainfall an area receives; e.g., in desert areas the ratio of dry to wet deposition is higher than in areas that receive significantly more rain each year.

(e) *1 point for correctly describing one positive environmental consequence of volcanic activity. 1 point maximum.*

Many land areas and formations were created directly from volcanic processes. A new lava flow may engulf and bury the land, but, over time, new soil and vegetation eventually develop through weathering processes. In warm, humid climates, soil development is more rapid, as volcanic ash and lava weathers to form new, rich, loamy soils.

(f) *1 point for providing an example in which evolution and/or speciation may have occurred as a consequence of volcanoes. 1 point maximum.*

Volcanic activity can create new landforms and/or drastically alter existing ones, destroying all or most life on an existing island(s). Over time, life forms begin to reappear as environmental conditions on the island become more hospitable through primary succession. Examples of life forms in which evolution and/or speciation may have occurred as a consequence of volcanoes include the following:

Galapagos Islands—These islands ($\sim$1 million years old and relatively young) are quite distant from the mainland of South America, have a volcanic origin, and provided an opportunity for new arrivals to "radiate" into new open niches. This isolation and the process of primary succession allowed the development of life forms on the islands.

Hawaiian Islands—These islands are relatively young ($<$5 million years old) and have a volcanic origin (hot spots). An example of speciation that may have occurred as a consequence of volcanoes are the Hawaiian honeycreepers, which branched into many different species and evolved a wide variety of colors and beak shapes.

5

Land and Water Use

Learning Objectives

In this unit, you will learn:

For added practice with this unit and topics, log onto Online Labs 5A–5C at online.barronsbooks.com

5.1 "The Tragedy of the Commons"

Garrett Hardin wrote "The Tragedy of the Commons" in 1968. The essay parallels what is happening worldwide in regards to resource depletion and pollution. The seas, air, water, animals, and minerals are all "the commons" and are for humans to use, but those who exploit them become rich. However, the price of depleting the resources of the commons is an external cost paid by all people on Earth.

Examples of current environmental problems that echo the issues of sustainability brought up in "The Tragedy of the Commons" include the following:

- Air pollution
- Burning of fossil fuels and consequential global warming
- Frontier logging of old-growth forests and the practice of "slash and burn"
- Habitat destruction and poaching
- Over-extraction of groundwater and wastewater due to excessive irrigation
- Overfishing
- Uncontrolled human population growth, leading to overpopulation

Limits to "The Tragedy of the Commons" include the following:

- Breaking a "commons" into smaller, privately owned parcels fragments the policies governing the entire "commons." Different standards and practices used on one parcel may or may not affect all other parcels.
- Economic decisions are generally short term, based on reactions in the world market, while environmental decisions are long term.
- Incorporating discount rates into the valuation of resources would be an incentive for investors to bear a short-term cost for a long-term gain.
- Land that is privately owned is subject to market pressure. For example, if privately owned timberland is increasing in value at an annual rate of 3% but interest rates on loans to purchase the land are 7%, this could result in the land's being sold or the timber's being harvested for short-term profits.
- Some "commons" are easier to control than others. Land, lakes, rangeland, deserts, and forests are geographically defined and easier to control than air or the open oceans, which do not belong to any one group.

5.2 Clearcutting

Clearcutting is when all of the trees in an area are cut at the same time. Environmental impacts of clearcutting include the following:

- A decrease in biodiversity caused by the loss of habitats
- Allows sunlight to reach the ground, making it warmer and drier—environment not suitable for many forest plants
- A temporary increase in the amount of available wood followed by very long times of no wood availability
- Microclimates within the forest are changed, affecting the associated food webs
- Reduction in both long-term (old-growth forests) and short-term carbon sinks, both of which result in more CO_2 eventually entering the atmosphere
- Water runoff (which results in soil erosion) increases

Edge Effects

An edge effect refers to how the local environment changes along some type of boundary or edge (Figure 5.1). Forest edges are created when trees are harvested, particularly when they are clear-cut. Tree canopies provide the ground below with shade and maintain a cooler and moister environment below.

As time passes and a stand of young trees emerges on a clear-cut, the environment changes and the edge begins to fade.

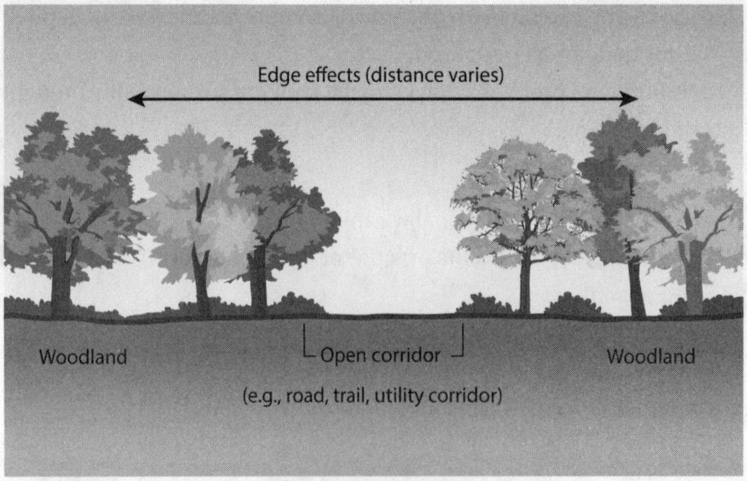

Figure 5.1 Edge effect
Source: U.S. Department of Agriculture

Deforestation—Tree Harvesting

Deforestation is the conversion of forested areas to non-forested areas, which are then used as grasslands for livestock, grain fields, mining, petroleum extraction, fuel wood cutting, commercial logging, tree plantations, or urban development. Natural deforestation can be caused by desertification, forest fires, glaciation, tsunamis, and volcanic eruptions (Figure 5.2).

Impacts of deforestation include the following:

- Allowing runoff into aquatic ecosystems
- Changing local climate patterns
- Decreasing soil fertility brought about by erosion. Forest soils are moist, but without protection from sun-blocking tree cover, they quickly dry out.
- Degrading environment(s) with reduced biodiversity and reduced ecological services. Eighty percent of Earth's land animals and plants live in forests.
- Increasing habitat fragmentation
- Increasing the amount of carbon dioxide released into the air from burning and tree decay
- Reducing the available habitats for migratory species of birds and butterflies
- Threatening the extinction of species with specialized niches

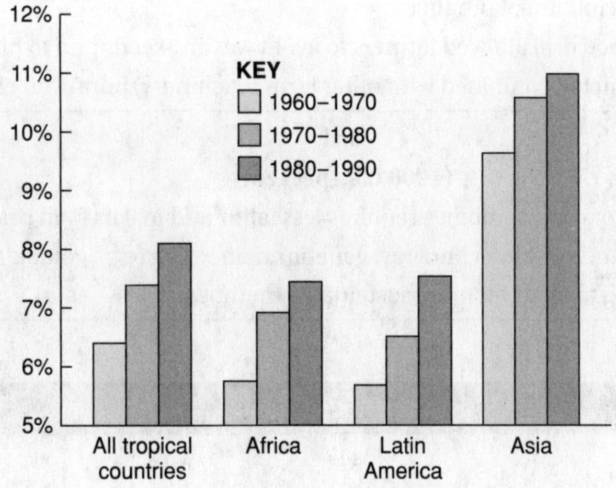

Figure 5.2 Rate of deforestation in tropical rainforests

5.3 The Agricultural and Green Revolutions

Agricultural Revolutions
1st **(2000+ B.C.E.)** People went from hunting and gathering to the domestication of plants and animals, which allowed people to settle in areas and create cities. Settled communities permitted people to observe and experiment with plants to learn how they grow and develop.
2nd **(1700–1900 C.E.)** Occurred at the same time as the Industrial Revolution—mechanization had a major role in this revolution and changed the way people farmed; e.g., went from hoes and shovels to plows and seed drills. ▪ Advances were made in breeding livestock. ▪ Increased agricultural output made it possible to feed large, urban populations. ▪ Methods of soil preparation, fertilization, crop care, and harvesting improved. ▪ New banking and lending practices helped farmers afford new equipment and seed. ▪ New crops came into Europe from trade with the Americas; e.g., corn and potatoes. ▪ Railroads allowed distribution of products. ▪ The invention of the seed drill allowed farmers to avoid wasting seeds and to plant in rows. ▪ The invention of the tractor, combined with other farm machinery, improved efficiency on farms.
3rd **(1900 C.E.–present)** Mechanization such as tractors and combines requires less labor and makes food prices more affordable. Scientific farming methods such as biotechnology, genetic engineering (e.g., golden rice), and the use of pesticides are now beginning to focus on more sustainable methods.

Green Revolutions
1st **(1940s–1980s)** The First Green Revolution started in the 1940s and involved the use of inorganic fertilizers, synthetic pesticides, new and more efficient methods of irrigation, and the beginning of developing high-yielding crop seeds that were also disease resistant and more tolerant of changing climatic conditions.
2nd **(1980s–Present)** The Second Green Revolution began in the mid-1980s with new engineering techniques and free-trade agreements involving property rights of food production, which helped to shape agricultural policies and food production and distribution systems worldwide. Examples of innovations during the Second Green Revolution include the development and expansion of genetically modified organisms (GMOs) (e.g., animals, plants, microorganisms) whose genetic makeup has been modified in a laboratory using genetic engineering or transgenic technology, resulting in genes that do not occur in nature. Examples include Golden Rice™, which is modified with daffodil genes to produce more beta-carotene (converts to Vitamin A) and Bt corn (corn modified with a bacterial insecticide gene that produces insect toxins within the cells of the corn).

Norman Borlaug

- Borlaug's work is estimated to have saved over 1 billion human lives from starvation and death in India, Mexico, and Pakistan.
- Borlaug's new, more productive strains require more fertilizer and pesticides.
- Borlaug's work has been criticized for large-scale monoculture and input-intensive farming techniques to countries that had previously relied on subsistence farming to support smaller populations.

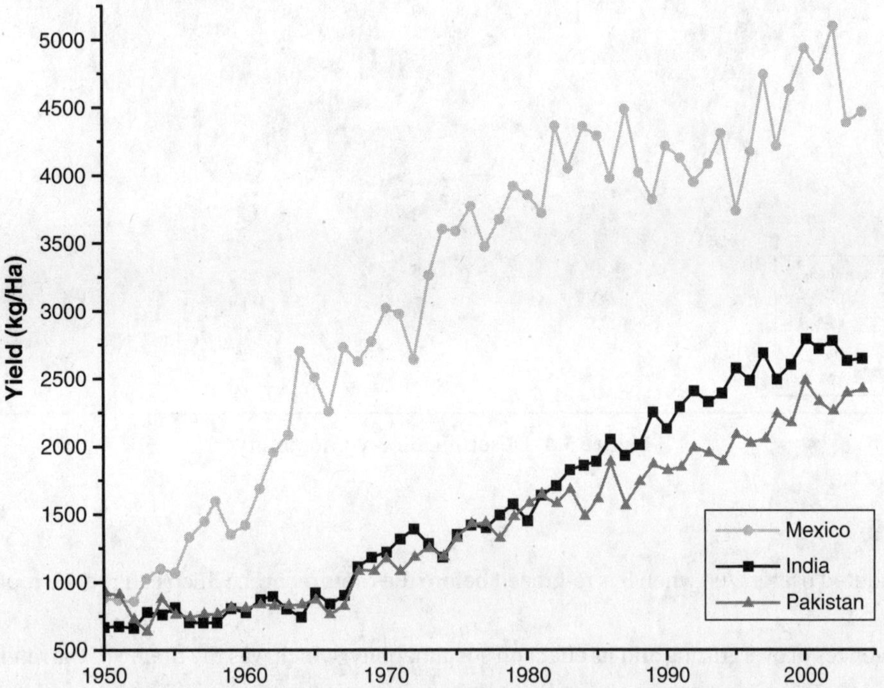

Figure 5.3 Wheat production in India, Mexico and Pakistan *(1950–2004) Source: United Nations Food and Agricultural Organization (FAO)*

- Created disease-resistant and high-yielding strains of food crops (e.g., wheat and rice) that could withstand harsh climates; e.g., wheat production increased 300% in Mexico.
- Pioneer in laying the groundwork for agricultural technological advances that helped alleviate world hunger.

5.4 Impacts of Agricultural Practices

Agricultural Productivity

Agricultural productivity implies greater output with less input—i.e., a more efficient use of scarce resources. As farms become more efficient, they are able to produce more products at a lower cost, which tends to stabilize food prices and make more food available to more people, which is vital for developing countries. For example, between 1950 and 2000, the amount of milk produced per cow rose an average of 250% and corn yields averaged an increase of 300%. However, increases in productivity are often linked with issues of sustainability, safety, pollution, and demands on finite resources.

Desertification

Desertification is the conversion of marginal rangeland or cropland to a more desert-like land type. It is often caused by overgrazing, soil erosion, prolonged drought, or climate change as well as the overuse of available resources, such as nutrients and water. Desertification proceeds with the following steps: First, overgrazing results in animals' eating all available plant life. Next, rain washes away the trampled soil, as there is no longer enough vegetation to hold back water. Wells, springs, and other sources of water dry up, and the vegetation that is left dies from drought or is taken for firewood. Then weeds that are unsuitable for grazing may begin to take over, and the ground becomes unfit for seed germination. Finally, wind and dry heat blow away the topsoil (Figure 5.4).

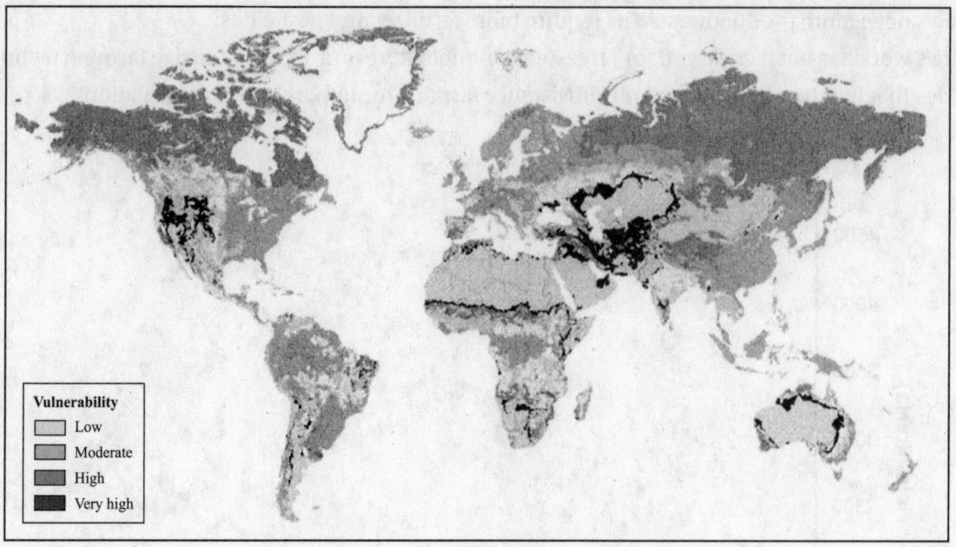

Figure 5.4 Desertification vulnerability

Overgrazing

A plant is considered overgrazed when it is re-grazed before the roots recover, which can reduce root growth by up to 90%.

The consequences of overgrazing and its effect on sustainability, which was the theme of Garrett Hardin's 1968 essay "The Tragedy of the Commons," occurs when plants are exposed to grazing for too long without sufficient recovery periods. When a plant is grazed severely, it uses energy stored in its roots to support regrowth. As this energy is used, the roots die back. The degree to which the roots die back depends on the severity of the grazing. Root dieback does add organic matter to the soil, which increases soil porosity, the infiltration rate of water, and the soil's moisture-holding capacity, and if sufficient time has passed, enough leaves will regrow and the roots will also regrow. Environmental consequences of overgrazing include the following:

- A decrease in soil porosity.
- Biodiversity decreases as a result of reducing native vegetation, which then leads to erosion.
- Desirable plants become stressed, while weedier species thrive, in these harsher conditions.
- Eutrophication increases due to cattle wastes' negatively impacting aquatic environments.
- Pastures become less productive.
- Predator–prey relationships and the balance achieved through predator-control programs are affected.
- Riparian areas are affected by cattle's destroying banks and streambeds, thereby increasing silting.
- Soils have less organic matter and become less fertile.
- The incidence of disease in native plants increases.
- The infiltration rate and moisture-holding capacity of the soil drops and susceptibility to soil compaction increases.

Reasons for Increases in Agricultural Productivity

Fertilizers

Fertilizers provide plants with the nutrients needed to grow healthy and strong. There are two main categories of fertilizer: inorganic and organic.

Inorganic Fertilizers

An inorganic fertilizer is a fertilizer mined from mineral deposits or manufactured from synthetic compounds.

Advantages of using inorganic fertilizers include the following:

- Controlled release—inorganic fertilizers accurately dispense plant nutrients over an extended period from 3 to 8 months, eliminating the need for repeated fertilizer applications throughout the growing season.
- Easier to use because inorganic fertilizers are more concentrated, which allows them to be applied with precise concentrations (e.g., dilution) that may be required.
- Fast acting as the nutrient-rich salts dissolve quickly and are immediately available to the plants.
- Lower cost as inorganic fertilizers are mass produced in factories.
- Precise content—plant nutrients are provided as a percentage by weight (N-P-K); e.g., 12-10-10 on a bag represents 12% nitrogen, 10% phosphorus, and 10% potassium.
- Soil organisms are not required to break down organic matter to release nutrients.

Disadvantages of inorganic fertilizers include the following:

- Consists primarily of only three of the 13 nutrients that plants need (nitrogen, phosphorus, and potassium).
- "Leached" chemicals are then sources of pollution for surface and subsurface water supplies.
- Over time, soils treated with only inorganic fertilizers lose organic matter and the all-important living organisms that help to build a quality soil. As the soil structure declines and the water-holding capacity diminishes, more and more of the chemical fertilizer applied will leach through the soil as water percolates downward, requiring ever-increasing amounts of chemicals to stimulate plant growth.
- Soluble inorganic chemical fertilizers contain mineral salts that plant roots can absorb quickly; however, these salts do not provide a food source for soil microorganisms or earthworms.

Organic Fertilizers

Any fertilizer that originates from an organic source, such as bone meal, compost, fish extracts, manure, or seaweed, is considered to be organic.

Pros	Cons
They add organic matter and humus to the soil, which provides a constant and balanced supply of nutrients and improves root growth.	They are often more expensive per unit than inorganic fertilizers.
They feed and sustain beneficial organisms that live in the soil, which inorganic fertilizers often destroy by increasing acid levels in the soil.	They are labor-intensive, as organic fertilizers can be bulky.
They improve moisture retention in the soil, making leaching less likely. In addition, ingredients in organic fertilizers are naturally biodegradable.	They have limited nutrient availability, as the release of nutrients from organic fertilizers can be dependent on both climate and the presence of microorganisms in the soil.
They improve the quality and structure of the soil, which improves water retention and drainage.	They are potentially pathogenic, as incomplete composting can leave certain pathogens in the organic matter, which can enter the water system or the food crops, causing human health and environmental problems.

Genetically Modified Organisms (GMOs)

Genetically modified foods are foods produced from organisms (both animal and plant) that have had changes introduced into their DNA.

Genetic engineering techniques allow for the introduction of new traits as well as greater control over traits when compared to previous methods; e.g., selective breeding. It is estimated that up to 80% of foods in grocery stores contain ingredients that come from genetically modified crops.

Pros	Cons
Faster growth = greater productivity = greater profit	GMO seeds are usually sterile, requiring farmers to purchase new seeds each year.
GMO crops lower the price of food and increase nutritional content, helping to alleviate world hunger.	GMO foods have not been proven safe (so far) for human consumption through long-term human clinical trials.
Growing GMO crops leads to environmental benefits such as reduced pesticide use, less water waste, and lower carbon emissions.	Modifying the genetic makeup of organisms may result in changes to the food supply that introduce toxins or trigger allergic reactions.
Higher crop yields	Less biodiversity
Less exposure to pesticides and herbicides	May result in pesticide-resistant strains
Less spoilage	May harm beneficial insects
May be able to grow crops in poorer-quality soil	May pose allergen risks
May require less water and fertilizer	May result in harmful mutations
More resistant to disease	Unknown ecological effects

Rangelands

Rangelands are native grasslands, woodlands, wetlands, and deserts that are grazed by domestic livestock or wild animals and make up about 40% of the landmass of the United States (~80% of the land in the western United States); they are the dominant type of land in arid and semiarid regions (Figure 5.5).

Rangelands are managed through livestock grazing and prescribed fire rather than more intensive agricultural practices of seeding, irrigation, and the use of fertilizers. Currently, livestock raised for meat on rangelands uses around 30% of global ice-free land and 8% of global freshwater, while producing about 18% of global greenhouse gas emissions.

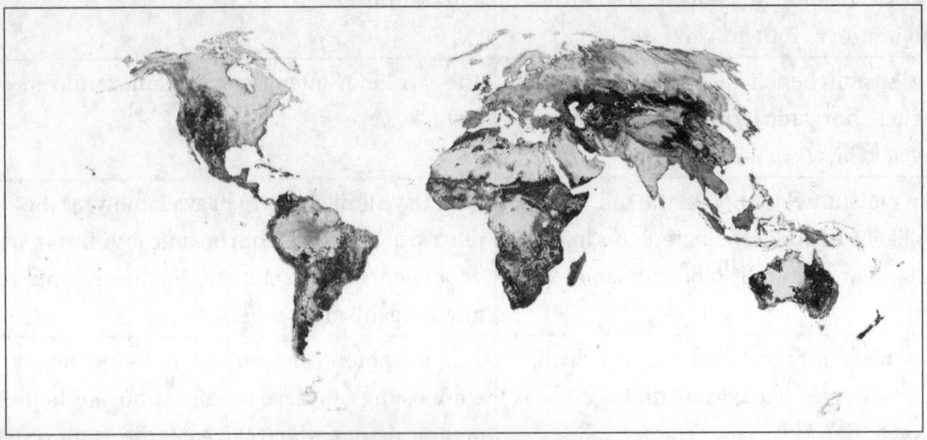

Figure 5.5 World rangelands
Source: National Aeronautics and Space Administration

Methods of rangeland management include the following:

- Controlling the number and distribution of livestock so that the carrying capacity is not exceeded
- Fencing off riparian (stream) areas to reduce damage to these sensitive areas
- Locating water holes, water tanks, and salt blocks at strategic points that do not degrade the environment
- Moving livestock from one area to another to allow the rangeland to recover
- Providing supplemental feed at selected sites
- Replanting barren rangeland with native grass seed to reduce soil erosion
- Restoring degraded rangeland
- Suppressing the growth of invasive plant species

Land administered by the Bureau of Land Management (BLM) is inhabited by 219 endangered species of wildlife, and livestock grazing is the fifth-rated threat to endangered plant species, the fourth leading threat for all endangered wildlife, and the number one threat to all endangered species in arid regions of the United States.

Slash and Burn

Slash-and-burn agriculture is a widely used method of growing food or clearing land in which wild or forested land is clear-cut and any remaining vegetation is burned.

Environmental consequences of slash-and-burn practices include the following:

- An increase in air pollution, including increasing the amount of atmospheric carbon dioxide, which leads to global warming
- An increase in dust storms
- An increase in soil erosion
- An increase in water pollution
- Loss of habitat for native species
- The resulting layer of ash resulting from slash and burn provides the newly cleared land with a nutrient-rich layer to help fertilize crops; however, the land is only fertile for a couple of years before the nutrients are depleted, which requires farmers to abandon the land and move to a new plot and clear more wild or forested land

Soil Erosion—Soil Degradation

Soil erosion is the movement of weathered rock or soil components from one place to another and is caused by flowing water, wind, and human activity (e.g., cultivating inappropriate land, burning of native vegetation, and deforestation). Soil erosion destroys the soil profile, decreases the water-holding capacity of the soil, and increases soil compaction. Because water cannot percolate through the soil, it runs off the land, taking more soil with it (positive feedback loop), and because the soil cannot hold water, crops grown in areas of soil erosion frequently suffer from water shortages. In areas of low precipitation, erosion can lead to significant droughts.

Poor agricultural techniques that lead to soil erosion include the following:

- Improper plowing of the soil
- Monoculture
- Overgrazing
- Removing crop wastes instead of plowing the organic material back into the soil
- Row cropping

Soil degradation is the decline in soil condition caused by its improper use or poor management, usually for agricultural, industrial, or urban purposes. Soils are a fundamental natural resource, and are the basis for all terrestrial life.

Soil Degradation

Desertification	Salinization	Waterlogging
DEFINITION: Productive potential of arid or semiarid land falls by at least 10% due to human activity and/or climate change.	**DEFINITION:** Water that is not absorbed into the soil evaporates, leaving behind dissolved salts in topsoil.	**DEFINITION:** Saturation of soil with water, resulting in a rise in the water table.
SYMPTOMS: Loss of native vegetation; increased wind erosion; salinization; drop in water table; reduced surface water supply.	**SYMPTOMS:** Stunted crop growth; lower yield; eventual destruction of plant life.	**SYMPTOMS:** Saline water envelops deep roots, killing plants; lowers productivity; eventual destruction of plant life.
REMEDIATION: Reduce overgrazing; reduce deforestation; reduce destructive forms of planting, irrigation, and mining. Plant trees and grasses to hold soil.	**REMEDIATION:** Irrigate or flush with freshwater to leach out salts; use drip irrigation, which requires less water; do not plant crops that require large amounts of water in areas prone to salinization; plant crops that remove salts from the soil (e.g., saltbush, barley, oats); improve drainage so water does not "sit" in one location; use organic or salt-free fertilizer; add humus to the soil; plant vegetation in coastal areas that would help block water surges resulting from storms and tsunamis.	**REMEDIATION:** Switch to less water-demanding plants in areas susceptible to waterlogging; utilize conservation-tillage farming; plant waterlogging-resistant trees with deep roots; take land out of production for a while; and/or install pumping stations with drainage pipes that lead to catchment-evaporation basins.

Tillage

Tillage is an agricultural method in which the surface is plowed and broken up to expose the soil, which is then smoothed and planted. Tillage is used for the following:

- Activating pesticides
- Burying heavy crop residue
- Incorporating manure and fertilizer into the root zone
- Leveling the soil
- Seedbed preparation
- Soil aeration
- Turning over cover crops
- Weed suppression

Environmental consequences of tillage include the following:

- Accelerating surface runoff, which affects soil erosion; e.g., reducing crop residue increases the impact of heavy rains, resulting in poor water infiltration
- Affecting soil organisms

- Deterioration in surface water quality due to increased concentration of sediment; e.g., sediments transport nitrogen and phosphorus from fields into lakes and streams, resulting in eutrophication
- Exposing the land to water and wind erosion
- Hardpan (a compacted and often clay-ey layer in soil that is impenetrable by roots) may develop.
- Topsoil loss exceeds the rate of topsoil replacement, resulting in a decrease in soil fertility as a result of the loss of organic matter and nutrient loss (e.g., ~half of plant-available phosphorus and ~all of plant-available potassium is concentrated in topsoil).

5.5 Irrigation Methods

Irrigation is the application of controlled amounts of water to plants at needed intervals and has been a necessary component of agriculture for over 5,000 years. Irrigation helps grow agricultural crops, maintain landscapes, and provide water for crops during periods of less than average rainfall.

Seventy-five percent of all freshwater used on Earth is used for agriculture, but approximately 75% of that water is lost through evaporation, leakage, and seepage. There are five basic types of irrigation used worldwide today. Refer to the chart on the following page.

Ditch	Ditches are dug and seedlings are planted in rows. The plantings are watered by placing canals or furrows in between the rows of plants. Siphon tubes are used to move the water from the main ditch to the canals.
Drip	Water is delivered at the root zone of a plant through small tubes that drip water at a measured rate.
Flood	Water is pumped or brought to the fields and is allowed to flow along the ground among the crops. Being simple and inexpensive, it is the method most widely used in less-developed countries.
Furrow (Channel)	Small parallel channels are dug along the field length in the direction of predominant slope. Water is applied to the top end of each furrow and flows down the field under the influence of gravity, which can result in more water infiltrating into the ground at the beginning of the furrow and less toward the end.
Spray	Uses overhead sprinklers, sprays, or guns to spray water onto crops.

Drip Irrigation

Since 2000, the area under drip irrigation has increased by 650%, with the most dramatic gains having occurred in China and India. However, most farmers in the world today still irrigate by flooding their fields or running water down furrows between their rows of crops, with less than half the water applied to the field actually irrigating a crop due to waste and inefficiency. Compared with conventional flood irrigation, drip methods can reduce the volume of water applied to fields by up to 70%, while increasing crop yields up to 90%. Keep in mind the following:

- Almost 20% of the world's farmland that is irrigated yields 40% of the world's food supply.
- Currently, California accounts for over 60% of the United States' agricultural land using drip irrigation.
- Less than 4% of the world's irrigated land utilizes drip irrigation.
- Most drip irrigation systems are used for fruits, vegetables, and other high-value crops that can provide a good return on investment.

Advantages/Disadvantages of Various Irrigation Techniques

Advantages	Type	Disadvantages
Very efficient, minimizes water loss due to evaporation as water directly reaches root zone, reduces weed growth, can be installed on difficult terrain (e.g., hillsides).	**Drip**	Requires precise installation, initial setup costs are high, requires constant maintenance (e.g., debris and mineral buildup).
Covers large areas, can be used on difficult terrain.	**Sprinkler**	Generally high initial costs, wind can scatter spray, evaporative loss, which adds to cost of water.
Inexpensive to install, land must be properly sloped.	**Surface**	Changes to land (e.g., plowing) may require re-sloping, uses large amounts of water, land subject to waterlogging

OGALLALA AQUIFER

The Ogallala Aquifer is a shallow water table aquifer surrounded by clay, gravel, sand, and silt located beneath the Great Plains of the United States. One of the world's largest aquifers, it occupies ~174,000 square miles (450,000 km^2) in the central United States.

During the 1940s, large-scale extraction of water from the aquifer was used for agricultural purposes. Today about 27% of the irrigated land in the entire United States lies over the aquifer, which yields about 30% of the groundwater used for irrigation in the United States. Currently, the aquifer is at risk for over-extraction and pollution, and it will take over 6,000 years to replenish naturally through rainfall.

5.6 Pest Control Methods

Controlling Pests

Pesticides can be used to control pests, but their use has drawbacks. Integrated Pest Management (IPM) is an ecologically based approach to control pests.

Types of Pesticides

Pesticides differ in several ways, including their chemistry, how long they remain effective in the environment (environmental persistence), their effect on the food web (bioaccumulation and biomagnification), what types of organism are affected, how the pesticides work (e.g., nervous system, reproductive cycles, blood chemistry), how fast they work, and their application.

Biological
Living organisms can be used to control pests. Examples include bacteria (e.g., Bt [*Bacillus thuringiensis*]), ladybugs, parasitic wasps, and certain viruses. The sterile insect technique is another method of biological control, whereby overwhelming numbers of sterile insects are released into the wild; the sterile males then compete with wild males to mate, ultimately reducing the next generation's population.

Carbamates

Carbamates, also known as urethanes, affect the nervous system of pests, which results in the swelling of tissue in the pest. One hundred grams of a carbamate have the same effect as 2,000 grams of a chlorinated hydrocarbon such as DDT. Carbamates are more water soluble than chlorinated hydrocarbons, which brings a greater risk of their being dissolved in surface water and percolating into groundwater.

CASE STUDY

On December 2, 1984, at the Union Carbide pesticide plant in Bhopal, India, a leak of chemicals from a plant manufacturing a carbamate pesticide occurred, which resulted in the exposure of hundreds of thousands of people. Roughly 8,000 people have since died from gas-related diseases, with 200,000 people having permanent injuries.

Fumigants

Fumigants are used to sterilize soil and prevent pest infestation of stored grain.

- **INORGANIC** pesticides are broad-based pesticides that include arsenic, copper, lead, and mercury. They are highly toxic and accumulate in the environment.
- **ORGANIC OR NATURAL** pesticides are natural poisons derived from plants such as tobacco or chrysanthemum.
- **ORGANOPHOSPHATES** are extremely toxic but remain in the environment for only a brief time. Examples include malathion and parathion to control mosquitoes and West Nile virus.

Persistent Organic Pollutants (POPS)

Persistent organic pollutants (POPS) are an organic (containing carbon) group of chemical compounds that do not break down through chemical or biological processes and which are able to pass through and accumulate in the fatty tissues of living organisms, and as such, POPS biomagnify (become more concentrated) in food pyramids. Studies have linked POP exposures to declines, diseases, or abnormalities in a number of wildlife species, including certain kinds of birds, fish, and mammals. An example of a POP is DDT—a pesticide.

Costs and Benefits of Pesticide Use

Despite the use of pesticides, many pests worldwide have increased in numbers due to the following:

- Genetic resistance
- Increased mobility of pests due to increased world trade
- Reduced crop rotation
- Reduction in crop diversity

Pros and Cons of Pesticides

Pros	Cons
Agriculture is more profitable.	Can accumulate in food chains.
Increase the food supply.	Estimates range from $5 to $10 in damage done to the environment for every $1 spent on pesticides. Pesticides are expensive to purchase and apply.
Kill unwanted pests that carry disease.	External costs of pesticide use may be high (e.g., healthcare [birth defects, cancer, and lower sperm count]).
More food means that food is less expensive (Law of Supply and Demand).	Inefficiency—only 5% of pesticides reach a pest.
Newer pesticides are safer, more specific (narrow spectrum), and less persistent (biodegradable).	Pesticide runoff and its effects on aquatic environments through biomagnification.
Reduce labor costs.	Pests develop resistance and create a pesticide treadmill.
Federal regulations provide mandatory guidelines for use and testing.	Threaten endangered species and pollinators; also affect human health.

The Pesticide Treadmill

Pesticide resistance describes the decreased susceptibility of a pest population to a pesticide that was previously effective at controlling the pest. Pest species evolve pesticide resistance via natural selection (i.e., the most resistant individuals survive and pass on their genetic traits to their offspring). In response to resistance, farmers may increase pesticide quantities and/or the frequency of pesticide applications, which magnifies the problem. In addition, some pesticides are toxic toward species that feed on or compete with pests, which can allow the pest population to expand, requiring more pesticides ("pesticide treadmill"). As a result, farmers progressively pay more for less and less benefit.

RELEVANT LAWS

DELANEY CLAUSE OF THE FOOD, DRUG, AND COSMETIC ACT (1958): Prohibits the U.S. Food and Drug Administration from approving any food additive found to cause cancer in animals or humans; however, currently it applies to only ~400 of the ~2,700 substances that are added to foods, many of which are determined to be "Generally Recognized as Safe" (GRAS). If any GRAS substance is found to be carcinogenic, it no longer is classified as GRAS and then becomes subject to the Delaney Clause.

5.7 Meat Production Methods

Concentrated Animal Feeding Operations (CAFOs)

Water and Air Quality

A concentrated animal feeding operation is an intensive animal feeding operation in which large numbers of animals are confined in feeding pens for over 45 days a year; e.g., 125,000+ chickens. The large amounts of animal waste from CAFOs present a risk to water quality and aquatic ecosystems. States with high concentrations of

CAFOs experience on average 20 to 30 serious water-quality problems per year as a result of manure management issues. Manure discharge from CAFOs can negatively impact water quality. Pollutants associated with CAFO waste principally include the following:

- Antibiotics
- Nitrogen and phosphorus, collectively known as nutrient pollution
- Odorous/volatile compounds such as ammonia (NH_3), carbon dioxide (CO_2), hydrogen sulfide (H_2S), and methane (CH_4)
- Organic matter (organic compounds that come from remains of organisms such as plants, animals, and their waste products in the environment)
- Pathogens (disease-causing organisms such as bacteria and viruses)
- Pesticides and hormones
- Salts
- Solids, including the manure itself and other elements mixed with it, such as spilled feed, bedding and litter materials, hair, feathers, and animal corpses
- Trace elements; e.g., arsenic

The two main contributors to water pollution caused by CAFOs are soluble nitrogen compounds and phosphorus. The eutrophication of water bodies from such waste is harmful to wildlife and water quality in aquatic systems like streams, lakes, and oceans.

Because groundwater and surface water are closely linked, water pollution from CAFOs can affect both sources if one or the other is contaminated. Surface water may be polluted by CAFO waste through the runoff of nutrients, organics, and pathogens from fields and storage and waste can be transmitted to groundwater through the leaching of pollutants.

CAFOs release several types of gas emissions—ammonia, hydrogen sulfide, methane, and particulate matter—all of which have varying human health risks. The primary cause of gas emissions from CAFOs is the decomposition of animal manure being stored in large quantities. Additionally, CAFOs emit strains of antibiotic resistant bacteria into the air.

Globally, ruminant livestock (e.g., cows) are responsible for about 35% of anthropogenic greenhouse gas emissions released per year, whereas in the United States, livestock are responsible for about 18% of greenhouse gas emissions globally and over 7% of greenhouse gas emissions through gases produced by digestion. Methane is the second most concentrated greenhouse gas contributing to global climate change, with livestock contributing nearly 30% of anthropogenic methane emissions. If no changes are made to the operations of CAFOs, global methane production is predicted to increase ~60% by 2030.

5.8 Impacts of Overfishing

Fishing is an important industry that is under pressure from growing demand and falling supply. Globally, fish provide more than 1.5 billion people with almost 20% of their average per capita intake of animal protein and 3 billion people with at least 15% of their protein intake. The global wild fish catch per person per year has dropped 30% since its peak in 1988, while, during the same period of time, the output from fish farming has increased 95%.

Overfishing

Marine life, including fisheries, as well as terrestrial life, depends upon primary producers. Aquatic plants require sunlight and are therefore largely restricted to shallow coastal waters, which make up less than 10% of the world's ocean area yet contain 90% of all marine species. Phytoplankton (the aggregate of plants and plantlike organisms in plankton) form the base of the marine food web and are responsible for approximately half of global carbon dioxide fixation, which is vital in controlling atmospheric CO_2 concentration and Earth's climate. However, the vast majority of marine surface waters are depleted in inorganic nutrients, including nitrogen, phosphorus, iron, and silica, shortages of which limit primary (plant) production, which in turn limits higher trophic marine animal populations, e.g., fish.

The world's oceans supply approximately 1% of all human food and represent approximately 10% of the world's protein sources, with China being responsible for the majority of ocean fish harvesting. About one-third of the total catch is used for purposes other than human consumption (e.g., fish oil, fish meal, animal feed) and about another third is classified as bycatch, including marine mammals, and other marine animals, such as sea turtles, birds, noncommercial fish, and shellfish, that are discarded because they were not the intended target of the fishing activity.

Methods to Manage Marine Fishing

- Eliminate government subsidies for commercial fishing.
- Increase the number of marine sanctuaries.
- Prevent importation of fish products from countries that do not adhere to sustainable fishing practices.
- Require and enforce labeling of fish products that were raised or caught according to sustainable methods.
- Require fishing licenses and open inspections, which limit the number and kind of fish caught per year, and trade sanctions should these limits be exceeded.

Methods to Restore Freshwater Fish Food Webs

- Control erosion.
- Control invasive species.
- Create or restore fish passages.
- Enforce laws that protect coastal estuaries and wetlands.
- Plant native vegetation on stream banks.

5.9 Impacts of Mining

The following table provides an overview of mining.

Steps	Descriptions	Environmental Effects and Issues
Mining	Removing mineral resource from the ground. Can involve underground mines, drilling, room-and-pillar mining, long-wall mining, open pit, dredging, contour strip mining, and mountaintop removal.	Mine wastes—acids and toxins. Displacement of native species. Reclamation of land and recycling.
Processing	Removing ore from gangue.	Pollution (air, water, soil, and noise). Involves transportation, processing, purification, smelting, and manufacturing. Human health concerns, risks, and hazards.
Use	Involves distribution to end user.	Air pollution involved in transporting final products to end user.

TYPES OF MINING

Type of Mining	Description
Surface	
Contour mining	Removing overburden from the seam in a pattern following the contours along a ridge or around a hillside
Dredging	A method for mining below the water table and usually associated with gold mining. Small dredges use suction or scoops to bring the mined material up from the bottom of a body of water.
In situ	Small holes are drilled into the Earth and toxic chemical solvents are injected to extract the resource.
Mountaintop removal	Removal of mountaintops to expose coal seams and disposing of associated mining overburden in adjacent "valley fills"
Open pit	Extracting rock or minerals from the Earth by their removal from an open pit when deposits of commercially useful ore or rocks are found near the surface
Strip mining	Exposes coal by removing the soil above each coal seam

Underground	
Blast	Uses explosives to break up the seam, after which the material is loaded onto conveyors and transported to a processing center
Longwall	Uses a rotating drum with "teeth," which is pulled back and forth across a coal seam—the material then breaks loose and is transported to the surface
Room and pillar	Approximately half of the coal is left in place as pillars to support the roof of the active mining area. Later, the pillars are removed and the mine collapses.

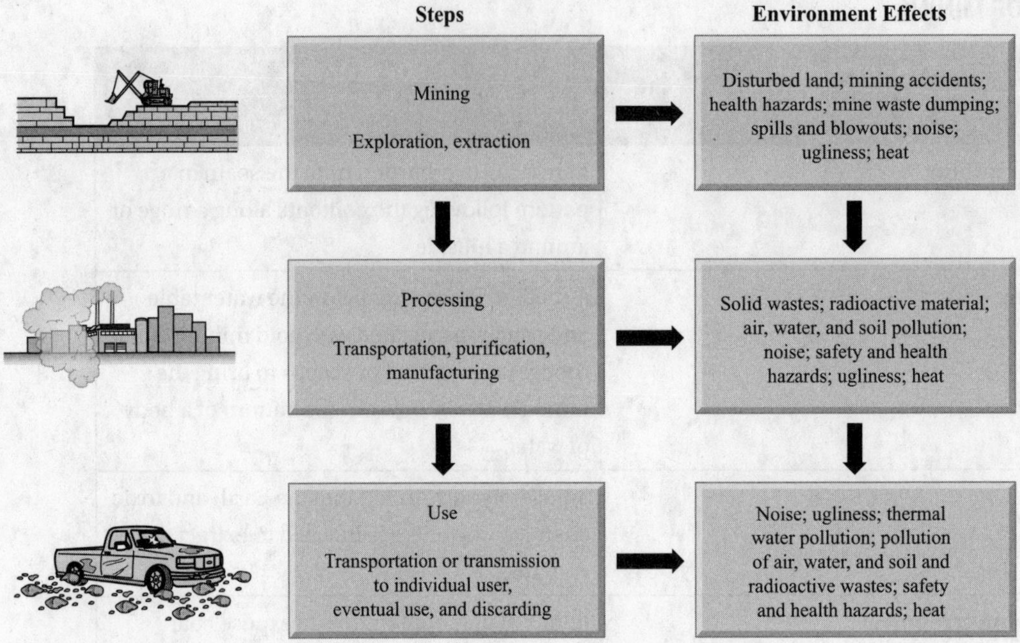

Figure 5.6 Steps required for manufacturing mining products and their environmental consequences

Environmental Damage from Mining

Environmental damage caused by mining includes the following:

- Acid mine drainage
- Chemicals from in situ leaching entering the water table
- Disruption of natural habitats
- Disruption of soil microorganisms and, consequently, nutrient cycling processes
- Dust released during the breakup of materials, causing lung problems and posing other health risks
- Land subsidence
- Large consumption and release of water

Mitigation of environmental damage caused by mining can take the following forms:

- Adding new topsoil and/or soil nutrients to improve soil fertility and soil structure
- Constructing wetlands for bioremediation
- Neutralizing the soil with limestone ($CaCO_3$) or lime (CaO)
- Recontouring the land to its former condition
- Replanting the area with fast-growing native vegetation
- Using sulfate-reducing bacteria (acidophiles) in catchment basin

5.10 Impacts of Urbanization

Urbanization

Urbanization refers to the movement of people from rural areas to cities and the changes that accompany it. Areas that are experiencing the greatest growth in urbanization are countries in Asia and Africa. Asia alone has close to half of the world's urban inhabitants even though 60% of its population still lives in rural areas. Africa, which is generally considered overwhelmingly rural, now has a larger urban population than North America. Reasons for this include access to jobs, higher standards of living, easier access to healthcare, mechanization of agriculture, and access to education. Nations with the most rapid increases in their urbanization rates are generally those with the most rapid economic growth (Figure 5.7).

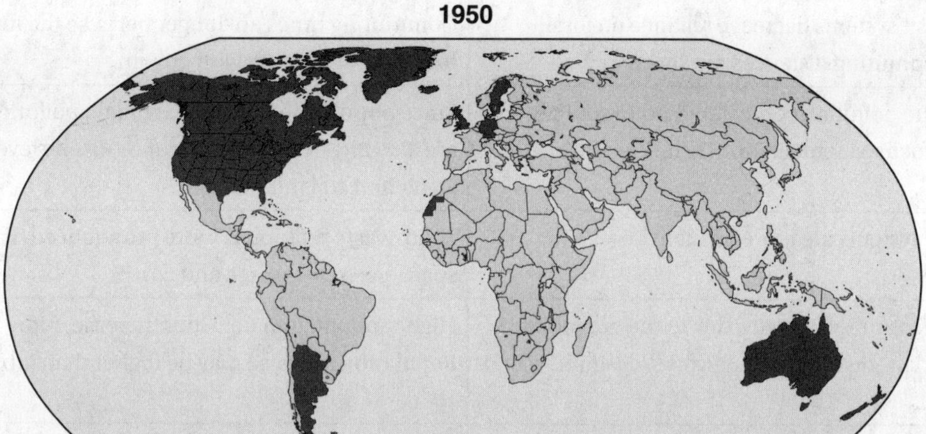

Figure 5.7 Projected worldwide urban growth from 1950 to 2030

Urbanization

Pros	Cons
Better educational delivery system.	Overcrowded schools.
Better sanitation systems.	Sanitation systems have greater volumes of wastes to deal with.
Large numbers of people generate high tax revenues.	Large numbers of poor people place strains on social services. This results in wealthier people's moving away from urban areas into suburbs and decreasing the tax base.
Mass transit systems decrease reliance on fossil fuels—commuting distances are shorter.	Commuting times are longer because the infrastructure cannot keep up with growth.
Much of the pollution comes from point sources, enabling focused remediation techniques.	Since population densities are high, pollution levels are also high (urban heat islands, ozone levels, and water and soil pollution).
Recycling systems are more efficient.	Solid-waste buildup is more pronounced. Landfill space becomes scarce and costly.
Urban areas attract industry due to the availability of raw materials, distribution networks, customers, and labor pool.	Higher population densities increase crime rates. Population increase may be higher than job growth.
Uses less land—less impact on the environment.	Impact on land is more concentrated and more pronounced. Examples include water runoff and flooding.

Urban (Suburban) Sprawl

Urban (or suburban) sprawl describes the expansion of human populations away from central urban areas into low-density and usually car-dependent communities. Reasons for urban sprawl include an increased need for and reliance on cars, higher family incomes (both parents working), tax advantages for buying a home, companies moving and/or expanding where there is more land available, with workers following, perception of better schools and lower crime, and aesthetics (more green lawns and less blight). Characteristics of urban sprawl include the following:

- Job sprawl—Job sprawl is characterized by low-density, geographically spread-out patterns of employment, where the majority of jobs in a given metropolitan area are located outside of the main city's central business district and increasingly in the suburbs.
- Land use conversion—Land for urban sprawl is often taken from agricultural lands, which are/were often located immediately surrounding cities. The extent of modern sprawl has consumed a large amount of the most productive agricultural land as well as forest, and other wilderness, areas.
- Leap frog development—Developments are typically separated by large greenbelts (i.e., tracts of undeveloped land), resulting in even lower population densities.
- Low-density housing—Most housing consists of single-family homes built on large lots characterized by having fewer stories than homes found in the city, spaced farther apart, and separated by lawns, landscaping, and/or roads.
- Single-use development—Commercial, residential, institutional, and industrial areas are separated from one another. As a result, the places where people live, work, shop, and seek recreation are far from one another and generally require a car.

Environmental consequences of urban sprawl may include the following:

- An increase in air temperatures because of the "heat island" effect
- Decreases in natural areas and forests
- More surface flooding because of more impervious, paved surfaces
- Water pollution increases because of urban runoff and rainwater picking up gasoline, motor oil, heavy metals, and other pollutants in runoff from parking lots and roads

Smart Growth

Smart growth is an urban planning and transportation plan designed to slow urban sprawl and concentrate growth in compact, walkable "urban villages" and advocates compact, transit-oriented, walkable, bicycle-friendly land use, neighborhood schools, and mixed-use development with a range of housing choices. Smart growth values long-range, regional considerations of sustainability. Sustainable development goals achieve a unique sense of community and place, and expand the range of transportation, employment, and housing choices, while equitably distributing the costs and benefits of development, preserving and enhancing natural and cultural resources, and promoting public health. Sustainable development strategies include the following:

- Adopting mixed-use planning—blending a combination of residential, commercial, cultural, institutional, and/or industrial uses in a specific location
- Developing greenbelts and other areas of largely undeveloped, wild, or agricultural land surrounding urban areas
- Providing property tax incentives to companies that locate in urban centers
- Providing subsidies for mass transit systems and riders
- Reducing urban blight by replacing abandoned buildings with green, open spaces

Urban or Planned Development

Urban development is the process of designing and shaping the physical features of cities and towns (e.g., blending commercial, industrial, and residential areas with mass transportation options) with the goal of making urban areas more attractive, functional, and sustainable.

There are more than 76 million residential buildings and approximately 5 million commercial buildings in the United States alone. Together, these buildings use one-third of all the energy and two-thirds of all the electricity consumed in the United States. Energy needs of buildings account for almost half of the sulfur dioxide emissions, one-fourth of the nitrous oxide emissions, and one-third of the carbon dioxide emissions. Some urban development strategies include the following:

- Conserving energy through government and private industry rebates and tax incentives for solar and other less-polluting forms of energy
- Conserving water through the use of xeriscaping (landscaping and gardening that reduces or eliminates the need for supplemental water from irrigation)
- Improving indoor air quality
- Placing buildings whenever possible near public transportation hubs that use a multitude of connections, such as light rail, subways, and park and rides
- Preserving historical and cultural aspects of the community while at the same time blending into the natural feeling and aesthetics of a community
- Using designs that minimize waste while utilizing recycled materials
- Using resource-efficient building techniques and materials

5.11 Ecological Footprints

An "ecological footprint" is a measure of human demand on Earth's ecosystems and is a standardized measure of demand for natural capital (world's stocks of natural assets, which include air, geology, soil, water, and all living things) that may be contrasted with the planet's ecological capacity to regenerate. Ecological footprints represent the amount of biologically productive land and sea area that is necessary to supply the resources a human population consumes, and to assimilate associated waste.

The figure below (Figure 5.8) shows the ecological footprints of various countries on Earth measured in global hectares per person. One global hectare represents the average productivity of all biologically productive areas measured in hectares on Earth in a given year. 1 ha = 2.5 acres = 10,000 m^2.

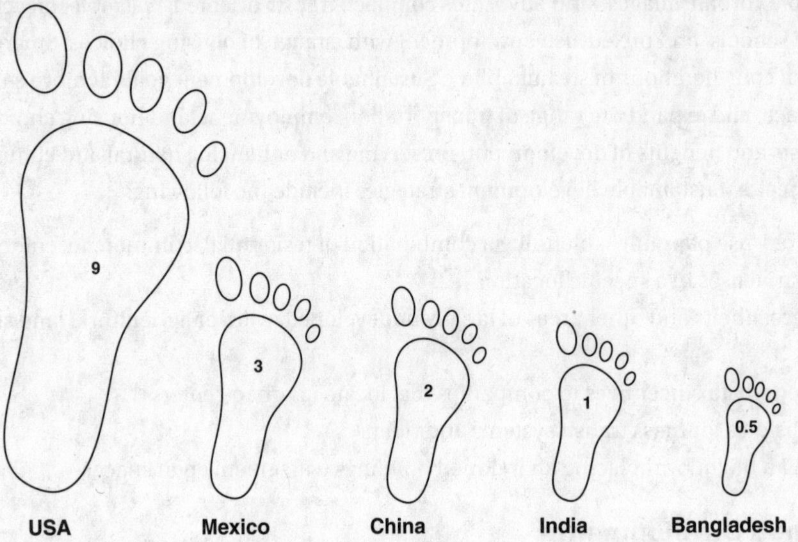

Figure 5.8 Ecological footprints of various countries on Earth measured in global hectares per person

5.12 Introduction to Sustainability

Healthy ecosystems and environments are necessary to the survival of all other living organisms. Sustainability refers to the capacity for the biosphere and human civilization to coexist through the balance of resources within their environment—in other words, to ensure that available resources are never depleted faster than those resources can be replaced. The impact on Earth systems is primarily due to the destruction of natural resources and the Earth's ecosystems by mankind.

One of the early attempts to understand the magnitude of human impact on finite natural resources (I) was developed in the 1970s and is known as the IPAT formula, which attempts to explain human consumption in terms of three variables: human population (P), levels of consumption (A for "affluence"), and the impact humans have on resources, symbolized by T (for "technology"):

$$I = P \times A \times T$$

It is estimated that by the year 2100, humans will have an environmental impact on nature seven times greater than they did in the year 2010. It is the combination of populations increasing in the developing world and unsustainable consumption levels in the developed world that poses a bleak outlook for sustainability.

The general trend is for higher standards of living to become less sustainable. Currently, each human on Earth uses ~2.8 global hectares of Earth resources per person worldwide, which is ~75% more than the 1.6 global hectares currently available per person. This deficit is being met by extracting renewable resources (e.g., forests, fisheries, clean water, etc.) faster than they can be replaced and extracting nonrenewable resources (e.g., fossil fuels, minerals, etc.) created in the distant past.

Solutions to the environmental problems created by unsustainable resource use and pollution production will require humanity to make transformative changes, including, but not limited to, sustainable agricultural practices, reducing consumption and waste, creating universal fishing quotas, and collaborative water management.

Threats to Sustainability

In 2009 scientists from the Stockholm Resilience Center and the Australian National University described nine planetary "sustainability boundaries." A portion of their results are shown in the table below.

Earth-System Process	Control Variable	Boundary Value	Current Value	Boundary Crossed	Preindustrial Value
Biodiversity loss	Extinction rate (per million per year)	10	> 100	yes	0.1–1
Climate change	Atmospheric carbon dioxide concentration (ppm by volume)	350	400	yes	280
Freshwater	Global human consumption of water (km^3/yr)	4,000	2,600	no	415
Land use	% land surface converted to cropland	15	11.7	no	low
Stratospheric ozone depletion	Dobson units	276	283	no	290

5.13 Methods to Reduce Urban Runoff

Urban runoff is surface runoff of rainwater created by urbanization. This runoff is a major source of urban flooding and water pollution in urban communities worldwide. Impervious surfaces, such as roads, rooftops, parking lots, and sidewalks, carry polluted storm water to storm drains instead of allowing the water to percolate through the soil, which causes a lowering of the water table (because groundwater recharge is lessened) and flooding, since the amount of water that remains on the surface is greater. Paved surfaces also may result in the following:

- An increase in algae growth due to the increased nutrients in the runoff
- An increase in groundwater depletion as water does not infiltrate into the soil to recharge aquifers
- An increase in the risk of infections and diseases through contaminated water supplies
- Creating microclimates due to the high heat capacity of asphalt, which causes heat to be more easily captured and stored longer than it would be in the normal landscape
- Fragmented habitats from urban runoff
- Reducing biodiversity and seriously impacting food webs in the area since there is less vegetation available for primary consumers. Most municipal storm sewer systems discharge untreated storm water to streams, rivers, and bays.

Urban runoff results in the following:

- Increased erosion and resulting sedimentation in the runoff with the sediments settling to the bottom of the water bodies and reservoirs, directly affecting the water quality and the storage capacity
- Increased temperature of the water in streams and waterways that significantly impacts fish and wildlife as heat is transferred from urban sources (asphalt, buildings, etc.)
- Runoff containing gasoline, motor oil, heavy metals, trash, fertilizers, and pesticides

Remediation steps include the following:

- Building constructed wetlands to naturally filter water before it's released into lakes, rivers, and oceans
- Building water retention-infiltration basins—shallow artificial ponds that are designed to infiltrate storm water through permeable soils into the groundwater aquifer
- Frequently using street-sweeping vacuums that can reduce the trash and other debris and pollutants that end up in runoff
- Planning and constructing more open green spaces and parks within urban communities to increase natural infiltration

5.14 Integrated Pest Management

Integrated pest management (IPM) is an ecological pest-control strategy that uses a combination of biological, chemical, and physical methods together or in succession and requires an understanding of the ecology and life cycle of pests. When used in combination, these methods can reduce or eliminate the use of traditional pesticides. The aim of IPM is not to eradicate pests but to control their numbers to acceptable levels. With IPM, chemical pesticides are the last resort.

Methods used in IPM include the following:

- Construction of mechanical controls such as traps, tillage, insect barriers, or agricultural vacuums equipped with lights
- Developing genetically modified crops that are more pest-resistant.
- Intercropping—a farming method that involves planting or growing more than one crop at the same time and on the same piece of land.
- Natural insect predators
- Planting pest-repellant crops
- Polyculture—the simultaneous cultivation or raising of several crops or types of animals
- Regular monitoring through visual inspection and traps followed by record keeping
- Releasing sterilized insects
- Rotating crops often to disrupt insect cycles
- Using mulch to control weeds
- Using pheromones or hormone interrupters
- Using pyrethroids or naturally occurring microorganisms (e.g., *Bt*)

When used effectively, IPM can reduce the following:

- Bioaccumulation and biomagnification of pesticides
- Pests' becoming resistant to a particular pesticide (genetic resistance—an inherited change in the genetic makeup of the pests that confers a selective survival advantage)
- The destruction of beneficial and non-targeted organisms (e.g., ladybugs, bees, and birds)

5.15 Sustainable Agriculture

Sustainable agricultural practices emphasize production and food systems that are profitable, environmentally sound, and energy efficient, and that improve the quality of life for both farmers and the public. Sustainable agriculture is based on enhancing the health of the land and rural communities and concentrating on long-term solutions rather than short-term treatment of symptoms. Examples of sustainable agricultural practices include the following:

- Developing ecologically-based pest management programs
- Diversifying farms to reduce economic risks
- Increasing energy efficiency in production and food distribution
- Integrating crop and livestock production
- Protecting the water quality
- Reducing or eliminating tillage in a manner that is consistent with effective weed control
- Rotating crops to enhance yields and facilitate pest management
- Using cover crops, green manure, and animal manure to build soil quality and fertility
- Using water and nutrients efficiently

Factors That Foster Sustainable Agriculture	Forces That Discourage the Adoption of Sustainable Agricultural Practices
Knowledge about the resources and processes provided by agriculture, such as clean water, soil conservation, and recreation (i.e., ecosystem services).	Agricultural subsidies that favor excessive production of a single commodity.
Policies or incentives that pay or reward producers for providing ecosystem services.	Political pressure to minimize environmental restrictions.
Policies that help alleviate pressure on marginal lands.	Consumers who are uninformed or wrongly informed about agricultural issues.
Public education to inform consumers and those involved in policy making about the environmental costs and benefits of alternative management practices.	Large populations seeking inexpensive food.
Understanding the impact of agricultural management practices on ecosystem services.	Economic incentives that reward growers for externalizing environmental costs to the rest of the society (e.g., policies that do not penalize water contamination due to pesticide runoff or soil erosion).

Soil Conversion Techniques and Sustainable Agriculture

- **Contour plowing**—plowing along the contours of the land in order to minimize soil erosion
- **No-till agriculture**—Soil is left undisturbed by tillage and the residue is left on the soil surface.
- **Planting perennial crops**—Perennials live for several years; e.g., fruit trees. Advantages of perennial crops over annual crops include the following:
  - Annual crops often are often harvested around the same time each year, whereas perennial crops can be planted at staggered times throughout the year.
  - Deeper root systems improve the structure of the soil as they create channels and spaces through which water can percolate, as well as pore spaces for aeration and less need for irrigation.
  - Energy is saved as there is no need for clearing areas for plowing, sowing seeds and composting, and mulching the soil to supply nutrients for plantings. Perennials also tend to need less care than annuals as they have established root systems.
  - Perennials can often be split into two or more separate plants, reducing the energy required for preparing for and planting new crops each year.
  - Perennial crops are also generally hardier than annuals and so are more likely to survive extreme weather events.
  - Perennial crops retain enough foliage to effectively become a cover crop, which protects the soil from evaporation by the sun and protects it from erosion by wind and rainfall, allowing for preserving nutrients to be available for plant growth.
- **Strip cropping**—cultivation in which different crops are sown in alternate strips
- **Terracing**—make or form (sloping land) into a number of level flat areas resembling a series of steps
- **Windbreaks**— rows of trees that provide shelter or protection from the wind

5.16 Aquaculture

Aquaculture, commonly known as mariculture or fish farming, includes the commercial growing of aquatic organisms for food and involves stocking, feeding, protecting from predators, and harvesting. Aquaculture is growing about 6% annually and provides 5% of the total food production worldwide, most of it coming from developing countries. Currently, the most popular products being produced through aquaculture include seaweeds, mussels, oysters, shrimp, and certain species of fish (primarily salmon, trout, and catfish). Kelp makes up about 17% of aquaculture output and is used as a food product and as a source of various products used in the food industry. Aquaculture is used to raise 80% of all mollusks, 40% of all shrimp, and 75% of all kelp.

Aquaculture offers several advantages over raising livestock in that cold-blooded organisms convert more feed to usable protein. For example, for every 1 million calories of feed required, a trout raised on a farm produces about 35 grams of protein whereas a chicken produces 15 grams of protein and cattle only produce 2 grams of protein.

For aquaculture to be profitable, the species must be marketable, inexpensive to raise, efficient at converting feed into fish biomass (achieving marketable size within 1 to 2 years), and disease resistant. Aquaculture creates dense monocultures that reduce biodiversity within habitats and requires and produces large levels of nutrients in the water that can be harmful to local water quality and other aquatic organisms.

Aquaculture offers possibilities for sustainable protein-rich food production and for economic development to local communities. However, aquaculture on an industrial scale may pose several threats to marine and coastal biological diversity, which creates wide-scale destruction and degradation of natural habitats and leaves nutrients and antibiotics in aquaculture wastes.

5.17 Sustainable Forestry

Steps that can be taken to reduce the effects of deforestation (Figure 5.9) include the following:

- Adopting uneven-aged forest management practices
- Educating farmers about sustainable forest practices and their advantages
- Growing timber on longer rotations
- Monitoring and enforcing timber-harvesting laws
- Reducing fragmentation in remaining large forests
- Reducing road building in forests
- Reducing or eliminating the practice of clear-cutting
- Relying on more sustainable tree-cutting methods, e.g., selective and strip cutting

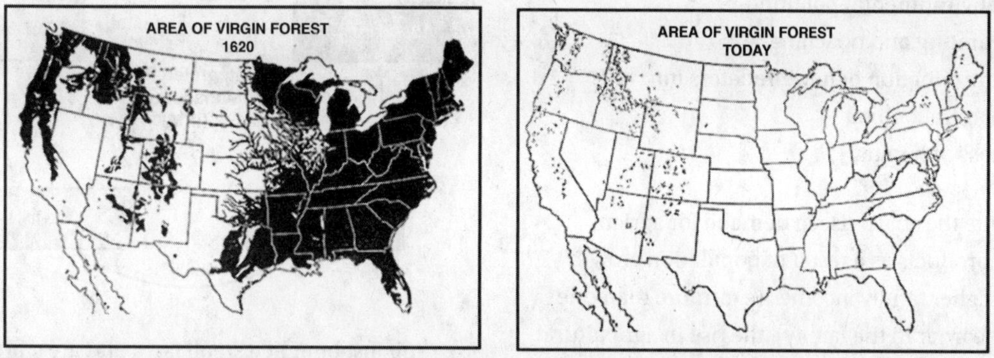

Figure 5.9 Deforestation in the United States (1620–present)

Practice Questions

1. Which of the following choices below is NOT consistent with the "First Green Revolution" that occurred between 1950 and 1970?

   (A) Genetically-modified organisms (GMOs)
   (B) Herbicides
   (C) Inorganic fertilizers
   (D) Pesticides

2. The greatest threat to the success of a species is

   (A) environmental pollution
   (B) hunting and poaching
   (C) introduction of new predators into the natural habitat
   (D) loss of habitat

3. Despite the sharp rise in demand for agricultural products as a result of population growth and higher family incomes (e.g., more than one wage earner in the family), the rise in agricultural productivity caused the average adjusted world price that farmers receive for their agricultural products to fall between 1900 and 2010.

   Which of the following statements are/is TRUE?

   I. Certain areas of the world (e.g., sub-Saharan Africa) have lower levels of agricultural productivity to meet demand compared with developed countries.
   II. Many places around the world are far behind the advances made during the first and second agricultural revolutions in terms of agricultural productivity, and there is considerable room for these developing countries to catch up. However, Southeast Asia, China, and Latin America are now approaching the land and labor productivity levels achieved by current industrialized nations of the 1960s.
   III. The annual increase in the rate of agricultural production since the "Green Agricultural Revolution" in the mid-1960s averages about 2% growth per year. To attain and sustain these increases, farmers have relied on higher use of inputs; e.g., fertilizer, irrigation, mechanization, and pest control.

   (A) I
   (B) II
   (C) III
   (D) All choices are true.

Refer to the following diagram to answer Question 4.

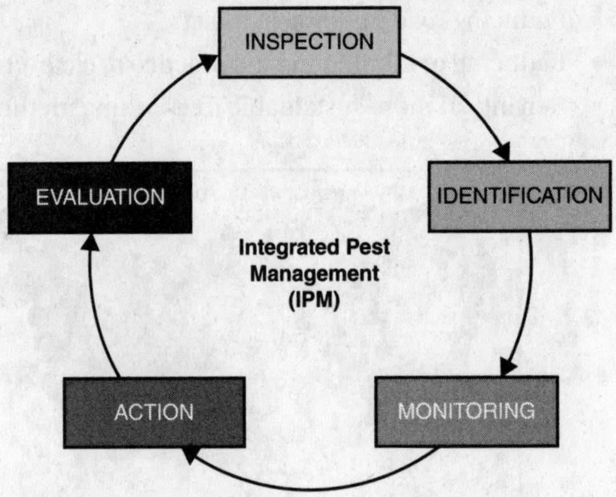

4. You just bought a small farm and are starting to grow some fruits and vegetables. Later in the season, you begin noticing that some of your fruits and vegetables are being destroyed by insects and the problem is only getting worse. Your neighbor suggested that you try an integrated pest management (IPM) approach first. Which of the choices below would be the most logical order of the steps you would follow to reduce the damage caused by the pests, starting with inspecting your crops for the type of pests causing the problem?

   (A) Inspection → Action → Identification → Monitoring ↔ Evaluation
   (B) Inspection → Evaluation ↔ Monitoring → Identification → Action
   (C) Inspection → Identification → Action ↔ Monitoring → Evaluation
   (D) Inspection → Monitoring ↔ Identification → Action → Evaluation

5. Planting trees and/or shrubs between rows of crops is known as

   (A) alley cropping
   (B) contour farming
   (C) planting windbreaks
   (D) strip cropping

6. The one area of the world that is NOT expected to increase food production soon is

(A) Asia
(B) India
(C) Latin America
(D) sub-Saharan Africa

7. Most foods derived from genetically modified crops contain

(A) hundreds of additional genes
(B) one or two additional genes
(C) the same number of genes as food produced from conventional crops
(D) the same number of genes as foods produced from hybrid crops

8. Which region of the developing world is currently as urbanized as the developed world?

(A) Asia
(B) Europe
(C) Latin America
(D) North America

9. The most lethal form of malnutrition is

(A) a deficiency of micronutrients in the diet
(B) a deficiency of vitamins in the diet
(C) lack of carbohydrates in the diet
(D) lack of protein in the diet

10. Adding more fertilizer does NOT necessarily increase crop production. This fact is an example of

(A) Liebig's Law of Minimum
(B) limiting factors
(C) the Law of Supply and Demand
(D) the Second Law of Thermodynamics

11. Soil that is transported by the wind is

(A) aeolian
(B) alluvial
(C) gangue
(D) r-ill

12. The following is an excerpt from an article written by John Parker in the February 24, 2011, issue of *The Economist* entitled "Feeding the World: The Nine Billion People Question."

> "An era of cheap food has come to an end. A combination of factors—rising demand in India and China, a dietary shift away from cereals towards meat and vegetables, the increasing use of corn as a biofuel, and falls in currency values—have brought to a close a period starting in the early 1970s in which the real price of staple crops (rice, wheat and maize) fell year after year due to an increase in productivity and technology.
>
> The end of the era of cheap food has coincided with growing concern about the prospects of feeding the world. As world population continues to increase, it stirs Malthusian fears. The consumer price rises have once again plunged into poverty millions of people who spend more than half their income on food which seems to suggest that the world cannot even feed its current population, let alone the 9 billion expected by 2050. Adding further to the concerns is climate change, of which agriculture is both a cause and a victim."

Which of the methods listed below are feasible to deal with the situation as outlined in the article above?

I. Close yield gaps—the difference between actual and potential crop yields.
II. Increase agricultural resource efficiency, e.g., switch from flooding fields or ditch irrigation to drip irrigation, providing sustainable crop nutrients.
III. Increase food supplies by shifting diets and reducing waste.

(A) I
(B) II
(C) III
(D) All choices are true

13. Countries with the largest population growth rates are found in

(A) Africa
(B) Asia
(C) Europe
(D) South America

14. Which of the following is NOT a concept that would be used in designing a sustainable city?

(A) Conserve natural habitats
(B) Design more affordable and fuel-efficient automobiles
(C) Focus on energy and resource conservation
(D) Provide ample green space

15. The Second Law of Thermodynamics would tend to support

(A) people eating a balanced diet from all food groups
(B) people eating about the same amount of grain as meat
(C) more people becoming vegetarians
(D) people eating more meat than grains and vegetables

16. Most of Earth's land area is

(A) agricultural
(B) desert
(C) forest
(D) rangeland

17. Most grain that is grown in the United States is used

(A) for cereals and baked goods
(B) for export
(C) to create fuel and liquor, e.g., gasohol, ethanol, beer
(D) to feed cattle

18. The complex environmental issue of global warming requires an understanding of the problem and cooperation by all countries of the world (e.g., international treaties), as global warming is caused by and affects all countries and life forms on Earth. Thus, international protocols, treaties, and agreements dealing with global warming

(A) are controversial, and international regulations include the United Nations Framework Convention on Climate Change, the Kyoto Protocols, and the Paris Agreement; however, not all countries have agreed to cooperate on how to solve the issue or to agree to the Paris Agreement
(B) establish specific figures for all countries of the world regarding acceptable concentrations of various greenhouse gases
(C) are addressed by the Montreal Protocol, which addresses various pollutants that contribute to global warming
(D) are generally not supported by developing countries of the world due to the financial impacts that international agreements and protocols would cause

19. What is the number one cause of soil erosion?

(A) chemical degradation
(B) physical degradation
(C) water erosion
(D) wind erosion

20. A process in which small holes are drilled into the earth and water-based chemical solvents are used to flush out desired minerals is known as

(A) chemical leaching
(B) *ex situ* leaching
(C) heap leaching
(D) *in situ* leaching

Free-Response Question

By Brian Palm, Catholic Memorial School, West Roxbury, MA
A.B., Dartmouth College, Hanover, NH
MSc, Oxford University, UK

(a) According to the U.S. Department of Agriculture, in 1967, 62 million acres of corn were planted. In 2016, that number rose to 93 million acres. What is the percent increase in planted acreage over this approximately 50-year period of time?

(b) Thomas Malthus described concerns about the different rates of growth for the human population and its food production. Use a graph and words to demonstrate your understanding of the trends he was predicting for both food availability and the human population.

(c) The recent era of farming and agricultural production is dramatically different from the other types of subsistence farming commonly utilized in the preceding 10,000 years. List TWO ways in which today's approach differs from earlier forms of agriculture AND describe TWO ways in which this approach can cause greater environmental harm.

(d) A range of sustainable or eco-friendly agricultural practices might be employed to reduce this sector's impact on our environment. IDENTIFY one method that would achieve this goal and DESCRIBE how it positively affects the local or global environment.

Answer Explanations

1. **(A)** The development of genetically modified organisms (GMOs) began during the Second Green Revolution (1950–1970s). For more information on the Agricultural and Green Revolutions, refer to page 178.

2. **(D)** There are many reasons why certain species decline and become endangered; however, a majority of studies show that the most significant factor is loss of habitat and land degradation.

3. **(D)** In situations where world agricultural prices paid to farmers decrease due to increased efficiency and higher output, farmers (especially small, independent farmers) try to increase agricultural productivity and lower inputs to remain competitive with larger, more efficient, and more cost-productive corporate farms. Worldwide, factors that compound this dynamic include the following:
   - A lack of governmental commitment to agricultural research and development due to competing financial interests and priorities
   - A lack of technology (compared to developed countries). Recent studies suggest that agricultural production would need to roughly double to keep pace with the projected demands of population growth, primarily in less-developed countries; dietary changes (e.g., meat consumption); and increasing bioenergy use. Compounding this challenge, agriculture must also address the significant environmental concerns of climate change, biodiversity loss, and degradation of land and freshwater.
   - Expected rates of future population growth in less-developed countries
   - Low incomes, which are typical of small farms (e.g., subsistence farming)

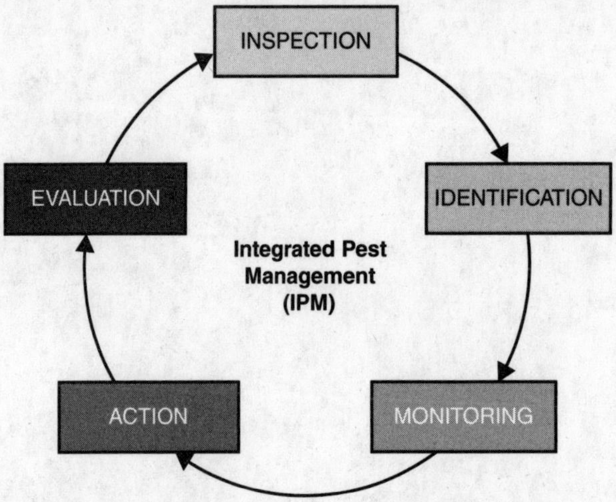

4. **(C)** Begin by first inspecting the crop to determine what is causing the problem. Is it a pest, a fungus, or a problem involving too much or the wrong type of fertilizer? Once you know that the problem is a pest, your next job would be to correctly identify the pest; this will determine the course of action or "action plan." Once the pest is identified, then you move into an action plan with constant monitoring (the reason for the double-headed arrows in the diagram below). Integrated Pest Management (IPM) emphasizes the growth of a healthy crop with the least possible disruption to agro-ecosystems (e.g., as little pesticides as possible) and encourages natural pest control mechanisms. Once enough time has passed, then you evaluate through inspection whether or not your action plan was adequate or not.

5. **(A)** See the "Types of Agriculture" in this unit for a more detailed explanation.

6. **(D)** Droughts, pestilence, AIDS, and civil strife in this area have had serious impacts on food production.

7. **(B)** Combining genes from different organisms is known as a recombinant DNA technology, and the resulting organism is said to be genetically modified, genetically engineered, or transgenic. Genetically modified products include medicines, vaccines, foods, feeds, and fibers.

8. **(C)** Despite the rapid growth of urbanization in Asian and African countries, their current percentage of people living in urban areas is still half that of Latin America. About 75% of Latin America's population currently lives in urban areas.

9. **(D)** Protein malnutrition, more commonly referred to as protein-energy malnutrition, is the most lethal form of malnutrition. Protein is necessary for key body functions, e.g., the development and maintenance of muscle tissue.

10. **(B)** There is a limit to crop yield vs. fertilizer application. Additional fertilizer past a certain point actually harms plants.

11. **(A)** Aeolian soils consist of sand-sized particles that can be transported by wind.

12. **(D)** Closing "yield gaps" could substantially increase global food supplies. Bringing yields to within 95% of their potential for 16 important food and feed crops could result in a 58% increase in dietary calories available from those crops. Additional gains in productivity, focused on increasing the maximum yield of key crops, are likely to be driven by genetic breakthroughs made during the third agricultural revolution, which we are currently in. Many "yield gaps" are due to insufficient nutrient availability as a result of soil degradation (desertification, salinization, and waterlogging); water shortages; continued use of inorganic fertilizers, which supply only a limited number of soil nutrients; or a failure to rotate crops or allow fields to go fallow (uncultivated). And, finally, shifting crop production away from livestock feed, bioenergy crops, and other non-food applications—i.e., shifting 16 major world food crops to 100% human food (not growing food for raising beef, pork, chicken, etc.)—could increase global food calories by 49% as approximately one-third of food that is currently produced is never consumed. In addition to adding more available food resources, this strategy lowers (1) greenhouse gas emissions, (2) water use, and (3) water pollution.

13. **(A)** Currently, the global population is approximately 7 billion people, with more than half living in Asia, while one quarter of the world's population lives in Africa. High fertility rates in Africa and Asia, a decline in mortality rates, and an increase in the median age of the world population all contribute to global population growth. The total world population is expected to increase by approximately 4 billion people by the year 2100. The increased global population growth rate is a direct result of people's living longer because of better living conditions and better nutrition. In terms of population, three out of five of the most populous countries in the world are located in Asia.

14. **(B)** Designing affordable and fuel-efficient cars would increase the number of automobiles already in cities, since more people could afford them. More private vehicles on the road results in more congestion and pollution.

15. **(C)** The Second Law of Thermodynamics states that whenever energy is transformed, there is a loss of energy through the release of heat. This occurs when energy is transferred between trophic levels throughout a food web. When one animal feeds off another, there is a loss of heat (energy) in the process, with additional energy loss occurring during respiration and movement; therefore, more and more energy is lost as one moves up through trophic levels. For example, approximately 50 pounds (1350 kg) of corn and soybeans is capable of supporting one person if converted to beef; however, that same amount of corn and soybeans would support 22 people if consumed directly.

16. **(C)** About one-third of Earth's surface is covered by forests; however, that number is declining at an ever-increasing rate because of deforestation, primarily converting the land to agricultural and forest-product use, e.g., timber. For example, between 1990 and 2016, the world lost 502,000 square miles (1.3 million square kilometers) of forest—an area larger than South Africa. Since humans started cutting down forests, 46% of all trees have been felled. Approximately 17% of the Amazon rainforest has been destroyed over the past 50 years, and losses are increasing.
17. **(D)** There are approximately 1.5 billion cattle on Earth, grazing on about 25% of all land and consuming enough grain to feed hundreds of millions of people. In the United States, cattle consume around 70% of all grain produced.
18. **(A)** The Paris Agreement is an agreement within the United Nations Framework Convention on Climate Change (UNFCCC) dealing with greenhouse-gas emissions mitigation, adaptation, and finance, starting in the year 2020. Terms of the international agreement include the following:
    - Holding the increase in the global average temperature to below 2°C above pre-industrial levels
    - Increasing the ability to adapt to the adverse impacts of climate change and foster climate resilience and low greenhouse gas emissions development in a manner that does not threaten food production
    - Making finance flows consistent with a pathway toward low greenhouse gas emissions and climate-resilient development. Under the Paris Agreement, each country shall determine, plan, and regularly report on the contribution that it makes to mitigate global warming. There is no mechanism to force a country to set a specific target by a specific date, but each target should go beyond previously set targets.
19. **(C)** Water is the primary cause of soil erosion—the detachment and transport of soil material.
20. **(D)** The process of in situ leaching can lead to permanent contamination of groundwater, which is often used for drinking water supplies, and can also contaminate land with chemicals that are used in the process.

Free-Response

10 Total Points Possible

(a) *Maximum 2 points total: 1 point for the correct calculation setup and 1 point for the correct answer.*

According to the U.S. Department of Agriculture, in 1967, 62 million acres of corn were planted. In 2016, that number rose to 93 million acres. What is the percent increase in planted acreage over this approximately 50-year period of time?

Responses must include a setup similar to the one below. While units are not required in the setup, the numbers should be consistent, and they should recognize the correct mathematical operation (in this case a variant of the percent change equation) where:

$$\%\,Change = \frac{New\ number - Old\ number}{Old\ number} \times 100 = \frac{93 - 62}{62} \times 100 = 50\%$$

(b) *Maximum 2 points total: 1 point for a correct representation of the graph and 1 point for a correct description and connection to Malthus' thinking.*

Thomas Malthus described concerns about the different rates of growth for the human population and its food production. Use a graph and words to demonstrate your understanding of the trends he was predicting for both food availability and the human population.

The two lines may be drawn on a single set of axes or they might be represented on two separate axes. You should demonstrate Malthus' prediction that food availability would grow arithmetically ("linear" or "constant" growth are also acceptable descriptions) over time and his prediction that the human population would grow exponentially. "Population Growth" and "Food Production" are the two quantities that should be used on the y-axes. The description of these predictions should include the recognition that a population demonstrating exponential growth has the potential to overtake a food source that is only growing arithmetically. At that time, food shortages would occur, putting downward pressure on the population, including famines.

(c) *Maximum 4 points total: 2 points for listing TWO ways in which today's agricultural practices (referred to as industrial agriculture) differ from earlier forms of agriculture and 2 points for listing TWO correct descriptions of the environmental harm caused by the associated agricultural practices. The description MUST align with the listed practice.*

The recent era of farming and agricultural production is dramatically different from the other types of subsistence farming commonly utilized in the preceding 10,000 years. List TWO ways in which today's approach differs from earlier forms of agriculture, AND describe TWO ways in which this approach can cause greater environmental harm.

The table below provides a range of possible answers.

Modern/Industrial Agricultural Practice	Associated Explanation of the Environmental Impact Practice
Monoculture/ single-crop planting	• The loss of diversity has the potential to make the system more vulnerable to disease and pests. The loss of diversity of seeds and crops reduces naturally occurring food webs.
Increased use of fossil fuels in production (fertilizers, herbicides, mechanization)	• Increased levels of carbon dioxide cause climate change, which can reduce and alter natural habitats, causing a loss of certain species. • Pollutants from combustion and manufacturing can cause habitat destruction and/or species death.
Abundant use of fertilizers	• Runoff from agricultural lands creates areas of higher than normal nitrogen/phosphorus levels in adjacent water (fresh/salt), which can cause massive food web changes, species death, or dead zones.

(d) *Maximum 2 points total: 1 point for IDENTIFYING one agricultural practice and 1 point for DESCRIBING how it positively affects the local or global environment. Answers may vary, but each correct answer should IDENTIFY one practice AND connect that specific practice to the environmental benefit.*

A range of sustainable or eco-friendly agricultural practices might be employed to reduce this sector's impact on our environment. IDENTIFY one method that would achieve this goal, and DESCRIBE how it positively affects the local or global environment.

Examples include the following:

- Crop rotation creates situations that allow crops to replenish soil nutrients that were extracted by the preceding crop. For example, a legume crop such as soybeans or alfalfa can return nitrogen to soils where corn farming has depleted nitrogen. This reduces demand for synthetic fertilizers, which require fossil fuels in their production. Avoided carbon production will reduce the risk of climate change.
- Intercropping, or the planting of adjacent crops that have different nutrient needs and harvesting/planting schedules, helps to reduce the need for large applications of synthetic fertilizer that can cause soil and water toxicity. Applying this practice also means that all areas of the agriculturally active area are NOT exposed to erosion during planting season, which reduces topsoil loss.
- Planting and maintaining vegetated riparian buffer zones between agriculturally active areas and local waterways helps to reduce erosion while filtering out fertilizers and agricultural products (such as pesticides or herbicides). Reductions in nitrates and phosphates prevent eutrophication while maintaining water quality for aquatic organisms.

6

Energy Resources and Consumption

Learning Objectives

In this unit, you will learn:

- ➔ 6.1 Renewable and Nonrenewable Resources
- ➔ 6.2 Global Energy Consumption
- ➔ 6.3 Fuel Type and Uses
- ➔ 6.4 Distribution of Natural Energy Resources
- ➔ 6.5 Fossil Fuels
- ➔ 6.6 Nuclear Power
- ➔ 6.7 Energy from Biomass
- ➔ 6.8 Solar Energy
- ➔ 6.9 Hydroelectric Power
- ➔ 6.10 Geothermal Energy
- ➔ 6.11 Hydrogen Fuel Cell
- ➔ 6.12 Wind Energy
- ➔ 6.13 Energy Conservation

For added practice with this unit and topics, log onto Online Problem Sets 6A and 6B along with Lab 6A at online.barronsbooks.com

Introduction to Energy

Energy is defined as the fundamental entity of nature that is transferred between parts of a system in the production of physical change within the system and is usually regarded as the capacity for doing work. The law of conservation of energy states that energy can be converted in form, but not created or destroyed.

The sun is the source of energy for most of life on Earth, and, as a star, the sun is heated to high temperatures by the conversion of nuclear energy to heat in its core by the process of nuclear fusion. Inside the sun, this process begins with protons, which are hydrogen nuclei. Through a series of steps, these protons fuse together and are turned into helium. This fusion process occurs inside the core of the sun, and the transformation results in a release of energy that keeps the sun hot. The resulting energy is radiated out from the core of the sun and moves across the solar system.

Living organisms require available energy to stay alive. Carnivores get energy from eating other animals, herbivores get energy from eating plants, plants get energy from the sun, and detritivores get energy from consuming dead organic matter.

Human civilization requires energy to function. Humans obtain energy from resources such as fossil fuels, nuclear fuel, or renewable energy. The processes of Earth's climate and ecosystem are driven by the radiant energy Earth receives from the sun and the geothermal energy contained within the Earth.

Forms of Energy

Form	Description
Chemical	Chemical energy is stored in bonds between atoms in a molecule.
Electrical	Electrical energy results from the motion of electrons.
Electromagnetic	Electromagnetic energy travels by waves.
Mechanical	There are two types of mechanical energy: potential energy (a book sitting on a table) and kinetic energy (a baseball flying through the air).
Nuclear	Nuclear energy is stored in the nuclei of atoms, and it is released by either splitting or joining atoms.
Thermal	Heat is the internal energy in substances—the vibration and movement of the atoms and molecules within substances.

Power and Units

Power is the amount of work done per unit of time and is measured in watts (1 watt = 1 joule/sec). A 100-watt lightbulb uses 100 joules of energy every second, as does a home computer or a person standing, and it takes 1 kilowatt, or 1,000 joules, of energy to keep your cell phone charged for a year. The average cost of 1 kW of electricity in the United States is about 12 cents.

The kilowatt-hour is the most common unit of electrical energy used in the United States. An electric dishwasher, for example, uses about 2 kWh per load. The unit of energy most common in the world, though, is the joule. It would take approximately 1 joule of energy to lift an apple 3 feet (1 m) vertically from the ground, while an average person sitting still produces 100 joules of heat every second.

Units of Energy/Power

Unit or Prefix	Description
British thermal unit (Btu)	Btu is a unit of energy used in the United States, although in most countries it has been replaced with the joule. A Btu is the amount of heat required to raise the temperature of 1 pound of water by 1°F. 1 watt is approximately 3.4 Btu/hr. 1 horsepower is approximately 2,540 Btu/hr. 12,000 Btu/hr is referred to as a "ton" in many air-conditioning applications.
Horsepower	Primarily used in the automobile industry. 1 horsepower (HP) is equivalent to 746 watts.
Kilowatt hour (kWh)	The kilowatt (kW) is a unit of power, a measure of energy used at a given moment, not over time (e.g., a car is traveling at a speed of 60 miles per hour). Kilo = 1,000 or 10^3. 1 kW = 10^3 watts. Mega = (M) 1,000,000 or 10^6. 1 MW = 10^6 watts. Unit of energy equal to 1,000 watt hours or 3.6 megajoules. The kilowatt hour is most commonly known as a billing unit for energy delivered to consumers by electric utilities, whereas the kilowatt hour (kWh) is a unit of total energy used over a specific period of time (e.g., a car used a gallon of gasoline to go 13 miles in 30 minutes). Examples: A heater rated at 1,000 watts (1 kilowatt), operating for one hour, uses one kilowatt hour (equivalent to 3.6 megajoules) of energy. Using a 60 watt lightbulb for one hour consumes 0.06 kilowatt hours of electricity.

TIP

Be sure to practice these types of conversion problems. They appear very frequently in the Free-Response Questions on the APES exam. Also, be sure you are comfortable with scientific notation and the factor-label method.

Conversions

Let's do a couple of sample conversion problems that you might experience on the APES exam. We will break them down into steps and show you how they might be graded.

EXAMPLE PROBLEM #1

Thorpeville is a rural community with a population of 8,000 homes. It gets its electricity from a small, municipal coal-burning power plant just outside of town. The power plant's capacity is rated at 20 megawatts, with the average home consuming 10,000 kilowatt hours (kWh) of electricity per year. Residents of Thorpeville pay the utility $0.12 per kWh. A group of entrepreneurs is suggesting that the residents support a measure to install 10 wind turbines on existing farmland. Each wind turbine is capable of producing 1.5 MW of electricity. The cost per wind turbine is $2.5 million dollars to purchase and operate for 20 years.

(a) The existing power plant runs 8,000 hours per year. How many kWh of electricity is the current plan capable of producing?

2 points. 1 point for correct setup. 1 point for correctly calculating the amount of electricity generated. You must correctly convert MW to kW. No points will be awarded without showing your work. Alternative setups are acceptable.

$$\frac{20\ \cancel{\text{MW}}}{1} \times \frac{(1 \times 10^6\ \cancel{\text{watts}})}{1\ \cancel{\text{MW}}} \times \frac{1\ \text{kW}}{10^3\ \cancel{\text{watts}}} = 2 \times 10^4\ \text{kW}$$

$$\frac{(2 \times 10^4\ \text{kW})}{1} \times \frac{8{,}000\ \text{hours}}{1\ \text{yr}} = 16{,}000 \times 10^4\ \text{kWh/yr}$$

$$= 1.6 \times 10^8\ \text{kWh/yr}$$

(b) How many kWh of electricity do the residents of Thorpeville consume in one year?

2 points. 1 point for correct setup. 1 point for correctly calculating the amount of electricity generated. No points will be awarded without showing your work. Alternative setups are acceptable.

$$\frac{8 \times 10^3\ \cancel{\text{homes}}}{1} \times \frac{1 \times 10^4\ \text{kWh}/\cancel{\text{home}}}{1\ \text{yr}} = 8 \times 10^7\ \text{kWh/yr}$$

(c) Compare answers (a) and (b). What conclusions can you make?

2 points, plus 1 possible elaboration point. 1 point for comparing answer (a) with answer (b) with an explanation of why the numbers in parts (a) and (b) would be the same or different (must be a viable reason).

OR

1 point for a solid or accurate explanation of why (a) and (b) are different even if the calculations were not attempted.

1 possible elaboration point for explanations that go into great detail about why the numbers are the same or different.

Note: If you say that (a) and (b) are the same, you must state that this can only occur if the households have backup systems that will produce energy for them if they exceed the power generated by the plant.

The power plant produces 1.6×10^8 kWh per year. The residents, however, only use 8×10^7 kWh per year. This leaves a surplus of $1.6 \times 10^8 - 8 \times 10^7 = 8 \times 10^7$ kWh in one year that can be sold to other towns. At a rate of $0.12 per kWh, this provides a surplus of 8×10^7 kWh × $0.12/kWh = $0.96 × 10^7 = $9,600,000.

Differences between Thorpeville's consumption and the power plant's output could be attributed to the following:

- The power plant needs to produce higher amounts of power to compensate for line loss.
- The power plant needs to produce higher amounts of power to supply energy to the town during peak hours, not just the average usage.
- The power plant needs to plan for possible future growth of the town.
- The power plant was built over capacity to provide a source of income to the town.

(d) Assuming that the population of Thorpeville remains the same for the next 20 years, and that electricity consumption remains stable per household, what would be the cost (expressed in $ per kWh) of electricity to the residents over the next 20 years if they decided to go with wind turbines?

2 points. 1 point for correct setup. 1 point for correct answer with calculations. Alternative setups are acceptable. If your answer in part (b) is incorrect but you appropriately use it as the basis for the calculations for answering the question in part (d), you will receive full credit for answering part (d) if the setup and calculations are correct, even if the answer is not correct.

Based on current community consumption of 8×10^7 kWh/ yr from part (b):

$$\text{kWh for 20 years} = \frac{8 \times 10^7 \text{ kWh}}{\cancel{\text{year}}} \times \frac{20 \cancel{\text{years}}}{1} = 1.6 \times 10^9 \text{ kWh}$$

$$\text{Direct cost for 20 years} = \frac{10 \cancel{\text{turbine}}}{1} \times \frac{\$2.5 \times 10^6}{\cancel{\text{turbine}}} = \$2.5 \times 10^7$$

$$\text{cost/kWh} = \frac{\$2.5 \times 10^7}{1.6 \times 10^9 \text{ kWh}} = \$0.016/\text{kWh}$$

(e) Describe three pros of the proposed wind farm vs. three cons of the existing coal-burning plant.

Pros (Wind Farm)	Cons (Coal)
■ The electricity produced from the wind turbines costs \$0.016 per kWh, but each homeowner would also have to pay \$25,000,000/8,000 homes = \$3,125.00 over 20 years (\$156.25/yr) to pay for the wind turbines. 10,000 kWh at \$0.016 per kWh for electricity produced from wind turbines = \$160 plus \$156.25 per year to pay for the wind turbines = \$316.25 per year. ■ The wind is free. ■ Zero emissions. Acid rain as well as photochemical smog would be reduced. ■ No heavy-metal or radioactive emissions. ■ No thermal pollution. ■ Multiple use of the land.	■ Electricity costs \$0.12 per kWh produced from the coal-burning plant. 10,000 kWh of electricity per year from the coal-burning plant at \$0.12 per kWh = \$1,200 per year. Clearly, electricity produced from wind turbines is much cheaper. ■ Produces air pollution, specifically SO_2 and NO_x. ■ As labor prices increase, the price of coal would be expected to increase over the next 20 years. ■ Coal-burning plants produce heavy metals such as mercury, lead, and cadmium pollution along with radioactive contaminants. ■ Can produce thermal pollution to local streams. However, cooling towers can be installed to reduce this form of pollution. ■ Cannot utilize the concept of multiple use of land.

EXAMPLE PROBLEM #2

(a) An electric water heater requires 0.30 kWh to heat a gallon of water. The thermostat is set to 150° F. The cost of electricity is $0.20 per kWh. A washing machine with a flow rate of 6.0 gallons per minute runs four times each Saturday. Each time it runs it takes in water for a total of 15 minutes. How much total water does the washing machine use in one year?

2 points. 1 point for correct setup. 1 point for correct answer with calculations. Alternative setups are acceptable.

$$\frac{4\ \cancel{\text{cycles}}}{\cancel{\text{Saturday}}} \times \frac{15\ \cancel{\text{minutes}}}{1\ \cancel{\text{cycle}}} \times \frac{6.0\ \cancel{\text{gallons}}}{1\ \cancel{\text{minute}}} \times \frac{52\ \cancel{\text{Saturdays}}}{1\ \text{year}} = 18{,}720\ \text{gallons/year}$$

(b) Calculate the annual cost of the electricity for the washing machine, assuming that 3.0 gallons per minute of the water used by the machine comes from the hot-water heater.

2 points. 1 point for correct setup. 1 point for correct answer with calculations. Alternative setups are acceptable. If your answer in part (b) is incorrect but you appropriately used information from part (a) as the basis for answering the question, you will receive full credit, even if the numerical answer is wrong.

18,720 gallons/2 = 9,360 gallons of hot water per year

$$\frac{9360\ \cancel{\text{gallons}}}{1\ \text{year}} \times \frac{0.30\ \cancel{\text{kWh}}}{1\ \cancel{\text{gallon}}} \times \frac{\$0.20}{\cancel{\text{kWh}}} = \$561.60/\text{year}$$

Laws of Thermodynamics

First Law—Energy cannot be created or destroyed.
Second Law—When energy is converted from one form to another, a less useful form results (energy quality); e.g., only 20% of the energy in gasoline is converted to mechanical energy, while the rest is lost as heat and is known as low-quality energy.

6.1 Renewable and Nonrenewable Resources

Renewable Energy Sources

Renewable energy is defined as energy that is collected from resources that are naturally replenished on a human time scale, e.g., sunlight, wind, rain, tides, waves, and geothermal heat. In 2018, renewable energy sources accounted for about 11% of total U.S. energy consumption and about 17% of electricity generation (Figure 6.1).

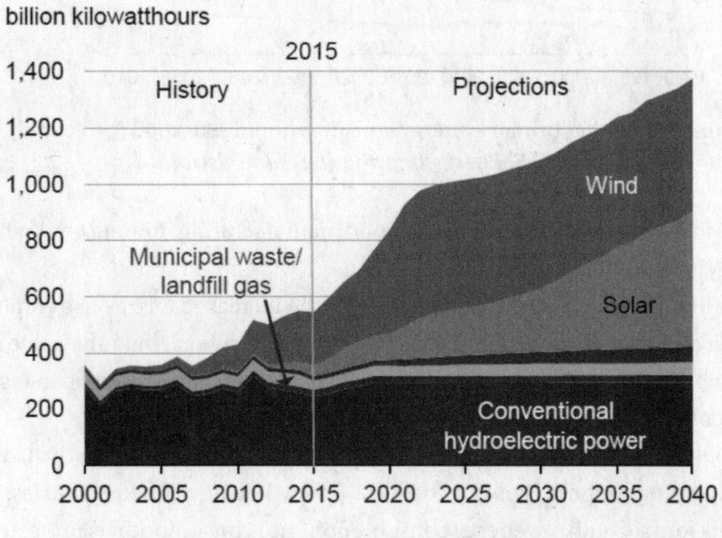

Figure 6.1 Projected United States renewable energy sources
Source: Energy Information Agency

Renewable energy resources exist over wide geographical areas, in contrast to other energy sources that are concentrated in a limited number of countries. Rapid deployment of renewable energy and energy efficiency is resulting in significant energy security, climate change mitigation, and economic benefits. At least 30 nations around the world already utilize renewable energy, contributing to more than 20% of their energy supply. As of 2021, renewable energy sources accounted for about 12.4% of total U.S. primary energy consumption.

Nonrenewable Energy Sources

Most fossil fuels, such as oil, natural gas, and coal, are considered nonrenewable resources in that their use is not sustainable because their formation takes billions of years. In the United States, most of the current energy demand comes from nonrenewable energy sources such as coal, natural gas, petroleum, propane, and uranium. Recent increases in the domestic production of oil and natural gas, due largely to fracking and increases in offshore drilling, have increased the supply and may continue to lower the price of these commodities. Arguments used to defend the continued use of fossil fuels include the following:

- Abundant supply, resulting in relatively low prices for consumers
- Concentrated fuel with a high net-energy yield
- Infrastructure already in place for extraction, processing, and delivery
- Politics (e.g., fear of losing existing jobs or concerns over unemployed coal miners)
- Technology already exists for their use (e.g., internal combustion engines running on gasoline).

6.2 Global Energy Consumption

The world's energy consumption will grow by nearly 50% between 2020 and 2050 and will be focused in regions where strong economic growth is driving demand—i.e., Asia (Figure 6.2).

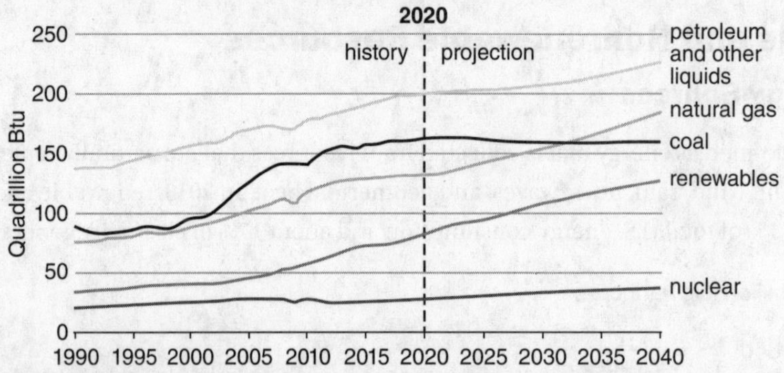

Figure 6.2 Global primary energy consumption, 1990–2040 (projected)
Source: U.S. Energy Information Administration

Industrial sectors, which include agriculture, construction, manufacturing, mining, and refining, account for the largest share of energy consumption of any end user.

People are using more energy than ever, with the majority of the increased energy use coming from developing countries. Today, coal, oil, and natural gas generate most of the world's energy, but renewable energy is growing fast, with the pace in renewables coming at the expense of coal, which has leveled off in response to concerns about the effect of its greenhouse gas emissions linked to climate change.

It is projected that a roughly 50% jump in energy consumption will occur by the year 2040, with 71% of that increase occurring in developing countries compared to ~18% growth in developed countries. Electricity is the world's fastest-growing form of end-use energy consumption, and consumption is projected to increase ~70% worldwide by 2040, with coal currently supplying the largest share of electricity generation (~40% of electricity in 2018 is produced by coal). Renewables are predicted to increase at a rate averaging ~3% a year through 2040 (Figure 6.3).

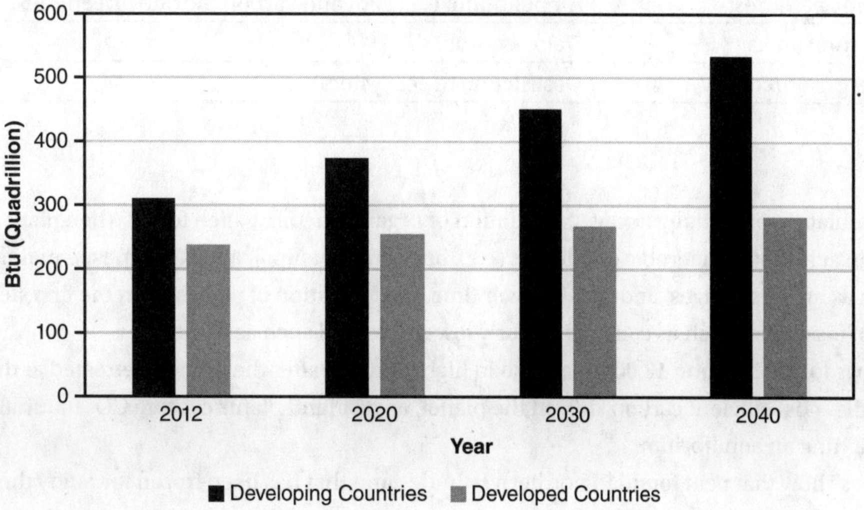

Figure 6.3 World energy consumption (developing vs. developed countries)

6.3 Fuel Types and Uses

The most widely used sources of global energy are fossil fuels, which are fuels formed from past geological remains of living organisms. (Figure 6.4)

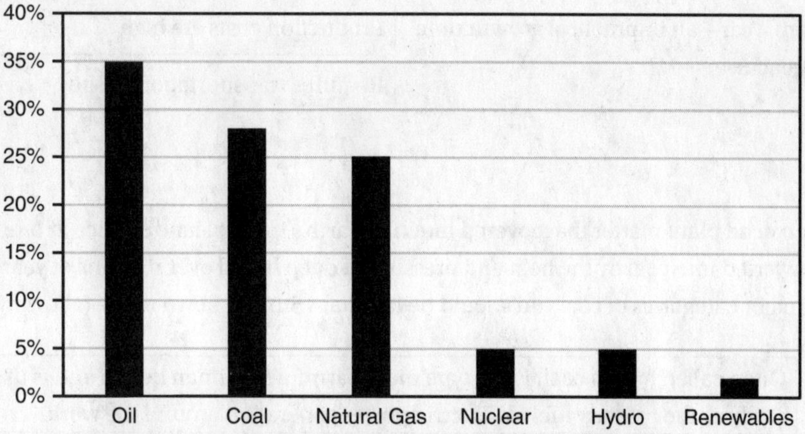

Figure 6.4 World energy consumption by fuel

Wood

Burning wood fuel creates the following by-products: carbon dioxide, heat, steam, water vapor, and wood ash. Another major component of wood smoke is fine particles that may account for a large portion of particulate air pollution and lead to asthma and heart disease.

Burning wood creates more atmospheric CO_2 than the decomposition of wood in a forest because by the time the bark of a dead tree has rotted, the log has already been occupied by other plants and microorganisms, which continue to sequester the CO_2 by integrating the hydrocarbons of the wood into their own life cycle.

Pros	Cons
Ash can be used as fertilizer	Accelerated soil erosion
Easily accessible near forests	Air pollution; e.g., soot and carbon dioxide (greenhouse gas)
Inexpensive in rural areas	Deforestation
Renewable energy source	Destruction of ecosystems

Peat

Peat is an accumulation of partially decayed vegetation or organic matter, which forms when plant material does not fully decay in acidic and anaerobic conditions (e.g., bogs and swamps), and is composed mainly of wetland vegetation such as mosses, sedges, and shrubs. Over time, the formation of peat is often the first step in the geological formation of fossil fuels such as coal, particularly low-grade coal such as lignite.

Most peat bogs formed around 12,000 years ago in high latitudes after the glaciers retreated at the end of the last ice age. Peat is the most efficient carbon sink on the planet, as peatland plants capture CO_2 naturally released from the peat, maintaining an equilibrium.

Recent studies show that peat found in northern latitudes and that has been frozen for many thousands of years is now thawing due to climate change. As the permafrost melts, there is the potential for the release of billions of tons of methane gas into the atmosphere.

Pros	Cons
Natural deposits are found around the world	Air pollution; e.g., soot and carbon dioxide (greenhouse gas)
Technically, a renewable energy source; however, peat accumulates ~1 mm/year—an impractical growth time period for widespread use	Destruction of ecosystems
	Production costs are high
	Requires transportation to sites

Coal

Coal is formed when dead plant matter that covered much of Earth's tropical land surface at one time decays into peat and is then converted into coal by the heat and pressure of deep burial over millions of years.

There are three major categories of coal you should be familiar with, as shown in the following table.

Lignite	Often called brown coal, is the type most harmful to human health and is used almost exclusively as the primary fuel for electric power generation around the world
Bituminous	Used primarily as fuel in steam-electric power generation
Anthracite	Used primarily for residential and commercial space heating

The United States has the largest proven recoverable coal reserves in the world, whereas China is the world's largest producer of coal.

Coal supplies ~25% of the world's energy (with Asia using ~75% of that) and ~40% of the world's electricity and is the largest anthropogenic source of carbon dioxide. Coal makes up ~40% of the total fossil fuel emissions and almost 25% of total global greenhouse gas emissions.

Millions of tons of ash from burning coal are released into the atmosphere each year. Carbon dioxide, nitrogen oxides (NO_x), sulfur oxides (SO_x), particulates (especially $PM_{2.5}$ particles, which are extremely dangerous), and heavy metals are some of the pollutants released into the air when coal is burned. It is estimated that almost one million premature deaths occur each year (primarily in China and India) due to the environmental effects caused by burning coal. In addition, asthma, heart attacks, and lung cancer have been attributed to the burning of coal.

Acid rain is another environmental by-product of burning coal, which releases sulfur dioxide (SO_2), and then combines with water vapor in the atmosphere to produce acid rain.

The U.S. Clean Air Act requires up to a 90% reduction in the release of sulfur-containing gases.

The following technologies can be used to remove pollutants from flue gases produced by burning coal:

- **Baghouse filters**—Fabric filters that can be used to reduce particulates.
- **Burning pulverized coal at lower temperatures**—Coal is crushed into a very fine powder and injected into a firebox.
- **Coal gasification**—A process that turns coal and other carbon-based fuels into gas known as "syngas." Impurities are removed from the syngas before it is combusted, which results in lower emissions of sulfur dioxide, particulates, and mercury.
- **Cyclone separator**—A method of removing particulates through rotational (spinning) effects and gravity.
- **Electrostatic precipitator**—A filtration device that removes fine particles, like dust and smoke, from a flowing gas using an electrostatic charge.
- **Fluidized-bed combustion**—A method of burning coal in which the amount of air required for combustion far exceeds that found in conventional burners. This process can be used to reduce the amount of NO_x, SO_x, and particulates.
- **Scrubbers**—Systems that inject chemical(s) into a dirty exhaust stream to "wash out" acidic gases. Scrubbers can also be used to reduce SO_x and particulates from burning coal.
- **Sorbents**—Activated charcoal, calcium compounds, or silicates can convert gaseous pollutants in the smokestacks, such as mercury, into compounds that can then be collected with baghouse filters, electrostatic precipitation, or scrubbers.

Pros

- Known world coal reserves will last approximately 300 years at the current rate of consumption (Figure 6.5).
- They have relatively high net-energy yield.
- U.S. government subsidies keep prices low, making coal's overall cost low compared to other fuels.

Cons

- Burning coal releases mercury, arsenic, sulfur (in the form of H_2S and a major contributor to both wet and dry acid precipitation), and radioactive particles into the air.
- Coal contributes to global warming from emissions of coal combustion (e.g., CO_2 and methane [CH_4]).

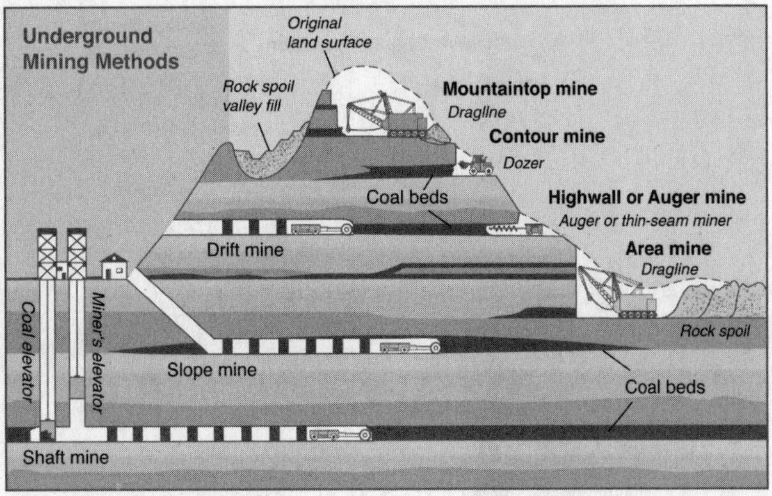

Figure 6.5 Surface mining methods

- Coal contributes to the formation of acid rain (wet deposition) caused by sulfuric acid (H_2SO_4).
- Dry deposition is caused by NO_x and/or SO_x.
- It contributes to the production of photochemical and industrial smog and releases particulates, all of which aggravate respiratory illnesses.
- It is expensive to process and transport and cannot be used effectively for transportation needs.
- Most extraction in the United States is done through either strip mining or underground mining, which cause disruptions to the land through erosion, leaching, runoff, and a decrease in biodiversity.
- Thirty-five percent of all CO_2 and 30% of all NO_x emissions released into the atmosphere are due to the burning of coal.
- Underground mining is dangerous and unhealthy.

Future Coal Consumption

Worldwide use of coal is expected to grow (but at a slower growth rate) in the immediate future due to:

- demand by developing countries;
- large amounts of reserves; and
- real or projected decreases in reserves of other nonrenewables, e.g., oil.
- the fact that infrastructure for mining, distribution, and transportation is already in place;

"Clean Coal"

"Clean coal' is a term for technology that attempts to mitigate emissions of carbon dioxide and other greenhouse gases that arise from the burning of coal for electrical power. Typically, the term "clean coal" is used by coal companies in reference to carbon capture and storage (CCS), which pumps and stores CO_2 emissions underground.

Large scale CCS is unproven (at this time) and may be decades away from being commercialized. Criticisms of this technology include:

- CCS is risky and expensive; a better option is renewable energy.
- CCS leaves behind dangerous waste material that has to be stored, just like nuclear power generating plants.
- CCS will actually increase CO_2 emissions and air pollutants, as CCS systems require 25% more energy than systems without CCS.
- Earthquakes and resulting land fissures could release the captured CO_2 being stored underground.
- Money spent on CCS will divert investments away from other solutions to mitigate climate change.
- Safe and permanent storage of CO_2 cannot be guaranteed, and even very low leakage rates could undermine any climate mitigation effects.

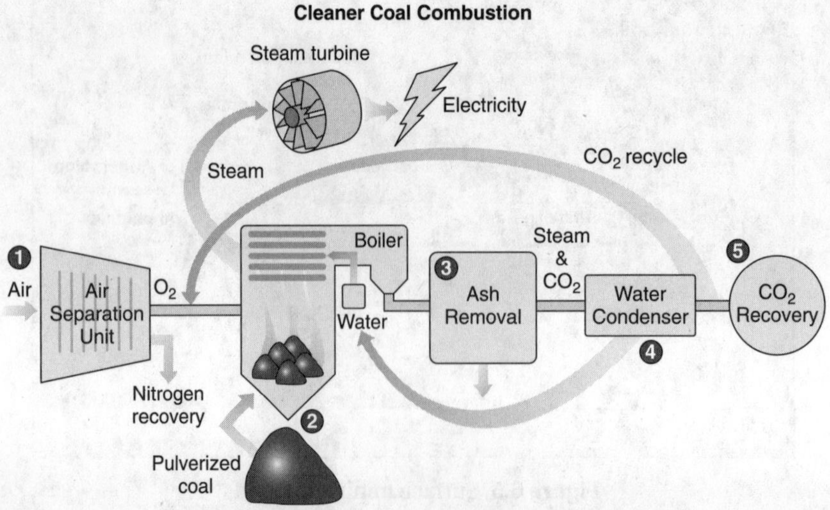

Figure 6.6 Steps used in producing "clean coal"

Natural Gas (Methane—CH_4)

Natural gas (which is primarily methane [CH_4]) is a fossil fuel formed when layers of buried plants and gases are exposed to intense heat and pressure over thousands of years. The energy that the plants originally obtained from the sun is stored in the form of chemical bonds in the methane molecules.

The energy that is released when natural gas is burned ($CH_4 + 2O_2 \rightarrow CO_2 + 2H_2O$ + heat energy) is used for cooking, to heat (or cool) homes and businesses, to generate electricity, and to power vehicles.

The increasing amount of carbon dioxide gas released into Earth's atmosphere over the last ~100 years by burning natural gas and other "fossil fuels" has dramatically risen, with human activity alone accounting for half of all methane emissions.

Agriculture, which includes enteric fermentation (methane gas released by cows) and manure management, is responsible for ~60% of all methane emissions into the atmosphere, with the remaining portion due to landfill waste and the oil and natural gas industry (Figure 6.7).

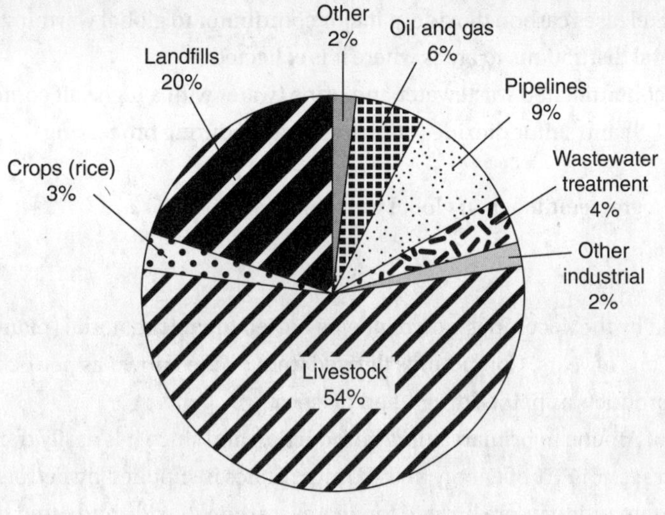

Figure 6.7 Methane emissions
Source: California Air Resources Board Methane Inventory

Methane has more than 80 times the warming potential (power) of carbon dioxide over the first 20 years after it reaches the upper atmosphere, and the burning methane is expected to increase ~2% annually for the foreseeable future.

Natural gas supplies about 30% of the world's current energy needs, with the Russian Federation and the Middle East supplying about 75%, and is expected to take on a greater role in the future due to the demand for cleaner fuels, the relatively low costs of building new natural gas power plants, and the use of natural gas as a transportation fuel and a source of hydrogen for fuel cells.

Natural gas typically flows from wells under its own pressure and is collected by small pipelines that feed into the large gas transmission pipelines. In the United States, about 20 trillion cubic feet (560 billion m^3) of natural gas is produced each year.

Two-thirds of the natural gas extracted in the United States is obtained through fracking, and at current production levels, there is enough natural gas to meet ~75 years of domestic production. There are approximately 50 years left of natural gas worldwide, with production expected to substantially increase due to demand coupled with projected long-term price increases due to the decrease in supply.

Pros

- Easily processed and transported as liquefied natural gas (LNG) over rail or by ship
- High net-energy yield
- Less CO_2, mercury, and particulates are released than when burning coal or oil.
- Less SO_2 and less NO_x are produced compared to burning coal or oil, resulting in less of an impact on acid rain and photochemical smog.
- Pipelines and distribution networks are already in place.
- Relatively inexpensive compared to other fossil fuels
- The extraction of natural gas is a much safer process than coal mining.
- Viewed by many as a transitionary fossil fuel as the world switches to renewable sources
- World reserves are estimated to be 125 years at the current rate of consumption, but may increase due to hydro-fracking.

Cons

- Burning natural gas releases carbon dioxide, a major contributor to global warming.
- Causes environmental disruptions to areas where it is collected.
- Extraction releases contaminated wastewater and brine (water with a high salt content).
- Hydrogen sulfide (H_2S) and sulfur dioxide (SO_2) are released during processing.
- Land subsidence
- Leakage of CH_4 has a greater impact on global warming than does CO_2.

Oil

Oil is a fossil fuel produced by the decomposition of deeply buried organic material (plants) under high temperature and pressure for millions of years. Compounds derived from oil are known as petrochemicals and are used in the manufacture of such products as paints, drugs, and plastics.

Oil accounts for ~28% of American primary energy production and since it is easily transportable, it is primarily used in the transportation sector; 71% of energy in the United States is supplied by petroleum.

As oil and other petroleum products are burned for energy, carbon dioxide and other greenhouse gases are released into the atmosphere. Atmospheric carbon dioxide levels were, up until ~150 years ago, about 180–300 ppmv (parts per million by volume) and are now currently ~415 ppmv, which has resulted in numerous environmental consequences; e.g., increased wildfires in California and increased frequency of hurricanes in the southern states and flooding in the Northeast (e.g., Hurricane Ida, 2021).

Oil Reserves

Up to 70% of Earth's global crude oil reserves have already been depleted, and it is estimated that there is ~50-year supply of oil left on Earth, with countries in the Middle East owning about half of what is left (Figure 6.8). The United States owns about 3% of the world's oil reserves but still uses about 30% of the oil extracted in the world each year. However, U.S. oil imports have decreased significantly due to energy-efficiency improvements, higher fuel economy standards, and advances in extractions (e.g., fracking). Keep in mind the following:

- It is estimated that approximately 13% of the world's oil reserves are in areas north of the Arctic Circle.
- Large amounts of undiscovered oil exist offshore throughout the world.
- The largest oil reserves are located in Venezuela, Saudi Arabia, and Canada.
- The United States still imports about 25% of the oil it needs, since it has a very large refining capacity, especially on the Gulf Coast, and needs to run at or near 100% capacity by importing crude oil and exporting refined petrochemical products; e.g., gasoline, diesel, heating oil, and lubricants.

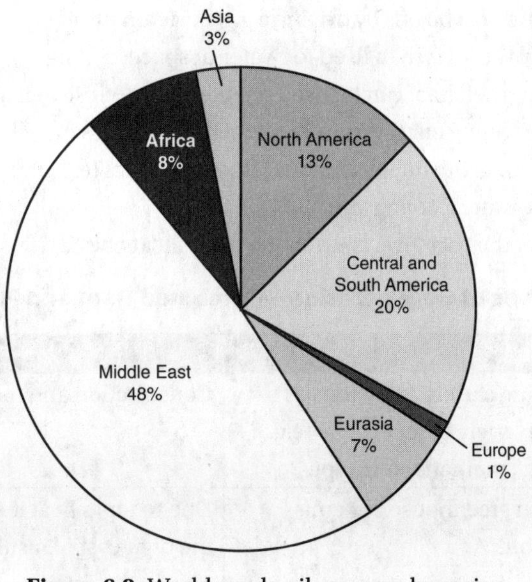

Figure 6.8 World crude oil reserves by region

Pros

- Ample supply for the immediate future
- Easily transported through established pipelines and distribution networks
- High net-energy yield
- Large U.S. government subsidies in place
- Versatile—used to manufacture many products (e.g., paints, medicines, plastics)

Cons

- Causes land disturbances in the drilling process, which accelerates erosion and increases earthquake activity
- Creates air pollution when burned (e.g., SO_2, NO_x, and CO_2)
- Oil spills, both on land and in the ocean from platforms, pipelines, and tankers, cause a disruption to wildlife (e.g., Deepwater Horizon oil spill in the Gulf of Mexico and the Exxon Valdez oil tanker leak in Alaska).
- Produces water pollution (e.g., wastewater and brine) when it is extracted
- World oil reserves are limited and declining.
- World supplies are politically unstable (e.g., 1973 U.S. oil embargo).

Cogeneration

Cogeneration, also known as combined heat and power (CHP), is an efficient technology to generate electricity and heat simultaneously at local facilities; otherwise, the heat produced from electricity generation is wasted.

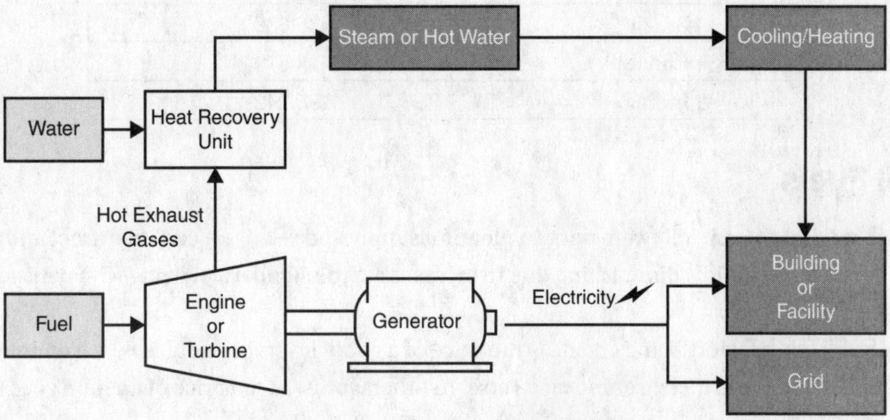

Figure 6.9 Cogeneration system

The supply of high-temperature heat begins by driving a gas or steam turbine-powered generator. The resulting lower temperature heat (formerly waste) is then used for water or space heating.

Cogeneration is most efficient when heat can be used on-site or very close to it and is recognized to be the most energy efficient method of transforming energy from fossil fuels or biomass into electric power. Cogeneration plants are commonly found in central heating systems of larger buildings (e.g., hospitals, hotels) and is commonly used in industry (e.g., processing water, cooling, etc.).

An example of cogeneration at the user level is when the waste heat of a car is used to heat the car's interior.

Pros and Cons of Cogeneration—Combined Heat and Power (CHP)

Pros	Cons
CHP is highly efficient in lowering carbon emissions as it is able to be fueled from a wide variety of renewable (e.g., biomass) or non-renewable (e.g., methane) fuel types.	Construction and maintenance costs are high.
CHP results in a net reduction in greenhouse gas emissions, and air and water pollution.	Due to rapid heat loss over longer distances and the high cost of insulating pipes, smaller CHP plants have higher costs and lower efficiency for producing electricity.
Cogeneration can generate the exact amount of electricity needed at any time or place allowing flexibility.	In order for CHP plants to run efficiently, a match between electricity and heat demand is required.
Cogeneration can meet the needs of a single household up to an industrial complex or city.	Regulatory barriers can be challenging.
Shorter transmission lines result in less transmission power loss and lower construction costs.	Some renewables (e.g., solar and wind) only produce power when the sun is shining or the wind is blowing and would store energy (batteries).

6.4 Distribution of Natural Energy Sources

Different regions of the Earth have access to different forms of renewable and nonrenewable natural energy resources based on their geographic location and past geologic processes. Access, or the lack thereof, to natural energy resources is a major factor that influences a country's economic development. For example, the United States has the world's largest coal reserves, with 491 billion tons, which accounts for ~27% of the world total.

The table below lists the top five countries in terms of the production of renewable energy.

Rank	Country	Total Renewable Energy (GWh)
1	China	1,398,207
2	United States	572,409
3	Brazil	426,638
4	Canada	418,679
5	India	195,242

6.5 Fossil Fuels

Fossil fuels (coal, natural gas, and oil) were once in plentiful supply and relatively easy to extract and transport; however, fixed supplies are rapidly diminishing due to increases in the standard of living with consequential increases in price.

Law of Supply: all other factors being equal, as the price of a good or service increases, the quantity of goods or services that suppliers offer will increase, and vice versa. In other words, as the price of an item goes up, suppliers will attempt to maximize their profits by increasing the quantity offered for sale.

Law of Demand: all other factors being equal, the quantity of the item purchased is inversely related to the price of the item; i.e., if the price of the commodity becomes higher, it has a lower demand.

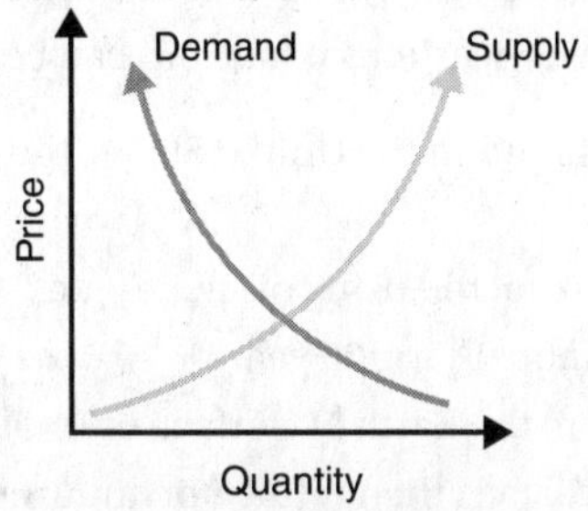

Figure 6.10 Law of Supply and Demand and its effect on price

Fossil fuels (coal, natural gas, and oil) are formed over time from deposits of once-living organisms and take thousands of years to form.

Coal originally comes from land vegetation, which over millions of years decays and becomes compacted. Compared to oil and natural gas, coal is quite abundant and is estimated to last at the current rate of extraction for 200 years or more. Since the mid-20th century, the use of coal has doubled; however, since the mid 1990s, it has declined as a fuel in more affluent countries but increased in less affluent countries.

Natural gas was formed from the remains of marine organisms and is relatively abundant and clean when compared to coal and oil. Natural gas consists primarily of methane (CH_4); it is highly compressed deep within the Earth and is brought to the surface by drilling. Natural gas reserves are more evenly distributed across the Earth than are coal or oil supplies.

Oil is a liquid fossil fuel that formed from the remains of marine organisms, and over millions of years, these deposits became trapped in small spaces in rock and sediment, which now can be accessed by drilling. As oil is extracted from the Earth (crude oil) it must later be refined or heated to separate it into its various components, a process known as refining.

Historically, fossil fuels were available in plentiful supply and were fairly easy to obtain and transport. But now, with diminishing supplies coupled with greater demand, fossil fuels are becoming more expensive and, at the same time, creating larger amounts of accumulating greenhouse gases and other pollutants (e.g., particulates, nitrogen [NO_x] and sulfur [SO_x] oxides) that result in air and water acidification, pollution, and global warming.

Other Fossil Fuel Nonrenewable Energy Resources

Methane Hydrates (Clathrates)

Methane hydrates (methane locked in ice) are a recently discovered source of methane that form at low temperature and high pressure. They are found (1) on land in permafrost regions; (2) beneath the ocean floor at water depths greater than 1,640 feet (500 m), where high pressures dominate and where the hydrate deposits themselves may be several hundred meters thick; and (3) on continental shelves.

Estimates are that there are enough methane hydrates to supply energy for hundreds or thousands of years.

Pros

- Expands natural gas reserves 160 times above the current available supplies; 1 m^3 of methane hydrate equals 164 m^3 of methane.
- Reserves are widely distributed across the globe.

Cons (in Addition to the "Cons" of Methane [CH_4]

- Could cause destabilization of the ocean floor (e.g., landslides and earthquakes).
- Releases methane into the atmosphere when hydrates are mined.

Oil Shale

Oil shale is an organic-rich, fine-grained sedimentary rock containing a solid mixture of organic chemical compounds (kerogen) from which liquid hydrocarbons (shale oil) can be produced.

- About 80% of the world's oil shale is found in the United States, the largest amount being in Wyoming's Green River Formation.
- Around 800 billion barrels could be extracted using oil shale—i.e., 300% as much as Saudi Arabia's proven reserves—meeting U.S. demand for more than 100 years.
- *Ex situ*—oil shale is first extracted from the Earth by surface or underground mining. Rock is then crushed and heated to release the shale oil, which is then refined of impurities, e.g., sulfur.
- *In situ*—oil shale is not mined or crushed. Rock is heated while it is still underground, where various components are separated and extracted.
- Large amounts of water and power are necessary for drilling, mining, and refining.
- Water is contaminated by toxic compounds and is costly to decontaminate.

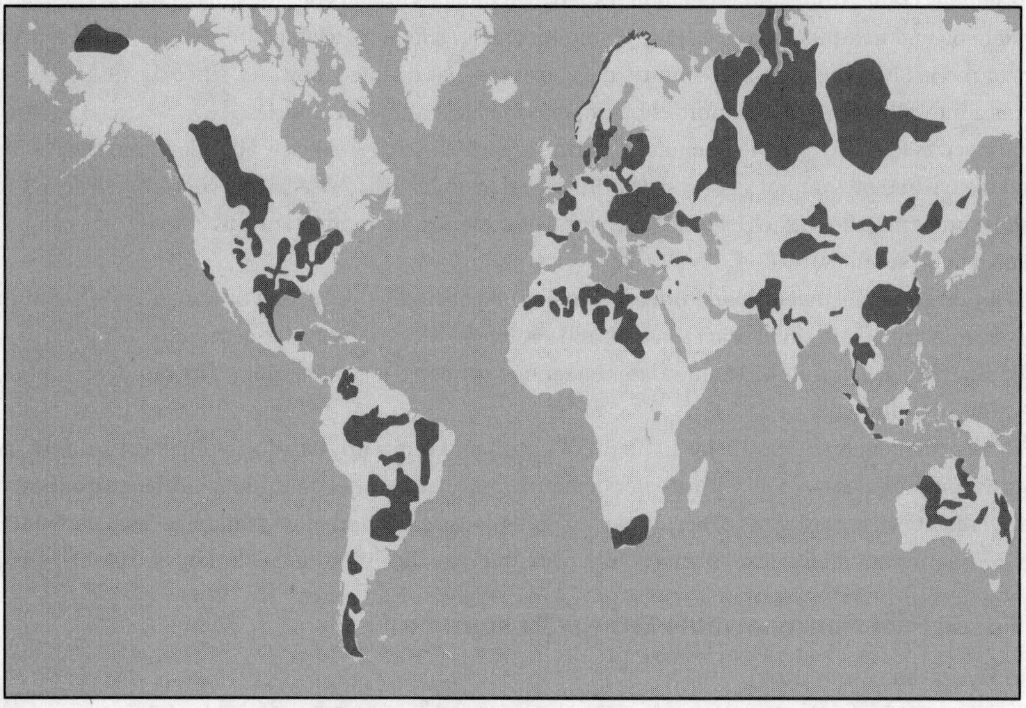

Figure 6.11 Major world oil shale deposits
Source: U.S. Energy Information Administration and U.S. Geological Survey

Pros

- Because there is less drilling involved in most instances, oil shale is generally less expensive to produce per barrel than regular oil products.
- Oil shale requires much less drilling or fracturing than other energy resources, resulting in less of an overall impact on the local environment.
- There will be less dependency on foreign countries and OPEC to meet future energy needs.

Cons

- Causes a negative impact on the local environment (e.g., increased truck traffic, more rigs traveling to sites, etc.).
- Causes earthquakes in local communities.
- Causes groundwater contamination due to extraction and processing. In situ methods have the potential of degrading aquifers.

- Disrupts local wildlife and flora.
- Oil (and other hydrocarbon by-products) derived from oil shale is a fossil fuel. Even though the world has large oil shale reserves, the problem remains that once the oil is obtained from shale, traditional issues of environmental pollution, acid rain, and global warming will continue.
- The net-energy yield of producing oil through oil shale is moderate since energy is required for blasting, drilling, crushing, heating the material, disposing of waste material, and environmental restoration.

Synfuels

A synthetic fuel (synfuel) is any fuel produced from coal, natural gas, or biomass (e.g., corn) through chemical conversion. The raw materials used to make synfuels have to be subjected to intense chemical and physical changes to be usable. An example is synthetic natural gas (SNG), which is chemically produced from coal.

Pros

- Can produce gasoline, diesel, or kerosene directly without having to be sent to a refinery.
- Easily transported through pipelines.
- Produces less air pollution.
- There are enough raw materials available worldwide to meet the current demand for hundreds of years.

Cons

- Low net-energy yield and requires energy to produce.
- Manufacturing plants are expensive to build.
- Product is more expensive than petroleum products.
- Still a carbon-based fuel, which, during combustion, produces gases that are harmful to the environment (e.g., global warming, air pollution, acid rain, etc.).
- Would increase the depletion of coal due to inherent inefficiencies.

Tar Sands

Tar sands contain bitumen—a semi-solid form of oil that does not flow. Specialized refineries are capable of converting bitumen to oil. Tar sand deposits are mined using strip mining techniques; in situ methods, using steam, can also be used to extract bitumen from tar sands. The sulfur content of oil obtained from tar sands is high, about 5%. Most of the tar sand deposits are located in Canada, the Russian Federation, and Venezuela, with those in Canada being the most concentrated and therefore the most economical to mine. The oil in tar sands represents about two-thirds of the world's total oil reserves.

The net-energy yield of producing oil through tar sands is moderate since energy is required for blasting, drilling, crushing, heating the material, disposing of waste material, and environmental restoration.

Pros

- Enormous growth potential since less than 5% has been produced so far.
- Provides jobs in extraction for Canadians and jobs in transportation and refining for Americans.
- Very large supply.
- Will help keep oil prices relatively low.

Cons

- Destructive to major boreal forests, an important carbon sink.
- Enormous greenhouse gas emissions.
- Large amounts of water required.
- Leads to deposition.
- Relatively low net-energy return.
- Requires expensive and risky pipelines (e.g., Keystone XL) to reach faraway markets.

- Water pollution is about 3 million gallons of toxic runoff per day.
- Widespread habitat destruction, both on land and in water.

Combustion

The combustion of any fossil fuel (e.g., coal, methane, oil) follows the following reaction:

$$\text{Fossil fuel} + \text{oxygen gas } (O_2) \rightarrow \text{carbon dioxide } (CO_2) + \text{water} + \text{energy}$$

Carbon dioxide produced during fossil fuel combustion for heat and electricity generation is a major contributor to global CO_2 emissions considered responsible for global warming due to its greenhouse gas effect. Fossil fuels such as coal are, however, expected to be continued to be used through the next several decades. About 2 billion tons of CO_2 are estimated to be emitted per year from existing coal-fired power plants around the world.

According to the United States Energy Information Agency, the consumption of coal and consequent CO_2 emissions are expected to be even greater in the future than they are today; e.g., global carbon dioxide emissions from all uses of coal are expected to increase from 13.0 billion metric tons in 2008 to 19.6 billion metric tons in 2035. In addition to the fossil fuels, CO_2 capture is also relevant to the utilization of "renewable" fuels such as biogas and syngas produced from biomass.

Steps Involved from Fuels to Electricity

Electrical energy generation using steam turbines as an example involves three energy conversions:

1. Extracting thermal energy from the fuel and using it to raise steam;
2. Converting the thermal energy of the steam into kinetic energy in the turbine; and
3. Using a rotary generator to convert the turbine's mechanical energy into electrical energy.

Hydraulic Fracturing ("Fracking")

Hydraulic fracturing, informally referred to as "fracking," is an oil and gas well development process that typically involves injecting water, sand, and chemicals under high pressure into a bedrock formation via a well. This process is intended to create new fractures in the rock as well as increase the size, extent, and connectivity of existing fractures. Hydraulic fracturing is commonly used in low-permeability rocks like sandstone, shale, and some coal beds to increase oil and/or gas flow to a well from petroleum-bearing rock formations.

Environmental issues related to hydraulic fracturing ("fracking") include:

- disruption to wildlife corridors and habitat
- forest fragmentation
- groundwater quality degradation
- night sky light pollution
- noise pollution that interferes with animal instinctive behaviors
- reduced air quality
- spills of chemicals at the surface
- surface water quality degradation from waste fluid disposal
- water availability

6.6 Nuclear Power

Nuclear Energy (Alternative Fuel)

Currently, there are 449 nuclear power plants in the world providing approximately 5% of the world's electricity. The United States produces the most nuclear power, providing around 20% of the U.S. electrical demand, with one in five households and businesses dependent on nuclear-generated electricity. Currently, radioactive wastes make up less than 1% of the total industrial toxic wastes in the United States.

Nuclear Fission

During nuclear fission, an atom splits into two or more smaller nuclei along with by-product particles (neutrons, photons, gamma rays, and alpha and beta particles). The reaction gives off heat (exothermic). If controlled, the heat that is produced is used to produce steam that turns generators that then produce electricity (Figure 6.12). If the reaction is not controlled, a "meltdown" can result. A nuclear meltdown is an informal term for a severe nuclear reactor accident that results in core damage from overheating, such as the Fukushima Daiichi nuclear disaster.

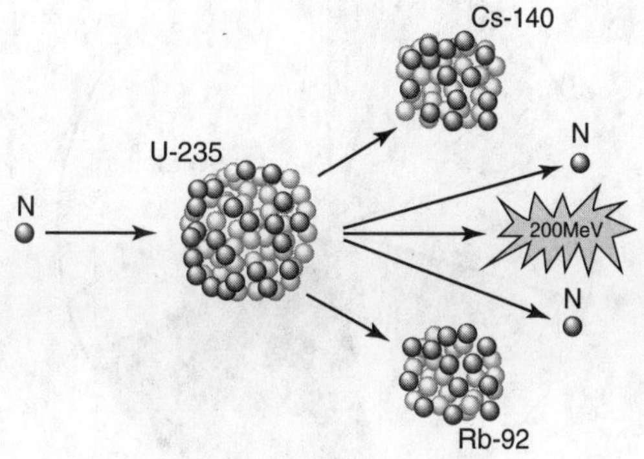

Figure 6.12 Nuclear fission

The amount of potential energy contained in nuclear fuel is 10 million times more than that of more traditional fuel sources such as coal or petroleum. The downside is that nuclear wastes remain highly radioactive for thousands of years and are difficult to dispose of.

The most common nuclear fuels are U-235, U-238, and Pu-239.

Nuclear Fuel

U-235

Less than 1% of all-natural uranium on Earth is U-235. U-235 differs from U-238 in its ability to produce a fission chain (self-sustaining) reaction. The minimum amount of U-235 required for a chain reaction is called the critical mass. Low concentrations of U-235 can be used if the speed of the neutrons is slowed down through the use of a moderator or control rods. Uranium that has been processed to separate out U-235 is known as enriched uranium. Nuclear weapons contain 85% or more of U-235, whereas nuclear power plants contain about 3%. The half-life (the time it takes for the radioactivity to fall to half its original value) of U-235 is 700 million years.

U-238

U-238 is the most common (99.3%) isotope of uranium and has a half-life of 4.5 billion years. When hit by a neutron, it eventually decays into Pu-239, which is used as a fuel in fission reactors. Most depleted uranium is U-238.

Pu (Plutonium)-239

Pu-239 has a half-life of 24,000 years and is produced in breeder reactors from U-238. Plutonium fission provides about one-third of the total energy produced in a typical commercial nuclear power plant. Control rods in nuclear power plants need to be changed frequently due to the buildup of Pu-239, which can be used for nuclear weapons. International inspections of nuclear power plants monitor the amount of Pu-239 produced by power plants.

Nuclear Power

As of 2017, 30 countries worldwide were operating 449 nuclear reactors for electricity generation, with 60 new nuclear power plants under construction, providing 11% of the world's energy requirements (Figure 6.13). The use of nuclear energy to produce electricity in the United States started during the 1960s and increased rapidly until the late 1980s. Reasons for its decline included cost overruns, higher-than-expected operating costs, safety issues, disposal of nuclear wastes, and the perception of its being a risky investment. However, due to electricity shortages, fossil fuel price changes and fluctuations, newer and safer technology, and global warming, there is a renewed interest in and demand for nuclear power plants.

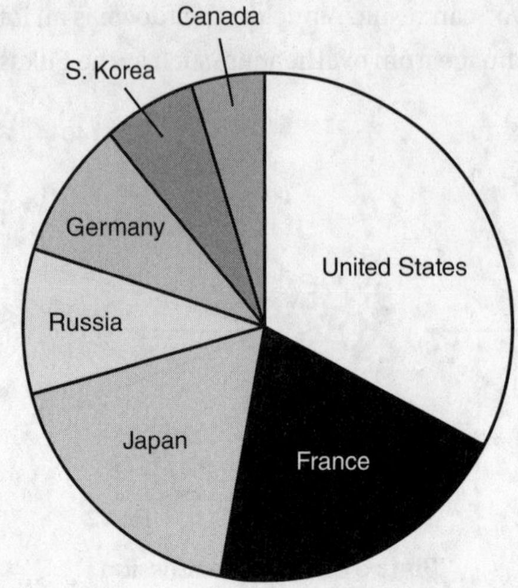

Figure 6.13 Countries that produce the most nuclear power (GW)

The United States is the world's largest producer of nuclear power (805 billion kWh in 2017, over 19% of the total electrical output), accounting for more than 30% of worldwide nuclear generation of electricity. France produces the highest percentage of its energy from nuclear reactors (78%), compared to 30% for the rest of the European Union.

Nuclear Reactor Components

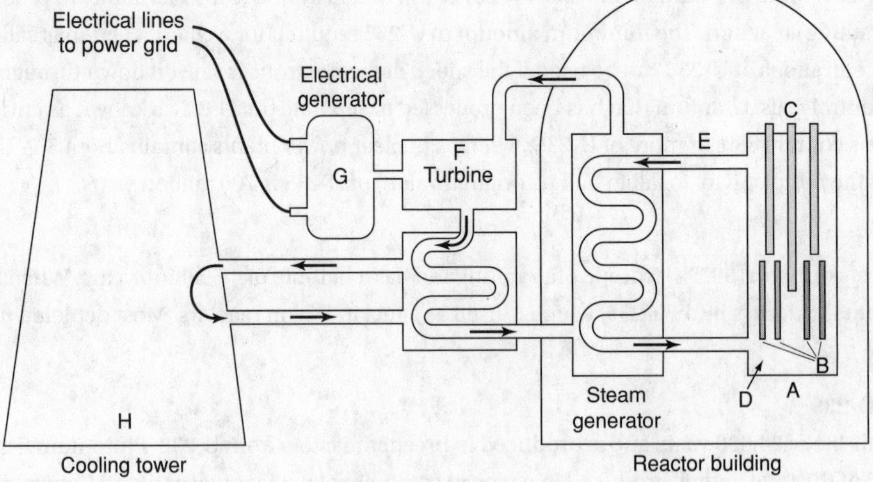

Figure 6.14 Diagram of a nuclear power plant

(A) Core—contains up to 50,000 fuel rods. Each fuel rod is stacked with many fuel pellets; each pellet has the energy equivalent of 1 ton (0.9 m.t.) of coal or 17,000 ft^3 (481 m^3) of natural gas or 149 gallons (564 L) of oil (Figure 6.14).

(B) Fuel—enriched (concentrated) U-235 is usually the fuel. The fission of an atom of uranium produces 10 million times the energy produced by the combustion of an atom of carbon from coal.

(C) Control rods—move in and out of the core to absorb neutrons and slow down the reaction.

(D) Moderator—reduces the speed of fast neutrons, thereby allowing a sustainable chain reaction. Moderators can be water, graphite (which can produce plutonium for weapons), or deuterium oxide (heavy water ^2H_2O or D_2O). Deuterium is an isotope of hydrogen whose nucleus comprises both a neutron and a proton, whereas the nucleus of a normal hydrogen atom consists of just one proton.

(E) Coolant—removes heat and produces steam to generate electricity.

(F) Turbine—a rotary mechanical device that extracts energy from a fluid flow and converts it into useful work. The work produced can be used for generating electrical power when combined with a generator.

(G) Electrical generator—a device that converts motive power (mechanical energy) or fuel-based power (chemical energy) into electric power.

(H) Cooling tower—a specialized heat exchanger in which air and water are brought into direct contact with each other in order to reduce the water's temperature. As this occurs, a small volume of water is evaporated, reducing the temperature of the water being circulated through the tower.

Pros

- Disruption of land is low to moderate.
- It is possible to fuel nuclear power plants with thorium, which is a greener alternative. China, Russia, and India have plans to start using thorium to fuel their reactors in the near future.
- It releases about one-sixth the CO_2 that fossil-fuel plants do, thus reducing global warming.
- Nuclear energy produces 62% of all U.S. emission-free electricity.
- Nuclear power plants provide a stable baseload of energy that can work synergistically with renewable energy sources such as wind and solar (e.g., the electricity production from the plants can be lowered when wind and solar resources are available and increased when the demand is high).
- There is enough uranium for the next 80 years.
- Water pollution is low.

Cons

- CO_2 and other wastes are released during the enrichment process.
- Decommissioning a nuclear power plant can cost over $300 million, which seriously impacts net profit.
- Exposure to radiation results in increases in genetic mutations and cancers and decreases in fertility rates.
- Low net-energy yield—the energy required for mining uranium, processing ore, building and operating the plant, decommissioning and dismantling the plant, and storing wastes.
- Mining uranium results in mine tailings that are radioactive, the alteration of wildlife habitats, and groundwater contamination.
- Nuclear wastes take millions of years to degrade—problems include where to store them and how to keep them out of the hands of terrorists.
- There can be safety and maintenance issues, including possible release of radioactive material into the environment (e.g., Fukushima Daiichi, Three Mile Island, and Chernobyl).
- Thermal pollution is the most persistent environmental threat—the effects can be diminished through the use of heat exchangers and cooling towers and ponds. It impacts the local environment by
  - decreasing biodiversity due to lower dissolved oxygen levels and higher water temperatures;
  - increasing bacterial growth in the water, which increases the risk of diseases;
  - increasing the growth of algae in waterways and decreasing the dissolved oxygen levels; and
  - increasing metabolic rates, which can result in thermal shock.
- Uranium is a nonrenewable resource.

CASE STUDIES

- **Chernobyl (1986)**
  - The Chernobyl disaster was a catastrophic nuclear accident that occurred in a light water graphite-moderated reactor in northern Ukraine and resulted in ~23,000 radiation-induced cancers and ~16,000 excess thyroid cancers as a result of iodine-131 exposure.
  - Over 572 million people in 40 different countries received some exposure to radioactivity—over 40 times more fallout than Hiroshima or Nagasaki.
- **Fukushima-Daiichi (2011)**
  - Following the 2011 Tohoku, Japan, earthquake, measured at a Richter scale reading of 9.0, a series of equipment failures, explosions, and release of radioactive materials occurred at the nuclear power plant.
  - The ~50-foot (15 m) tsunami that followed the earthquake flooded the emergency generators that circulated coolant, causing the reactor to overheat and explode.

6.7 Energy from Biomass

Biomass

Biomass is biological material derived from living, or recently living, organisms (e.g., forest residues, yard clippings, and wood chips) that can be burned in large incinerators to create steam that is used for generating electricity. Industrial biomass can be grown from numerous types of plants, such as switch grass, hemp, corn, poplar, willow, sugarcane, and bamboo. Approximately 50% of the world's renewable energy supply is derived from biomass.

Pros

- Biomass can be grown on marginal land that is not suitable for agriculture.
- Crop residues are available.
- Reduces the impact on landfills.

Cons

- Can result in the destruction of native habitats and the concurrent decrease in biodiversity
- Causes an increase in CO_2 production, resulting in a greater impact on global warming
- Causes severe air pollution if not burned in a centralized facility with air pollution control devices (e.g., burning wood in a home to cook meals)
- Could deplete the soil of nutrients if the same biomass continuously grows in the same location
- Land use for growing a variety of food crops in a sustainable manner
- Net-energy yield is low to moderate. Energy is required for drying and transporting material to a centralized facility.

Biogas

Anaerobic digestion is a collection of processes by which microorganisms break down biodegradable material, in the absence of oxygen, to produce methane gas, which is then burned to produce energy. Biodegradable waste materials that are used in anaerobic digesters include waste paper, grass clippings, leftover food, sewage sludge, and animal waste (Figure 6.15).

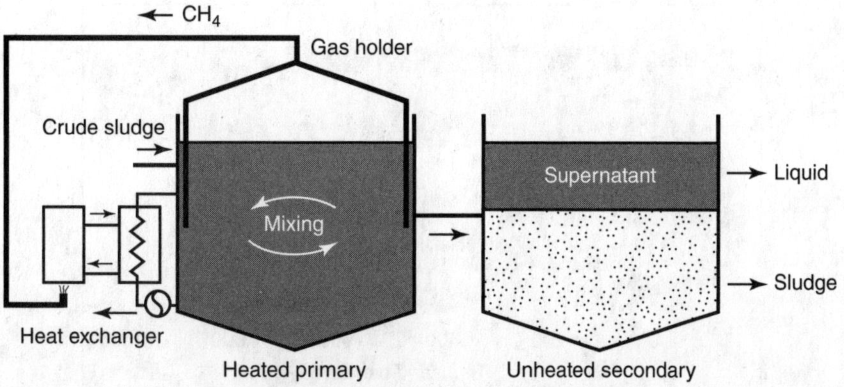

Figure 6.15 Anaerobic digester
Source: United Nations Environment Program
Division of Technology, Industry, and Economics

Pros

- Reduces the amount of animal waste that needs to be disposed of.
- Reduces the impact of land disturbances required for coal mining, oil, and natural gas drilling and extraction, including fracking.
- Reduces the impact of water pollution coming from feedlots, thereby reducing fecal bacterial contamination and nutrient loading that leads to eutrophication.
- Reduces the methane emissions from landfills that contribute to global warming.
- Reduces the reliance on coal and oil, thereby reducing the amount of mercury, sulfur, and particulates released into the atmosphere.
- There is no net increase in CO_2 emissions (i.e., the CO_2 released by burning methane came from plants that recently removed it from the atmosphere through photosynthesis rather than from fossil fuels that were formed millions of years ago).

Cons

- Contains some gases as impurities, which are corrosive to the metal parts of internal combustion engines.
- Difficult to enhance the efficiency of biogas systems.
- Not feasible to locate at all locations.
- The process is not as efficient or as economical on a large industrial scale as compared to other biofuels.

Biofuels

A biofuel is a liquid fuel (e.g., ethanol) produced from living organisms (e.g., corn, sugarcane) and is most commonly used in the United States and developing nations such as Brazil, China, and India. The United States and Brazil account for 90% of all biofuels produced (Figure 6.16).

Pros

- Biofuel is a renewable energy source as long as it is used sustainably (e.g., addressing soil erosion and deforestation).
- Biofuels are biodegradable.
- Biofuels can be converted into biodiesel or bioethanol to power vehicles.
- Biofuels can be produced anywhere, as opposed to fossil fuels, which only occur in specific locations (e.g., coal, oil, and natural gas).

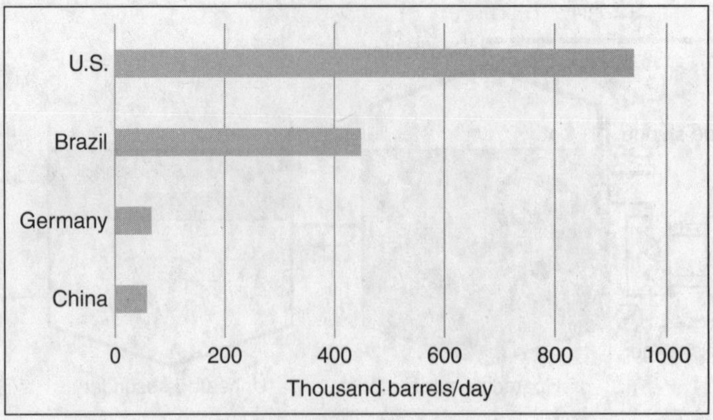

Figure 6.16 Countries that produce the most biofuels
Source: World Economic Forum

- Biofuels could supply half of the world's demand for electricity.
- If algae is used as the source for the biofuel, several unique advantages are possible:
  - Algal biofuel can yield up to 100 times more fuel per unit area when compared with other biofuels, such as corn or sugarcane, and it can be grown on marginal land that is unsuitable for traditional crops.
  - Algae can be grown with minimal impact on freshwater resources. It can also recover phosphorus from wastewater as a method to reduce the impact of fertilizer runoff. Ammonia, nitrates, and phosphates that normally render water unsafe serve as nutrients for algae.
  - Algae cultivation requires significantly less pesticides and fertilizers, resulting in less environmental impacts due to runoff.
  - Algae has a faster growth rate than terrestrial crops and can convert a higher fraction of their biomass to oil than traditional crops can (e.g., 60% for algae compared to about 3% for terrestrial plants).
  - The cultivation of algae does not significantly impact the land, resulting in less of an impact on wildlife, habitats, biodiversity, changes to the soil that can result in erosion, salinization, desertification, loss of soil nutrients, and groundwater depletion.
- The combustion of biofuels does not produce SO_x, CO, or unburned hydrocarbons.
- When biofuels are burned for energy, the carbon they release was only recently removed from the atmosphere via photosynthesis, not from long-standing carbon sinks; therefore, it does not contribute to a net increase in the amount of CO_2 in the atmosphere.

Cons

- Requires adequate water and fertilizer, sources of which are declining.
- The corn being diverted to produce ethanol would raise food prices (Law of Supply and Demand: if all other factors remain equal, the higher the price of a good, the less people will demand that good; Figure 6.10).
- There is a risk of a loss of jobs in current and traditional fossil-fuel producing and supporting industries.
- There is the potential for massive deforestation and a loss of habitat, resulting in a decrease in biodiversity.

6.8 Solar Energy

Active and Passive

Solar energy consists of collecting and harnessing radiant energy from the sun to provide heat and/or electricity. Electrical power and/or heat is generated at home and at industrial sites through photovoltaic cells, solar collectors, or at a central solar-thermal plant.

A passive solar heating design does not include any type of mechanical heating device and functions by incorporating building features that absorb heat and then release it slowly to maintain the temperature throughout the building. These building features, often referred to as thermal mass, may include large windows, stone flooring, and brick walls. The energy collected moves through the building according to the Second Law of Thermodynamics, which states that heat moves from warm to cool areas.

Additional steps that homeowners of passive solar designs can use (depending on the local solar and weather conditions) include the following:

- In areas of high solar input, planting deciduous trees to block sunlight (leaves fall in the winter to allow sunlight in); in areas of little solar input, decreasing light-blocking vegetation around windows
- Increasing insulation in attics, walls, and crawl spaces or basements
- Installing awnings and thermal blinds on all windows
- Using double- or triple-paned windows with heat-reflective coatings
- Using reflective coatings on roofs and exterior walls
- Using skylights to allow free sunlight in
- When designing a home, positioning the home to maximize or minimize (depending on the location) solar input

Active solar heating is a much more involved process, generates more heat than passive systems, and relies on three components: a solar collector to absorb the solar energy, a solar storage system, and a heat transfer system (e.g., pumps and blowers) that uses an external source of power to disperse the heat to the appropriate places in the building (Figure 6.17).

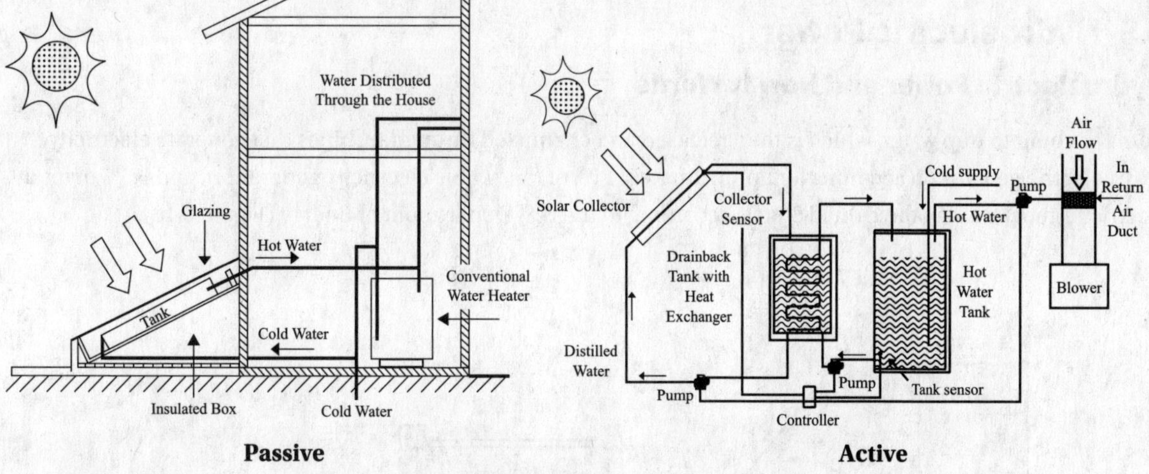

Figure 6.17 Passive vs. active solar systems

A residential photovoltaic system consists of solar panels to absorb and convert sunlight into electricity, a solar inverter to change the electric current from DC to AC, and a battery storage and backup system. Systems range from small, residential rooftop-mounted systems with capacities from a few to several tens of kilowatts, to large utility-scale power stations of hundreds of megawatts. Nowadays, most residential systems are connected to the electrical grid, with any extra electricity generated sold back to the local power company. A residential rooftop system produces about 95% of net clean renewable energy over its expected lifetime of about 30 years. Because of the exponential growth of photovoltaics, prices for PV systems have rapidly declined in recent years.

Federal, state, and municipal governments can encourage the adoption of more solar energy systems by

- providing tax incentives, credits, subsidies, and/or low-interest loans to manufacturers and/or homeowners; and
- requiring all new construction to include some type of solar energy system that is appropriate for the area.

Pros

- Pollution only comes from the manufacturing and disposal of collectors. No air, water, nuclear, or thermal pollution is produced during the generation of electricity.
- The supply of solar energy is limitless.
- There is very little impact on wildlife or local habitats as collectors are usually mounted on rooftops.
- When connected to the grid, home solar photovoltaic systems can be smaller and thus less expensive, no back-up system is required, and excess energy can be sold back to the energy companies.

Cons

- Commercial facilities require large amounts of land, resulting in changes to native habitats.
- Current efficiency is between 10% and 25% and is not expected to increase anytime soon.
- It is inefficient where sunlight is limited or seasonal, thus requiring backup systems.
- Systems deteriorate and must be replaced.
- Toxic materials are required in the manufacturing of solar collectors and photovoltaic systems, including battery backup systems, and safe disposal procedures and practices are required.

Worldwide Growth in Solar Power

Photovoltaic (PV) systems are gaining momentum on a worldwide scale. Asia is the fastest-growing region in the world for the installation of PV systems, with more than 60% of the global installations in China and Japan alone, accounting for half of all worldwide installations. Cumulative photovoltaic capacity is currently sufficient to supply about 1% of global electricity demands, and by 2050, solar power is anticipated to become the world's largest source of electricity.

6.9 Hydroelectric Power

Hydroelectric Power and How It Works

Dams are built to trap water, which is then released and channeled through turbines that generate electricity. Hydroelectric generation accounts for approximately 44% of renewable electricity generation, and 6.5% of total electricity generation in the United States, according to the U.S. Department of Energy (Figure 6.18).

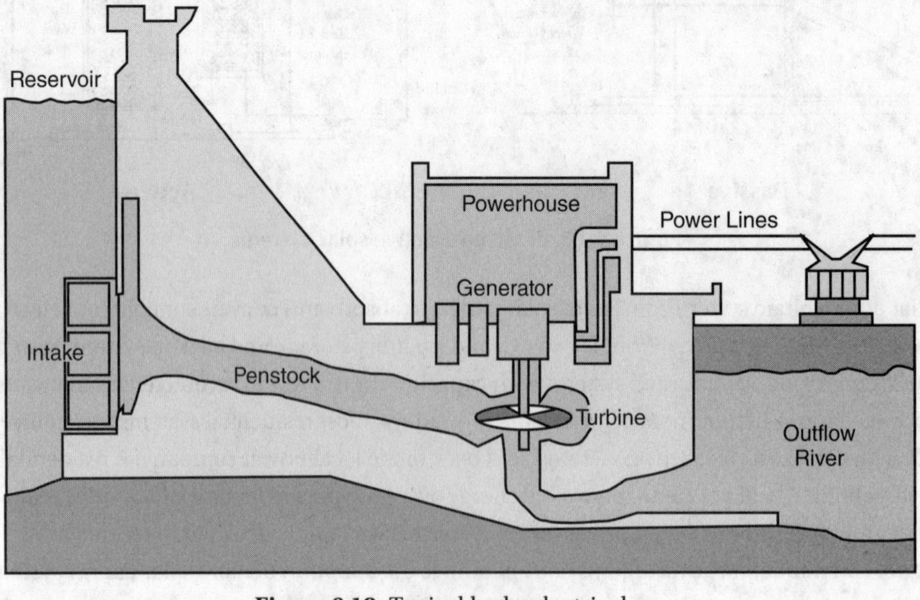

Figure 6.18 Typical hydroelectric dam

There are about 75,000 dams in the United States that block ~600,000 miles (~1 million km) of what had once been free-flowing rivers. Salmon are migratory fish that hatch in streams and rivers and then swim downstream to the ocean to live most of their lives, returning to the rivers and streams from which they hatched to spawn (anadromous). Dams now block almost every major river system in the West and have destroyed important spawning and rearing habitats for salmon. As a result of habitat destruction, at least 100 major U.S. West Coast salmon runs are extinct. Dams also change the character of rivers, creating slow-moving, warm-water pools that are ideal for predators of salmon. Low water velocities in large reservoirs can also delay salmon migration and expose fish to higher water temperatures and disease.

To reduce the impacts of dams on fish, fish passage facilities and fish ladders have been built to help juvenile and adult fish migrate over or around many dams. Spilling water at dams over the spillway can also help pass juvenile fish downstream because it avoids sending the fish through turbines. Water releases from upstream storage reservoirs have been used to increase water velocities and to reduce water temperatures to improve migration. Juvenile fish are also collected and transported downstream in barges and trucks.

Pros

- Dams help control flooding.
- Long life spans.
- Low operating and maintenance costs, which result in affordable electricity.
- Moderate to high net-useful energy.
- No polluting waste products.
- Provide water storage for municipal and agricultural use and serve as areas for water recreation (e.g., boating, fishing, and swimming).

Cons

- Dams are expensive to build.
- Dams create large flooded areas behind the dam from which people are displaced.
- Dams destroy wild rivers.
- Dams destroy wildlife habitats and keep fish (e.g., salmon) from migrating.
- Dams reduce the amount of land available for agriculture.
- Sedimentation behind the dam requires dredging. Dredging prevents sediments and nutrients from reaching downstream and enriching the farmland.
- There is a risk of flooding from catastrophic dam failure (e.g., from earthquakes).
- Water behind the dam is slow-moving and can be a breeding ground for mosquitoes, snails, and flies—the vectors that carry malaria, river blindness, schistosomiasis, and the Zika virus.

Flood Control

Floods can be caused by the following:

- Failures of dams, levees, and pumps
- Fast snowmelt
- Increased amounts of impervious surfaces, e.g., asphalt or concrete
- Natural hazards, such as wildfires, that reduce the supply of vegetation that absorbs rainfall
- Prolonged heavy rainfall
- Severe winds over water
- Tsunamis
- Unusually high tides and storm surges

The effects of flooding can be diminished by the following:

- Barrier islands, levees, reservoirs, and sea walls
- Catchment basins, diversion canals, and stream channels
- Estuaries and floodplains
- Groundwater replacement
- Surface runoff
- Water being absorbed by grass and vegetation

6.10 Geothermal Energy

Geothermal

Heat contained in underground rock and fluids from molten rock (magma), hot dry-rock zones, and warm-rock reservoirs produces pockets of underground steam and hot water that can be used to drive turbines, which can then generate electricity. Geothermal energy is being used in California, Hawaii, Iceland, Japan, Mexico, New Zealand, and Russia, and geothermal resources tend to follow tectonic plate boundaries (Figure 6.19).

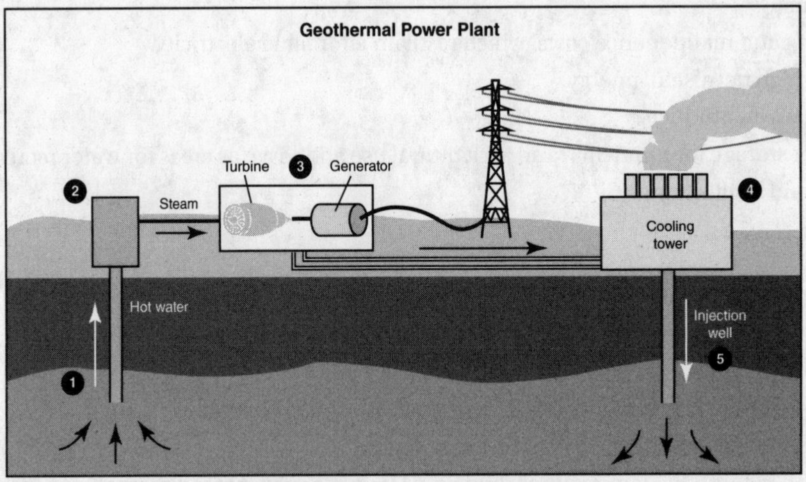

Figure 6.19 Geothermal power plant
Source: U.S. Environmental Protection Agency

Pros

- Competitive cost; no significant cost fluctuations determined by gas and oil prices
- Limitless and reliable source, if managed properly
- Little air pollution; small carbon footprint
- Moderate net-energy yield
- Not weather dependent like solar or wind are
- Potential 60% savings on business and residential heating and cooling costs
- The disruption to wildlife and land is minimal.

Cons

- Can degrade the ecosystem due to corrosive, thermal, or saline wastes
- Causes noise, odor (hydrogen sulfide—H_2S), and land subsidence
- Discharges into the Earth would include sulfur dioxide and silica, potentially contaminating the groundwater.
- High upfront costs for building plants
- Reservoir sites are scarce.
- Source can be depleted if not managed properly.
- Uses a large amount of water

6.11 Hydrogen Fuel Cells

Hydrogen Fuel Cells and How They Work

Nine million tons of hydrogen are produced in the United States each day—enough to power 20 to 30 million cars or 5 to 8 million homes. Most of this hydrogen is used by industries in refining and treating metals and processing foods. The hydrogen fuel cell operates similarly to a battery with two electrodes—oxygen passes over one and hydrogen passes over the other. The hydrogen reacts with a catalyst to form negatively charged electrons and positively charged hydrogen ions (H^+). The electrons flow out of the cell to be used as electrical energy. The hydrogen ions then move through a membrane, where they combine with oxygen and the electrons to produce water. Unlike batteries, fuel cells never run out (Figure 6.20).

NOTE

Remember, hydrogen fuel is not a renewable energy source.

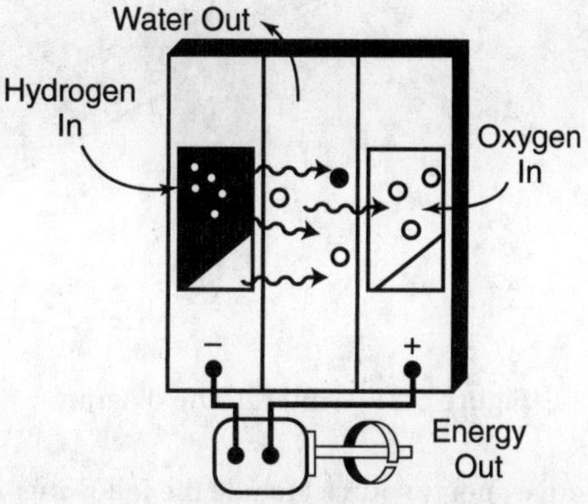

Figure 6.20 Hydrogen fuel cell

Pros

- Hydrogen does not destroy wildlife habitats and has minimal environmental impact.
- Hydrogen can be stored in compounds to make it safe to handle. Hydrogen is explosive, but so are methane, propane, butane, and gasoline.
- Hydrogen is easily transported through pipelines.
- Ordinary water (either ocean water or freshwater) can be used to obtain hydrogen.
- The energy to produce hydrogen could eventually come from fusion reactors or other less-polluting sources.
- The waste product is pure water.

Cons

- As of now, it is difficult to store hydrogen gas for automobiles.
- Changing from a current fossil fuel–based system to a hydrogen-based system would be very expensive.
- Energy is required to produce the hydrogen from either water or methane.
- Hydrogen gas is explosive.

6.12 Wind Energy

Wind

Wind turbines work very simply: instead of using electricity to make wind—like a fan—wind turbines use wind to make electricity. Wind turns the giant turbine blades, and then that motion powers generators (Figure 6.21). Wind turbines clustered together are often called wind farms. Using wind power is by far the most efficient method of producing electricity (Figure 6.22).

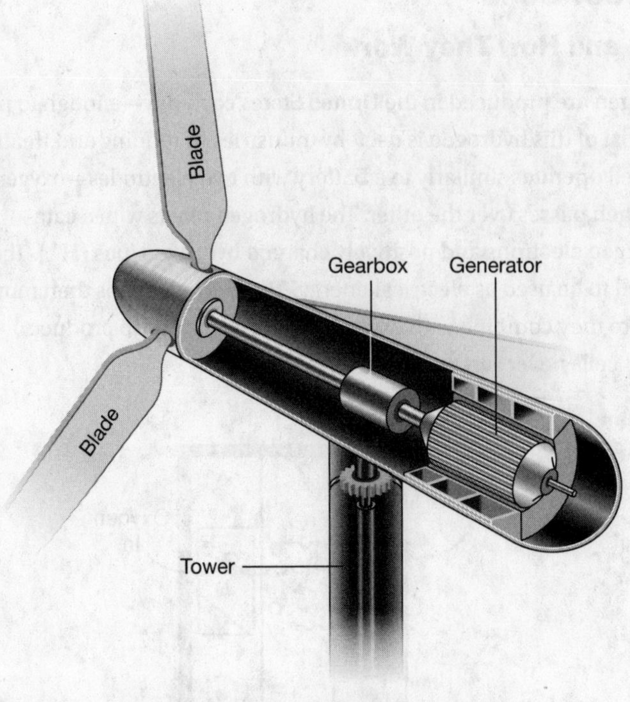

Figure 6.21 Wind turbine diagram

Facts about wind power as an alternative energy source include the following:

- About 6% of the electrical demand in the United States is now produced from wind energy.
- Offshore wind represents a major opportunity to provide power to highly populated coastal cities.
- One megawatt of wind energy can offset approximately 2,600 tons of CO_2.
- The country with the largest wind energy installed capacity is China, followed by the United States.
- The current capacity of wind power in the United States powers approximately 20 million homes.
- The largest turbines can harness energy to power 600 American homes.
- There has been a 25% increase in wind turbine use in the last decade, but wind energy only provides a small percentage of the world's energy.

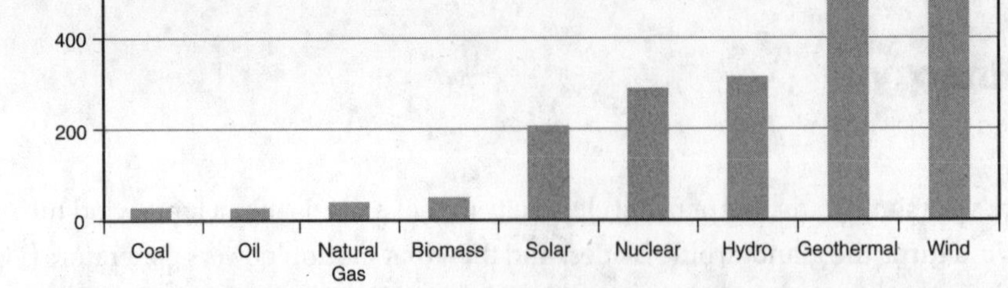

Figure 6.22 Energy efficiency of various methods that produce electricity

Pros

- All electrical needs of the United States could be met by wind in North Dakota, South Dakota, and Texas.
- Destruction and changes in the natural habitat are minimal.
- Maintenance is low, and wind farms are automated.
- The land underneath wind turbines can be used for agriculture (multiple-use).
- There is moderate to high net-energy yield.
- There is no pollution. Wind farms are in remote areas, so noise pollution is minimal to humans.
- Wind farms can be built quickly and can also be built out on sea platforms.

Cons

- Backup systems need to be in place when the wind is not blowing.
- Causes visual and noise pollution
- May interfere with communication, such as microwaves, TVs, and cell phones
- May interfere with the flight patterns of birds
- Steady wind is required to make the investment in wind farms economical; few places are suitable.

6.13 Energy Conservation

How We Conserve

Energy Star® is a joint program of the U.S. Environmental Protection Agency and the U.S. Department of Energy and is designed to protect the environment through energy-efficient products and practices. Each product, building, or home that is Energy Star® certified is independently certified to use less energy and cause fewer of the emissions that contribute to climate change. Since its inception in 1992, Energy Star® has saved American families and businesses more than $450 billion in energy costs and over 3.5 trillion kWh of electricity, and has decreased greenhouse gas emissions by 3.5 billion tons (3,200 kg). There are hundreds of ways to conserve energy. Below are just a few examples:

- Add extra insulation and seal air leaks. Improving attic insulation and sealing air leaks can save 10% or more on annual energy bills.
- Change to a programmable HVAC (heating, ventilation, and air conditioning) thermostat. A programmable thermostat can save as much as 15% on heating and cooling costs.
- Change to more efficient LED lighting—unlike fluorescent lights, LED lights do not contain mercury and can be disposed of with the regular household trash.
- Minimize phantom loads. "Phantom load" refers to the energy that an appliance or an electronic device consumes when it is not actually turned on. Seventy-five percent of the electricity used to power home electronics is consumed while the products are turned off.
- Use more energy-efficient appliances. Energy Star® appliances use between 10% and 50% less energy and water than their conventional counterparts.

Energy Consumption (Past)

Wood (a renewable energy source) served as the primary form of energy until the mid- to late-1800s, even though water mills were important to some early industrial growth. Coal became dominant in the late 19th century before being overtaken by oil in the middle of the 20th century, a time when natural gas usage also rose quickly. From the late 19th century until the early 21st century, coal and oil were the world's primary sources of energy with the largest growth (Figure 6.23).

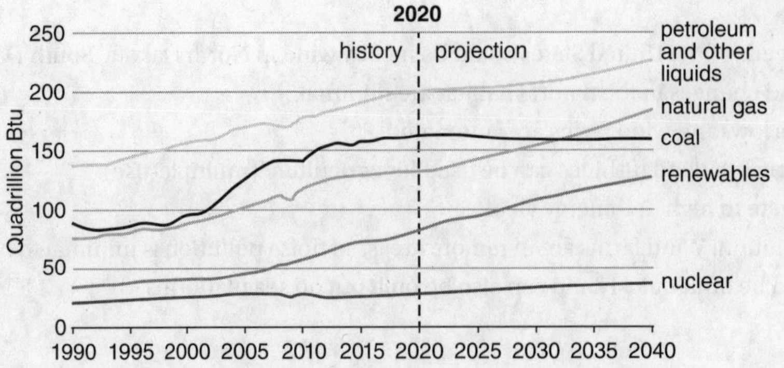

Figure 6.23 Global primary energy consumption 1990–2040 (projected)
Source: U.S. Energy Information Administration

The three major fossil fuels—petroleum, natural gas, and coal, which together provided 87% of the total U.S. primary energy over the past decade—have dominated the U.S. fuel mix for over 100 years.

The United States was energy independent until the late 1950s. At that time, energy consumption began to outpace domestic production, which then led to oil imports (Figure 6.24).

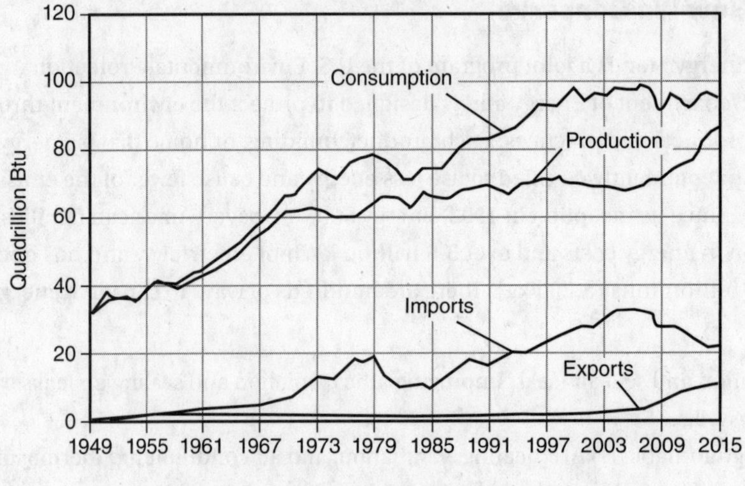

Figure 6.24 U.S. primary energy overview (1949–2015)
Source: U.S. Energy Information Administration

Energy Use (Present)

Currently, about 40% of the world's energy is consumed in the Asia Pacific region (mostly China), 25% is consumed in Europe, and 20% is consumed in North America. However, the energy demand projected for the year 2035 shows marked increases in the energy demand for Africa, the Middle East, and India due to projected population growth (Figure 6.25).

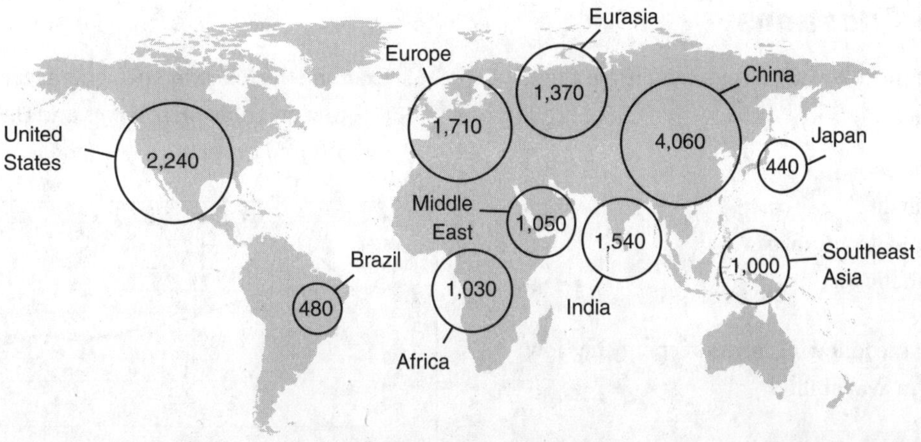

Figure 6.25 Projected energy demand for 2035 (million tons of oil equivalent)
Source: International Energy Agency, "World Energy Outlook"

Energy Crisis

In a free-market economy, the price of energy is driven by the principles of supply and demand, and sudden changes in the price of energy can occur if either the supply or the demand changes.

In some cases, an energy crisis can be brought on by a failure of world markets to adjust prices in response to shortages. For example, the oil supply is largely controlled by nations with significant reserves of easily extractable oil, such as Saudi Arabia, which belongs to an association of oil-producing countries known as OPEC (Organization of Petroleum Exporting Countries). When OPEC reduces the output quotas of its member countries, the price of oil increases as the supply diminishes. Similarly, OPEC can boost oil production to increase supplies, which drives down the price. When OPEC raises the price of oil too high, the demand decreases and the production of oil from alternative sources becomes profitable.

LAW OF SUPPLY AND DEMAND

The Law of Supply and Demand explains the interaction between the supply of a resource and the demand for that resource. It also defines the effect the availability of a particular product and the desire (or demand) for that product have on price. Generally, a low supply and a high demand increases the price, while in contrast, a greater supply and a lower demand causes the price to fall.

Practice Questions

1. Which of the following forms of energy is a renewable resource?
   (A) Biomass
   (B) Oil shale
   (C) Synthetic natural gas
   (D) Synthetic oil

2. Which of the following forms of energy has low short-term availability?
   (A) Coal
   (B) Nuclear energy
   (C) Petroleum
   (D) Solar energy

3. Which of the following is a source of moderate to high net energy?
   (A) Fission
   (B) Geothermal
   (C) Tar sands
   (D) Wind

4. Which of the following alternatives would NOT lead to a sustainable energy future?
   (A) Assess penalties or taxes on continuous use of coal and oil
   (B) Create policies to encourage governments to purchase renewable energy devices
   (C) Create tax incentives for independent power producers
   (D) Decrease fuel-efficiency standards for cars, appliances, and HVAC (heating, ventilation, and air conditioning) systems

5. Examine the following two graphs that show U.S. proven crude oil reserves and U.S. crude oil imports and exports.

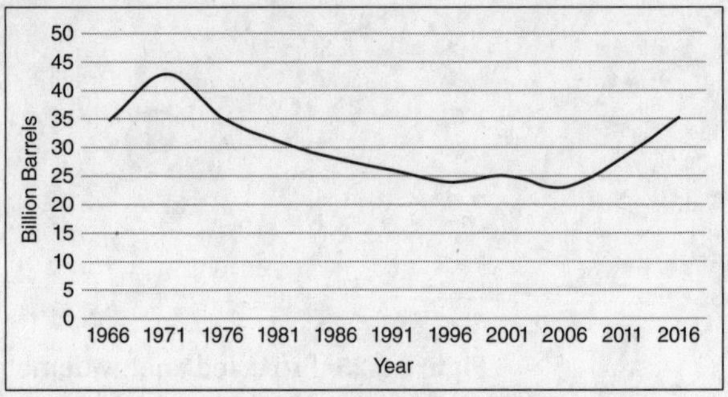

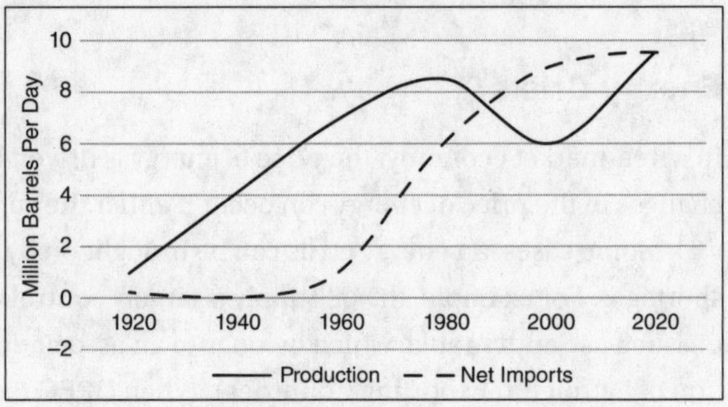

Which of the statements listed below are reasonable answers as to why the United States continues to import crude oil and yet has a sizeable proven crude oil reserve?

I. Venezuela and Canada export less expensive, asphalt-laden oil. U.S. refineries can import this oil at a discount, as few other countries have the necessary technology and refining equipment to produce gasoline and diesel from it, which frees up a lot of high-quality U.S. domestic oil to be shipped overseas and sold at a premium to countries with lower-tech refineries; i.e., the United States imports cheap oil and exports expensive oil. It's profitable because the transport cost is less than the price difference between high- and low-quality oil.

II. Transport options within the United States are not fully capable of handling all the necessary movement of oil; e.g., Oklahoma may have a surplus of oil, which frees up some of the Gulf Coast oil production to go overseas, while New England doesn't have enough pipelines connecting it to the rest of the country to domestically import all its oil, and has relatively older, lower-tech refineries, with the result that New England must import oil.

III. The price of oil fluctuates, and if the price of oil is low, there is a strong financial incentive to store oil and wait until it can be sold at a higher price.

(A) I
(B) II
(C) III
(D) All statements are reasonable and true.

6. Which country currently ranks number one in both coal reserves and the use of coal as an energy source?

(A) China
(B) India
(C) Russian Federation
(D) United States

7. The lowest average sustainable generating cost (cents per kWh) comes from what energy source?

(A) Coal
(B) Large hydroelectric facilities
(C) Nuclear
(D) Solar photovoltaic

8. The fastest-growing renewable energy resource today is

(A) geothermal
(B) large-scale hydroelectric
(C) nuclear
(D) wind

9. The least-efficient energy conversion device listed below is a (an)

(A) fluorescent lightbulb
(B) incandescent lightbulb
(C) internal-combustion engine
(D) steam turbine

10. Which is NOT an advantage of using nuclear fusion?

(A) Abundant fuel supply
(B) No air pollution
(C) No high-level nuclear waste
(D) All are advantages

11. Only about 10% of the potential energy of gasoline is used in powering an automobile. The remaining energy is lost as low-quality heat. This is an example of the

(A) First Law of Efficiency
(B) First Law of Thermodynamics
(C) Law of Conservation of Energy
(D) Second Law of Thermodynamics

12. The Law of Conservation of Mass and Energy states that matter can neither be created nor destroyed and that the total energy of an isolated system is constant despite internal changes. Which society offers the best long-term solution to the constraints of this law?

(A) Free-market economy
(B) Global market economy
(C) High-throughput society
(D) Low-throughput society

13. Which of the following methods CANNOT be used to produce hydrogen gas?

(A) Coal gasification
(B) Producing it from plants
(C) Reforming
(D) All of the above are methods of producing hydrogen gas.

14. Examine the graph below to see U.S. electricity net generation by fuel.

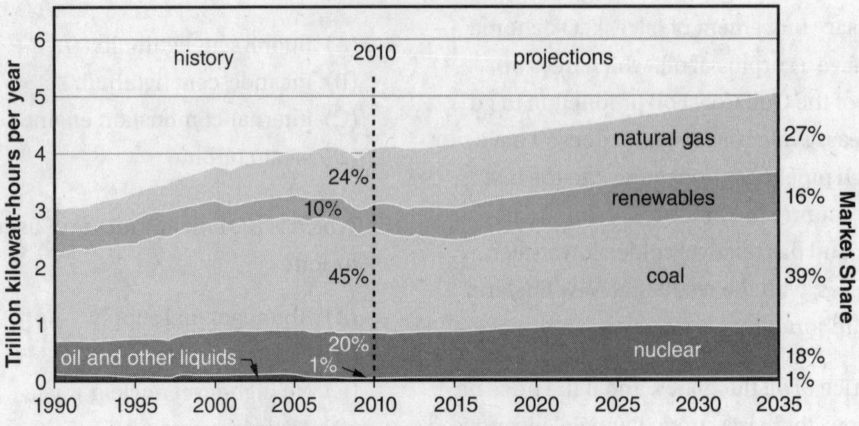

U.S. electricity net generation by fuel

Which of the following statement(s) is/are TRUE?

(A) Annual generation from coal decreases from 45% to 39% between 2010 and 2035.
(B) Annual generation from natural gas increases by 66% from 1990 to 2035.
(C) Renewable energy generation grows.
(D) All statements are true.

15. Which of the following terms is NOT a unit of power?

(A) Horsepower
(B) Joule
(C) Kilowatt
(D) All of the above

16. Automobile manufacturers make money by selling cars. Cars pollute and use fossil fuels. Traditionally, American auto manufacturers have encouraged customers to buy bigger and more powerful cars. This lack of incentive to improve energy efficiency is known as a

(A) beneficial, negative feedback loop
(B) harmful, negative feedback loop
(C) harmful, positive feedback loop
(D) mutualistic, positive feedback loop

17. If you were designing a house in the United States, you lived in a cold climate, and you wanted it to be as energy efficient as possible, you would place large windows to capture solar energy on which side of the house?

(A) East
(B) North
(C) South
(D) It makes no difference on which side.

18. Which of the following is NOT an advantage of building a hydroelectric power plant?

(A) Control flooding
(B) High construction cost
(C) Moderate-to-high net-useful energy
(D) Relatively low operating cost

19. Examine the graph below showing energy consumption per capita.

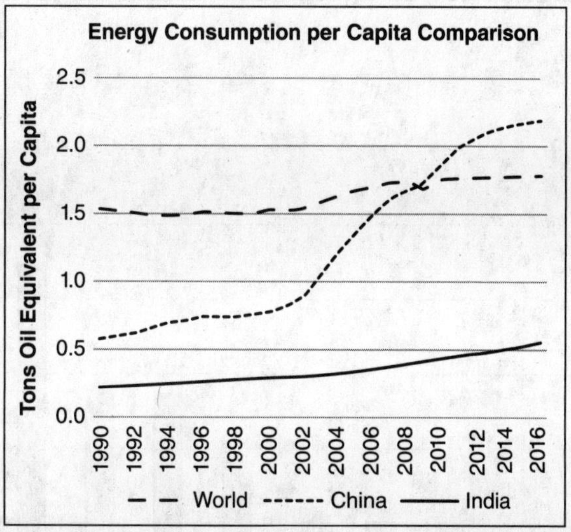

What was the percentage difference between the energy consumption per capita in China as compared to the energy consumption per capita in India in 2016?

(A) 27%
(B) 85%
(C) 170%
(D) 340%

20. Which of the following nonrenewable energy sources has the LEAST environmental impact when burned?

(A) Coal
(B) Gasoline
(C) Natural gas
(D) Oil shale

Free-Response Question

By Brian Palm, Catholic Memorial School, West Roxbury, MA
A.B., Dartmouth College, Hanover, NH
MSc, Oxford University, UK

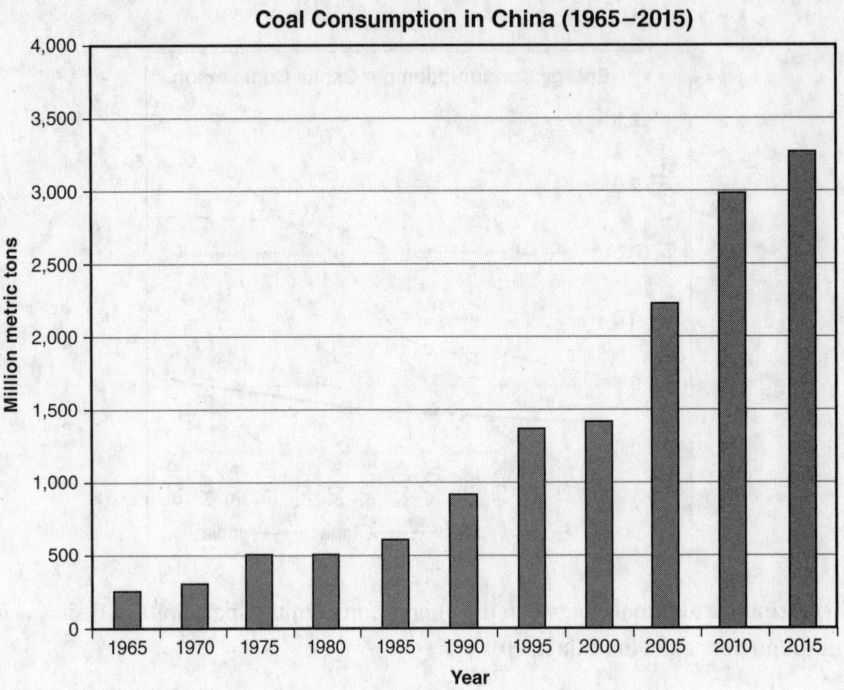

The graph above depicts coal consumption in China from 1965 to 2015.

(a) According to this graph, China's coal consumption increased from 1.5 billion tons of coal in 2003 to 3 billion tons in 2010. Using the Rule of 70, calculate the percentage growth in China's coal consumption during this time.

(b) One ton of coal produces 30 million Btu (British thermal units, a measure of energy), and there are 3,000 Btu in 1 kWh. Assume that all of the coal in China in 2010 was combusted in coal-fired power plants that are 50% efficient. How many kWh were provided to the electric grid through the combustion of this type of coal during that year?

(c) Describe the process of converting coal into usable electricity, and describe ONE pollution control device that could be used to reduce emissions of a particular air pollutant.

(d) As an environmental scientist, you might have concerns about the amount of coal being consumed by China, so you have been hired by the nuclear lobby to convince China to begin constructing nuclear reactors to provide their electricity. Describe one environmental benefit that China could realize by making this transition.

(e) Mr. Lorax was hired by a local environmental group to recommend ways that the region might avoid building ALL forms of large-scale electricity-generating plants (nuclear, coal, or other). IDENTIFY one strategy that Mr. Lorax might recommend, AND EXPLAIN how this strategy reduces the demand for more electricity-generating plants.

Answer Explanations

1. **(A)** Renewable resources are those resources that theoretically will last indefinitely either because they are replaced naturally at a higher rate than they are consumed or because their source is essentially inexhaustible.

2. **(C)** At the current rate of consumption, global oil reserves are expected to last another 45–50 years. With projections of increased consumption in the near future, this figure will be even lower.

3. **(D)** Net-useful energy is defined as the total amount of useful energy available from an energy resource over its lifetime minus the amount of energy used in extracting it and delivering it to the end user.

4. **(D)** The key word in this question is "NOT." To foster a sustainable energy future, fuel efficiency standards would have to increase.

5. **(D)** All statements are reasonable and true.

6. **(D)** More than 80% of the world's total proven coal reserves are located in just ten countries, with the United States topping the list with more than a quarter of the proven coal reserves, while China, which ranks third, is the biggest producer and consumer of coal.

7. **(B)** Nonrenewable resources of energy, such as natural gas or oil, can have dramatic price increases that may result in major increases in the cost of electricity. Renewable resources of energy, especially hydroelectric and wind, provide the least expensive method of producing electricity.

8. **(D)** During the 1990s, wind power experienced an annual growth of 22%. Wind power supplies less than 2% of the energy used in the United States. The country with the largest sector of its energy needs met through wind power is Denmark at 8%.

9. **(B)** An incandescent lightbulb is only 5% efficient as compared with a fluorescent lightbulb at 22%. The typical internal combustion engine is 10% efficient, and a hydrogen fuel cell is 60% efficient. The United States wastes as much energy each day as two-thirds of the world consumes.

10. **(D)** The major fuel used in fusion reactors, deuterium, could be readily extracted from ordinary water. The tritium required would be produced from lithium, which can be extracted from seawater and land deposits.

11. **(D)** The Second Law of Thermodynamics states that when energy changes from one form to another, some of the useful energy is always degraded into a lower quality, more dispersed (higher entropy), and less useful form.

12. **(D)** A low-throughput society, also known as a low-waste society, focuses on matter and energy efficiency through reusing and recycling, using renewable resources at a rate no faster than can be replenished, reducing unnecessary consumption, emphasizing pollution prevention rather than waste reduction, and controlling population growth.

13. **(D)** Reforming is a chemical process that splits water molecules. Hydrogen gas can be produced from algae by depriving the algae of oxygen and sulfur. Coal gasification is the conversion of coal into synthetic natural gas (SNG), which can then be converted into hydrogen.

14. **(D)** To calculate a percentage increase: First, calculate the difference (increase) between the two numbers you are comparing:

$$\text{Increase} = \text{New Number} - \text{Original Number}$$

Then, divide the increase by the original number and multiply the answer by 100%.

$$\%\ \text{Increase} = (\text{Increase} \div \text{Original Number}) \times 100\%$$

If your answer is a negative number, then this represents a percentage decrease.

(a) Factors that negatively impact the future of coal include projected low natural gas prices, rising costs of new coal-fired plant construction costs and the costs to meet environmental air quality emission standards, and concerns over potential greenhouse gas emissions policies.

(b) Eighty-five gigawatts of new natural gas capacity are projected to be added through 2035 as stable prices and relatively low prices are expected to make natural gas the most attractive energy source.

(c) Non-hydro renewables account for a majority of this growth, with wind, solar, biomass, and geothermal generation all significantly larger by 2035.

15. **(B)** A joule is a unit of energy.

16. **(C)** Pollution is harmful, which is the first part of the answer. A negative feedback loop tends to slow down a process, while a positive feedback loop tends to speed it up. Speeding up or increasing sales would be a positive feedback, which is the second half of the answer.

17. **(C)** In cold climates, large south-facing windows allow significant solar energy into the house and also provide daylighting. Properly sized overhangs can prevent overheating in the summer. In hot climates, north-facing windows can provide daylighting without heating the house. East- and west-facing windows generally cause excessive heat gains in the summer and heat losses in the winter, so they are usually small. Although overhangs are impractical for east- and west-facing windows, vertical shading can be used, or trees and shrubs can be strategically located, to shade the windows.

18. **(B)** The construction of hydroelectric plants is initially expensive, but the operating costs are relatively low since there are no fuel costs.

19. **(D)** First, calculate the difference (increase) between the two numbers you are comparing:
India (N_1) = ~ 0.5 tons oil equivalent

China (N_2) = ~ 2.2 tons oil equivalent

Difference = $N_2 - N_1 = 2.2 - 0.5 = 1.7$

Then, divide the increase by N_1 and multiply the answer by 100%.

$(1.7 / 0.5) \bullet 100\% = 340\%$

20. **(C)** The combustion of natural gas releases very small amounts of sulfur dioxide and nitrogen oxides, virtually no ash or particulate matter, and lower levels of carbon dioxide, carbon monoxide, and other reactive hydrocarbons than coal or oil. The following chart compares emission levels of various pollutants for three forms of energy: natural gas, oil, and coal.

Fossil Fuel Emission Levels*

Pollutant	Natural Gas	Oil	Coal
Carbon dioxide (CO_2)	117,000	164,000	208,000
Carbon monoxide (CO)	40	33	208
Nitrogen oxides (NO_x)	92	448	457
Sulfur dioxide (SO_2)	1	1,122	2,591
Particulates (PM_x)	7	84	2,744
Mercury (Hg)	0.000	0.007	0.016

**Pounds of pollutant per billion Btu of energy input*

Free-Response

10 Total Points Possible

(a) *Maximum 2 points total: 1 point for the correct calculation setup and 1 point for the correct answer.*

According to this graph, China's coal consumption increased from 1.5 billion tons of coal in 2003 to 3 billion tons in 2010. Using the Rule of 70, calculate the percentage growth in China's coal consumption during this time.

Units are not required in this setup. Students should use the correct mathematical operation (in this case an appropriate use of the Rule of 70 equation) where:

$$\frac{70}{\text{growth rate \%}} = 7 \text{ years to double}$$

A sample calculation follows:

$$\frac{70}{x} = 7$$

$$70 = 7x$$

$$10 = x$$

Therefore, the percentage growth in China's coal consumption during this time was 10%.

(b) *Maximum 2 points total: 1 point for the correct calculation setup and 1 point for the correct answer.*

One ton of coal produces 30 million Btu (British thermal units, a measure of energy), and there are 3,000 Btu in 1 kWh. Assume that all of the coal in China in 2010 was combusted in coal-fired power plants that are 50% efficient. How many kWh were provided to the electric grid through the combustion of this type of coal during that year?

While units are not required in the setup, the numbers should be consistent, and they should demonstrate that the correct mathematical operations were used. Note that the application of the 50% efficiency can be accomplished in a number of places in the calculation, but it will simply reduce the value by one-half (e.g., from 3 to 1.5).

$$\frac{3.0 \times 10^{9}\ \cancel{\text{tons}}}{1} \times \frac{\cancel{3} \times 10^{\cancel{7}\,4}\ \cancel{\text{Btu}}}{\cancel{1\ \text{ton}}} \times \frac{1\ \text{kWh}}{\cancel{3}\ 10^{3}\ \cancel{\text{Btu}}}\ 3 \times 10^{13}\ \text{kWh} \times 0.5 = 1.5 \times 10^{13}\ \text{kWh}$$

(c) *Maximum 3 points total: 2 points for adequately describing at least TWO steps (in the correct order) in the electricity process AND 1 point for describing a pollution control device that is commonly found in coal-fired power plants.*

Describe the process of converting coal into usable electricity, and describe ONE pollution control device that could be used to reduce emissions of a particular air pollutant.

Electricity generation using coal includes all of the following steps:

- Coal is combusted in the boiler, a structure that is surrounded by tubes containing water.
- Coal is commonly pulverized prior to being injected into boilers, where it is combusted.
- Coal is mined and transported to a coal-fired power plant.
- The flue gases are sent up the stack after pollution control devices remove specific pollutants.
- The high-temperature, high-pressure steam flows through a series of pipes that connect to a turbine.
- The steam is then condensed, cooled, and returned as water to the boiler.
- The steam spins the turbine at very high speeds.

- The turbine is connected to the generator, which then moves wires around a magnet, creating current or electricity.
- The water in the tubes surrounding the boiler is turned into steam and often super-heated after this conversion.

There are a number of pollution control devices that are used to clean the flue gases in a coal-fired power plant:

- Ammonia is sprayed into the flue gases to remove NO_x gases.
- A spray of a calcium carbonate solution is used to remove SO_x gases.
- Baghouse filters remove particulate matter from the gases as the gases are passed through fabric filters.
- Charcoal, or wet carbon, is introduced at high temperatures to remove mercury.
- Electrostatic precipitators may also be used to remove particulate matter by charging and then attracting the particles to collectors.

(d) *Maximum 2 points total: 2 points for adequately identifying AND connecting the environmental benefit to this transition.*

As an environmental scientist, you might have concerns about the amount of coal being consumed by China, so you have been hired by the nuclear lobby to convince China to begin constructing nuclear reactors to provide their electricity. Describe one environmental benefit that China could realize by making this transition.

Nuclear energy has a range of potentially positive environmental outcomes. Of primary importance is the fact that no carbon dioxide is produced after the plant is constructed. Nuclear energy production also does not release sulfur oxides, particulates, or mercury into the atmosphere during electricity generation.

A correct response to this question would describe one of these factors AND connect that factor to the appropriate environmental benefit. For example, describing the carbon-free source of electricity would need to be connected to a reduction in the potential for climate change, which would reduce the likelihood of habitat destruction and species loss. Fewer sulfur oxides could be connected to a reduction in acid rain and the associated habitat destruction.

(e) *Maximum 2 points total: 1 point for IDENTIFYING one strategy and 1 point for connecting this strategy to a decreased demand for large-scale electricity plants. Answers may vary, but each correct answer should IDENTIFY one strategy AND connect that specific strategy to EITHER a reduced electricity demand OR an alternative way of creating electricity supply.*

Mr. Lorax was hired by a local environmental group to recommend ways that the region might avoid building ALL forms of large-scale electricity-generating plants (nuclear, coal, or other). IDENTIFY one strategy that Mr. Lorax might recommend, AND EXPLAIN how this strategy reduces the demand for more electricity-generating plants.

Examples include the following:

- Conservation programs, education, and outreach to the community increase awareness about electricity use in the community. Similar to an efficiency improvement program, the reduced demand from electricity users in homes and businesses will reduce the need to develop new, large-scale supply plants.
- Creating demand-reduction incentives is one strategy to reduce demand for large-scale generating plants. Programs that provide rebates for high-efficiency lighting or appliances increase the likelihood that electricity consumers will purchase these more efficient items, install them in their homes, and subsequently reduce the amount of electricity needed by each home. As old and inefficient stock is replaced, the community avoids the need for electricity and reduces the necessity for new generating capacity.
- Rooftop solar panels will allow customers to produce more of their own electricity, reducing the amount of electricity required from the larger grid and reducing the need for new centralized generating capacity.

7

Atmospheric Pollution

Learning Objectives

In this unit, you will learn:

For added practice with this unit and topics, log onto Online Labs 7A–7B at online.barronsbooks.com

7.1 Introduction to Air Pollution

Air pollution occurs when harmful or excessive quantities of substances are introduced into Earth's atmosphere. With over 50,000 deaths in the United States, 750,000 deaths in China, and over 6 million deaths worldwide per year projected by 2050 (compounded by indoor air pollution's being 2–5 times worse than outdoor levels), air pollution remains a primary cause of death in the world today.

Measurement Units

The most common form of expressing air pollutants is parts per million (ppm); 1 ppm represents one particle of a pollutant for every 999,999 particles of air. The symbol "m" is often used to represent one millionth and is equivalent to one drop of ink in 40 gallons (150 L) of water or one second in 280 hours.

Primary Pollutants—Emitted directly into the air (Figure 7.1)

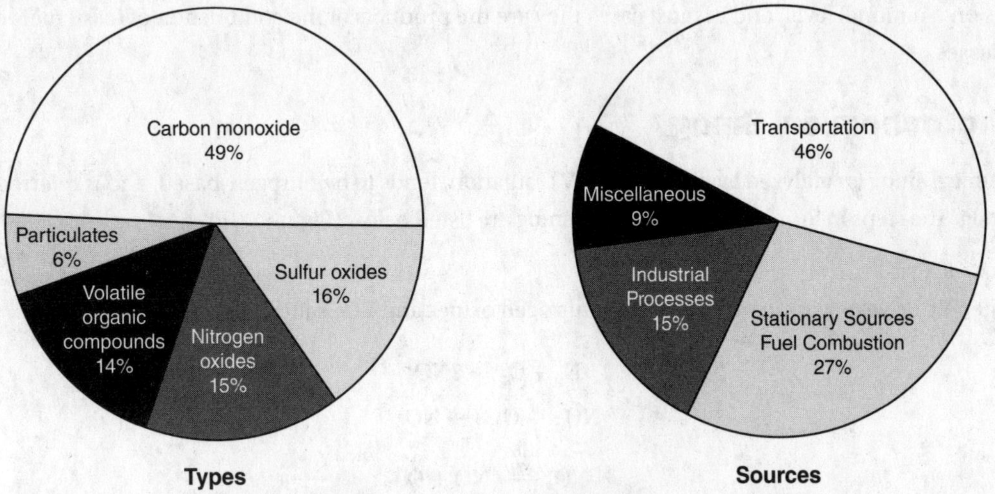

Figure 7.1 Primary pollutants

Secondary Pollutants—Result from primary air pollutants' reacting together and forming new pollutants. Examples include ozone, which is formed when hydrocarbons and nitrogen oxides (NO_x) combine in the presence of sunlight; NO_2, which is formed when NO combines with oxygen in the air; and acid rain, which is formed when sulfur dioxide or nitrogen oxides react with water.

Point Source Pollution

Point source air pollution occurs when the contaminant comes from an obvious source, such as a certain factory that produces certain poisonous chemicals and toxic gases (e.g., benzene) that are released directly into the air as part of the manufacturing process.

Non-Point Source Pollution

Non-point source air pollution occurs when the contaminant comes from a source that is not easily identifiable or from a number of sources spread over a large, widespread area. For example, car tailpipes are considered non-point source pollution because their sheer numbers make the emitted pollution appear as though it is being emitted from a diffuse source. Water pollution can come from non-point sources as we consider agricultural lands and residential areas producing fertilizers, herbicides, and insecticides that end up in rivers or bodies of water.

Point Source Pollution vs. Non-Point Source Pollution

Point Source Pollution	Non-Point Source Pollution
Discharge is usually easy to identify.	Many small diffuse sources from many different locations.
Fairly easy to monitor above and below the discharge point.	Difficult to monitor; many stations required.
Relatively easy to control since the type of contaminant and location are known.	Individual contributions are smaller, but cumulative effects may be significant.
The industry can be fined if it does not obey the terms of the permit.	Difficult to locate and fine offenders.

Criteria Air Pollutants

Criteria air pollutants are a set of eight air pollutants that cause smog, acid rain, and other health hazards and are typically emitted from many sources in industry, mining, transportation, power generation, and agriculture. They were the first set of pollutants recognized by the U.S. Environmental Protection Agency (EPA) as needing standards on a national level, and in most cases they are the products of the combustion of fossil fuels or industrial processes.

7.2 Photochemical Smog

Photochemical smog is catalyzed by ultraviolet (UV) radiation, tends to be nitrogen-based, and is referred to as brown smog. The steps in forming photochemical smog are listed below (Figure 7.2):

- 6 A.M.–9 A.M.

  As people drive to work, concentrations of nitrogen oxides and VOCs increase:

$$N_2 + O_2 \rightarrow 2NO$$

$$NO + VOCs \rightarrow NO_2$$

$$NO_2 \xrightarrow{uv} NO + O$$

- 9 A.M.–11 A.M.

  As traffic begins to decrease, nitrogen oxides and VOCs begin to react, forming nitrogen dioxide (NO_2):

$$2NO + O_2 \rightarrow 2NO_2$$

- 11 P.M.–4 P.M.

  As the sunlight becomes more intense, nitrogen dioxide is broken down and the concentration of ozone (O_3) increases:

$$NO_2 \xrightarrow{uv} NO + O$$

$$O_2 + O \rightarrow O_3$$

  Nitrogen dioxide also reacts with water vapor to produce nitric acid (HNO_3) and nitric oxide (NO):

$$3NO_2 + H_2O \rightarrow 2HNO_3 + NO$$

  Nitrogen dioxide can also react with VOCs released by vehicles, refineries, and gas stations to produce toxic PANs (peroxyacyl nitrates):

$$NO_2 + VOCs \rightarrow PANs$$

- 4 P.M.–Sunset

  As the sun goes down, the production of ozone is halted.

$$\text{Net result: } NO + VOCs + O_2 + uv \rightarrow O_3 + PANs$$

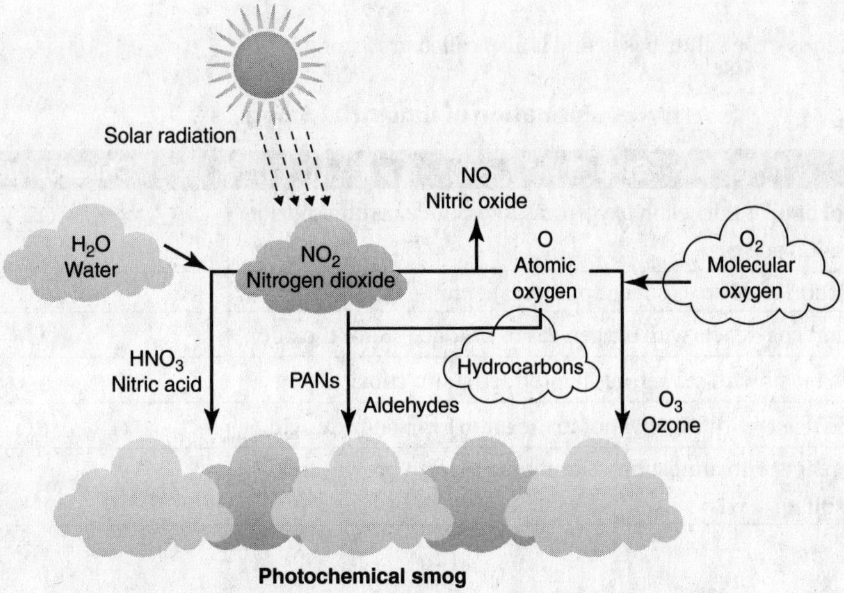

Figure 7.2 Steps in the formation of photochemical smog

7.3 Thermal (Temperature) Inversion

Thermal or temperature inversions occur when the air temperature increases with the height above the ground, as opposed to the normal decrease in temperature with height (Figure 7.3). This effect can lead to pollution such as smog being trapped close to the ground with possible adverse effects on human health (e.g., asthma, emphysema, and increases in lung cancer).

Thermal inversions commonly occur at night when solar heating ceases and the surface cools, which then cools the atmosphere immediately above it. A warm air mass then moving over a colder one keeps the cooler air mass trapped below, and the air becomes still, which then results in dust and pollutants being trapped and their concentrations increasing.

A nearly permanent, naturally occurring temperature inversion occurs over Antarctica.

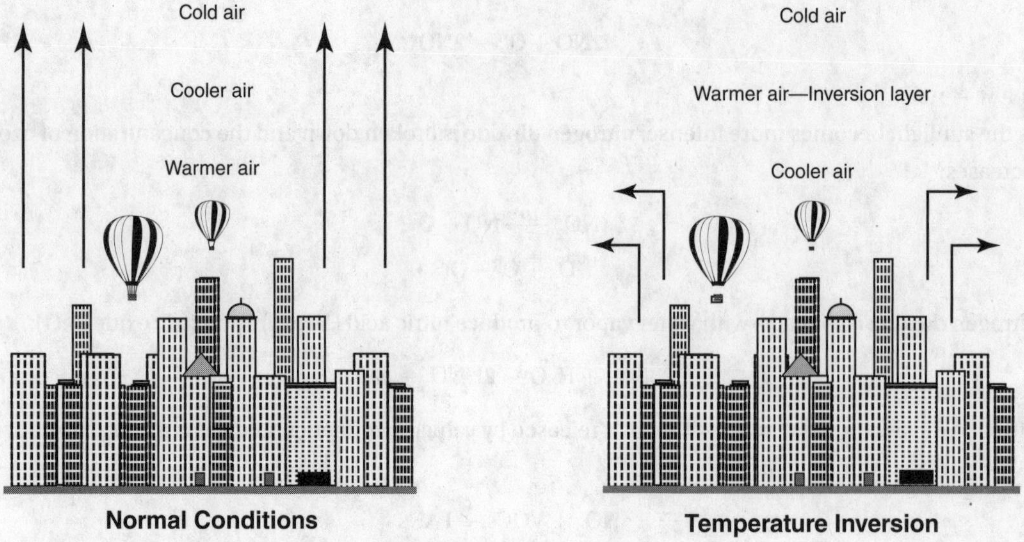

Figure 7.3 Thermal temperature inversion

7.4 Atmospheric CO_2 and Particulates

Industrial Smog

Industrial smog tends to be sulfur-based and is also called gray smog.

Formation of Industrial Smog

Step	Chemical Reaction
1. Carbon in coal or oil is burned in oxygen gas to produce carbon dioxide and carbon monoxide gas.	$C + O_2 \rightarrow CO_2$ $2C + O_2 \rightarrow 2CO$
2. Unburned carbon ends up as soot or particulate matter (PM).	C
3. Sulfur in oil and coal reacts with oxygen gas to produce sulfur dioxide.	$S + O_2 \rightarrow SO_2$
4. Sulfur dioxide reacts with oxygen gas to produce sulfur trioxide.	$2SO_2 + 3O_2 \rightarrow 2SO_3$
5. Sulfur trioxide reacts with water vapor in the air to form sulfuric acid.	$2SO_3 + 2H_2O \rightarrow 2H_2SO_4$
6. Sulfuric acid reacts with atmospheric ammonia to form brown, solid ammonium sulfate.	$H_2SO_4 + 2NH_3 \rightarrow (NH_4)_2SO_4$

Carbon Monoxide (CO)

Carbon monoxide is a colorless, odorless, and tasteless gas that is slightly less dense than air and is produced from the partial oxidation of carbon-containing compounds. Carbon monoxide forms when there is not enough oxygen to produce carbon dioxide, such as when operating a stove, a water heater, or an internal-combustion engine in an enclosed space.

Carbon monoxide is present in small amounts in the atmosphere, primarily as a product of the following:

- Natural and man-made fires, such as wildfires and burning crop residues
- Photochemical reactions in the troposphere (largest single source)
- The burning of fossil fuels
- Volcanic activity

Carbon monoxide is a temporary atmospheric pollutant in some urban areas, primarily produced from the exhaust of internal-combustion engines but also from incomplete combustion of various other fuels (e.g., charcoal, coal, natural gas, oil, propane, trash, and wood). Carbon monoxide, in the presence of oxygen, nitrogen dioxide, and ultraviolet radiation, also produces tropospheric (ground level) ozone, a component of photochemical smog:

$$CO + 2O_2 + uv^* \rightarrow CO_2 + O_3$$

**ultraviolet radiation*

Methods to reduce carbon monoxide pollution include the following:

- Building more public transportation infrastructure
- Requiring catalytic converters on all cars worldwide; however, this only converts carbon monoxide to carbon dioxide—a greenhouse gas
- Switching to renewable energy sources

Lead (Pb)

Lead is used in building construction, lead-acid batteries for vehicles, bullets and shot (which are ingested by wildlife), fishing weights, solder, and shields for radiation. Tetraethyl lead (a gasoline additive) was phased out in the United States starting in 1975 to meet tighter emissions regulations in an effort to reduce toxic pollution.

Exposure to lead, which generally occurs from lead-based paint chips from older buildings, mining, smelters, and municipal waste incineration, can occur from inhalation of polluted air and dust and from the ingestion of lead in food and/or water. The United States banned lead-based paint in 1977 in residential properties and public buildings, along with toys and furniture that contained lead paint. However, in 2017, the Environmental Protection Agency moved to weaken these regulations.

Symptoms of lead poisoning include failure of the blood to make hemoglobin, which results in anemia (low red blood cell count), endocrine (hormone) disruptors, mental retardation and disabilities, hypertension (high blood pressure), miscarriages and/or premature births, and even death at relatively low concentrations.

Nitrogen Oxides (NO_x)

NO_x is a generic term for NO (nitric oxide) and NO_2 (nitrogen dioxide), which are produced from the reaction of nitrogen and oxygen gases in the air. NO_x gases are formed whenever nitrogen occurs in the presence of high-temperature combustion; e.g., forest fires, internal-combustion engines, lightning, power station boilers, and volcanoes. NO_x gases react to form acid rain and are essential for the formation of tropospheric ozone.

When NO_x and volatile organic compounds (VOCs) react in the presence of sunlight, they form photochemical smog, a significant form of air pollution, especially in the summer. Agricultural fertilization and the use of nitrogen-fixing plants also contribute to atmospheric NO_x by promoting nitrogen fixation by microorganisms.

NO_x should not be confused with nitrous oxide (N_2O), which is a greenhouse gas. Nitrous oxide is a major air pollutant, with levels of N_2O having increased by more than 15% since 1750. Considered over a 100-year period, N_2O is calculated to have about 300 times more global-warming potential than carbon dioxide.

Agricultural croplands contribute about 70% of the total nitrous oxide (N_2O) emissions. Nitrous oxide (N_2O) is produced naturally in the following ways:

- By denitrification—the process whereby ammonia (NH_3) is oxidized to nitrite (NO_2^-) by bacteria of the genus *Nitrosomonas*
- In the soil during the process of nitrification (the oxidation of ammonia [NH_3] to nitrite [NO_2^-] and nitrate [NO_3^-])
- Nitrite (NO_2^-) is further reduced to nitric oxide (NO), molecular nitrogen (N_2), and nitrous oxide (N_2O).

Nitrous oxide also causes ozone depletion whereby nitrous oxide (N_2O) gives rise to nitric oxide (NO) on reaction with oxygen atoms: $N_2O + O \rightarrow 2NO$ and NO in turn reacts with ozone:

$$NO + O_3 \rightarrow NO_2 + O_2$$

Nitrous oxide is one of the most significant ozone-depleting substances and is expected to remain so throughout the 21st century.

Ozone (O_3)

Ozone is an inorganic molecule with the chemical formula O_3, and tropospheric (ground-level) ozone is a secondary air pollutant. Examples include (1) UV reacting with the NO_x released by vehicles, which causes oxygen atoms to react with O_2 gas, resulting in ozone O_3:

$$NO_2 + uv \rightarrow NO + O$$
$$O + O_2 \rightarrow O_3$$

and (2) when volatile organic compounds (VOCs) react with NO_x, producing peroxyacyl nitrates (PANs), which are powerful respiratory and eye irritants that are present in photochemical smog:

$$NO_x + VOCs \rightarrow PANs + O_3$$

Human activities that can lead to the formation of tropospheric ozone include burning fossil fuels (e.g., by using power plants, motor vehicles, etc.) and releasing VOCs. Ozone is a short-lived greenhouse gas that decays in the atmosphere more quickly than carbon dioxide does, and because of its short-lived nature, tropospheric ozone does not have strong global effects, but instead is more influential in its effects on smaller, more localized areas.

Tropospheric ozone can have the following effects:

- Cause asthma and bronchitis
- Harm lung function and irritate the respiratory system
- Result in heart attacks and other cardiopulmonary problems
- Suppress the immune system

Tropospheric ozone impacts ecosystems by affecting animal respiratory systems in much the same manner as it does human respiratory systems. In addition, it negatively affects plant life by stressing plants to the point of making them less vigorous (e.g., producing less chlorophyll) and more susceptible to disease and pests. This results in a decrease in primary productivity, which negatively impacts food webs and decreases crop yields.

Peroxyacyl Nitrates (PANs)

PANs are secondary pollutants. Because they break apart quite slowly in the atmosphere into radicals (a group of atoms that are unstable and highly reactive) and NO_2, PANs are able to move far away from their urban and industrial origin. PANs can cause the following:

- Eye irritation
- Impaired immune systems
- Inhibited photosynthesis
- Reduced crop yields by damaging plant tissues (e.g., cells, leaves, etc.)
- Respiratory problems

Methods to reduce PANs include the following:

- Limiting wood-burning fireplaces and stoves in new home construction
- Reducing smokestack emissions through baghouse filters, cyclone precipitators, scrubbers, and/or electrostatic precipitators
- Reducing the incineration of municipal and industrial wastes
- Reducing the reliance on fossil fuels, especially oil and coal

Sulfur Oxides (SO_x)

The most common sulfur oxide is sulfur dioxide (SO_2), a colorless gas with a penetrating, choking odor that readily dissolves in water to form an acidic solution. The main emission source of sulfur dioxide is the burning of fossil fuels used by power stations, oil refineries, and large industrial plants. Motor vehicles, tar-sand extraction, and ore processing, as well as natural sources such as active volcanoes, marshes, hot springs, and forest fires, release sulfur dioxide.

Sulfur dioxide is toxic to a variety of plants and reduces crop yields. Sulfur dioxide, emitted in sufficient quantities at low or ground level, can combine with air moisture to cause gradual damage to some building materials (e.g., limestone) by forming an acid solution that gradually dissolves the stonework.

Sulfur dioxide irritates the throat and lungs, and, if there are fine dust particles in the air, can damage the respiratory system.

Steps that can be taken to reduce the amount of SO_2 in the atmosphere include the following:

- Fluidized gas combustion (i.e., crushed limestone is added to crushed coal to produce calcium sulfite, calcium sulfate, or gypsum)
- Using only low-sulfur coal
- Using scrubbers in the smokestacks—in the scrubber, there is a water spray that produces a cloud of fine water droplets, which mix with crushed limestone and react with the sulfur and "pull" it out of the exhaust
- "Washing" the coal—by mixing crushed coal with water, impurities with different densities are able to be separated and removed

Suspended Particulate Matter (PM_x)

Suspended particulate matter (PM_x) is microscopic solid or liquid matter suspended in Earth's atmosphere. The "*x*" refers to the size of the particles; e.g., PM_{10} represents particles with an average size of 10 microns (ten-millionths of a meter; the average diameter of a human hair is 50 microns).

Naturally Occurring PM_x	Anthropogenic PM_x
Dust storms	Burning of fossil fuels—power plants
Forest and grassland fires	Incineration of wastes
Sea spray	Soil erosion—desertification, deforestation
Volcanoes	Vehicle exhaust

In general, the smaller and lighter a particle is, the longer it will stay in the air. Larger particles (greater than 10 micrometers in diameter) tend to settle to the ground by gravity in a matter of hours, whereas the smallest particles (less than 1 micrometer) can stay in the atmosphere for weeks and are mostly removed by precipitation.

Particulate matter

- affects the diversity of ecosystems;
- changes the nutrient balance in coastal waters and large river basins;
- damages sensitive forests and farm crops;
- depletes the nutrients in soil;
- increases health issues with humans and animals including asthma, lung cancer, cardiovascular disease, respiratory diseases, birth defects, and premature death. Particulate matter pollution is estimated to cause 3 million deaths per year worldwide and up to 50,000 deaths per year in the United States; and
- makes lakes and streams more acidic.

Airborne particulate matter can be reduced by

- conserving energy to reduce demands on power plants;
- increasing air-quality standards for emissions of particulate matter from smokestacks;
- increasing automobile emission standards;
- limiting the use of household and personal products that cause fumes;
- not burning leaves and other yard waste;
- not using wood in fireplaces (gas logs only) and changing building codes to outlaw future construction of wood-burning fireplaces;
- taking steps to reduce the incidence of wildfires; and
- using public transportation whenever possible.

Volatile Organic Compounds (VOCs)

VOCs are organic chemicals that have a high vapor pressure (easily evaporate) at ordinary room temperature. Their high vapor pressure results from a low boiling point, which causes large numbers of molecules to evaporate and enter the surrounding air.

VOCs include both human-made and naturally occurring chemical compounds and are not acutely toxic, but they have compounding long-term health effects. Since many people spend much of their time indoors, long-term exposure to VOCs in the indoor environment can contribute to "sick building" syndrome. Health effects of "sick building" syndrome include

- cancer;
- damage to the liver, kidney, and central nervous system;
- eye, nose, and throat irritation; and
- headaches, loss of coordination, and nausea.

7.5 Indoor Air Pollutants

"Sick building" syndrome (SBS) is a term used to describe a combination of ailments associated with an individual's place of work or residence, with many people spending the majority of their lives indoors sleeping, working, eating, and relaxing, where air circulation may be restricted and where indoor air pollutant levels may be two to five times greater than outdoor levels. Up to a third of new and remodeled buildings worldwide may be linked to symptoms of SBS. The following sections discuss some examples of specific indoor air pollutants.

Asbestos

Asbestos became a major part of manufacturing in America starting in the early 1800s and was commonly used until the mid-1970s. Its first popular use was as the lining in steam engines, and the construction, shipbuilding, and manufacturing industries used asbestos-containing products whenever possible. Asbestos is inexpensive, durable, and flexible and naturally acts as an insulating and fireproofing agent. Examples of where asbestos was commonly used include brake linings, bakeware, ceiling and flooring tiles, cigarette filters, drywall, fireproofing materials, insulation, plastics, roofing shingles, and vinyl products.

Breathing asbestos fibers can lead to lung cancer, mesothelioma (a cancer of the lining of the chest and the abdominal cavity), and asbestosis (lungs are scarred with fibrous tissue). The symptoms of breathing asbestos fibers usually do not appear until about 20–30 years after the first exposure, and ~15,000 Americans die each year due to asbestos-related diseases.

In June 2018, the U.S. Environmental Protection Agency (EPA) announced that it will not consider the health risks and impacts of asbestos already in the environment (e.g., asbestos used in tiles, piping, and adhesives in homes and businesses throughout the United States) when evaluating the dangers associated with the chemical compound.

More on Carbon Monoxide and How It Affects the Environment

In closed environments, the concentration of carbon monoxide can easily rise to lethal levels. Carbon monoxide poisoning is the most common type of fatal indoor air poisoning in many countries because it easily combines with hemoglobin to block the blood's oxygen-carrying capacity. Worldwide, carbon monoxide poisoning is responsible for more than half of all poisoning deaths, with about 2,500 people dying from it each year in the United States.

Around 50% of people, almost all in developing countries, rely on coal and biomass in the form of wood, dung, and crop residues for domestic energy. These materials are typically burned in simple stoves with very incomplete combustion. Consequently, women and young children are exposed to high levels of indoor air pollution every day, and there is consistent evidence that this daily exposure increases the risk of chronic obstructive pulmonary disease and acute respiratory infections in childhood, the most significant cause of death among children under five years of age in developing countries. Evidence also exists of carbon monoxide contributing to low birth weight, increased infant mortality, tuberculosis, and lung cancer.

Formaldehyde

Formaldehyde is an organic chemical (a molecule that contains carbon) that is prevalent in the indoor environment and is a carcinogen that is linked to lung cancer. The following table shows consumer products that contain formaldehyde.

Products That Contain Formaldehyde

Products	Examples
Combustion	Natural gas, kerosene, tobacco smoke
Insulation	Foam insulation
Paper products	Grocery bags, facial tissues, paper towels, disposable sanitary products
Pressed-wood products	Plywood, particle board, decorative paneling
Stiffeners, wrinkle resisters, and water repellents	Floor covering (rugs, linoleum, varnishes, plastics), carpet adhesive, fire retardants, permanent press clothes
Other	Cosmetics, deodorants, shampoos, fabric dyes, disinfectants

Persistent Organic Pollutants (POPs)

Radon

Radon is an invisible radioactive gas that results from the radioactive decay of radium, which can be found in rock formations (typically granite containing uranium) beneath buildings or in certain building materials and is probably the most pervasive serious hazard for indoor air in the United States and Europe (Figure 7.4). It is responsible for tens of thousands of deaths from lung cancer each year. Radon is a heavy gas and thus will tend to accumulate at floor level, and is more prevalent in well-insulated homes. In areas where natural radon levels occur, mitigation methods include the following:

- Increasing indoor ventilation with outdoor air
- Sealing basement foundations
- Sealing concrete slab floors

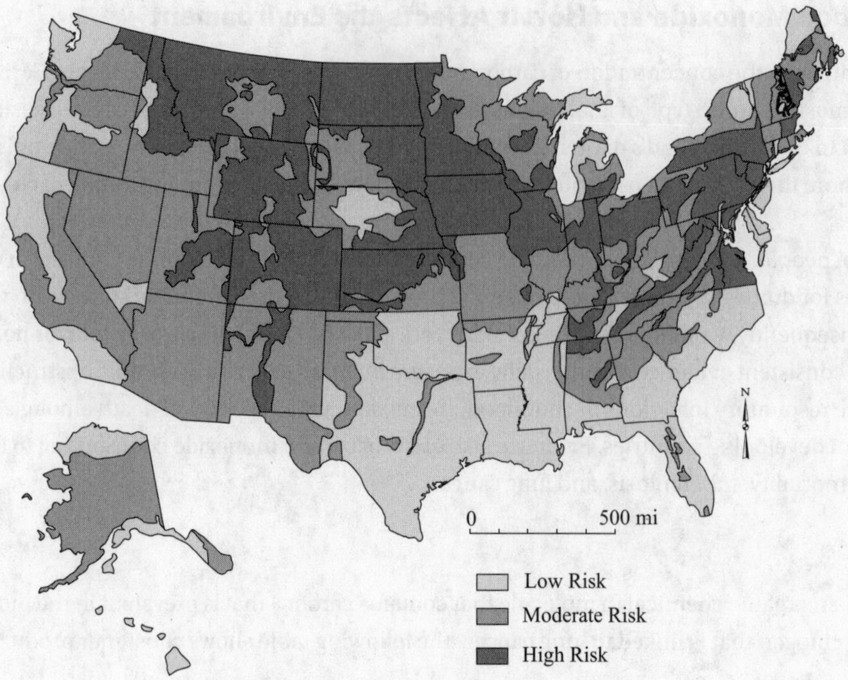

Figure 7.4 Radon risk in the United States

Tobacco Smoke

Cigarette smoke contains almost 5,000 chemical compounds, including 60 known carcinogens (cancer-causing chemicals), one of which is dioxin. It is responsible for 85% of lung cancers and is also associated with cancers of the mouth, larynx, esophagus, stomach, pancreas, kidney, bladder, and colon. Tobacco smoke has also been linked to leukemia, strokes, and heart attacks.

Cigarettes are smoked by about 20% of the world's population—over one billion people, the majority of whom are men. While smoking rates have leveled off or declined in developed nations, in developing nations, tobacco consumption continues to rise. More than 80% of all smokers now live in countries with low or middle incomes, and 60% are in just ten countries, with China ranked number one, with about three trillion cigarettes consumed per year (Figure 7.5).

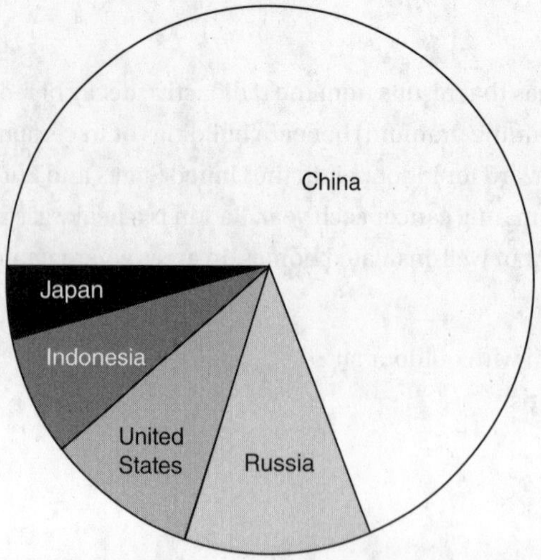

Figure 7.5 Cigarette consumption by country

Remediation Steps to Reduce Indoor Air Pollutants

- Add plants that absorb toxins; e.g., philodendron, spider plant, English ivy.
- Do not allow smoking indoors.
- Install air purification systems and ensure adequate fresh air ventilation when temperatures permit.
- Maintain all filters and vents.
- Monitor humidity levels to reduce mold and mildew.
- Test for radon gas and other dangerous indoor pollutants.
- Use "green" cleaning products.
- Use natural pest-control techniques.

7.6 Reduction of Air Pollutants

Catalytic Converters

Controlling Air Pollution/Pollution-Control Devices

A catalytic converter is an exhaust emission control device that converts toxic chemicals in the exhaust of an internal-combustion engine into less harmful substances. Inside a catalytic converter, a catalyst stimulates a chemical reaction in which by-products of combustion are converted to less toxic substances by way of catalyzed chemical reactions. Most present-day vehicles that run on gasoline are fitted with a "three way" converter, since it converts the three main pollutants:

1. Oxidation of carbon monoxide to carbon dioxide:

$$CO + O_2 \rightarrow CO_2$$

2. Oxidation of unburned hydrocarbons to carbon dioxide and water:

$$C_xH_{2x} + O_2 \rightarrow CO_2 + H_2O$$

3. Reduction of nitrogen oxides to nitrogen and oxygen:

$$NO_x \rightarrow O_2 + N_2$$

Although catalytic converters are effective at removing hydrocarbons and other harmful emissions, they do not reduce the emission of carbon dioxide produced when fossil fuels are burned.

Additionally, the Environmental Protection Agency (EPA) has stated that catalytic converters are a significant and growing cause of global warming because of their release of nitrous oxide (N_2O), a greenhouse gas over 300 times more potent than carbon dioxide. The EPA states that motor vehicles contribute approximately 50% of nitrous oxide emissions, which make up 7% of greenhouse gases.

Remediation Steps to Reduce Air Pollution

- Ban open burning of waste.
- Buy smaller cars and energy-efficient appliances.
- Decrease unnecessary travel.
- Distribute solar cook stoves to developing countries to replace wood and coal.
- Drive within the speed limit and keep tires inflated.
- Institute flexible work shifts.
- Maintain vehicle properly with regular tune-ups and oil changes.
- Reduce idling and turn off engines while waiting.
- Use mass transit systems or carpool when possible.
- Toughen Corporate Average Fuel Economy (CAFE) standards.
- Toughen legislation to reduce sulfur content in fuel.

- Use fans instead of air conditioners.
- Use fluorescent or LED lighting.
- When buying a car, consider its fuel efficiency.

RELEVANT LAWS

CLEAN AIR ACT (1963)—The Clean Air Act is designed to control air pollution on a national level and requires the Environmental Protection Agency (EPA) to develop and enforce regulations to protect the general public from exposure to airborne contaminants that are known to be hazardous to human health. The Act also

- addressed acid rain, ozone depletion, and toxic air pollution;
- established new auto gasoline reformulation requirements;
- expanded federal enforcement authority;
- required comprehensive federal and state regulations for both stationary (industrial) and mobile sources of air pollution; and
- served as the first major environmental law in the United States to include a provision for citizen lawsuits.

7.7 Acid Rain (Deposition)

Acid deposition is a complex chemical and atmospheric phenomenon that occurs when emissions of sulfur and nitrogen compounds and other substances are transformed by chemical processes in the atmosphere and then deposited on Earth in either wet forms (e.g., fog, rain, snow) or dry forms (e.g., acidic gases or particulates).

Dry Deposition

In areas where the weather is dry, acidic chemicals in the air may become incorporated into dust or smoke, sticking to the ground, buildings, homes, cars, and trees, and can then be washed from these surfaces by rainstorms, leading to increased acidic runoff. About half of the acidity in the atmosphere falls back to Earth through dry deposition.

Wet Deposition (Acid Rain)

Wet deposition refers to acid rain, fog, and snow. As this acidic water flows over and through the ground, it affects a variety of plants and animals. The strength of the effects depends on many factors, including how acidic the water is; the chemistry and buffering capacity of the soils involved; and the types of organisms, such as fish, macroinvertebrates, trees, and other living things, that rely on the water and soil.

Acid rain causes acidification of lakes and streams. It contributes to the damage of trees at high elevations and many sensitive forest soils through nitrogen saturation and by creating acidic conditions that are unhealthy for decomposers and mycorrhizal fungi. Acid shock, which is caused by the rapid melting of snow pack that contains dry acidic particles, results in acid concentrations in lakes and streams that are five to ten times higher than that of acidic rainfall. In addition, acid rain accelerates the decay of building materials and paints, including irreplaceable statues and sculptures. Acid rain also leaches essential plant nutrients from the soil, such as Ca^{2+}, K^{+}, and Mg^{2+}, as well as heavy metal ions, such as Pb^{2+}, Cd^{2+}, and Hg^{2+}.

Acid deposition due to sulfur dioxide (SO_2) begins with sulfur dioxide's being introduced into the atmosphere by burning coal and oil, smelting metals, organic decay, and ocean spray. It then combines with water vapor to form sulfurous acid ($SO_2 + H_2O \rightarrow H_2SO_3$), which then reacts with oxygen to form sulfuric acid ($H_2SO_3 + \frac{1}{2}O_2 \rightarrow H_2SO_4$).

Acid deposition due to nitrogen oxides (NO_x) begins with nitrogen oxides formed by burning oil, coal, or natural gas. They are also found in volcanic vent gases and are formed by forest fires, bacterial action in the soil, and lightning-induced atmospheric reactions. Nitrogen monoxide, also known as nitric oxide (NO), reacts with oxygen gas to produce nitrogen dioxide gas ($NO + \frac{1}{2}O_2 \rightarrow NO_2$), which then reacts with the water vapor in the atmosphere to produce nitrous and nitric acids ($2NO_2 + H_2O \rightarrow HNO_2 + HNO_3$).

Effects of Acid Deposition

Acid deposition results in the following:

- Acid shock, caused by the rapid melting of the snow pack that contains dry acidic particles, which may result in acid concentrations in lakes and streams that are up to ten times higher than that of acidic rainfall
- An increase in fish kills (at all stages of development) due to the increased acidity of the water and the increase in the levels of the toxins released into the water
- Changes in animal life due to changes in the amount and types of vegetation available
- Changes in vegetation (e.g., vigor, disease and pest susceptibility, abundance, and biodiversity) because of changes in the soil pH and in the soil ecosystems. This results in changes in the food webs.
- Increased leaching of soil nutrients (e.g., Ca^{2+}, K^{+}, and Mg^{2+})
- Increased solubility of toxic metals, including methyl mercury, lead, and cadmium
- Reduced buffering capacity of the soil

Remediation and Reduction Strategies

Remediation and reduction strategies that are designed to reduce the effects of acid deposition include the following:

- Continued support and enforcement of the Clean Air Act
- Creating allowance trading systems
- Requiring continuous emission monitoring (CEM) systems for industries that are historically heavy polluters (power companies, refineries, mining companies, etc.)
- Requiring that only Energy Star® appliances be sold in the United States
- Requiring the use of pollution control technology for existing coal-burning power plants
- Tax incentives to encourage buyers of new vehicles to switch to green vehicles
- Tax incentives to power companies to encourage them to convert to less-polluting sources of energy

Heat Islands and Air Pollution

Urban heat islands occur in metropolitan areas that are significantly warmer than their surroundings; e.g., urban air can be 108°F (68°C) warmer than the surrounding area. Since warmer air can hold more water vapor, rainfall can be as much as 30% greater downwind of cities when compared with areas upwind.

Reasons for higher urban temperatures (which are usually more pronounced at night) are as follows:

- Additional heat is produced by burning the fuels necessary for air conditioning, transportation, lighting, etc.
- Additional impervious materials in urban areas reduce the natural cooling effect provided by the evaporation of water from soil and leaves and the shading provided by trees.
- Buildings interfere with the outgoing thermal radiation emitted by Earth's surface.
- There is a lack of vegetation and standing water.
- There is more black asphalt and other building surfaces that absorb heat and reduce the reflectivity of sunlight (reduces the albedo).

Human activities that increase the heat-island effect include the operation of automobiles, air conditioners, and industry. Since the demand for air conditioning rises during summer months, problems associated with energy availability and pricing become compounded.

A "street canyon" is a place where the street is flanked by buildings on both sides, creating a canyon-like environment. Classic examples of these human-built canyons are made when streets separate dense blocks of structures, especially skyscrapers. These urban street canyons have an impact on various local conditions, including temperature, wind, and air quality.

High levels of pollution in urban areas can also create a localized greenhouse effect. Urban heat islands can directly influence the health and welfare of urban residents who cannot afford air conditioning, with as many as 1,000 people dying in the United States each year due to excessive temperatures. Urban heat islands can produce secondary effects on local meteorology, including altering local wind patterns, the development of clouds and fog, the number of lightning strikes, and the rate of precipitation. To develop sustainability, the heat-island effect can be slightly reduced by using white or reflective building and paving materials and increasing the amount of landscaping and parks.

Mitigation techniques to deal with urban heat islands include the following:

- Create urban parks or green spaces.
- Reduce height of buildings and increase the space between them.
- Use reflective and light-colored building and paving materials.

7.8 Noise Pollution

Noise pollution is unwanted human-created (anthropogenic) sound that disrupts the environment. The dominant form of noise pollution is from transportation sources (e.g., motor vehicles, aircraft noise, and rail transport noise). Noise regulation by governmental agencies effectively began in the United States with the 1972 Federal Noise Control Act.

Effects of Noise Pollution

Sensory hearing loss is caused by damage to the inner ear and is the most common form associated with noise pollution. In addition to contributing to hearing loss, excessive noise can cause

- a decrease in alertness and the ability to memorize;
- anxiety and nervousness;
- cardiovascular problems, which manifest as an accelerated heartbeat and high blood pressure; and
- gastrointestinal problems.

Noise Control Measures

Below are some mitigation techniques to reduce noise.

Techniques to Reduce Roadway Noise

- Create computer-controlled traffic flow devices that reduce braking and acceleration, and implement changes in tire designs.
- Create noise barriers.
- Introduce newer roadway surface technologies.
- Limit times for heavy-duty vehicles.
- Place limitations on vehicle speeds.

Techniques to Reduce Aircraft Noise

- Develop quieter jet engines.
- Reschedule takeoff and landing times.

Techniques to Reduce Industrial Noise

- Create new technologies in industrial equipment.
- Control residential noise, such as power tools, garden equipment, and loud entertainment.
- Install noise barriers in the workplace.

Practice Questions

1. ______ contributes to the formation of _________ and thereby compounds the problem of ________.

   (A) Carbon dioxide, carbon monoxide, ozone depletion
   (B) Nitric oxide, ozone, photochemical smog
   (C) Ozone, carbon dioxide, acid rain
   (D) Sulfur dioxide, acid deposition, global warming

2. Photochemical smog does NOT require the presence of

   (A) nitrogen oxides (NO_x)
   (B) peroxyacyl nitrates (PANs)
   (C) ultraviolet radiation (UV)
   (D) volatile organic compounds (VOCs)

3. Which reaction is NOT involved in the formation of acid deposition?

   (A) $2H_2SO_3 + O_2 \rightarrow 2H_2SO_4$
   (B) $2NO_2 + H_2O \rightarrow HNO_2 + HNO_3$
   (C) $O_3 + C_xH_y \rightarrow$ peroxyacyl nitrates (PANs)
   (D) $SO_2 + H_2O \rightarrow H_2SO_3$

4. Which letter on the graph below represents the optimum level of pollution balanced with the most efficient use of available resources to control it?

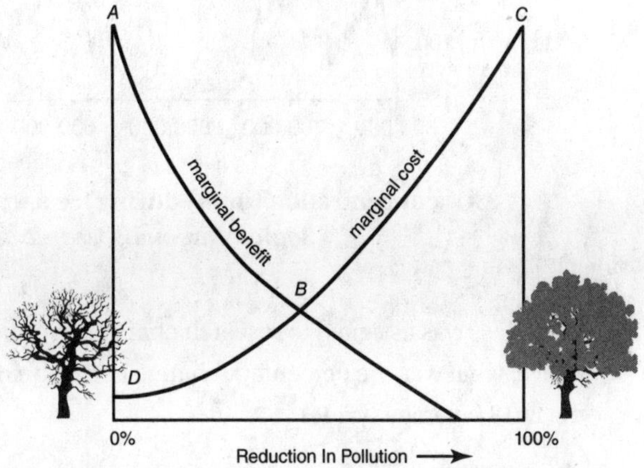

   (A) A
   (B) B
   (C) C
   (D) D

Examine the following diagram to answer Question 5.

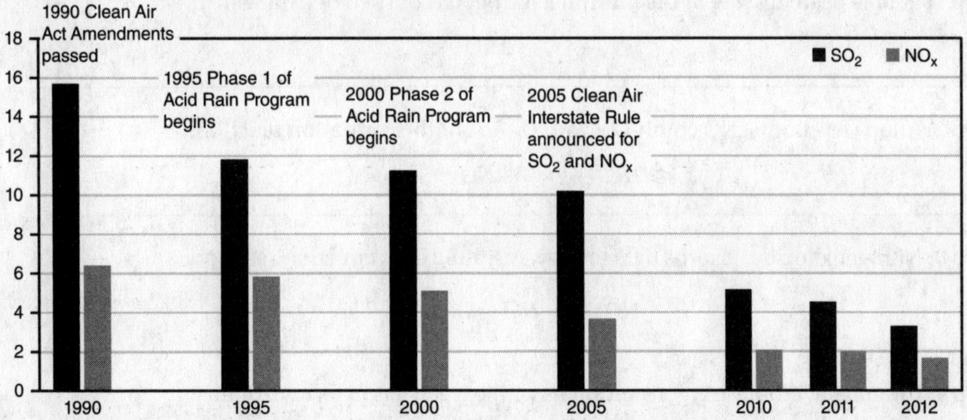

SO_2 and NO_2 emissions from the electric power sector (measured in million short tons)
Source: U.S. Environmental Information Agency

5. What are/is the primary reason(s) for the trends in nitrogen dioxide (NO_2) and sulfur dioxide (SO_2) emissions from American power plants from 1990 to 2012?

   (A) Passage of the Clean Air Act in 1990
   (B) Phasing in of power plants powered by natural gas
   (C) Retrofitting of power plants with scrubbers, switching to lower sulfur coal, and technology
   (D) All of the above

Use the following diagram to answer Question 6.

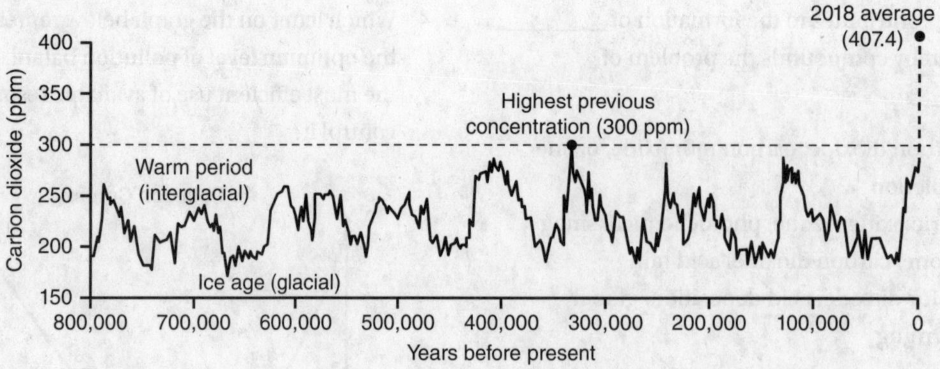

CO_2 concentration (ppm) during ice ages and warm periods for the last 800,000 years
Source: National Oceanic and Atmospheric Administration

6. Your class was assigned to research changes in atmospheric carbon dioxide levels over time. Using the graph above, what was the percentage change in carbon dioxide levels from the highest previous concentration to the 2018 (average) value?

(A) ~3.6%
(B) ~26%
(C) ~36%
(D) ~360%

7. Below are possible steps involved in the formation of photochemical smog. Which of the choices below represents the correct order of the process and each step having the correct chemical reaction?

Choice 1

Step 1: People begin driving in the morning, nitrogen is burned or oxidized: $N_2 + O_2 \rightarrow 2NO$
Step 2: After a few hours, NO combines with O_2, in another oxidation reaction: $2NO + O_2 \rightarrow 2NO_2$
Step 3: Nitrogen dioxide absorbs light energy, resulting in a reduction reaction: $NO_2 \xrightarrow{uv} NO + O$
Step 4: In sunlight, atomic oxygen combines with oxygen gas to form ozone: $O + O_2 \rightarrow O_3$
Step 5: Reaction is temperature and sunlight dependent: $O_3 + NO \rightleftharpoons NO_2 + O_2$

Choice 2

Step 1: People begin driving in the morning, nitrogen is burned or oxidized: $O_3 + NO \rightleftharpoons NO_2 + O_2$
Step 2: After a few hours, NO combines with O_2, in another oxidation reaction: $NO + O_2 \rightarrow NO_3$
Step 3: Nitrogen dioxide absorbs *light* energy, resulting in a reduction reaction: $N_2O_2 + \xrightarrow{\text{light}} 2NO$
Step 4: In sunlight, atomic oxygen combines with oxygen gas to form ozone: $O + O_2 \rightarrow O_3$
Step 5: Reaction is temperature and sunlight dependent: $O_2 + 2NO_3 \rightleftharpoons 2NO + 3O_2$

Choice 3

Step 1: People begin driving in the morning, nitrogen is burned or oxidized: $N + O_2 \rightarrow NO_2$
Step 2: In sunlight, atomic oxygen combines with oxygen gas to form ozone: $O + O_2 \rightarrow O_3$
Step 3: Nitrogen dioxide absorbs light energy, resulting in a reduction reaction: $N_2O_2 \xrightarrow{uv} N_2 + O_2$
Step 4: Reaction is temperature and sunlight dependent: $O_3 + NO \rightleftharpoons NO_2 + O_2$
Step 5: After a few hours, NO combines with O_2 in another oxidation reaction: $2NO + O_2 \rightarrow 2NO_2$

(A) Choice 1
(B) Choice 2
(C) Choice 3
(D) None of the choices are correct

8. In 1970, Congress passed the Occupational Safety and Health Act (OSHA), which granted OSHA the authority to adopt national standards for acceptable workplace exposure known as Permissible Employee Exposure Limits (PELs). A PEL of a particular hazard constitutes the acceptable level that an employee can be exposed to during a given workday. If an employee's PEL is at or above the accepted level, his or her employer must provide them with adequate protection equipment; e.g., ear plugs, HEPA facemasks, etc. To determine if an employee has been exposed above the PEL, OSHA requires the contractor to monitor its employees and analyze the data by applying an eight-hour time-weighted average calculation (TWA) to the collected data. The table below specifies the maximum number of hours per day a worker may be exposed to certain sound levels (measured in decibels [dB]); e.g., an eight-hour exposure to 90 dBA (decibel average) is considered a 100% "dose," the maximum allowable. Likewise, three hours at 97 dB is also considered a 100% dose. Less-than-maximum times at a certain level would correspond to a less-than-maximum dose: 4 hours at 90 dB would be a 50% dose.

Sound Level (dB)	Duration per Day (hours)
85	16
90	8
92	6
95	4
97	3
100	2
102	1.5
105	1
110	.5
115	.25

Typically, workers may be exposed to different noise levels during the course of an eight-hour shift. To calculate the dosage in such scenarios, the partial times at the different sound levels would be added up. The following formula can be used to determine the percentage dose for a day with different sound levels:

$$D = 100 \times o(C_n/T_n) \text{ where}$$

D = The percentage dose – anything at 100% or less is considered acceptable
C_n = The actual hours at a certain sound level
T_n = The allowable time for that sound level

Using the chart provided above, what is the percentage dose of sound exposure for a construction worker who was exposed to four hours of sound measured at 90 dB and four hours of sound measured at 85 dB?

(A) 25%
(B) 50%
(C) 75%
(D) 120%

9. Which of the following pollutants would NOT contribute to acid rain?

(A) carbon dioxide (CO_2)
(B) carbon monoxide (CO)
(C) nitrogen dioxide (NO_2)
(D) sulfur dioxide (SO_2)

10. With respect to indoor air pollution, the average American spends about __________% of their time indoors.

(A) 0%–25%
(B) 26%–50%
(C) 51%–75%
(D) 76%–100%

11. Which of the figures shown below correctly illustrates both normal atmospheric conditions and atmospheric conditions that occur during a thermal inversion?

(A)

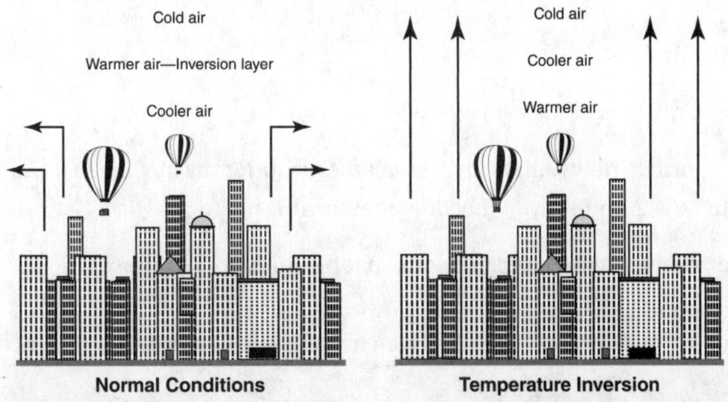

(B)

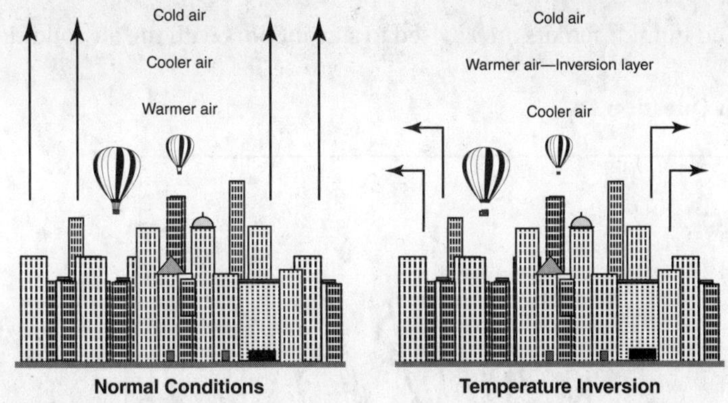

(C)

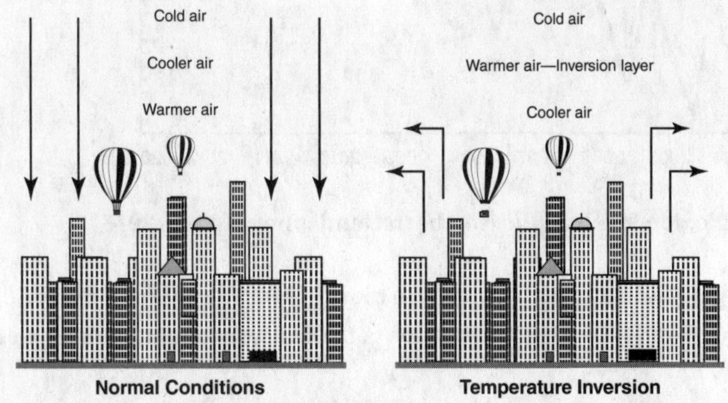

(D)

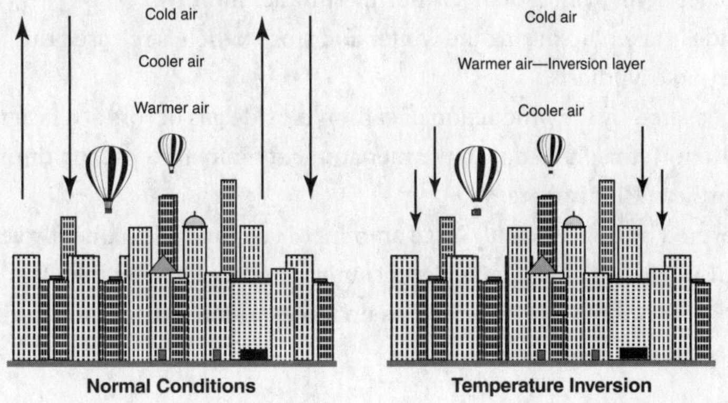

12. What is the single largest cause of worldwide air pollution?

    (A) Farms
    (B) Factories
    (C) Vehicles burning carbon-based fuels
    (D) Volcanoes

13. Sarah and her dad were watching the morning news and heard the weather reporter say that the AQI (Air Quality Index) was 235 and that it was going to be a "purple day." What does that mean?

    (A) Air quality is moderate; however, there may be a risk for some people, particularly those who are unusually sensitive to air pollution.
    (B) Air quality is very unhealthy and a "Health Alert" is being given as the risk of health effects is increased for everyone.
    (C) The day will be cold and windy.
    (D) The day will be extremely hot and humid; seniors are advised to stay indoors with the air conditioner on.

Examine the following graph to answer Question 14.

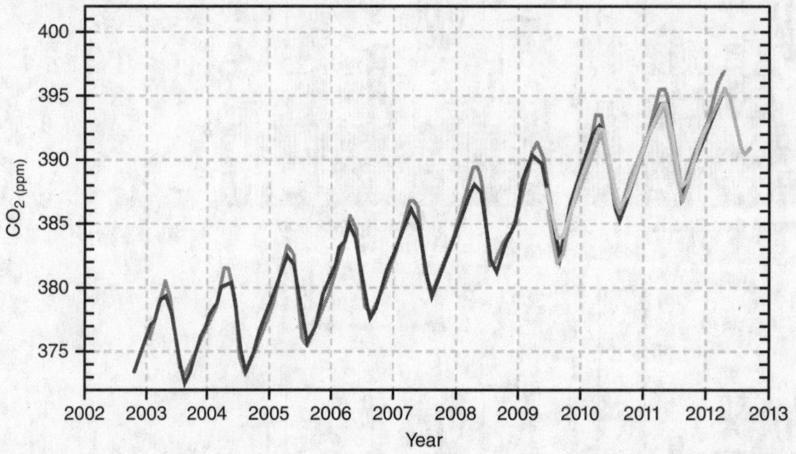

Tropospheric carbon dioxide levels in the Northern Hemisphere (2002–2013)

14. The graph shown above shows fluctuations in atmospheric carbon dioxide concentrations in the Northern Hemisphere from 2002 to 2013. Regular and repeating fluctuations in the CO_2 concentration might best be explained as which of the following?

    (A) The Northern Hemisphere is colder in the winter than it is during summer months.
    (B) More atmospheric carbon dioxide is taken up during the winter and stored, then is released during the summer through the process of photosynthesis.
    (C) The rate of photosynthesis as measured by the production of carbon dioxide gas occurs at a faster rate during the summer in the Northern Hemisphere due to warmer ambient temperatures than during the colder, winter months in the Northern Hemisphere.
    (D) The Northern Hemisphere is covered by a significantly large amount of land and consequently vegetation compared to the Southern Hemisphere, which contains significantly more ocean water than land. During the summer, this vegetation in the Northern Hemisphere takes up more carbon dioxide, and during the winter, when it decays, it releases more carbon dioxide.

Refer to the following choices to answer Questions 15–17.

(A) Clean Air Act
(B) Kyoto Protocol
(C) Montreal Protocol
(D) United Nations Air Pollution Control Act

15. Designed to protect the ozone layer by phasing out the production of numerous substances that are responsible for ozone depletion.

16. First international agreement on climate change with the objective to keep the global temperature rise below 28°C by stabilizing the level of greenhouse gases in the atmosphere.

17. Required the U.S. Environmental Protection Agency (EPA)

- to establish a cap-and-trade program to limit acid rain;
- to mandate emission controls for sources of hazardous air pollutants;
- to set national emission standards for air pollution, including motor vehicles, power plants, and other industrial sources;
- to set deadlines for the achievement of those standards by state and local governments; and
- to set health-based standards for ambient air quality.

18. Mr. Five's AP Environmental Science class was discussing the impact of cars and trucks on air pollution. During the discussion, one of the students mentioned that he had been told that with everything being equal, before catalytic converters were required

(1) cars used to get more horsepower;
(2) cars used to run quieter; and
(3) gas mileage was better.

Which of the following statements above is FALSE?

(A) #2
(B) #1 and #3
(C) All statements are TRUE
(D) All statements are FALSE

19. Which of the possible four answer choices below (A, B, C, or D) correctly describes (in order) the formation of ground-level or tropospheric ozone using the numerical codes shown below?

1 = Carbon dioxide (CO_2)
2 = Nitrogen oxides (e.g., NO_2)
3 = Oxygen gas (O_2)
4 = Strong spring or summer sun along with low relative humidity and light winds
5 = Sulfur oxides (e.g., SO_2)
6 = Tropospheric ozone (O_3)
7 = Volatile organic compounds (VOCs)
8 = Water vapor (H_2O)
9 = Weak winter or fall sun along with high relative humidity and strong winds

(A) #1 + #2 + #3 → #6
(B) #4 + #5 → #7
(C) #2 + #7 + #4 → #6
(D) #3 + #5 + #8 + #9 → #6

20. Mr. Fitzpatrick was discussing various methods used to reduce air pollutants with his AP Environmental Science class. Use the following choices to fill in the blanks in Mr. Fitzpatrick's lecture.

(1) catalytic converter
(2) electrostatic precipitators
(3) scrubbers
(4) vapor recovery nozzles

"Good morning class. Today we are going to discuss various methods that are used to reduce air pollutants. We'll discuss ____, which are used by coal-burning power plants to reduce the gas mists and particulate matter from burning coal and which simultaneously provide cooling of the hot gases released. We'll also be discussing ____, which are also used by coal-burning power plants in an effort to remove fine particles and smoke from the effluent flowing gases with an electrostatically charged filter. Then we'll switch to talking about ____, which are federally mandated to capture and collect gasoline vapors so that they do not escape into the air. And finally, before we head off for lunch, we'll wrap up talking about ____, which are required by many states to reduce toxic gases and pollutants in exhaust gases coming from an internal combustion engine into less-toxic pollutants by catalyzing an oxidation and a reduction reaction."

(A) 2, 3, 4, 1
(B) 1, 3, 2, 4
(C) 3, 2, 4, 1
(D) 4, 1, 3, 2

Free-Response Question

The following graph shows the levels of atmospheric sulfur dioxide and nitrogen dioxide gases in Paris, France, from 1960 to 2013.

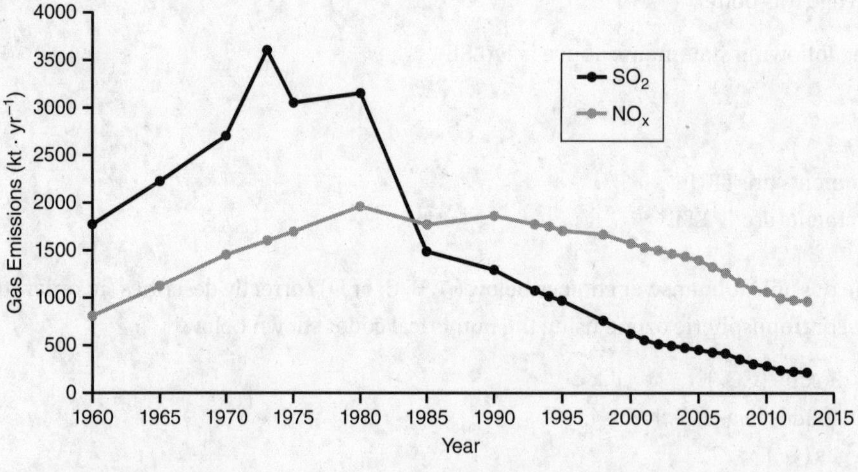

(a) Describe and compare the atmospheric concentration **trends** for either nitrogen dioxide (NO_2) or sulfur dioxide (SO_2) in France from 1960 to 2013, comparing its largest % increase during those years to its final concentration in 2013.

(b) Calculate the percentage change in each gas from 1960 until the year in which it reached its maximum value.

(c) Identify the major source(s) of air pollution caused by nitrogen dioxide and sulfur dioxide and describe the primary health and environmental effect(s) they cause.

(d) Describe ONE method that can be used to reduce the concentration of one of the pollutants.

Answer Explanations

1. **(B)** As sunlight becomes more intense during the day, nitrogen dioxide (NO_2) is broken down into nitric oxide (NO) and oxygen atoms (O), which results in an increase in ozone (O_3).

$$NO_2 + uv \rightarrow NO + \mathbf{O}$$

$$O_2 + \mathbf{O} \rightarrow O_3$$

2. **(B)** Peroxyacyl nitrates (PANs) are secondary pollutants (not directly emitted as exhaust from power plants or internal combustion engines) and are powerful respiratory and eye irritants present in photochemical smog and are good indicators for the source of volatile organic compounds (VOCs).

3. **(C)** The key to answering this question is to look at the products of each reaction and remember that most acids generally start with a hydrogen (H) atom; e.g., $\mathbf{H}_2SO_4$—sulfuric acid or $\mathbf{H}NO_3$—nitric acid), but not all compounds that begin with H are necessarily acids; e.g., water—H_2O. All choices except (C) have at least one product that starts with "H."

4. **(B)** Point B represents the point at which society has reached the maximum benefit of controlling air pollution at the least cost. Any more money spent on controlling air pollution would not be worth the cost for the benefit received (law of diminishing returns) and would take money away from other needs.

5. **(D)** The decline in emissions of NO_x and SO_2 from electricity-generating plants is primarily due to the 1990 Clean Air Act, retrofitting existing power plants with scrubbers (reduces sulfur from flue gas), switching to coal with lower sulfur content, and adding after-burners to the exhaust flue gas stream to reduce SO_2. The Clean Air Act also allowed a cap and trade provision—a system for controlling carbon emissions and other forms of atmospheric pollution by which an upper limit is set on the amount a power plant can release into the environment but which allows further capacity to be bought from other organizations that have not used their full allowance. Furthermore, as time progressed and natural gas became more abundant and affordable due to higher supply from fracking technology, older plants were either retrofitted to be able to burn natural gas or were rebuilt to do so.

6. **(C)** Percentage increase and percentage decrease are measures of percentage change, which is the extent to which a variable (something that changes) gains or loses value. The figures are arrived at by comparing the initial (or before) and final (or after) quantities. Percentage change is calculated by using the following formula: % change = $(x_2 - x_1) / x_1 * 100\%$, where x_2 represents the final value and x_1 represents the initial value. In this case, x_1 or the initial value of carbon dioxide that was the highest prior to the 2018 average was ~300 ppm around 350,000 years ago, and the 2018 value was ~400 ppm:

$$\textbf{\% change} = \mathbf{100\ (x_2 - x_1) / x_1} = \frac{100\ (407 - 300)}{300} \times 100\% = \mathbf{\sim 36\%}$$

7. **(A)** Photochemical smog is a type of air pollution that results from the reaction of solar radiation with airborne pollutant mixtures of nitrogen oxides (NO_x) and volatile organic (hydrocarbon) compounds. Due to industry and the number of motor vehicle emissions, photochemical smog is more of a problem in large cities that have a dry, sunny, and warm climate. Nitric oxide (NO) and nitrogen dioxide (NO_2) can also react with hydrocarbons instead of ozone to form other volatile compounds known as PANs (peroxyacyl nitrates).

8. **(C)** $D = 100 \times (4/8 + 4/16) = 75\%$. The dose of sound the worker received over his shift (75%) is acceptable since it is less than or equal to 100%.

9. **(B)** Carbon monoxide (CO) is produced when hydrocarbon-based fuels burn in a limited amount of oxygen and at high temperatures (e.g., a faulty home furnace: $CH_{4(g)}$ (methane) + $O_{2(g)} \rightarrow \mathbf{CO}_{(g)} + H_{2(g)}$).

   The Fischer-Tropsch process is a series of chemical reactions that converts a mixture of carbon monoxide and hydrogen into liquid hydrocarbons (fuels) and is one of the most important industrial reactions; it is used for the production of ammonia (e.g., agricultural fertilizer). $CO + H_2O \rightleftharpoons CO_2 + H_2$ does produce carbon dioxide BUT only at very high temperatures, far beyond those found in the atmosphere. The main contributors of acid rains are carbon dioxide (CO_2), nitrogen dioxide (NO_2), and sulfur dioxide (SO_2). Normal rainwater has an acidity of pH 5.6 due to naturally dissolved CO_2 in the atmosphere; however, excess CO_2 produced from combustion (both naturally [e.g., wild fires] or from anthropogenic [human-caused] sources; e.g., automobiles and factory emissions) combining with water forms higher levels of carbonic acid (H_2CO_3), which adds additional H^+ to rainwater, increasing the acidity. Important chemical reactions to know for the production of hydrogen ions (**H^+**), which make a solution acidic, include the following:

   1. $CO_2 + H_2O \rightarrow H_2CO_3$ (carbonic acid)
      $H_2CO_3 \rightarrow \mathbf{H^+} + HCO_3^-$
   2. $N_2 + O_2 \rightarrow 2NO$
      $NO + \frac{1}{2}O_2 \rightarrow NO_2$
      $3NO_2 + H_2O \rightarrow 2HNO_3$ (nitric acid) $+ NO$
      $HNO_3 \rightarrow \mathbf{H^+} + NO_3^-$
   3. $2SO_2 + O_2 \rightarrow 2SO_3 + 2H_2O \rightarrow 2H_2SO_4$ (sulfuric acid)
      $H_2SO_4 \rightarrow \mathbf{H^+} + HSO_4^-$
      $HSO_4^- \rightarrow \mathbf{H^+}\ SO_4^{2-}$

10. **(D)** In a recent survey of 9,000 Americans conducted by the United States Environmental Protection Agency (EPA), it was revealed that people in the United States spend ~87% of their time in an indoor environment. Of this, 69% of their time was spent at home, and 18% was spent in some other type of indoor venue; e.g., motor vehicle, shopping, work, etc. A poll of people living in the United Kingdom found that the average UK citizen spends 17 minutes (~1%) out of the 1,440 minutes in a day outdoors (weather was a major factor). And, according to the EPA, indoor levels of pollutants can be as much as 100 times the levels found outside.

11. **(A)** A thermal inversion, also known as a temperature inversion, is a reversal of the normal behavior of temperature in the troposphere (the region of the atmosphere nearest Earth's surface), in which a layer of cool air at the surface is overlain by a layer of warmer air (under normal conditions air temperature usually decreases with height).

12. **(A)** The most harmful pollutant to human health is called $PM_{2.5}$, a particulate matter smaller than 2.5 microns in diameter that's found in dust, smoke, and soot; a micron is 1/1,000 of a millimeter or one-millionth of a meter. $PM_{2.5}$ is especially dangerous because it can get lodged in the lungs and cause long-term health problems like asthma and chronic lung disease. According to a recent report from Columbia University, the largest source of deadly air pollution in many parts of the world is farms, which are major contributors to fine particulate matter. Agriculture requires fertilizers, which release ammonia (NH_3) into the air. As ammonia encounters pollutants produced by factories and vehicles powered by carbon-based fuels, including nitrogen oxides (NO_x) and sulfates (SO_x), through a series of chemical reactions, these molecules combine to generate $PM_{2.5}$. Recent studies have shown that almost 6 million premature deaths each year and more than 80% of all people living in urban areas (especially in the developing world) are breathing air contaminated by $PM_{2.5}$.

	Top 10 Countries for HIGH $PM_{2.5}$ Concentrations	Top 10 Countries for LOW $PM_{2.5}$ Concentrations
#1	Pakistan 115.7	Australia 5.7
#2	Qatar 92.4	Brunei 6.6
#3	Afghanistan 86	New Zealand 6.8
#4	Bangladesh 83.3	Estonia 7.2
#5	Egypt 73	Finland 7.3
#6	United Arab Emirates 64	United States 7.4
#7	Mongolia 61.8	Canada 7.5
#8	India 60.6	Iceland 8.2
#9	Bahrain 56.1	Sweden 8.7
#10	Nepal 50.0	Ireland 8.8

Note: Of the top ten cities in the United States with the highest $PM_{2.5}$ levels, California had nine.

13. **(B)** The AQI or Air Quality Index is the Environmental Protection Agency's index for reporting air quality. Based on federal air quality standards, the AQI includes measures for six major air pollutants: carbon monoxide, nitrogen dioxide, ozone, sulfur dioxide, and two sizes of particulate matter. AQI values at or below 100 are generally thought of as satisfactory. When AQI values are above 100, air quality is unhealthy. Below is a chart showing the six categories of air quality used by the EPA.

AQI Color	Level of Concern	AQI	Description of Air Quality
Green	Good	0–5	Air quality is satisfactory, and air pollution poses no risk.
Yellow	Moderate	51–100	Air quality is acceptable; however, there may be some risk to people sensitive to air pollution.
Orange	Unhealthy for sensitive groups	101–150	Members of sensitive groups may experience health effects; e.g., asthma.
Red	Unhealthy	151–200	Some members of the general public may experience more serious health effects.
Purple	Very unhealthy	201–300	Health alert. Risk of health effects is increased for everyone.
Maroon	Hazardous	300 +	Emergency conditions.

14. **(D)** Begin by remembering what occurs during the process of photosynthesis:

$$CO_2 + H_2O \rightarrow C_6H_{12}O_6 + O_2$$

During the fall and winter in the Northern Hemisphere, when there are lower ambient light levels and colder temperatures and when trees and plants begin to lose their leaves and decay, carbon dioxide is released into the atmosphere along with carbon dioxide produced from the combustion of fossil fuels used for heating purposes. This, combined with there being fewer trees and plants removing carbon dioxide from the atmosphere due to dormancy during winter months and lower light levels, allows concentrations of carbon dioxide to climb all winter, reaching a peak by early spring. During the Northern Hemisphere's spring and summer months, however, plants absorb a substantial amount of carbon dioxide through photosynthesis, thus removing it from the atmosphere. As there is more water surface area in the Southern Hemisphere compared to land surface area, and since the Northern Hemisphere's winter is the Southern Hemisphere's summer, this pattern is reversed in the Southern Hemisphere.

15. **(C)** Explanation incorporated into the question.

16. **(B)** Explanation incorporated into the question.

17. **(A)** Explanation incorporated into the question.

18. **(A)** Once the catalytic converter is removed from a vehicle, some vehicles do experience an increase in power as a catalytic converter creates a back-pressure on the engine that slightly reduces its performance. In regards to better gas mileage, since the catalytic converter places a strain on the engine with its constrictive design, it must work harder to achieve the same energy than it does without the device in place. This means there is a slight improvement in fuel consumption without a catalytic converter, which translates to slightly better mileage. In addition, the catalytic converter works like a muffler on the average vehicle, as it reduces the impact of the gases emerging from the engine due to the combustion and burning of fuel. Along with the muffler, the catalytic converter dampens the sound of the exhaust.

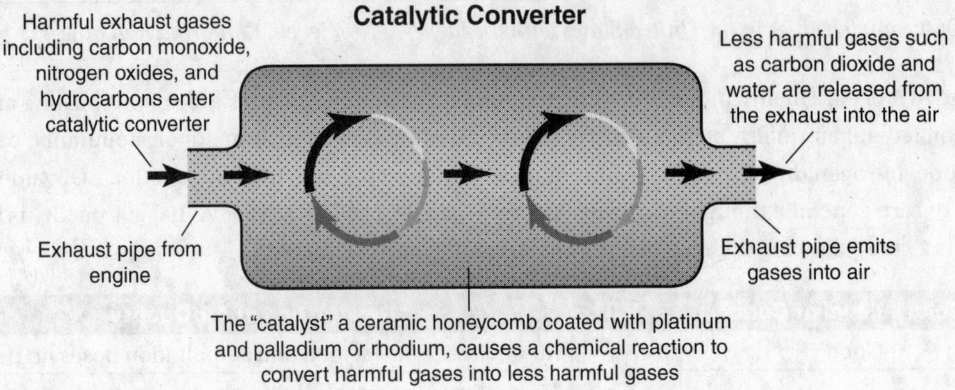

19. **(C)**

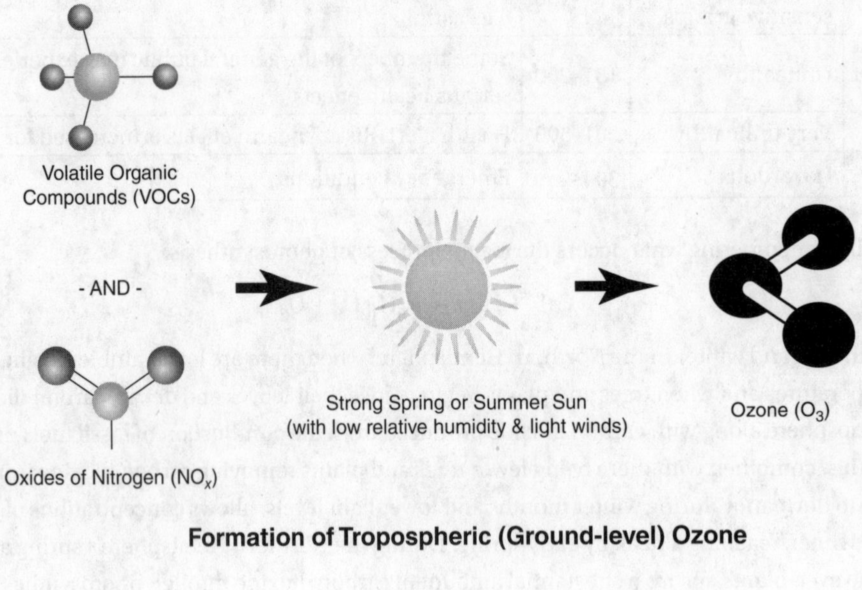

20. **(C)** No explanation required as the explanations are incorporated into the scenario.

Free-Response

10 Total Points Possible

(a) *Maximum 2 points: 1 point for correctly describing the trend of NO_2 concentration and 1 point for correctly describing the trend for sulfur dioxide concentration.*

*Describe and compare the atmospheric concentration **trends** of nitrogen dioxide (NO_2) and sulfur dioxide (SO_2) in France from 1960 to 2013.*

Nitrogen dioxide emissions increased from ~900 kilotons per year in 1960 until ~1980, in which the amount produced reached a maximum of ~1900 kilotons per year but at a slightly lower rate than sulfur dioxide was being produced at, based on a shallower slope for NO_2 during those years. Nitrogen dioxide continued to decrease until ~2013, when it again reached a value of about 900 kilotons per year. Sulfur dioxide gradually increased in atmospheric concentration from 1960 at ~1900 kilotons per year until ~1974, when it reached a maximum of ~3600 kilotons per year and then began to dramatically decrease from 1974 to ~2013 to about 50 kilotons per year.

(b) *Maximum 2 points: 1 point for correctly calculating the change in nitrogen dioxide gas from 1960 until it reached its maximum and 1 point for correctly calculating the change in sulfur dioxide gas from 1960 until it reached its maximum.*

Calculate the percentage change in each gas from 1960 until the year in which it reached its maximum value.

NO_2:

x_1: 900 kilotons/yr

x_2: 1900 kilotons/yr

$$\text{Percentage change} = \frac{(x_2 - x_1)}{x_1} \times 100\% = \frac{1900 \text{ kt/yr} - 900 \text{ kt/yr}}{900 \text{ kt/yr}}$$

= **111% increase**

SO_2:

$x_1 = 1900$ kilotons/yr

$x_2 = 3600$ kilotons/yr

$$\text{Percentage change} = \frac{(x_2 - x_1)}{x_1} \times 100\% = \frac{3600 \text{ kt/yr} - 1900 \text{ kt/yr}}{1900 \text{ kt/yr}} \times 100\%$$

= **~89% increase**

(c) *Maximum 2 points: 1 point for correctly describing the source of nitrogen dioxide and 1 point for correctly describing the source of sulfur dioxide.*

NO_2: Nitrogen dioxide forms when fossil fuels such as coal, diesel, natural gas, or oil are burned at high temperatures. NO_2 and other nitrogen oxides in the outdoor air contribute to particle pollution and to the chemical reactions that make ozone. Cars, trucks, and buses are the largest sources of emissions, followed by power plants.

SO_2: Sulfur dioxide forms when sulfur-containing fuel is burned for motor vehicles, trains, ships, and off-road diesel equipment. In addition, SO_2 is emitted from some industrial processes—e.g., natural gas and petroleum extraction and oil refining—and is also released during volcanic activity.

(d) *Maximum 2 points for correctly describing the primary health effect(s) of nitrogen dioxide and 1 point for correctly describing the primary health effect(s) of sulfur dioxide.*

NO_2: Health effects of exposure to nitrogen dioxide include the following:

- Cardiopulmonary effects
- Coughing and wheezing
- Increased asthma attacks
- Premature death
- Reduced lung function

SO_2: Health effects of exposure to sulfur dioxide include the following:

- Bronchitis or emphysema
- Shortness of breath and chest tightness, especially during physical activity
- Wheezing

(e) *Maximum 2 points: 1 point for correctly describing at least one environmental effect of nitrogen dioxide and 1 point for correctly describing at least one environmental effect of sulfur dioxide.*

NO_2: Environmental effects of exposure to nitrogen dioxide include the following:

- Acid rain, which affects crops and habitats (both land and water)
- Contributes to the reduction of visibility both directly, by selectively absorbing the shorter blue wavelengths of visible light, and indirectly, by contributing to the formation of nitrate aerosol haze that decreases visibility

SO_2: Environmental effects of exposure to sulfur dioxide include the following:

- Acid rain, which harms both aquatic and terrestrial ecosystems
- Promotes chemical reactions that facilitate the accumulation of mercury in soil and water

8

Aquatic and Terrestrial Pollution

Learning Objectives

In this unit, you will learn:

- ➔ 8.1 Sources of Pollution
- ➔ 8.2 Human Impacts on Ecosystems
- ➔ 8.3 Endocrine Disruptors
- ➔ 8.4 Human Impacts on Wetlands and Mangroves
- ➔ 8.5 Eutrophication
- ➔ 8.6 Thermal Pollution
- ➔ 8.7 Persistent Organic Pollutants (POPs)
- ➔ 8.8 Bioaccumulation and Biomagnification
- ➔ 8.9 Solid Waste Disposal
- ➔ 8.10 Waste Reduction Methods
- ➔ 8.11 Sewage Treatment
- ➔ 8.12 Lethal Dose (LD_{50})
- ➔ 8.13 Dose Response Curve
- ➔ 8.14 Pollution and Human Health
- ➔ 8.15 Pathogens and Infectious Diseases

For added practice with this unit and topics, log onto Online Labs 8A–8C at online.barronsbooks.com

8.1 Sources of Pollution

Pollution is defined as the introduction of substances (or energy) that cause adverse changes in the environment and living entities. There are four major types of pollution which will be outlined here.

Air Pollution

Air pollution refers to the release of harmful contaminants (chemicals, particulates, toxic gasses, etc.) into the earth's atmosphere and is is now the world's fourth-largest risk factor for early death (e.g., 4.5 million deaths were linked to outdoor air pollution exposures in 2019, and another 2.2 million deaths were caused by indoor air pollution) with over 90% of all humans currently breathing air that exceeds the World Health Organization's (WHO) guidelines.

Contributors to air pollution include:

- Burning fossil fuels
- Exhaust gasses from industries and factories
- Mining operations

Environmental impacts of air pollution include, but are not limited to:

- Acid rain
- Global warming
- Hazards to wildlife

- Increased risk of cancer
- Increased risk of cardiovascular and respiratory illness
- Ozone depletion

The air pollutants methane (CH_4) and carbon dioxide (CO_2) ultimately come from the production and use of energy in an especially destructive positive feedback loop, which ultimately raises the earth's air temperature; e.g., the 2023 northern hemisphere summer experienced the hottest summer temperatures ever experienced on Earth.

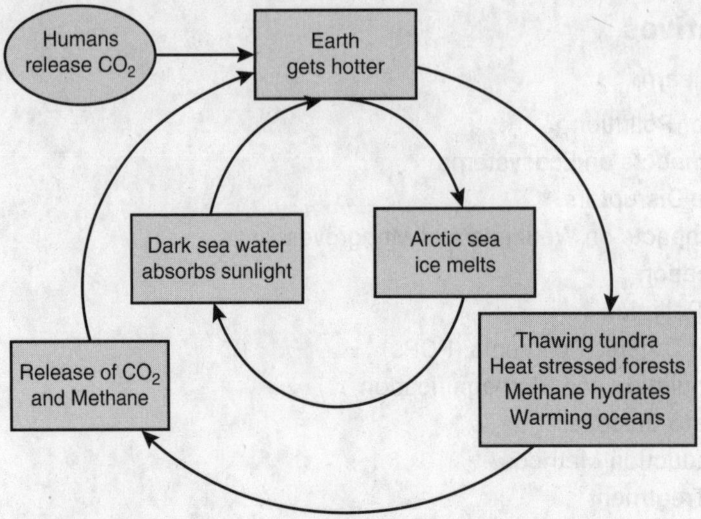

Figure 8.1 Climate Feedback Loop Influenced by CO_2

The effects of air pollution on the human body vary depending on the type of pollutant and the length and level of exposure, as well as other factors including a person's individual health risks and the cumulative impacts of multiple pollutants.

Smog (fog or haze combined with smoke and other atmospheric pollutants; e.g. ground level ozone) and soot (carbon particles resulting from the incomplete combustion of hydrocarbons) are the two most prevalent types of air pollution, both of which come from cars and trucks, engines, factories, forest fires, incinerators, and power plants: basically anything that combusts fossil fuels; e.g., coal, gasoline, natural gas (methane), or oil.

A number of air pollutants pose severe health risks and can sometimes be fatal, even in small amounts. Almost 200 of them (e.g., benzene—a carcinogen which can cause eye, skin, and lung irritation in the short term and blood disorders in the long term) are regulated by law in the United States. They can contain dioxins, which can affect the liver in the short term and harm the endocrine, immune, reproductive, and nervous systems, and lead and mercury that can harm the central nervous system (damaging children's brains and affecting their IQ and ability to learn), kidneys, endocrine system, and reproductive system.

Noise Pollution

Noise pollution results from an excessive amount of noise in the surrounding environment and can have many sources, some which include:

- Industrial noise, e.g., construction sites, factories, heavy machinery, etc.
- Noise from social events, e.g., fireworks, music concerts, sporting events, etc.
- Transportation noises from jets, trucks, etc.

Noise pollution is now very common due to dense urbanization and industrialization and can cause:

- Hearing loss
- High blood pressure (hypertension)
- Ringing in the ears, i.e., tinnitus
- Sleeping disorders

Soil Pollution

Soil pollution refers to the degradation of land due to the presence of chemicals in the soil. Over time, these toxic chemicals will get absorbed by the plants and will pass up through the food chain.

Some common causes of soil pollution are:

- Acid rain caused by air pollution
- Improper industrial waste disposal
- Intensive farming and agrochemicals (like fertilizers and pesticides)
- Mining activities
- Oil Spills

Environmental effects of soil pollution include:

- Degradation of vegetation due to the increase of salinity of the soil
- Impact to the natural flora and fauna residing in the soil
- Loss of soil nutrients, which renders the soil unfit for agriculture
- Toxic dust which can cause respiratory problems, e.g., lung cancer

Water pollution

Water pollution is the contamination of aquifers, groundwater, lakes, oceans and rivers. This form of environmental degradation occurs when pollutants are directly or indirectly discharged into water bodies without adequate treatment to remove harmful compounds.

Water pollution occurs when particulate matter and/or toxic pollutants are introduced into lakes, rivers, or oceans through human activity such as oil spills and improper sewage treatment or through natural processes, e.g., eutrophication.

These toxic pollutants can bioaccumulate and move up the food chain, ultimately reaching humans. Additional environmental consequences of water pollution include:

- Disruption of ecosystems
- Increased levels of toxic chemicals (e.g., mercury) in water bodies
- Increased risk of water-borne diseases
- **Non-point source pollution**—a combination of pollutants from a large area rather than from specific identifiable sources, e.g., discharge pipes. Runoff is generally associated with non-point source pollution, as water is emptied into streams or rivers after accumulating contaminants from specific sources, e.g., parking lots, roads, feed lots, construction sites, agricultural runoff, etc.
- **Point source pollution**—release pollutants from known locations, such as discharge pipes, that are regulated by federal and state agencies. Major point sources of wastewater discharge are factories and sewage treatment plants.
- Threats to marine life

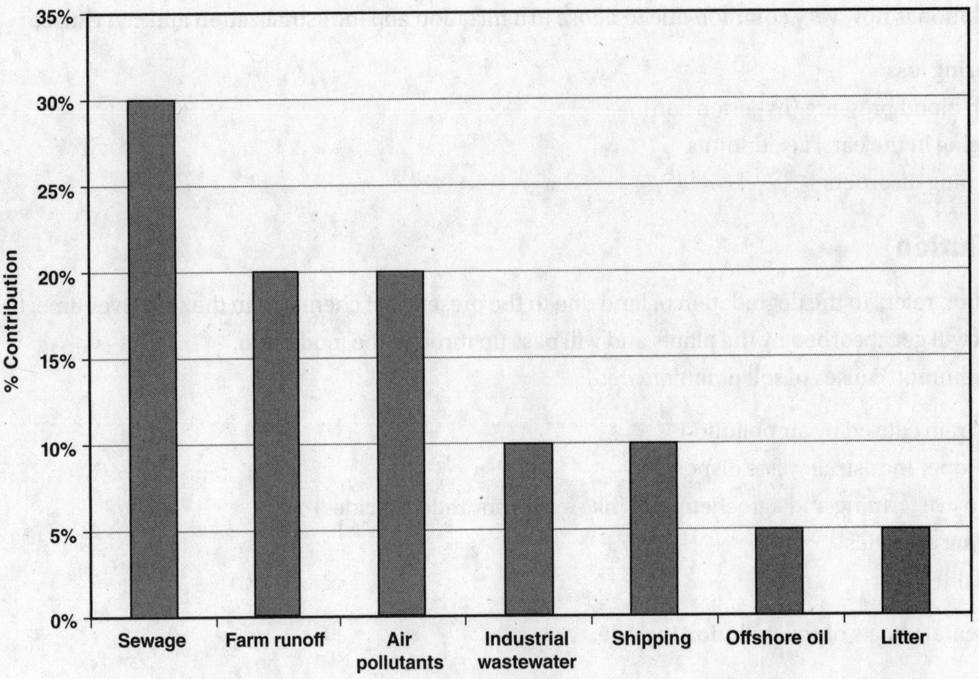

Figure 8.2 Source and percentage of ocean pollutants

- **Thermal Pollution**—the degradation of water quality by any process that changes ambient water temperature. A common cause of thermal pollution is the use of water as a coolant by power plants and industrial manufacturers.

8.2 Human Impacts on Ecosystems

As mentioned previously, airborne pollutants can enter water bodies and cause acidification and/or eutrophication. Examples include mercury, which results in higher mercury levels in fish through bioaccumulation and biomagnification, and nitrogen and sulfur oxides, which can lead to water acidification and/or eutrophication.

Assorted Chemicals

A variety of chemicals from industrial, residential, and agricultural sources can cause water pollution, e.g., metals, solvents, oils, detergents, and pesticides. These pollutants can accumulate in fish and shellfish, poisoning the people, animals, and birds that consume them.

On a square-foot basis, homeowners apply more chemicals to their lawns than farmers apply to their fields. Each year, road runoff and other non-spill sources impart an amount of oil to the oceans that is more than five times greater than the *Exxon Valdez* spill—about 21 million barrels.

Discharge of oily wastes and oil-contaminated ballast water and wash water are all significant sources of marine pollution. Drilling and extraction operations for oil and gas can also contaminate coastal waters and groundwater. The Environmental Protection Agency (EPA) estimates that over 100,000 gasoline storage tanks are leaking chemicals into the groundwater. New evidence strongly suggests that components of crude oil, called polycyclic aromatic hydrocarbons (PAHs), persist in the marine environment for years and are toxic to marine life at concentrations in the low parts per billion (ppb) range. Chronic exposure to PAHs can affect the development of marine organisms, increase susceptibility to disease, and jeopardize normal reproductive cycles in many marine species.

Studies have also shown that up to 90% of drug prescriptions pass through the human body unaltered. Animal farming operations that use growth hormones and antibiotics also send large quantities of these chemicals into the water, with most wastewater treatment facilities not equipped to filter out personal care products, household products, or pharmaceuticals, and as a result, a large portion of these chemicals pass directly into local waterways. Studies on the effects of these chemicals have discovered

- fragrance molecules inside fish tissues;
- ingredients from birth control pills cause gender-bending hormonal effects in frogs and fish; and
- remnants of detergents, which disrupt fish reproduction and growth.

8.3 Endocrine Disruptors

A gland is an organ (e.g., heart, brain, leaf, stem) that secretes particular chemical substances for use in the body or for discharge into the surroundings.

The endocrine system is a network of glands that make the hormones that help cells communicate with each other and is responsible for almost every cell, organ, and function in both humans and animals. Examples of common glands in humans that make up the endocrine system include the pituitary gland, which is located in the brain; the thyroid gland in the neck; the pancreas near the stomach; and the ovaries or testes in the pelvic region.

Endocrine disruptors are chemicals that can interfere with endocrine or hormonal systems and can cause behavior, learning and developmental disorders, birth defects, cancerous tumors, and loss of fertility.

Endocrine disruptors are contained in many common household and industrial products, with many of them being slow to break down in the environment, making them accumulate over time. Examples include cosmetics, detergents, flame-retardants, liners of metal food cans, pesticides, some plastic bottles and containers, and toys. The following table gives examples of specific endocrine disruptors.

Bisphenol A (BPA)	Used in plastic manufacturing and epoxy; e.g. food containers
Dioxins	By-product of herbicide production and paper bleaching, and released during burning wastes and wildfires
Phthalates	Used to make plastics more flexible; e.g., some food packaging, cosmetics, children's toys, and medical devices
Polychlorinated biphenyls (PCBs)	Used to make electrical equipment (transformers), heat transfer fluids, and lubricants

Unlike being able to remove some of the oil from an oil spill, or trash off a beach, or reducing sulfur or nitrogen oxides from combustion, there is currently no cost-effective or practical method to remove endocrine disruptors from the environment.

8.4 Human Impacts on Wetlands and Mangroves

A wetland is a place where the land is covered by water, which can be freshwater, saltwater, or brackish water (having more salinity than freshwater, but not as much as seawater). Wetlands include marshes, ponds, the edge of a lake or ocean, the delta at the mouth of a river, and low-lying areas that frequently flood.

The destruction of wetlands is of major environmental concern because wetlands are some of the most productive habitats on Earth, as they support high concentrations of animals (e.g., birds, fish, invertebrates, and mammals) and serve as breeding grounds and nurseries for many of these species. Wetlands also allow for the cultivation of rice, a food source for half of the world's population. And, finally, wetlands provide a wide range of ecosystem services; e.g., flood control, recreation, storm protection, and water filtration.

A mangrove is a shrub or small tree that grows in slightly salty (brackish) water formed by seawater's mixing with freshwater in estuaries. Mangroves have evolved a complex salt filtration system and complex root system to be able to survive in environments in which their roots are exposed to salt water, low-oxygen mud, and the stress of wave action.

Mangroves are found worldwide in the tropics and subtropics and are found in ~120 countries worldwide, occupying a total area on Earth of ~53,000 square miles (~140,000 km^2) (Figure 8.3).

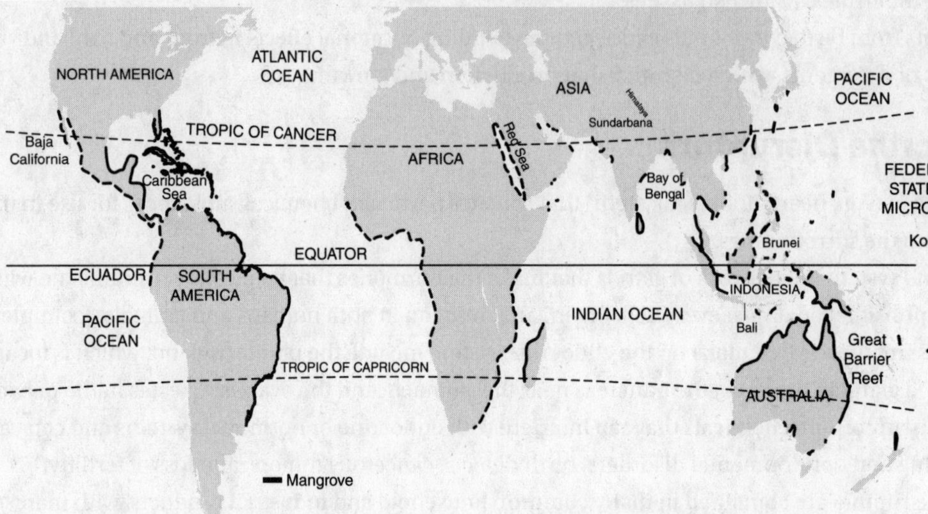

Figure 8.3 Location of mangrove forests around the world
Source: NASA (National Aeronautics and Space Administration)

However, since the Industrial Revolution, humans have had an increasingly negative impact on one of most productive ecosystems on Earth through

- an increase in hurricanes, which may in part be due to climate change, resulting in more frequent sea surges (a destructive rise in sea level above normal high-tide level in a coastal area that is caused by a combination of low atmospheric pressure and strong onshore winds, which also causes massive damage);
- causing water pollution from chemical herbicides and oil spills, which coat the above-ground roots, limiting the transport of oxygen to underground roots;
- diking and dredging, which allows standing water to cover the aerial roots, making it impossible for oxygen to reach these specialized roots as well as their underground root systems; and
- filling in these sensitive areas for urban development, impacting water quality and runoff. These land and coastal activities result in increased erosion as well as the reduction of nursery areas supporting commercial and game fisheries.

8.5 Eutrophication

Eutrophication is the process whereby human activity increases the amount of nutrients (especially nitrates and phosphates) entering surface waters (Figure 8.4).

Ecological Effects of Cultural Eutrophication

- Changes in species composition and dominance
- Decomposed algae by anaerobic bacteria release toxic gases, e.g., hydrogen sulfide
- Decreased biodiversity in water bodies
- Dissolved oxygen (DO) depletion (hypoxia), resulting in fish kills
- Increase in algal blooms
- Increase in turbidity
- Increased phytoplankton biomass
- Toxic phytoplankton species, a cause of neurotoxic shellfish poisoning

Human Activities That Contribute to Cultural Eutrophication

- Discharge from water treatment facilities that do not have the capacity to handle nutrient and biodegradable waste discharge
- Fertilizers and pesticides from residential and agricultural runoff (e.g., feedlots, golf courses, lawns, etc.)
- Sewer and drainage overflows that can occur when the rainfall amount exceeds the wastewater treatment capacity
- Use of household products that contain phosphates (e.g., detergents)

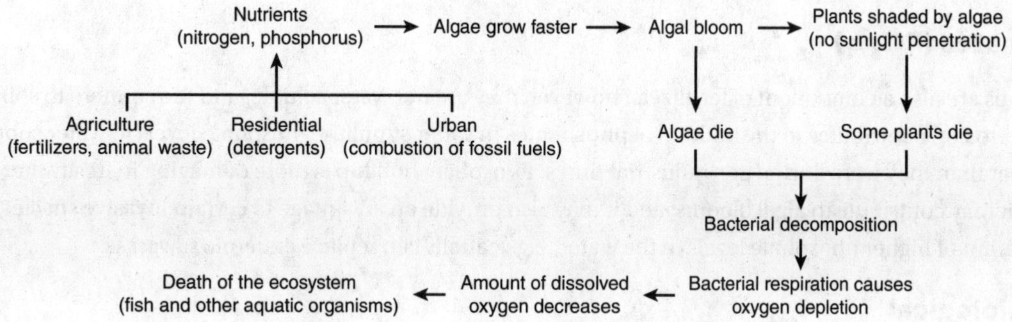

Figure 8.4 Steps in cultural eutrophication

Steps for Controlling Cultural Eutrophication

- Controlling runoff from feedlots
- Controlling the application and timing of applying fertilizer
- Constructing wastewater lagoons and retention ponds near agricultural areas
- Planting vegetation (buffer zones) along streambeds, which slows erosion and absorbs some of the excess nutrients
- Updating building codes to utilize permeable pavement to absorb the excess urban runoff
- Upgrading existing water treatment plants to better control nitrate and phosphate pollution through tertiary standards and other advanced technologies
- Using monetary and tax incentives to convert existing watering systems to drip irrigation and to replace landscaping with native vegetation that is less water-demanding

Steps Involved in Algal Blooms

- An increase in algae occurs due to increased nitrate and/or phosphate concentrations and results in decreased light penetration and the killing off of deeper plants and their supply of oxygen to the water.
- Decaying fish and algae produce toxins in the water.
- Lower oxygen concentrations cause fish and other aquatic organisms to die and contaminate the water at a high rate.
- The oxygen concentration decreases in the water due to the consequences of there being increased material for decomposers (consumers of dissolved oxygen).

Biodegradable Wastes

The release of biodegradable wastes into water supplies is also a major component of cultural eutrophication. These wastes are used by bacteria and other microorganisms and can result in oxygen depletion (hypoxia) in the water, causing anaerobic bacteria to increase, which results in "dead zones." Most dead zones occur in bottom, and near bottom, water near inhabited coastlines, where aquatic life is most concentrated. Major dead zones include the northern Gulf of Mexico region surrounding the outfall of the Mississippi River, areas of the Chesapeake Bay, and the coastal regions of the Pacific Northwest.

Nitrates (NO_3^-)

Nitrates are water soluble and are found in fertilizers, which can remain on fields and accumulate, leach into groundwater, or end up in surface runoff and cause algal blooms in surface waters, resulting in decreased dissolved oxygen levels. They can also contaminate drinking water supplies by reducing the effectiveness of hemoglobin ("blue baby" syndrome), and they contribute to acid rain by forming nitrous acid (HNO_2) and nitric acid (HNO_3). The bacterial decomposition of nitrogen-based fertilizers produces nitrous oxide (N_2O), which contributes to global warming and the depletion of stratospheric ozone.

Phosphates (PO_4^{3-})

Phosphates are also a component of fertilizers; however, they are not water soluble, and they adhere to soil particles. Soil erosion contributes to the buildup of phosphates in water supplies. Phosphate levels in water supplies are 75% higher than they were during pre-industrial times. Phosphate buildup is more damaging in freshwater systems because it may contribute to algal blooms, and it may also provide an advantage to certain invasive species that are more tolerant of higher phosphate levels in the water; e.g., cattails can replace endemic sawgrass.

Microbiological

Disease-causing (pathogenic) microorganisms, such as bacteria, viruses, and protozoa, can result in swimmers getting sick and shellfish becoming contaminated. Examples of waterborne diseases include cholera, typhoid, shigella, polio, meningitis, and hepatitis. In developing countries, an estimated 90% of the wastewater is discharged directly into rivers and streams without treatment. In the United States, almost a trillion gallons (4 trillion liters) of raw sewage is dumped into rivers, lakes, and bays each year by leaking sewer systems and inadequate combined sewer/storm systems that overflow during heavy rains. Leaking septic tanks and other sources of sewage can also cause groundwater and stream contamination.

Beaches also suffer the effects of water pollution from sewage; e.g., ~25% of all beaches in the United States have annual water pollution advisories or are closed each year due to bacterial buildup caused by sewage.

Mining

Mining causes water pollution in a number of ways:

- In the case of gold mining, cyanide is intentionally poured onto piles of mined rock (a leach heap) to extract the gold from the ore chemically. Some of the cyanide ultimately finds its way into nearby water. Additionally, huge pools of mining waste slurry are often stored behind contaminant dams that often leak or infiltrate groundwater supplies.
- Mining companies in developing countries often dump mining waste directly into rivers or other bodies of water as a method of disposal.
- The mining process exposes heavy metals and sulfur compounds that were previously locked away in the earth. Rainwater leaches these compounds out of the exposed earth, resulting in acid mine drainage and heavy-metal pollution that can continue long after the mining operations have ceased.
- The rainwater falling on piles of mining waste (tailings) transfers pollution to freshwater supplies.

Noise Pollution on Marine Life

Many marine organisms, including marine mammals, sea turtles, and fish, use sound to communicate, navigate, and hunt. Because of oceanic water noise pollution caused by commercial shipping, military sonar, and recreational boating, some species may have a harder time hunting, detecting predators, or navigating properly.

Oil Spills

Oil is one of the world's main sources of energy, but because it is unevenly distributed worldwide, it must be transported by ship across oceans and by pipelines across land, which can result in accidents when transferring oil to vessels, when transporting oil, and when pipelines break, as well as when drilling for oil. Oil that is accidentally released into a marine environment drastically affects wildlife in the following ways:

- As the seabirds attempt to preen, they typically ingest oil that covers their feathers, causing kidney and liver damage. This, along with their limited foraging ability, quickly results in dehydration.
- Since oil floats on top of water, less sunlight penetrates into the water, limiting the photosynthesis of marine plants and phytoplankton and affecting the food web in the ecosystem.
- The oil penetrates the feathers of seabirds, reducing the feathers' insulating ability and making the birds more vulnerable to temperature fluctuations and making them much less buoyant in the water. Recovering spilled oil is difficult and depends on many factors, including the type of oil spilled, the temperature of the water, and the types of shorelines and beaches involved.

Methods for cleaning up oil include the following:

- Chemical agents, such as dispersants, sorbents, and detergents, that act to disperse the oil, absorb it, or cause it to clump into gel-like agglomerations that sink
- Controlled burning and booming, skimming, and/or vacuuming the oil from the surface or shoreline
- The use of microorganisms to break down the oil

Plastic

The "Great Pacific Garbage Patch" is a large system of rotating ocean currents (gyres) of marine litter in the central North Pacific Ocean and is characterized by high concentrations of floating plastics, chemical sludge, and other debris that have been trapped by the currents of the North Pacific Gyre (Figures 8.5 and 8.6).

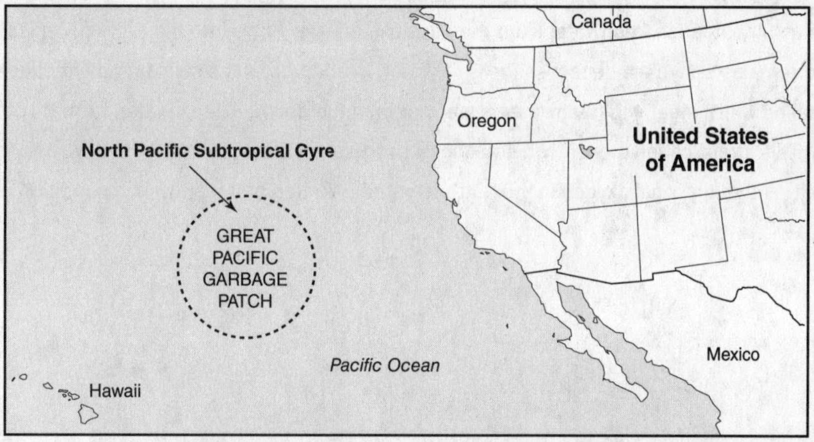

Figure 8.5 The "Great Pacific Garbage Patch"

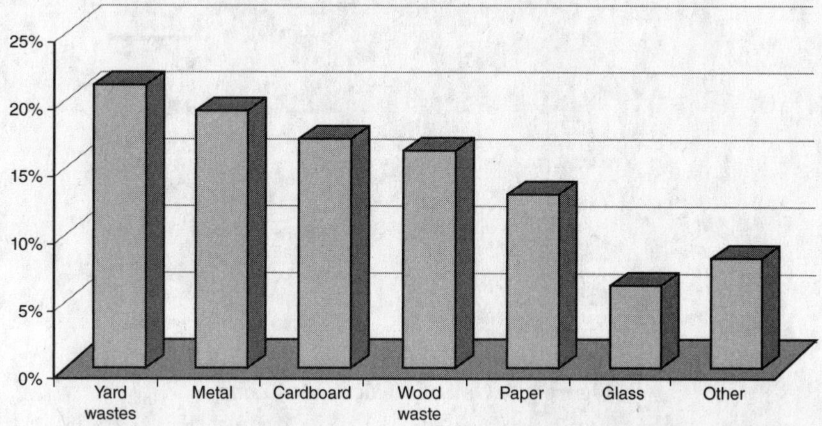

Figure 8.6 Types of material found in the "Great Pacific Garbage Patch"

The "garbage patch" formed gradually as a result of marine pollution gathered by oceanic currents as the gyre's rotational pattern drew in waste material from across the North Pacific Ocean, including coastal waters off North America and Japan. As materials are captured in the currents, the wind-driven surface currents gradually move floating debris toward and trap it in the center of the gyre. Plastics and other plastic-like substances (such as nylon from fishing nets) can entangle fish, sea turtles, and marine mammals, causing injury and death, while other types of plastic can break down into extremely small particles and become ingested.

As the plastic photodegrades into smaller and smaller pieces, it remains as plastic polymers leaching toxic chemicals into the upper water column. As the plastic further disintegrates, it becomes small enough to be ingested by aquatic organisms and birds near the ocean's surface and eventually enters the marine food chain. The floating debris can also absorb organic pollutants from seawater, which, through bioaccumulation, distributes these toxins throughout the food chain.

Marine plastics also facilitate the spread of invasive species that attach to floating plastic. On the macroscopic level, the plastic kills birds and sea turtles as the animals' digestion systems cannot break down the plastic inside their stomachs. Furthermore, the large amount of debris makes it much more difficult for animals to see and detect their normal sources of food through the water column.

CASE STUDY

GULF OF MEXICO DEAD ZONE: Currently, the Gulf of Mexico's dead zone, off the coast of Louisiana and Texas, is the largest hypoxic (low-oxygen) zone in the United States (about the size of the state of Connecticut) (Figure 8.7). The hypoxic zone in the Gulf of Mexico forms every summer and is a result of excess nutrients from the Mississippi River (which is the drainage area for about 40% of the continental United States) and seasonal stratification (layering) of waters in the Gulf. Nutrient-laden freshwater from the Mississippi River flows into the Gulf of Mexico. This freshwater is less dense and remains above the denser saline Gulf water. In addition to the saline gradient caused where the freshwater and saline water meet, the freshwater is warmer than the deeper ocean water, further contributing to the stratification. This stratification prevents the mixing of oxygen-rich surface water with oxygen-poor water on the bottom of the Gulf. Without mixing, oxygen in the bottom water is limited and the hypoxic condition continues.

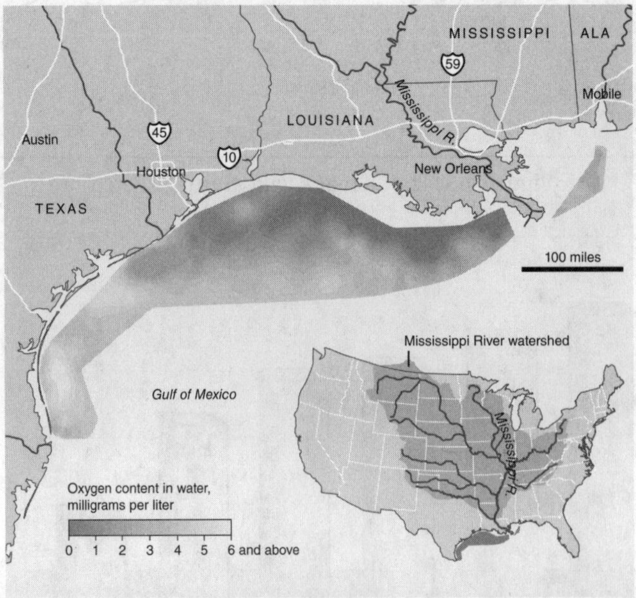

Figure 8.7 Gulf of Mexico Dead Zone
Source: National Oceanic and Atmospheric Administration (NOAA)

Groundwater Pollution

About 50% of the people in the United States depend on groundwater for their water supplies, and in some countries, it may reach as high as 95%. Almost half of the water used for agriculture in the United States comes from groundwater.

The EPA estimates that each day, over a trillion gallons (4.5 trillion L) of contaminated water seeps into groundwater supplies in the United States, where over 9 billion gallons (34 billion L), or 60%, of the most hazardous liquid waste solvents, heavy metals, and radioactive materials are injected directly into deep groundwater via thousands of injection wells. Although the EPA requires that these effluents be injected below the deepest sources of drinking water, some pollutants have already entered underground water supplies in Florida, Ohio, Oklahoma, and Texas.

Water that enters an aquifer remains there for approximately 1,400 years, compared with 16 days for water entering a river system. Once an aquifer is contaminated, it is practically impossible to remove the pollutants. For example, in Denver, Colorado, 20 gallons (~80 L) of organic solvents contaminated a trillion gallons (~4 trillion L) of groundwater. Initial cleanup of contaminated groundwater locations in the United States could cost up to $1 trillion over the next 30 years.

8.6 Thermal Pollution

Thermal pollution is the discharge of heated liquid into natural waters at a temperature harmful to the environment. In the United States, about 80% of thermal pollution is generated by power plants, with the remainder coming from chemical plants, oil refineries, pulp and paper mills, smelters that process ore, and steel mills. Additional sources of thermal pollution include soil erosion, which often causes water bodies to rise, thus making them more exposed to sunlight; deforestation, which allows more sunlight to reach water bodies; runoff from hot paved surfaces; and natural causes, such as volcanoes, geothermal vents, and hot springs.

When water is used as a coolant and returns to its natural source at a higher temperature, the warmer water does not contain as much oxygen, which ultimately results in a condition known as "thermal shock," which affects amphibians, aquatic plants, fish, etc. Thermal shock also increases the metabolic rate of aquatic organisms, which ultimately reduces the finite amount of resources available. As a result, food webs are altered and subsequently the dynamics and biodiversity of the environment are changed.

The dissolved oxygen (DO) in water is typically expressed as a percentage of the oxygen that would dissolve in the water at the prevailing temperature and salinity, both of which affect the solubility of oxygen in water. An aquatic system with 0% dissolved oxygen is called anaerobic, whereas at low DO concentrations (1%–30%), it is called hypoxic and results in impaired reproduction of fish via endocrine disruption (see Section 8.3 above). A "healthy" aquatic environment should rarely experience DOs less than 80%.

Aquatic plant growth rates are increased by warmer water, resulting in a shorter life span and ultimately in species overpopulation, which often leads to algae blooms, which further reduce DO concentrations in the water.

At the cellular level, higher water temperatures may also cause cell walls in plants to become less permeable during osmosis, coagulation of certain proteins, and disruption to enzymes and their role in metabolism.

Thermal pollution can also involve the release of colder water, as when, for example, reservoirs release extra water in anticipation of impending flood conditions, with biological and subsequent environmental consequences.

Solutions to thermal pollution can include the following:

- Installing efficient cooling and spray towers near the primary source(s) of thermal pollution
- Planting more shade trees near waterways and coastal locations
- Raising public awareness of the problem and possible solutions

8.7 Persistent Organic Pollutants (POPs)

Chlorinated hydrocarbons, such as the pesticide DDT, are synthetic organic compounds that belong to a group of chemicals known as persistent organic pollutants, or POPs.

POPs are organic (carbon) compounds that are resistant to environmental degradation through chemical or biological processes or decomposition due to light. As pesticides, they affect the nervous systems of pests.

POPs have been observed to persist in the environment, be capable of long-range transport, bioaccumulate in human and animal tissue, biomagnify in food chains, and have potential significant impacts on human health and the environment.

Chemical characteristics of POPs include the following:

- Ability to travel long distances through the atmosphere before being deposited on Earth. Thus, POPs can be found all over the world, including in areas where they have never been used and remote regions such as the middle of oceans and Antarctica
- A tendency to evaporate easily in hot regions and accumulate in cold regions, where they tend to condense and persist
- High molecular masses
- High fat solubility, which makes them able to pass through biological membranes and bioaccumulate in the fatty tissues of living organisms
- Low water solubility

As mentioned previously, urban runoff is surface runoff of rainwater created by urbanization. This runoff is a major source of urban flooding and water pollution in urban communities worldwide. Impervious surfaces, such as roads, rooftops, parking lots, and sidewalks, carry polluted storm water to storm drains instead of allowing the water to percolate through the soil, which causes a lowering of the water table (because groundwater recharge is lessened) and flooding, since the amount of water that remains on the surface is greater.

Paved surfaces also tend to

- create microclimates due to the high heat capacity of asphalt, which causes heat to be more easily captured and stored longer than it would in the normal landscape;
- fragment habitats;
- increase groundwater depletion as water does not infiltrate into the soil to recharge aquifers; and
- reduce biodiversity and seriously impact food webs in the area since there is less vegetation available for primary consumers.

Most municipal storm sewer systems discharge untreated storm water to streams, rivers, and bays. Urban runoff results in the following:

- An increase in algae growth due to the increased nutrients in the runoff
- An increase in the risk of infections and diseases through contaminated water supplies
- Increased erosion and resulting sedimentation in the runoff, with the sediments settling to the bottom of the water bodies and reservoirs, directly affecting the water quality and the storage capacity
- Increased temperature of water in streams and waterways, which significantly impacts fish and wildlife as heat is transferred from urban sources (asphalt, buildings, etc.)
- Runoff containing gasoline, motor oil, heavy metals, trash, fertilizers, and pesticides

Runoff remediation steps include the following:

- Building constructed wetlands to naturally filter water before it's released into lakes, rivers, and oceans
- Building water retention–infiltration basins, which are artificial ponds that are designed to infiltrate storm water through permeable soils into the groundwater aquifer

- Frequently using street-sweeping vacuums that can reduce the trash and other debris and pollutants that end up in runoff
- Planning and constructing more open green spaces and parks within urban communities to increase natural infiltration

RELEVANT LAWS

CLEAN WATER ACT (1972)—primary federal law that governs water pollution

- Ensures that surface waters meet standards necessary for human sports and recreation.
- Prohibits the release of high amounts of toxic substances into water.
- Regulates point sources of water pollution.
- Regulates the discharge of dredged or fill material into waters of the United States, including wetlands.
- Restores and maintains the chemical, physical, and biological integrity of the nation's water.

SAFE DRINKING WATER ACT (1974)—primary federal law intended to ensure safe drinking water (primarily from ground and underground sources)

- Does not apply to private wells.
- Requires the Environmental Protection Agency (EPA) to set universal standards for drinking water.
- Sets maximum levels for microorganisms, disinfectants, chemicals, and radionuclides.

8.8 Bioaccumulation and Biomagnification

Bioaccumulation

Bioaccumulation is the increase in the concentration of a pollutant within an organism. The rate at which a given substance bioaccumulates depends upon the following:

- The mode of uptake (along with food, through gills, contact with skin, etc.)
- The degree of fat solubility of the pollutant
- The rate at which the substance is eliminated from the organism
- The transformation of the substance by metabolic processes
- The lipid (fat) content of the organism

Biomagnification

Biomagnification is the increasing concentration of a substance in the tissues of organisms at successively higher trophic levels within a food chain. As a result of biomagnification, organisms at the top of the food chain generally suffer greater harm from a persistent toxin or pollutant than those at lower levels (Figure 8.8).

For biomagnification to occur, the following must be true:

- The pollutant must be long-lived. If it is short-lived, it will be broken down before it can become dangerous.
- The pollutant must be mobile. If it is not mobile, it will stay in one place and be less likely to be taken up by many organisms.
- The pollutant must be soluble in fats. If the pollutant is soluble in water, it will be excreted. In mammals, milk that is produced by females is often tested since the milk is high in fat and because the young are often more susceptible to damage from toxins.
- The pollutant must be biologically active (chemicals that have adverse effects in small amounts).

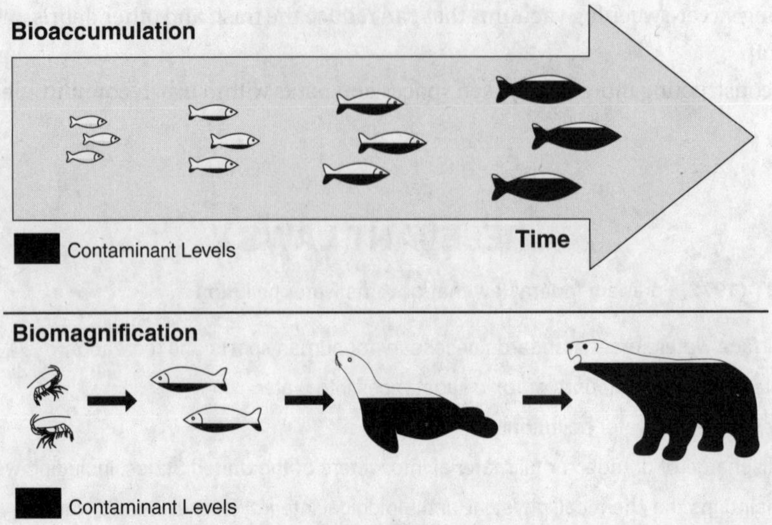

Figure 8.8 Bioaccumulation vs. biomagnification

RELEVANT LAWS

RESOURCE CONSERVATION AND RECOVERY ACT (RCRA) (1976)—Encouraged states to develop comprehensive plans to manage hazardous and nonhazardous industrial solid and municipal wastes. Set criteria for municipal landfills and disposable facilities.

COMPREHENSIVE ENVIRONMENTAL RESPONSE, COMPENSATION AND LIABILITY ACT (CERCLA-SUPERFUND) (1980)—Provided authority for the federal government to respond to releases or possible releases of hazardous substances that could threaten public health and/or the environment. Established rules for closed and abandoned hazardous waste sites and created a trust fund for cleanup if responsible parties for contaminated sites could not be located.

8.9 Solid Waste Disposal

Municipal Solid Waste

Municipal solid waste (MSW)—more commonly known as trash or garbage—consists of everyday items that are used and then thrown away, e.g., product packaging, grass clippings, furniture, clothing, bottles, food scraps, newspapers, appliances, paint, and batteries (Figures 8.9 and 8.10).

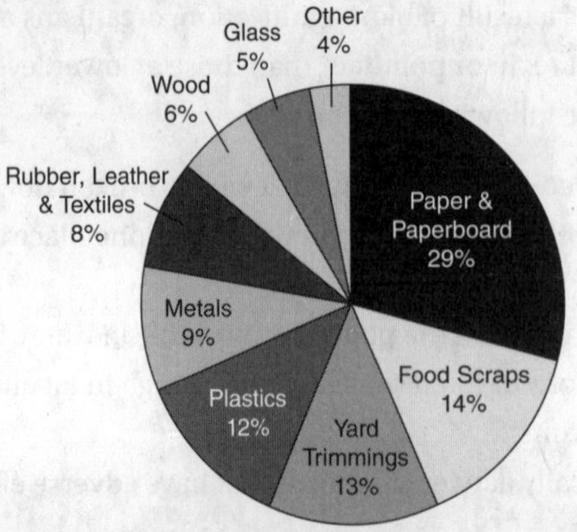

Figure 8.9 Amounts and types of municipal solid wastes (MSWs) in the United States

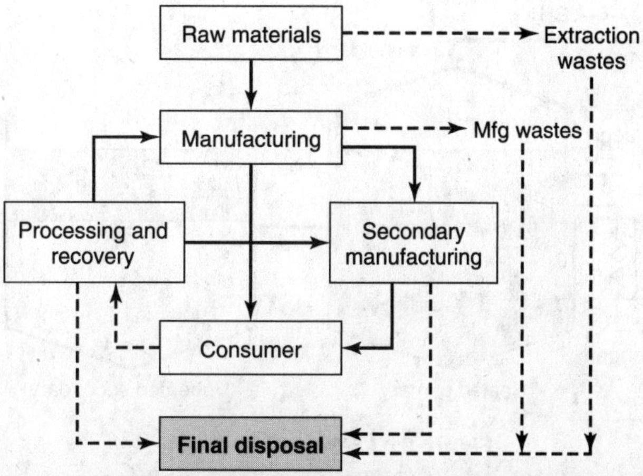

Figure 8.10 Solid waste flow diagram

Types of solid wastes include the following:

- **Hazardous**—paints, chemicals, pesticides, etc. These wastes can take hundreds of years to decompose.
- **Organic**—kitchen wastes, vegetables, flowers, leaves, or fruits. Usually decompose within two weeks. Wood can take 10 to 15 years to decompose.
- **Radioactive**—spent fuel rods and smoke detectors. Radioactive wastes can take hundreds of thousands of years to decompose.
- **Recyclable**—glass, metals, paper, and some plastics. Paper decomposes in 10 to 30 days, while glass does not decompose. Metals decompose in 100 to 500 years. Some plastics can take up to 1 million years to decompose.
- **Soiled**—hospital wastes. Cotton and cloth can take two to five months to decompose.

Disposal and Reduction Strategies

The collection, transportation, and disposal of municipal solid waste presents considerable costs, poses threats to health, and can result in a permanent loss of valuable materials. Local governments are addressing these challenges through a variety of strategies to divert waste from landfills and recover and repurpose valuable and/or toxic materials.

Anaerobic Digestion

Microorganisms are used to break down biodegradable material and sewage sludge in the absence of oxygen (Figure 8.11).

Pros

- It is a renewable energy source.
- It reduces the amount of organic matter, which might otherwise be destined to be dumped at sea (by some countries) or in landfills or burned in incinerators.
- It reduces or eliminates the energy footprint of waste treatment plants.
- It reduces the methane emission from landfills.
- It is best suited for organic material and is commonly used for industrial effluent, wastewater, and sewage sludge treatment.
- Nutrient-rich digestate (leftovers) can be used as fertilizer.
- Systems can be located at either sources of biowaste (e.g., large farms) or centralized facilities.

Cons

- Nutrient conservation may be undesirable on a farm with excess nutrients to manage.
- Systems are expensive if installed on individual farms.
- There are transportation and additional labor costs if farmers need to transport biowastes to a centralized facility.

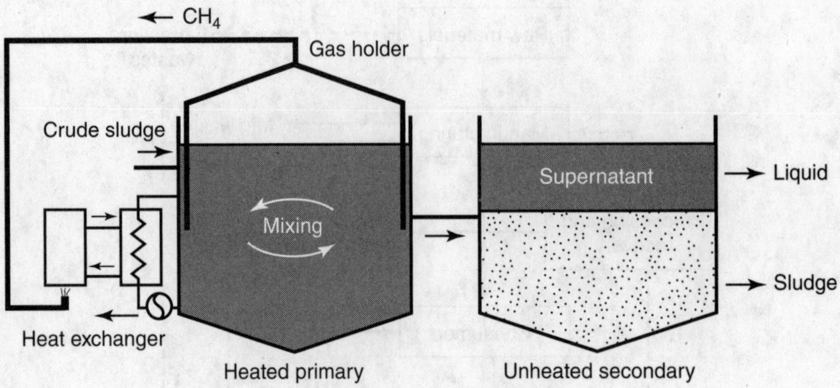

Figure 8.11 Anaerobic digester
Source: United Nations Environment Program Division of Technology, Industry, and Economics

Waste Solutions and Energy Recovery

Incinerating

Incineration is a waste treatment process that involves the combustion of substances contained in waste materials and conversion of the waste into ash, flue gas, and heat. The United States incinerates 15% of its waste, whereas France, Japan, Sweden, and Switzerland incinerate more than 40% of their waste and use the heat to generate electricity.

Burning hazardous wastes does reduce the volume of, and may detoxify, some hazardous pollutants, but it can release air pollutants and toxic ash (e.g., lead, mercury, and dioxins). Cities have been burning municipal solid waste since the 1880s. Most incinerators operating today use the heat from burning trash to produce steam that can generate electricity in "waste-to-energy" plants. Communities and environmental groups have long opposed these facilities, arguing that they are serious polluters and undermine recycling efforts.

Newer incinerators utilize a process called co-incineration, or co-firing, in which wastes are burned alongside traditional fossil fuels like coal in facilities such as cement kilns, coal-fired power plants, and industrial boilers. As of 2018, there were 86 incinerators across 25 states burning about 30 million tons of garbage annually—about 12% of the total U.S. waste stream—and producing about 0.4% of total U.S. electricity generation.

The Environmental Protection Agency is currently considering a plan to allow states to classify waste incineration and co-incineration as carbon-neutral forms of energy production. However, emissions typically associated with incineration include particulate matter, lead, mercury, and dioxins.

Pros

- Incineration reduces the impact on landfills.
- Mass burning is inexpensive.
- Released heat energy can be used to generate electricity.
- What is left is 10%–20% of the original volume.

Cons

- Air pollution—including cadmium, lead, mercury, NO_x, SO_2, and chemicals that accumulate in fat cells, e.g., polychlorinated biphenyls (PCBs) and dioxins—is carcinogenic, and these chemicals degrade very slowly in the environment.
- Ash is more concentrated with toxic materials.
- Initial costs of incinerators are high.
- Sorting out noncombustible items, such as batteries and plastics, is expensive.
- There is no way of knowing the toxic consequences.
- This process adds to acid deposition and global warming.

Exporting

The global waste trade is the international trade of waste between countries for further treatment, disposal, and/or recycling. Toxic or hazardous wastes are often exported from developed countries to developing countries.

Pros

- Exporting gets rid of the problem immediately for one country.
- It can be a source of income for poor countries.

Cons

- Environmental "racism"—placing landfills in low-income areas—may occur.
- The cost of transportation is high.
- There is a possibility of polluting roadways and waterways during transport of the solid waste.
- There may be unknown long-term effects.

Ocean Dumping

Ocean dumping is the deliberate disposal of municipal and/or hazardous wastes at sea.

Pros

- It is convenient.
- It is inexpensive.

Cons

- Debris may float to unintended areas.
- It is illegal in the United States. The Ocean Dumping Act (1972) gives the EPA the power to monitor, enforce, and regulate the dumping of sewage, sludge, and industrial, radioactive, biohazardous, or medical waste materials into U.S. territorial waters.
- Marine organisms can be negatively impacted—vast numbers of seabirds, sea turtles, and other sea animals die each year by ingesting and/or getting tangled in plastic.

Recycling

See Section 8.9.

Remanufacturing

In 1995, the EPA implemented a program to promote waste reduction and resource conservation through the use of materials recovered from solid waste, and to ensure that the materials collected in recycling programs will be used again in the manufacture of new products.

Pros

- Remanufacturing is beneficial to inner cities as an industry because materials are available and jobs are needed.
- Remanufacturing recovers materials that would have been discarded.

Cons

- Toxic materials may be present (e.g., CFCs, heavy metals, toxic chemicals, etc.).

Reuse

Reuse means to use again, especially in a different way or after reclaiming or reprocessing.

Pros

- Cloth diapers do not impact landfills.
- Industry models are already in place (e.g., auto salvage yards, building materials, etc.).
- Refillable glass bottles can be reused up to 15 times.
- Reuse is the most efficient method of reclaiming materials.

Cons

- Only when items are expensive and labor is cheap is reuse economical.
- The cost of collecting materials on a large scale is expensive.
- The cost of washing and decontaminating containers is also expensive.

Sanitary Landfills as a Strategy

Sanitary landfills are a method of waste disposal where the waste is buried either underground or in large piles, and where waste is isolated from the environment until it is safe. It is considered safe when it has completely degraded biologically, chemically, and physically.

Pros

- Geologic studies and environmental impact studies are performed prior to building the landfills.
- Plastic liners, drainage systems, and other methods help control leaching material into groundwater.
- The collection of methane and the use of fuel cells can be used to supplement the energy demand.
- The use of anaerobic methane generators reduces the energy dependency on other energy sources.
- Waste is covered each day with dirt to help prevent insects and rodents.

Cons

- Degradable plastics do not decompose completely and take a long time to decompose.
- Methane is released as materials decay.
- NIMBY (not in my backyard).
- Rising land prices may make it too expensive to build sanitary landfills near populated areas.
- Suitable areas are limited.
- There are high costs for running and monitoring the landfill.
- Transportation costs to the landfill are high.

Reducing

Reducing is lessening the amount of hazardous wastes by substituting and using products that are more "Earth friendly" (e.g., substituting Puron® for Freon® in air conditioning systems). Freon's® molecular structure contains chlorine, which seriously degrades the stratospheric ozone layer. Puron® substitutes fluorine for chlorine, and has less of an impact on the stratospheric ozone layer. Another method is replacing mercury thermometers with alcohol-based thermometers.

Hazardous Wastes

Radioactive Wastes

Radioactive wastes are usually a by-product of nuclear power generation and other applications of nuclear fission, such as research and medicine. Radioactive waste is hazardous to most forms of life and the environment, and is regulated by government agencies to protect human health and the environment.

Low-level radioactive wastes contain low levels of radiation and remain dangerous for a relatively short time (i.e., hundreds of years or less). They can be stored in shielded barrels on-site until they can be disposed of at a hazardous waste landfill.

High-level radioactive wastes contain high levels of radiation and remain dangerous for a very long time (i.e., tens of thousands of years). Storage typically involves sealing the radioactive waste in a steel cylinder, which is then placed in a concrete cylinder, which acts as a radiation shield and must be cooled before it can be sent to hazardous waste sites that are especially designed to handle radioactive material. These sites must meet the following criteria:

- Be located in areas that do not experience much rain (to minimize the chance of infiltration into the water table should there be leakage)
- Be at least 2,000 feet (about 600 meters) below Earth's surface
- Be fairly close to the source of the waste so that transportation times and risks are minimized
- Be located in remote locations so as not to be near areas of human habitation
- Not be located near areas that could be problematic, such as areas near volcanoes or areas that are prone to earthquakes, hurricanes, or tornadoes

Some of the effects of being exposed to sub-lethal amounts of radiation are cancer, sterility, DNA and chromosomal damage, damage to the immune system, birth defects in children, internal bleeding, and, with high-level exposure, death.

Reactive Wastes

Reactive wastes are wastes that are unstable under normal conditions. Reactive wastes can cause explosions, gases, toxic fumes, or vapors when heated, compressed, or mixed with water. Examples are lithium batteries and explosives.

Source-Specific Wastes

Source-specific wastes are wastes from specific industries (e.g., petroleum refining or pesticide manufacturing). Examples of source-specific wastes include certain sludges and wastewater from treatment and production processes.

Teratogens

Teratogens are substances found in the environment (e.g., drinking alcohol [ethanol], radioactive compounds, dioxin, mercury, tobacco, and excessive caffeine) that can cause birth defects.

Toxic Wastes

Toxic wastes are wastes that are harmful or fatal when ingested or absorbed (e.g., lead or mercury). When toxic wastes are disposed of on land, these toxins may leach and pollute the groundwater. The following are strategies for handling hazardous wastes.

Capping

Landfill capping is a containment technology that forms a barrier between the contaminated media and the surface, thereby shielding humans and the environment from the harmful effects of its contents and hopefully limiting the migration of the contents. A cap must restrict surface water infiltration into the contaminated subsurface to reduce the potential for contaminants to leach from the site. In 1976, Congress passed the Resource Conservation and Recovery Act (RCRA), tightening the regulatory oversight of existing landfills and establishing basic standards for covering landfills (e.g., landfill caps and containing leachate; Figure 8.12).

Hazardous waste caps consist of three layers:

1. An upper topsoil layer
2. A compacted soil barrier layer
3. A low-permeability layer made of a synthetic material, covering at least two feet of compacted clay

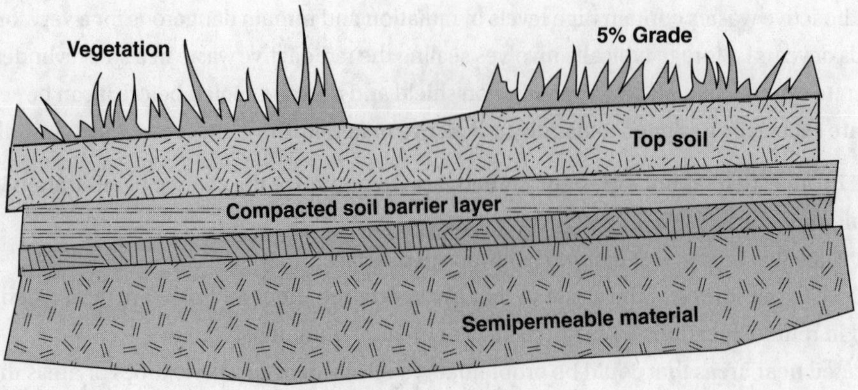

Figure 8.12 Landfill cap
Source: Sandia National Laboratory

Hazardous Waste Landfills

Hazardous waste landfills are excavated or engineered sites where non-liquid hazardous waste is deposited for final (permanent) disposal, and are selected and designed to minimize the chance of a release of hazardous waste into the environment. Design standards for hazardous waste landfills require a double liner, a double leachate collection and removal system, a leak detection system, and wind dispersal controls. They also may utilize sealed drums, as liquid wastes may not be placed in a hazardous waste landfill. Operators of hazardous waste landfills must also comply with inspections, monitoring, release response plans, and closure and post-closure plans (e.g., monitoring leachate, preventing storm water run-on and runoff).

Perpetual Storage

Perpetual storage attempts to place the hazardous waste material in a highly condensed or concentrated form, such that the wastes are isolated from the environment for extended periods of time.

The following section lists the current methods used to isolate and store hazardous wastes.

Salt Formations, Caves, and Mines

Salt dome and bed formations, underground caves, and mines are geologic repositories. The absence of flowing water within natural salt formations prevents the wastes from dissolving and spreading. Rooms and caverns in the salt can be sealed, thus isolating the waste from the biosphere.

Surface Impoundments

Surface impoundments are natural topographic depressions, man-made excavations, or diked areas that are used for temporary storage and/or for the treatment of liquid hazardous waste, e.g., holding, storage, settling, aeration pits, ponds, and lagoons. Hazardous waste surface impoundments are required to be constructed with a double liner system, a leachate collection and removal system, and a leak detection system.

Pros

- Built quickly, wastes can be retrieved, and, if lined, can store wastes for long periods of time
- Low initial startup and operating costs

Cons

- Earthquake issues
- Groundwater contamination
- Overflow if flooding occurs

- Promotes waste production
- VOCs (volatile organic compounds) pollution

Underground Deep Well Injection

An injection well stores fluid deep underground in geologically stable, porous rock formations, such as sandstone or limestone, or into or below the shallow soil layer. The fluid may be water, wastewater, brine (saltwater), or water mixed with chemicals.

Pros

- Low cost
- Requires simple technology
- Wastes can be retrieved.

Cons

- Can cause earthquake issues
- Can cause groundwater contamination
- Can cause leaks

Waste Piles

Waste piles are non-containerized piles of solid, non-liquid hazardous waste that are used for temporary storage or treatment. When closing a waste pile, all waste residue and contaminated soils and equipment must be removed or decontaminated.

Pros

- Fairly easy to identify leaks
- Provides a temporary storage area to remove wastes from sensitive areas

Cons

- Shipping materials to facilities can result in accidents

Treatment, Disposal, and Cleanup of Contaminated Sites

Reduction and cleanup of hazardous wastes can occur by producing less waste, converting the hazardous material to less hazardous or nonhazardous substances, and placing the toxic material into perpetual storage.

Brownfields

A "brownfield" is land that was previously used for industrial or commercial purposes, may have been contaminated with hazardous wastes, and is commonly found in large urban areas. However, once cleaned up, the land may be available for housing, retail, or industrial uses again. Reclamation of brownfields has the following advantages:

- Property values increase in the area, which results in increased tax revenues that can support further conservation efforts.
- There may be a reduction in crime as job opportunities become more available.
- There is a reduction in vehicle miles traveled when development occurs at a brownfield site, resulting in reduced pollution emissions.
- There is a reduction of stormwater runoff.
- Threats to the environment and human health are reduced.

8.10 Waste Reduction Methods

Recycling

Recycling is the process of collecting and processing materials that would otherwise be thrown away as trash and turning them into new products. There two forms of recycling:

1. **Closed-loop recycling**—a production system in which the waste or by-product of one process or product is used in making a similar product (e.g., recycling waste newspaper to make paper-board or other types of paper).
2. **Open loop recycling**—the conversion of material from one or more products into a new and different product, involving a change in the inherent properties of the material itself—typically to a lower-quality product material; e.g., recycling flexible plastic bottles to make rigid plastic drainage pipes.

Pros

- Conserves natural resources such as minerals, timber, and water
- Creates jobs
- Reduces air and water pollution
- Reduces the impact on landfills
- Reduces energy requirements to produce products (e.g., recycling aluminum cuts energy use by 95%, and producing steel from scrap reduces energy requirements by 75%)
- Reduces the amount of waste sent to incinerators (reduces air pollution) and landfills (reduces aquifer pollution)
- Reduces the need for raw materials and the costs associated with them
- Reduces the dependence on foreign oil
- Turns waste into an inexpensive resource

Cons

- Recycling has poor regulation.
- There are fluctuations in market prices, which may make recycling unprofitable.
- Throwaway packaging is more popular and convenient.

Composting

Composting is a natural biological process that is carried out under controlled aerobic conditions. In this process, various microorganisms, including bacteria and fungi, break down or biodegrade organic matter (e.g., food waste, manure, leaves, paper, wood, crop residues, etc.) into simpler substances and turn it into a valuable organic fertilizer.

Pros

- Aids in water retention
- Creates nutrient-rich soil additives
- No major toxic issues

- Reduces impact on landfills
- Reduces greenhouse gas production—mixing the layers of organic materials helps to provide air for oxygen-requiring bacteria to decompose material without producing greenhouse gases
- Slows down soil erosion

Cons

- NIMBY (not in my backyard)
- Public reaction to odor, vermin, and insects

e-Waste

Electronic waste, commonly referred to as e-Waste, includes discarded computers; electronic home, office, and entertainment equipment; mobile phones; television sets; appliances; and so on. These items are eventually destined for reuse, resale, salvage, recycling, disposal in landfills or the ocean, or for scrap (copper, steel, plastic, etc.). Electronic waste makes up 5% of all municipal solid waste worldwide, nearly the same amount as all plastic packaging and solid-waste streams, because people are upgrading their electronics more frequently than ever before. As much as 80% of the electronic waste meant to be recycled in the United States does not get recycled in the United States, but is sent to China, India, and developing countries for salvage or disposal.

Reasons for exporting e-Waste include the following:

- By exporting e-Waste, the exporters do not have to comply with stringent laws and regulations.
- Exporters of e-Waste are not responsible for the health costs of the workers in foreign countries.
- It's less expensive to export e-Waste due to the less-expensive labor costs abroad.
- There are no NIMBY (not in my backyard) issues by exporting the problem.

Environmental effects of e-Waste include the following:

- Air pollution that is caused by burning unsalvageable items, such as plastics
- Groundwater contamination
- Human and wildlife health issues (e.g., birth defects, learning disabilities, and reproductive problems) caused by breathing in and consuming heavy metals
- Surface runoff containing heavy metals

Methods to reduce e-Waste include the following:

- Adopting laws that outlaw exporting wastes
- Providing tax incentives to companies that adopt "trade-in" allowances to consumers who have older models
- Requiring a "disposal tax" on products

Landfill Mitigation

Landfills are the oldest and still most common form of waste disposal and treatment in most places around the world. These sites produce large amounts of combustible and toxic gases that seep up from the ground to lower-pressure areas, such as ground surface and off-site areas.

Many landfill sites are subsequently reclaimed as sites for new commercial, industrial, recreational (e.g., golf courses, parks), and residential facilities. While winds may dilute the gas emanating from landfills, it can still pose a significant environmental hazard to neighboring communities as toxic and combustible gases *can* build up in the vicinity or within the complexes themselves, potentially compromising the health and safety of the occupants.

Combustion of Gases from Organic Decomposition Landfills

Anaerobic bacteria in the landfill break down the trash in the absence of oxygen-producing ~50% methane (CH_4) and ~50% carbon dioxide (CO_2) gases. The methane that is produced can be either burned off (flared) at the site or piped to a power facility that burns it to produce steam that powers turbines on electrical generators, which deliver electricity to the local electrical grid (Figure 8.13). A benefit of this process is that methane, a strong greenhouse gas, is prevented from entering the atmosphere. Methane is a more potent greenhouse gas than CO_2 but has a shorter atmospheric life span; thus, its relative climate impact reduces significantly over time.

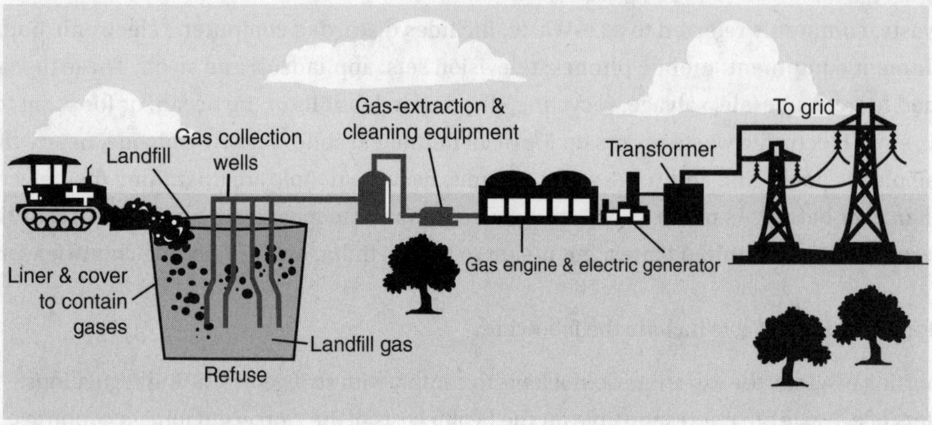

Figure 8.13 Steps involved in landfill gas conversion

8.11 Sewage Treatment

Sewage treatment incorporates physical, chemical, and biological processes to remove contaminants from wastewater.

Septic Systems

A septic system consists of an underground tank and a drain field. Wastewater enters a tank to allow solids to settle, while anaerobic bacteria digest the settled solids. The excess liquid leaves the tank and moves through a pipe to a leach field, where the water then percolates into the soil. About 25% of the population of North America relies on septic tank systems (Figure 8.14).

Pros

- Energy and chemicals are generally not needed to pump and treat the sewage, which impacts rivers.
- Municipal treatment plants can overflow in times of intense downpour or overuse, which is not an issue here.
- There are no monthly sewer charges.

Cons

- Maintenance of a municipal sewage system is managed and paid for by the city (taxpayers), while the costs of septic system installation and maintenance are paid for by the homeowner.
- Many pollutants do not decompose in a septic tank and may contaminate the groundwater.
- Septic systems do not remove suspended solids, control nitrate or phosphate levels, or disinfect the effluent.
- Septic tanks must be periodically pumped out, usually about once every 3 to 5 years.

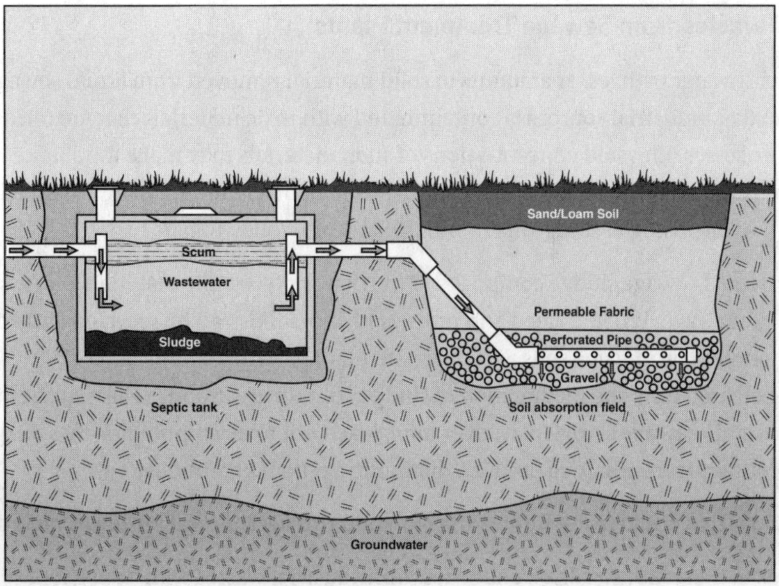

Figure 8.14 A typical septic system

Sewage (Wastewater) Treatment Facilities

Wastewater treatment is a process that converts wastewater, which is water that is no longer needed or suitable for its most recent use, into an effluent that can be either returned to the water cycle with minimal environmental issues or reused. If the wastewater is predominantly from municipal sources (households and small industries), it is called sewage, and its treatment is called sewage treatment (Figure 8.15).

- **First stage**—physical separation. Inorganic solids (e.g., trash and sand) are separated from wastewater by screens.
- **Second stage**—biological treatment. Dissolved biological matter is progressively converted into a solid mass by using microorganisms.
- **Third stage**—chemical treatment. Biological solids are neutralized, then disposed of or reused. Phosphorus is removed through coagulation by using lime, $Ca(OH)_2$. Nitrates may be removed using a variety of techniques, the most common involving the use of alcohols or sugars and denitrifying bacteria. The treated water is then disinfected chemically (using ozone and/or chlorine), through UV radiation, and/or physically through lagoons or microfiltration.
- **Fourth stage**—advanced treatment. Pollutants may not be eliminated by the previous treatment processes. Activated carbon filters, ozone, and enzymes are utilized to eliminate micro-pollutants such as pharmaceuticals, chemicals, and pesticides. Introduction of specific fungi may also be used.

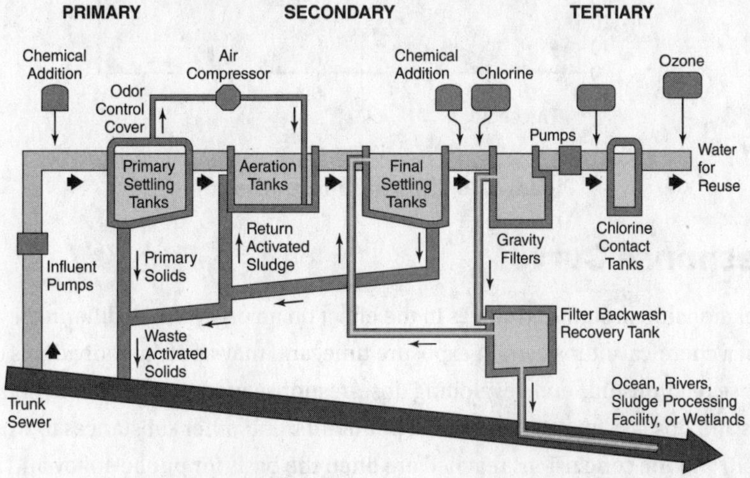

Figure 8.15 Sewage treatment plant

Disposing of Solid Wastes from Sewage Treatment Plants

Sewage sludge is mostly water with lesser amounts of solid material removed from liquid sewage. Sludge that originates from commercial or industrial sources is contaminated with toxic materials that are released into the sewers from the industrial processes. Elevated concentrations of such materials may make the sludge unsuitable for agricultural use, and it may then have to be incinerated or disposed of in a landfill. Environmental issues that result from the incineration or disposing of sludge in landfills include the following:

- Bio-solids are treated sewage sludge composed of nutrient-rich organic materials that result from the treatment of domestic sewage. When treated and processed, bio-solids can be recycled and applied as fertilizer to improve and maintain productive soils and stimulate plant growth.
- Gases that are released from the incineration of sludge contain greenhouse gases and/or toxins.
- Landfills that contain sewage sludge may, over time, leach out toxic compounds into the groundwater.
- Landfills that contain sewage sludge produce methane, a potent greenhouse gas.

Environmental issues that could result from using bio-solids for agricultural purposes include the following:

- Sewage systems in most industrialized countries mix industrial wastes with household sewage.
- Some pharmaceutical drug compounds, birth control chemicals, and cosmetic ingredients remain in bio-solids even after sophisticated removal treatments. These micro-pharmaceutical pollutants may leach out and end up polluting the surface and groundwater sources and ultimately the food chain.

8.12 Lethal Dose (LD_{50})

Toxicity

The mean lethal dose (LD_{50}) for a substance is the dose (milligrams of substance per kilogram of body mass) required to kill half the members of a tested population after a specified time, and is frequently used as an indicator of a substance's acute toxicity. The lower the LD_{50}, the greater the toxicity. The x-axis in Figure 8.16 is the dose, typically in a logarithmic scale, that allows comparison of a very wide range of doses, and the y-axis is the response, or the percentage of subjects that show a response. Dose-response curves normally take the form of an S (sigmoid) curve. Limitations on LD_{50} values include the fact that there can be wide variability between species; e.g., most humans can eat chocolate, but chocolate can be lethal to dogs.

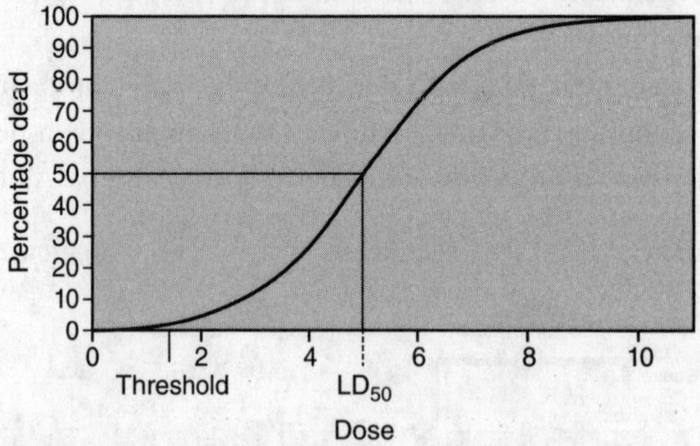

Figure 8.16 LD_{50} dose-response curve

8.13 Dose-Response Curve

The dose-response relationship describes changes in the effect on an organism of differing levels of exposure (or doses) to a drug or a chemical after a certain exposure time, and may apply to individuals or to a population. Studying dose-response relationships and developing dose-response models is central to determining "safe" vs. "hazardous" levels and dosages for drugs, potential pollutants, and other substances to which humans or other organisms are exposed, and the conclusions reached are often the basis for public policy and law.

Interpreting Dose-Response Curves

When the dose of a substance is plotted against the percentage of a population that responds to that dose, the result is called a dose-response curve. For most effects, small doses are not toxic; however, the point at which toxicity first appears is known as the threshold dose level. From that point, the curve increases with higher dose levels (Figure 8.17).

Major differences among the substances being tested may exist not only in the slope, or potency, but also in the point at which the threshold is reached. A curve with a steep slope indicates the chemical has a high potency, or toxic strength, compared to other chemicals. The greater the slope, the greater the potency.

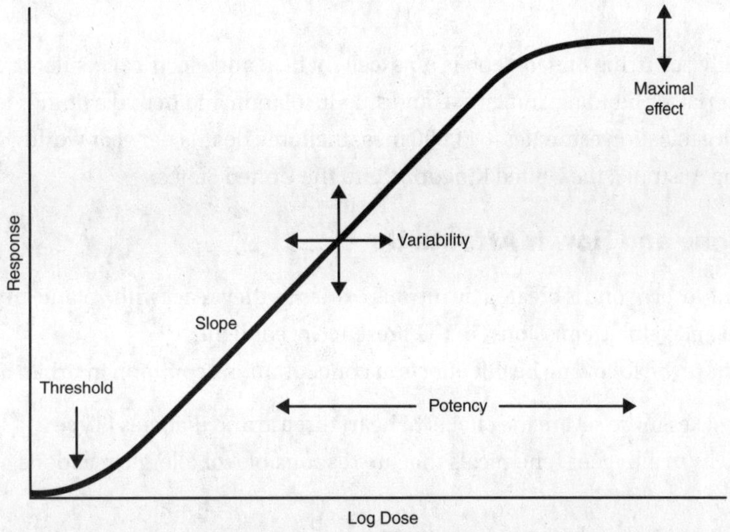

Figure 8.17 Dose-response curve

Acute vs. Chronic Effects

Acute—Acute health effects are caused by rapid, severe, and sudden exposures to a substance. Normally, a single large exposure is involved, as with exposure to dioxin or carbon monoxide.

Chronic—Chronic health effects are caused by prolonged and repeated exposures over time. Symptoms may not be immediately apparent and are often irreversible, as is the case with smoking cigarettes, and asbestos, lead, and mercury poisoning.

	Acute	Chronic
Onset	Rapid	Gradual
Cause	Usually one	May be several
Duration	Short	Indefinite
Diagnosis	Usually accurate	Often uncertain
Diagnostic tests	Often decisive	May not be of much value
Treatment	Cures are often available	Cures are often not available

8.14 Pollution and Human Health

Pollution is the leading cause of death around the globe and kills three times more people (~9 million) than AIDS, malaria, and tuberculosis combined. Although nearly 92% of pollution-related deaths occur in developing countries, developed countries are also affected since pollution spreads around the world. For example, ~30% of all shellfish consumed in the United States comes from China; however, the U.S. Food and Drug Administration tested the imports and discovered that about 10% of the imports were unfit for human consumption due to the presence of heavy metals and antibiotic residues.

We will now look at two specific examples of pollution and their effects on human health.

Mesothelioma

Mesothelioma is a cancer found in the lining surrounding the lungs, the stomach, the heart, or the testicles. There are about 3,000 new cases of mesothelioma in the United States every year, with the only known cause of mesothelioma being exposure to asbestos. Inhaling or ingesting asbestos causes these microscopic asbestos fibers to get lodged in the lining of the chest or abdomen, and over time the fibers cause tumors to develop, as the body is incapable of expelling the fibers.

Symptoms include shortness of breath, cough, wheezing, and tightness in the chest. The lag time between initial exposure to asbestos and the development of mesothelioma can be as high as ~58 years. There is no cure for mesothelioma.

Asbestos is a naturally occurring mineral and is an excellent heat and electrical insulator, and was once widely used as a building material for building, industrial, and ship insulation and in brake linings for vehicles.

The World Health Organization estimates ~43,000 mesothelioma deaths per year worldwide, with the highest-rate countries including Australia, the United Kingdom, and the United States.

Tropospheric Ozone and How It Affects Us

Tropospheric or ground-level ozone is created by nitrous oxides as they react with volatile organic compounds (VOCs) (e.g., industrial and vehicle emissions) in the presence of sunlight.

Ozone is known to have the following health effects at concentrations common in urban air:

- Aggravates and makes more severe the effects of heart disease and diabetes (Type 2)
- Increases sensitivity to allergens (chemicals that are responsible for allergic reactions, which can increase asthma attacks)
- Inflammation of lung tissue, which can be permanent
- Irritation of the respiratory system. Affects people with preexisting respiratory conditions to a greater degree, e.g., asthma, COPD (chronic obstructive pulmonary disease), and those that engage in strenuous outdoor activity, e.g., contractors
- Suppression of the immune system

8.15 Pathogens and Infectious Diseases

A pathogen is a microorganism that causes disease and that lives and survives within a host. Once the pathogen invades the host, it manages to avoid the host's immune responses and uses the host's resources to replicate before exiting and spreading to a new host through either airborne particles, bodily fluids (e.g., through sneezing or coughing), or contact with surfaces that contain the pathogen (e.g., clothing, doorknobs, skin, etc.).

It is important to note that pathogens

- are more common in equatorial-type climate zones, and as world climates change and become warmer and potentially more humid, it is expected that pathogens will increase in areas in which they had not been known to previously occur;
- are more likely to occur in areas that may not have adequate health services, lack adequate waste disposal services, and have contaminated water supplies; e.g., low income areas;
- can occur anywhere, despite what seem to be clean and sanitary conditions; and
- have the capacity to adapt and evolve, which over time may make them resistant to current drug treatments.

Three common types of pathogens include the following:

1. **Bacteria**—a group of diverse single-cell microorganisms. It is important to note that not all bacteria cause infections, but those that do are called pathogenic bacteria. Illnesses caused by harmful bacterial infections trigger the body's immune system to produce antibodies, which may or may not be successful in combatting the bacterial infection. Overuse of antibiotics over time and through the law of natural selection allows resistant

bacteria populations to increase, making it more and more difficult to treat bacterial infections. Specific beneficial bacteria are used in the food industry in the manufacturing of cheese, chocolate, and sourdough bread.

2. **Parasites**—any group of organisms that live in or live on a host to the detriment of the host. Parasitic infections are common in tropical and subtropical regions but can occur anywhere. Parasites can spread through contaminated blood, food, sexual contact, soil, or water. Three common types of parasites that can cause disease in humans include the following:
   - Ectoparasites—multi-celled organisms that live on or feed off of skin; e.g., mosquitoes, ticks
   - Protozoa—a group of living, single-celled eukaryotes (possessing a well-defined nucleus within a membrane), either free-living or parasitic, which feed on organic matter; e.g., microorganisms, organic material
   - Worms—able to live on (e.g., leech) or within (e.g., tapeworm) a host
3. **Viruses**—made up of a piece of genetic code from either DNA or RNA and which are protected by a coating of protein. Once the pathogen enters the host, it uses the host's cells and body's resources to replicate and produce more viruses, which are then released back into the host. It is important to note that antibiotics do not destroy viruses; however, there are some antiviral drugs that have been developed that have proven effective.

Specific Infectious Pathogens

Dysentery

Dysentery is a general term used to describe a series of contagious and serious infections of the stomach and small intestine that can involve amoebas, bacteria, and parasites. Symptoms may include abdominal pain, vomiting, and diarrhea containing large amounts of blood.

Each year dysentery affects over 165 million people worldwide, resulting in over a million deaths. Diarrheal diseases account for one in nine children's deaths worldwide, making it the second leading cause of death among children under the age of five years old in developing countries (e.g., Africa, Asia, and Central and South America) that have little or no infrastructure for providing clean water.

In the United States, approximately 30,000 dysentery cases occur each year. Most of these cases are non-fatal due to access to medical care.

Efforts to reduce the incidence of dysentery worldwide include improving the water quality by eliminating the discharge of contaminated human and animals wastes and wastewater into the drinking water supply.

Cholera

Cholera is an acute (severe and sudden in onset vs. chronic, which is a long-developing syndrome) watery diarrhea infection caused by ingestion of food or water contaminated with a specific bacterium. Cholera remains a global threat to public health and an indicator of wealth inequity and lack of social development. There are ~1.3 to 4.0 million cases and up to 143,000 deaths worldwide each year due to cholera.

Transmission of cholera is closely linked to inadequate access to clean water and sanitation facilities. At-risk areas include urban slums and refugee camps where minimum requirements of clean water and sanitation have not been met.

Infection rates can be reduced by ensuring adequate clean water and health education, including the need for hand-washing, safe preparation and handling of food, proper disposal of human wastes, and quick recognition of the disease and follow-up treatment.

Plague

A plague is an infectious disease caused by a bacterium. Symptoms include fever, weakness and headache, and possible swelling of lymph nodes, and in some forms, there may be shortness of breath, chest pain, or skin dying.

Certain forms of plague can be spread by flea bites or by handling infected animals, while the pneumonic form is generally spread between people through the air via infectious droplets.

Transmission of plague can occur by

- consuming contaminated food or water;
- coughing or sneezing on another person;
- direct physical contact involving touching or sexual contact with an infected person;
- exposure to infected insects and animals (including humans); and
- indirect contact; e.g., handling contaminated soil or a surface.

Worldwide, there are about 600 cases reported a year, generally occurring in Africa, Asia, and South America. The current risk of death without treatment is ~70%, while with treatment it is about 10%. The worst outbreak of plague occurred in Europe in the 14th century and resulted in 50 million deaths.

Malaria

Malaria is caused by a parasite that is spread to people through the bites of infected female *Anopheles* mosquitoes. Nearly half of the world's population (with children making up ~67% of all cases) is currently at risk of malaria, with ~93% of malaria cases and deaths occurring in sub-Saharan Africa.

Transmission of the parasite by mosquitos is heavily dependent on climatic conditions in the area, such as rainfall patterns, temperature, and humidity.

Currently, the most widely used methods to control the mosquitos is widespread spraying of insecticides and the use of mosquito nets. However, increasing resistance to insecticides by the mosquito populations compromises this method of control. Research into releasing sterilized female mosquitoes into the wild populations is showing promise. Resistance to antimalarial medicines is also a recurring problem.

Middle East Respiratory Syndrome (MERS)

Middle East respiratory syndrome (MERS) is an illness caused by a virus and linked to the Arabian Peninsula. Most MERS patients develop severe respiratory illness, with symptoms of cough, fever, and shortness of breath; about three out of every ten cases are fatal. Most people confirmed to have MERS-CoV infection have had severe respiratory illness with symptoms of cough, fever, and/or shortness of breath.

Severe Acute Respiratory Syndrome (SARS)

Severe acute respiratory syndrome (SARS) is a viral respiratory illness caused by a coronavirus and is associated with the coronavirus. The SARS virus was first identified in 2003, and is believed to be an animal virus stemming from bats that affected 26 countries and resulted in more than 8,000 cases in 2003. Symptoms of SARS include fever, headache, and diarrhea.

The main way that SARS seems to spread is by close person-to-person contact. The virus that causes SARS is thought to be transmitted most readily by respiratory droplets (droplet spread) produced when an infected person coughs or sneezes. Droplet spread can happen when droplets from the cough or sneeze of an infected person are propelled a short distance (generally up to 3 feet) through the air and deposited on the mucous membranes of the mouth, nose, or eyes of persons who are nearby. The virus also can spread when a person touches a surface or object contaminated with infectious droplets and then touches his or her mouth, nose, or eye(s). In addition, it is possible that the SARS virus might spread more broadly through the air (airborne spread) or by other ways that are not now known.

Practice Questions

1. ________ contributes to the formation of ________ and thereby compounds the problem of ________.

(A) Carbon dioxide, carbon monoxide, ozone depletion
(B) Nitric oxide, ozone, photochemical smog
(C) Ozone, carbon dioxide, acid rain
(D) Sulfur dioxide, acid deposition, global warming

2. Photochemical smog does NOT require the presence of

(A) nitrogen oxides (NO_x)
(B) peroxyacyl nitrates (PANs)
(C) ultraviolet radiation (UV)
(D) volatile organic compounds (VOCs)

3. Which of the following is generally NOT considered to be a teratogen?

(A) Benzene
(B) Ethanol (drinking alcohol)
(C) Radiation
(D) All of the above are considered to be teratogens

4. Which of the following steps is NOT involved in the production of industrial smog?

(A) $2C + O_2 \rightarrow 2CO$
(B) $NO_2 \rightarrow NO + O$
(C) $S + O_2 \rightarrow SO_2$
(D) $2SO_2 + O_2 \rightarrow 2SO_3$

5. Household water is most likely to be contaminated with radon in homes that

(A) are served by public water systems that use a surface water source
(B) are served by public water systems that use a groundwater source
(C) are served by water agencies that use ozone to disinfect the water
(D) have private wells

6. Which reaction is NOT involved in the formation of acid deposition?

(A) $2H_2SO_3 + O_2 \rightarrow 2H_2SO_4$
(B) $2NO + O_2 \rightarrow 2NO_2$
(C) $O_3 + C_xH_y \rightarrow$ peroxyacyl nitrates (PANs)
(D) $SO_2 + H_2O \rightarrow H_2SO_3$

7. You are an emergency room doctor working the night shift. A 67-year-old man and his wife come into the ER by ambulance. The wife is unconscious and the husband is complaining of their both having dull headaches, dizziness, shortness of breath, confusion, blurred vision, and nausea for the last two days. During your interview with the husband as to what may have caused this incident, he said the wife was washing clothes in the basement this evening near the hot water heater when she passed out, and both of them have not been feeling well for the last two days. Which of the following diagnoses would be most consistent with the couple's recent history?

(A) Asbestosis
(B) Bacterial infection from drinking from a nearby stream the week before
(C) Carbon monoxide poisoning
(D) Mercury poisoning

8. The following dose-response curve shows that

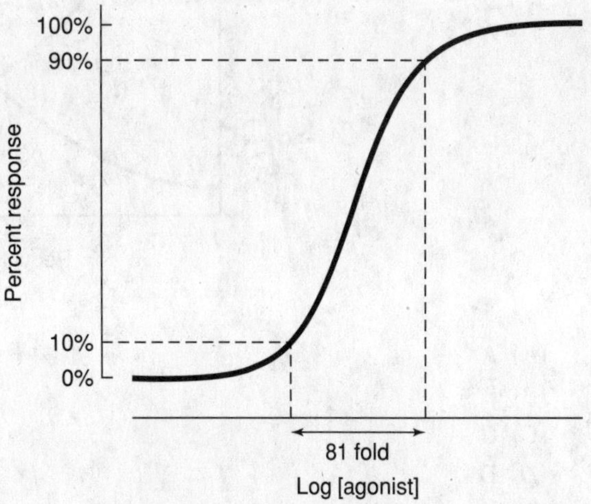

(A) an 80% response is achieved with a tenfold increase in agonist (a drug or other chemical that produces a reaction typical of a naturally occurring substance)
(B) 80% more agonist is required to achieve 80% more response
(C) 81 times more agonist is needed to achieve a 90% response than a 10% response
(D) larger amounts of agonist produce a corresponding increase in response

9. In developing countries, the most likely cause of respiratory disease would be

(A) industrial smog
(B) photochemical smog
(C) PM_{10}
(D) smoking

10. Humans who are LEAST susceptible to the effects of air pollution are

(A) adult males
(B) newborns
(C) teenagers
(D) the elderly

11. Which letter in the diagram below represents the optimum level of pollution balanced with the most efficient use of available resources to control it?

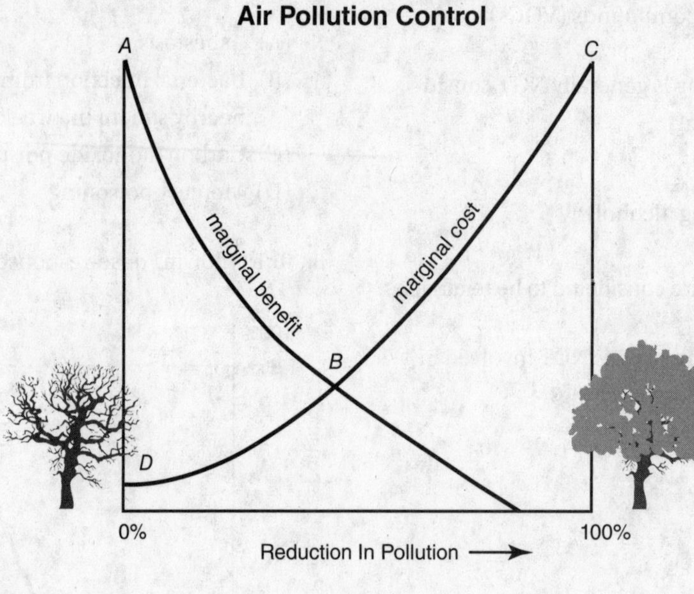

(A) A
(B) B
(C) C
(D) D

12. Which pollutant listed below best illustrates the effectiveness of pollution-control legislation?

(A) NO_2 (nitrogen dioxide)
(B) O_3 (ozone)
(C) Pb (lead)
(D) SO_2 (sulfur dioxide)

13. The diagram below shows the range of organisms found within certain sections of a river in an industrial area. Which section of the river most likely has the LOWEST level of dissolved oxygen?

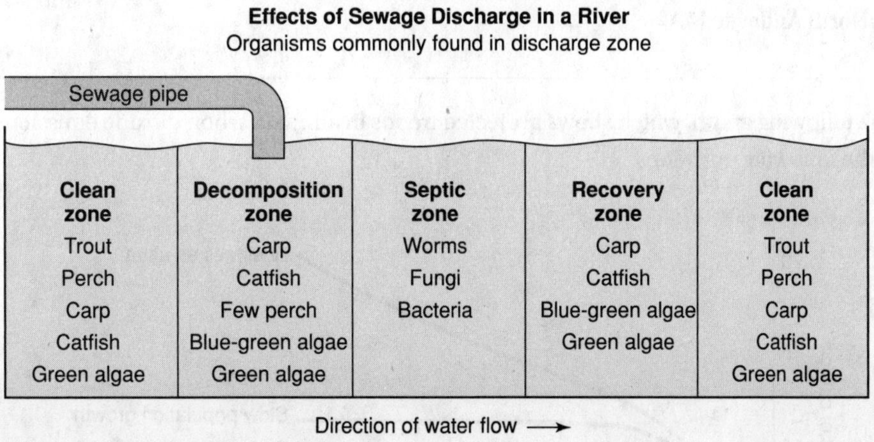

(A) Clean zone
(B) Decomposition zone
(C) Recovery zone
(D) Septic zone

14. Biochemical oxygen demand (BOD) is most often associated with which type of pollution listed below?

(A) Organic matter
(B) Thermal pollution
(C) Too much oxygen in the water
(D) Toxic chemicals

15. The major source of solid waste in the United States is

(A) agriculture
(B) factories
(C) homes
(D) mining wastes

16. What is the largest category of domestic solid waste in the United States?

(A) Metals
(B) Paper
(C) Plastic
(D) Yard waste

17. Which of the following is most readily recyclable?

(A) Glass
(B) Metal
(C) Paper
(D) Plastic

18. Which of the following statements is TRUE?

(A) Landfills and incinerators are more cost-effective and environmentally sound than recycling options.
(B) Landfills are significant job creators for rural communities.
(C) Recycling is more expensive than trash collection and disposal.
(D) None of the above.

19. Which act established federal authority for emergency response and cleanup of hazardous substances that have been spilled, improperly disposed of, or released into the environment?

(A) Hazardous Materials Transportation Act
(B) Resource Conservation and Recovery Act
(C) Solid Waste Disposal Act
(D) Superfund—Comprehensive Environmental Response, Compensation, and Liability Act

20. Which of the following methods of handling solid wastes is against the law in the United States?

(A) Dumping it in open landfills
(B) Dumping it in the ocean
(C) Exporting it to foreign countries
(D) Incinerating it

Free-Response Question

By Dr. Ian Kelleher
Brooks School, North Andover, MA

(a) Study the following graph, which shows projected trends in annual carbon dioxide emissions, and then answer the following questions.

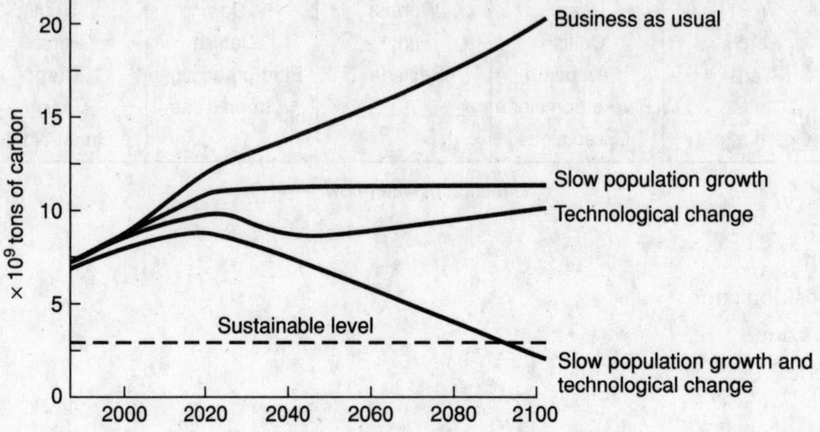

(i) In the "business as usual" model, identify TWO factors you think might contribute to the increase in carbon dioxide emissions.

(ii) Given the shape of the graph, what do you think is meant in this case by "technological change"?

(iii) What is meant by a "sustainable level" of carbon dioxide emissions? According to these predictions, what needs to happen for this level to be brought about?

(b) Explain the difference between remediation and alleviation of an environmental problem. Use the example of acid deposition to illustrate the difference.

(c) Look at the graph of CFC production below and account for the trends you observe.

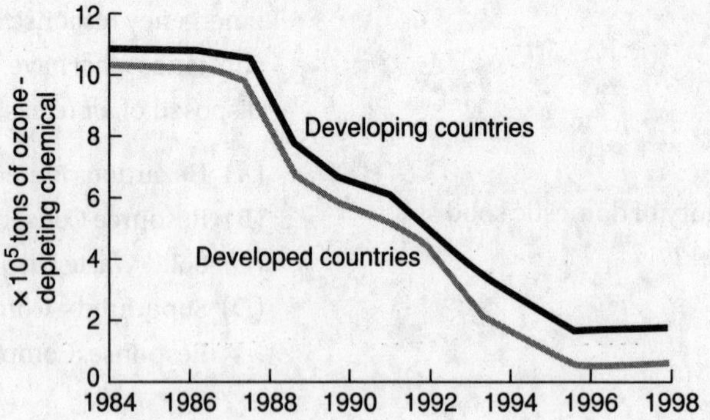

(d) In 1992, the World Bank designated indoor air pollution in developing countries as one of the four most critical global environmental problems. Provide and explain ONE example of a major source of an indoor pollutant to illustrate how the issue of indoor air quality in developing countries differs from that in developed countries.

Answer Explanations

1. **(B)** As sunlight becomes more intense, nitrogen dioxide (NO_2) is broken down into nitric oxide (NO) and oxygen atoms, and the concentration of ozone increases.

$$NO_2 + uv \rightarrow NO + O$$

$$O_2 + O \rightarrow O_3$$

2. **(B)** Peroxyacyl nitrates (PANs) are powerful respiratory and eye irritants present in photochemical smog and are good markers for the source of volatile organic compounds (VOCs). PANs can cause skin cancer and are irritating, as they dissolve more readily in water than ozone. PANs are secondary pollutants (not directly emitted as exhaust from power plants or internal combustion engines) and are formed from other pollutants by chemical reactions in the atmosphere, transporting NO_x to regions where it can more efficiently produce ozone.

$$NO_2 + VOCs \rightarrow PANs$$

3. **(D)** Teratogens are substances or environmental agents that cause abnormalities in a fetus during pregnancy. All choices are teratogens.

4. **(B)** This reaction is necessary in the formation of photochemical smog, but is not involved in the production of industrial smog.

5. **(D)** Radon is a naturally occurring radioactive gas that causes cancer and may be found in drinking water and indoor air. The Safe Drinking Water Act requires the Environmental Protection Agency (EPA) to develop regulations to reduce radon in drinking water.

6. **(C)** Carbon dioxide reacts with water to form carbonic acid (Equation 1).

$$CO_2 + H_2O \rightarrow H_2CO_3 \text{ (carbonic acid)} \quad \textit{(Equation 1)}$$

Carbonic acid then dissociates into hydrogen ions (H^+) and hydrogen carbonate ions (HCO_3^-) (Equation 2). The ability of H_2CO_3 to deliver H^+ is what classifies this molecule as an acid, thus lowering the pH of a solution.

$$H_2CO_{3(l)} \rightarrow H^+_{(aq)} + HCO_3^-{}_{(aq)} \quad \textit{(Equation 2)}$$

Nitric oxide (NO), which also contributes to the natural acidity of rainwater, is formed during lightning storms by the reaction of nitrogen and oxygen, two common atmospheric gases.

$$N_{2(g)} + O_{2(g)} \xrightarrow{\textit{lightning}} 2NO_{(g)} \quad \textit{(Equation 3)}$$

In air, NO is oxidized to nitrogen dioxide (NO_2),

$$2NO_{(g)} + O_{2(g)} \rightarrow 2NO_{2(g)} \quad \textit{(Equation 4)}$$

which in turn reacts with water to give nitric acid (HNO_3).

$$3NO_{2(g)} + H_2O_{(g)} \rightarrow 2HNO_{3(aq)} + NO_{(g)} \quad \textit{(Equation 5)}$$

Nitric acid then dissociates in water to yield hydrogen ions and nitrate ions (NO_3^-) in a reaction analogous to the dissociation of carbonic acid shown in Equation 2, again lowering the pH of the solution.

Although sulfuric acid may be produced naturally in small quantities from biological decay and volcanic activity, it is produced almost entirely by human activity, especially the combustion of sulfur-containing fossil fuels in power plants. When these fossil fuels are burned, the sulfur contained in them reacts with oxygen from the air to form sulfur dioxide (SO_2), accounting for approximately 80% of the total atmospheric SO_2 in the United States. Sulfur dioxide, like the oxides of carbon and nitrogen, reacts with atmospheric oxygen and water vapor to form sulfuric acid.

$$2SO_{2(g)} + O_{2(g)} + 2H_2O_{(g)} \rightarrow 2H_2SO_{4(l)} \quad \textit{(Equation 6)}$$

Sulfuric acid is a strong acid, so it readily dissociates in water to give an H^+ ion and an HSO_4^- ion (Equation 7).

$$H_2SO_{4(l)} \rightarrow HSO_4^-{}_{(aq)} + H^+_{(aq)} \quad \textit{(Equation 7)}$$

The HSO_4^- ion may further dissociate to give hydrogen and sulfate ions H^+ and SO_4^{2-}.

$$HSO_4^-{}_{(aq)} \rightarrow H^+_{(aq)} + SO_4^{2-}{}_{(aq)} \quad \textit{(Equation 8)}$$

Thus, the presence of H_2SO_4 causes the concentration of H^+ ions to increase dramatically, and so the pH of the rainwater drops to harmful levels.

7. **(C)** The most likely diagnosis would be carbon monoxide (CO) poisoning, which occurs when carbon monoxide builds up in the bloodstream. When too much carbon monoxide is in the air, the

body replaces the oxygen in the red blood cells with carbon monoxide, which can lead to serious tissue damage or even death. Carbon monoxide is a colorless, odorless, tasteless gas produced by burning a fossil fuel (e.g., gasoline, wood, natural gas, charcoal, or other fuel) in a tightly sealed or enclosed indoor space without proper ventilation and venting of the exhaust fumes. All of the patients' symptoms are consistent with CO poisoning.

8. **(C)** "81 fold" means 81 more times.
9. **(C)** Developing countries around the world are experiencing increased levels of particulate matter as a result of rapid increases in energy consumption, motor vehicle use, and population and economic growth.

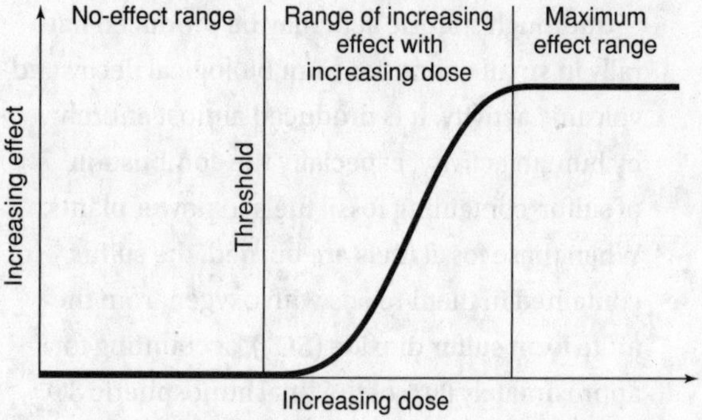

10. **(A)** The elderly are the most affected group since they are more likely to have compromised respiratory systems, based on many years of breathing air pollutants. Newborns are a very close second. Teenagers are highly impacted due to high activity levels and increased cases of asthma.
11. **(B)** Point B represents the point where society has reached the maximum benefit of controlling air pollution at the least cost. Any more money spent on controlling air pollution would not be worth the cost for the benefit received (Law of Diminishing Returns) and would take money away from other needs.
12. **(C)** From 1976 to 1980, the average amount of lead found in the blood of children was 15 μg/dL. But due to the Clean Air Act, the average level of lead had dropped to 2.7 μg/dL—an 82% decrease.

$$[(V_2 - V_1) \div V_1] \times 100\%$$
$$= [(2.7 - 15) \div 15] \times 100\%$$
$$= 82\% \text{ decrease}$$

13. **(D)** The level of dissolved oxygen in the septic zone is too low to support organisms that live by aerobic respiration. In this region, anaerobic (needing little to no oxygen) bacteria flourish and produce waste products such as hydrogen sulfide (H_2S) and methane (CH_4).
14. **(A)** Biochemical oxygen demand (BOD) is the amount of dissolved oxygen that is needed by aerobic organisms in a body of water to break down the organic matter present. BOD is a widely used indicator of the quality of water and is commonly expressed in milligrams of oxygen consumed per liter of water sample. BOD is often used as a gauge of the effectiveness of wastewater treatment plants.
15. **(D)** Mining wastes, along with oil and gas production wastes, constitute approximately 75% of all solid waste generated in the United States.
16. **(B)** Paper accounts for approximately 35% of domestic solid waste in the United States.
17. **(C)** Paper can be broken down into fibers and reused, generally without requiring a sorting process.
18. **(D)** Choice (A) is incorrect because properly designed recycling programs are cost-competitive with landfills and incinerators. Choice (B) is incorrect because recycling creates many more jobs in rural and urban communities than landfill employment does. Choice (C) is incorrect because when designed properly, recycling programs are cost-competitive with trash collection and disposal.
19. **(D)** The 1980 Superfund program, also known as CERCLA (Comprehensive Environmental Response, Compensation, and Liability Act), established federal authority for emergency responses to and cleanups of hazardous substances that have been spilled, improperly disposed of, or released into the environment.
20. **(B)** The 1972 Marine Protection, Research, and Sanctuaries Act, also known as the Ocean Dumping Act, and the 1976 Resource Conservation and Recovery Act (RCRA) prohibit the dumping of wastes into territorial ocean waters of the United States.

Free-Response

10 Total Points Possible

(a) *Maximum 5 points distributed as follows:*

(i) *In the "business as usual" model, identify TWO factors you think might contribute to the increase in carbon dioxide emissions. (2 points maximum for correctly identifying two factors [worth 1 point each] that contribute to an increase in CO_2 emissions.)*

Increased carbon dioxide emissions primarily come from increased burning of fossil fuels worldwide. The "business as usual" model would include an increasing global population, which would mean an increase in the demand for energy. In the "business as usual" model, this need would be met primarily by an increased use of fossil fuels, by far the most common source of energy in the world today. Increasing rates of development, particularly of developing countries in this model, would also lead to an increase in fossil fuel use.

Notable points:

Cover all of the relevant aspects of "business as usual," such as increasing size and development of populations.

(ii) *Given the shape of the graph, what do you think is meant in this case by "technological change"? (1 point maximum for correctly describing "technological change." The "technological change" curve shows a dramatic reduction in carbon dioxide emissions compared to the "business as usual" curve. Therefore, your answer needs to focus on a technology that can reduce fossil fuel use.)*

"Technological change" reduces carbon dioxide emissions on this graph. Increased use of alternative sources of energy, such as nuclear power, hydroelectric, solar, and tidal energy, would all decrease fossil fuel emissions. Technologies that increase energy efficiency and save power, such as more efficient car engines, would also decrease the use of fossil fuels and thus carbon dioxide emissions.

Technological change is a vague phrase that could mean many things. In this case, the curve shows a dramatic reduction in carbon dioxide emissions, so the question is asking about technologies that reduce fossil fuel use.

(iii) *What is meant by a "sustainable level" of carbon dioxide emissions? According to these predictions, what needs to happen for this level to be brought about? (1 point maximum for correctly explaining what is meant by a "sustainable level" of carbon dioxide emissions and 1 point maximum for describing what possibly needs to happen for this to occur.)*

A "sustainable level" of carbon dioxide emissions is one in which the amount absorbed by Earth's natural systems, such as oceans and plants, equals the amount released into the atmosphere. A combination of both technological change and slowed population growth is needed to bring carbon dioxide emissions to sustainable levels. As the graph shows, neither factor can produce sustainable levels on its own.

(b) *Maximum 3 points.*

Explain the difference between remediation and alleviation of an environmental problem. Use the example of acid deposition to illustrate the difference. (1 point maximum for correctly defining remediation. 1 point maximum for correctly defining alleviation. 1 point maximum for correctly explaining the difference using acid deposition as an example.)

Alleviation of an environmental problem means stopping or lessening its cause. Remediation means cleaning up the effects of the problem. Acid deposition forms in the atmosphere mainly from sulfur oxides (SO_x) and nitrogen oxides (NO_x). Both are present primarily as a consequence of the combustion of fossil fuels.

Anything that reduces the use of fossil fuel or limits the emissions of sulfur and nitrogen oxides would be an example of alleviation. Anything that neutralizes the effects of acid in the environment would be an example of remediation. Be careful in giving just an appropriate level of detail.

Methods of alleviation may concentrate on decreasing the amount of sulfur and nitrogen oxides released into the atmosphere. Any method of reducing the rate of consumption of fossil fuels, such as increased use of nuclear power and alternative energy sources, or laws and education to help conserve energy, would decrease the amount of sulfur and nitrogen oxides released. Clean-fuel technologies would result in less of these gases' being produced on combustion. Scrubbers in smokestacks can remove much of what is produced. Perhaps the most important step in reducing emissions in the United States, however, was the passing of the Clean Air Act in 1963 and its subsequent amendments.

Methods of remediation may concentrate on neutralizing the acid deposited in an environment. Acids can be neutralized by adding a base. Since an abundant, low-cost, and nontoxic material is often needed, limestone, $CaCO_3$, is commonly used. For example, limestone might be added to a lake to increase its pH. Powdered limestone could also be spread over agricultural lands to increase the pH of the soil. Many nutrients are more soluble in acidic soils and therefore might be washed away by rain. As a result, the addition of fertilizer is also required. Deforestation caused by acid deposition may be addressed by treating the soil and then replanting.

(c) *Maximum 1 point.*

Look at the graph of CFC production and account for the trends you observe. (1 point for correctly accounting for one possible reason for the trends as shown in the graph.)

When asked to explain the trends shown in a graph, a good starting point is to describe what they are. Writing this down first should also help you compose your explanation.

Between 1984 and 1988, both developed and developing countries used ozone-depleting chemicals at a fairly consistent rate of approximately 1 million tons per year. After 1988, the amount used by both developed and developing countries decreased sharply and at a fairly constant rate for the next seven years. By 1996, developed countries used 50,000 tons per year and developing countries used 150,000 tons. Use remained at approximately these levels for the next two years.

The reduction is likely to be due to countries' implementing technologies to comply with the Montreal Protocol (1989), which set limits for the emission of chemicals that cause depletion of the ozone layer. The biggest reduction has come from using alternatives to chlorofluorocarbons (CFCs)—the principal ozone-depleting agents. Alternatives include HFCs and HCFCs.

(d) *Maximum 1 point.*

In 1992, the World Bank designated indoor air pollution in developing countries as one of the four most critical global environmental problems. Provide and explain ONE example of a major source of an indoor pollutant to illustrate how the issue of indoor air quality in developing countries differs from that in developed countries. (1 point maximum for providing a correct example that includes a correct explanation.)

This sample may provide more detail than necessary. The important point is that the amounts for both developed and developing countries fell dramatically at the same time and basically at the same rate. This suggests that the drops come as a result of the implementation of new laws. Since we are dealing with a global situation, it is likely to be in the form of an international agreement.

Homes are perhaps the most important factor when considering indoor air quality since people tend to spend more time there than anywhere else. The majority of people in developing countries live in homes with much simpler technologies. Many of the pollutants found in houses in developed countries are not present. The major source of indoor air pollutants in developing countries, therefore, is the combustion of poor-quality fuels for heating, lighting, and cooking. Such dirty fuels include animal wastes, kerosene, and low-grade coal that may release large amounts of particulates, carbon monoxide, sulfur oxides, and other toxins on combustion. As fires are often burned indoors in places with inadequate ventilation, the levels of these pollutants and carbon monoxide can be greatly concentrated.

Energy sources are much more technologically advanced in developed countries, so this is not such an important source of indoor air pollution. For example, poorly maintained furnaces may produce some carbon monoxide. However, this problem is not nearly as widespread or generally as serious as the energy issue in developing countries. Instead, major sources of air pollution include lead (from lead paints), asbestos, and fumes from volatile organic compounds (VOCs) in paints, glues, plastics, and furniture. These things will be less common in houses in developing countries. Houses in developed countries are often tightly sealed, more so than in developing countries. This leads to increased levels of concentration. For example, radon gas, which occurs naturally from radioactive decay in certain rocks, might accumulate to dangerous levels in a modern air-conditioned house but not in a simpler hut with no glass in the windows.

The wording of the World Bank designation suggests that the question is referring to the general masses of the population in developing countries rather than the small, technologically developed percentage. The answer is thus strongly focused on this difference in lifestyle.

9

Global Change

Learning Objectives

In this unit, you will learn:

- ➔ 9.1 Stratospheric Ozone Depletion
- ➔ 9.2 Reducing Ozone Depletion
- ➔ 9.3 The Greenhouse Effect
- ➔ 9.4 Increase in the Greenhouse Gases
- ➔ 9.5 Global Climate Change
- ➔ 9.6 Ocean Warming
- ➔ 9.7 Ocean Acidification
- ➔ 9.8 Invasive Species
- ➔ 9.9 Endangered Species
- ➔ 9.10 Human Impacts on Biodiversity

For added practice with this unit and topics, log onto Online Labs 9A–9C at online.barronsbooks.com

9.1 Stratospheric Ozone Depletion

Stratospheric Ozone

The stratosphere contains approximately 97% of the ozone in the atmosphere, and most of it lies between 9 and 25 miles (15–40 km) above Earth's surface. Most ozone is formed over the tropics; however, slow circulation currents carry the majority of it to the poles. It also varies somewhat due to season, being somewhat thicker in the spring and thinner during the fall. Steps involved in the formation of stratospheric ozone include the following:

- Atomic oxygen can combine with oxygen molecules to form ozone:

$$O + O_2 \rightarrow O_3$$

- Atomic oxygen can combine with ozone to form oxygen molecules:

$$O + O_3 \rightarrow 2O_2 \text{ (back to first step)}$$

- Ultraviolet radiation (*uv*) can strike ozone and create oxygen molecules and atomic oxygen:

$$O_3 + uv \rightarrow O_2 + O$$

- Ultraviolet radiation (*uv*) strikes an oxygen molecule, creating atomic oxygen:

$$O_2 + uv \rightarrow O + O$$

Ultraviolet Radiation

The sun emits a wide variety of electromagnetic radiation, including infrared, visible, and ultraviolet (Figure 9.1). Ultraviolet radiation is subdivided into three forms: UVA, UVB, and UVC, listed in order of increasing energy (shorter wavelength).

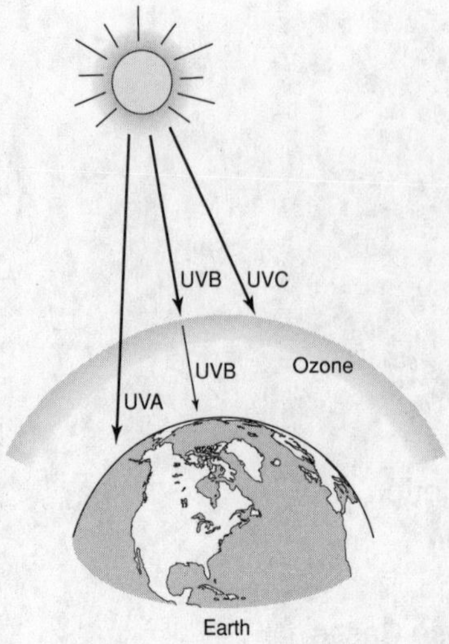

Figure 9.1 Ultraviolet radiation reaching Earth

UVA

UVA is closest to blue light in the visible spectrum and is the form of ultraviolet radiation that usually causes skin tanning. UVA radiation is 1,000 times less effective than UVB in producing skin redness, but more of it reaches Earth's surface than UVB. Birds, reptiles, and bees can see UVA since many fruits, flowers, and seeds stand out more strongly from the background in ultraviolet wavelengths. Many birds have patterns in their plumage (feathers) that are not visible in the normal spectrum (white light) but become visible in ultraviolet. The urine of some animals is also visible in the UVA spectrum.

UVB

UVB causes blistering sunburns and is associated with skin cancer.

UVC

UVC is found only in the stratosphere and is largely responsible for the formation of ozone.

Ozone Depletion

The ozone (O_3) layer is a belt of naturally occurring ozone gas that sits between 9 and 19 miles (15–30 km) above Earth and serves as a shield from the harmful ultraviolet B radiation emitted by the sun. Ozone is a highly reactive molecule and is constantly being formed and broken down in the stratosphere.

Causes of Ozone Depletion

Thinning of the ozone layer was first discovered over Antarctica in 1986 due to the presence of extremely cold polar stratospheric clouds. Ozone depletion occurs seasonally (in the winter) and is due to the presence of both natural and human-made compounds that contain halogens (bromine, chlorine, fluorine, or iodine). Measurements indicate that the ozone over the Antarctic has decreased as much as 60% since the late 1970s, with an average net loss of about 3% per year worldwide.

The main cause of degradation of the ozone layer is the presence of chlorofluorocarbons (CFCs) and halocarbons (halons), very stable compounds that, as a result of their chemical stability (persistence), are able to reach the stratosphere—there are no natural reservoirs of CFCs or halons in nature.

First manufactured during the 1920s, chlorofluorocarbons are nonflammable chemicals that contain atoms of carbon, chlorine, and fluorine. They are used in the manufacture of aerosol sprays, as blowing agents for foams and packing materials, as solvents, and as refrigerants.

By 1974, nearly 1 million tons of CFC gases were produced each year, with the largest single source of CFCs in the atmosphere coming from leakage from air conditioners. The average residence time for CFCs in the environment is 200 years.

When a CFC molecule enters the stratosphere, ultraviolet radiation causes it to decompose and produce atomic chlorine (Cl).

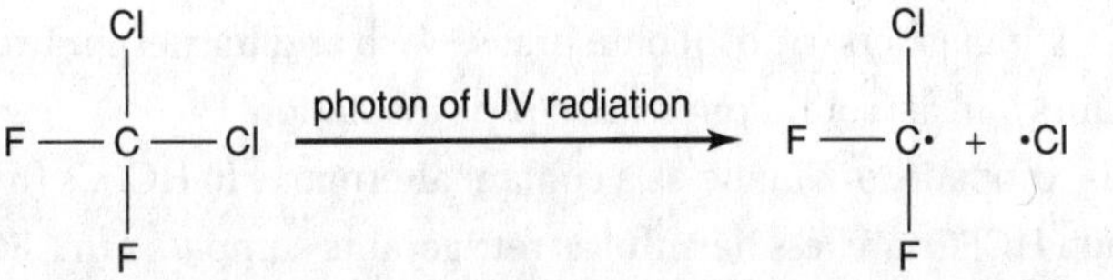

This atomic chlorine then reacts with the ozone in the stratosphere to produce chlorine monoxide (ClO):

$$Cl + O_3 \rightarrow ClO + O_2$$

The chlorine atom thus prevents the formation of ozone by sequestering oxygen atoms to form ClO. The chlorine monoxide then reacts with more ozone to produce even more chlorine in what essentially becomes a chain reaction:

$$ClO + O_3 \rightarrow Cl + 2O_2.$$

Thus, one chlorine atom released from a CFC can ultimately destroy over 100,000 ozone molecules. Much of the destruction of the ozone layer that is occurring now is the result of CFCs that were produced many years ago, since a CFC molecule takes about eight years to reach the stratosphere, and the residence time in the stratosphere for a CFC molecule is over 200 years.

Halocarbons (halons) are organic chemical molecules that are composed of at least one carbon atom with one or more halogen atoms; the most common halogens in these molecules are fluorine, chlorine, bromine, and iodine, which are used in fire extinguishers, soil fumigants and pesticides (e.g., methyl bromide), solvents, and foam-blown insulation. The first synthesis of halocarbons was achieved in the early 1800s. However, production began accelerating when their useful properties as solvents and anesthetics, and their uses in plastics and pharmaceuticals, were discovered.

A large amount of the naturally occurring halocarbons are created by forest fires and other forms of biomass burning, volcanic activities, and marine algae, which produce millions of tons of methyl bromide annually. Bromine, which is found in much smaller quantities than chlorine, is about 50 times stronger than chlorine in its effect on stratospheric ozone depletion.

Effects of Ozone Depletion

During the onset of the 1998 Antarctic spring, a hole three times the size of Australia (over 3,500 miles [5,600 km] in diameter) developed in the ozone layer over the South Pole.

Harmful effects of increased UV radiation include the following:

- A reduction in crop production
- A reduction in the effectiveness of the human body's immune system
- A reduction in the growth of phytoplankton and the cumulative effect on food webs
- Climatic changes
- Cooling of the stratosphere
- Deleterious effects on animals (since they don't wear sunglasses or sunscreen)
- Increases in cataracts
- Increases in mutations, since UV radiation causes changes in the DNA structure
- Increases in skin cancer
- Increases in sunburns and damage to the skin (premature aging)

9.2 Reducing Ozone Depletion

Strategies that can be used to help reduce ozone depletion include the following:

- Introduce tariffs on products produced in countries that allow the use of chlorofluorocarbons (CFCs).
- Offer tax credits or rebates for turning in old refrigerators and air conditioners.
- Support legislation that reduces ozone-destroying products found in certain aerosol propellants used in medical inhalers, fire extinguishers, aerosol hairsprays, wasp and hornet sprays, certain types of foam insulation used in refrigerators and air conditioners, and pipe insulation.
- Use helium, ammonia, propane, or butane as a coolant alternative to HCFCs (hydrochlorofluorocarbons) and CFCs. Even though HCFCs are less harmful as refrigerants compared to CFCs, they are 1,725 times greater in their greenhouse warming potential compared to carbon dioxide. Following the United Nations Montreal Protocol on Substances that Deplete the Ozone Layer, the use of HCFCs has been phased out in the majority of developed countries. Due to the fact that HCFCs are not considered to be as damaging as CFCs, HBFCs, or halons, the process of phasing out their use is not as strict and requires that, internationally, they no longer be used after 2030.

9.3 The Greenhouse Effect

When sunlight strikes Earth's surface, some of it is reflected back toward space as infrared radiation (heat). Greenhouse gases absorb this infrared radiation and trap the heat in the atmosphere (Figure 9.2).

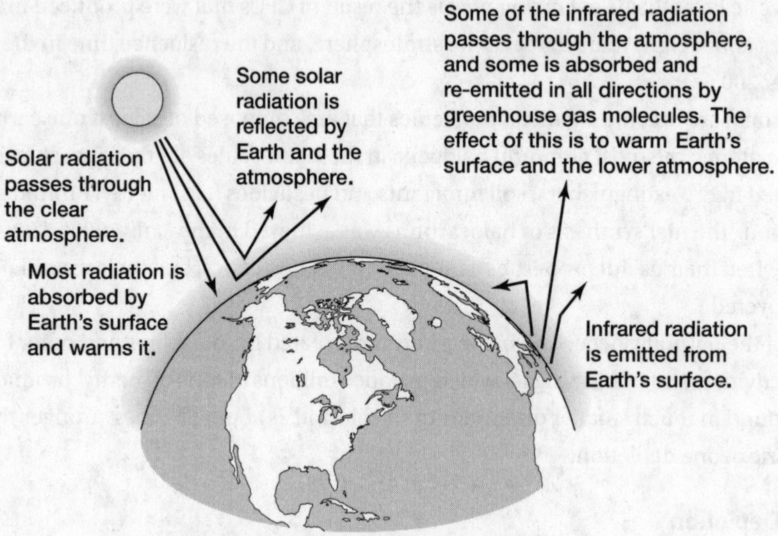

Figure 9.2 The greenhouse effect

9.4 Increases in Greenhouse Gases

Trends in Global Greenhouse Gas Emissions

The concentration of carbon dioxide in the atmosphere is naturally regulated by numerous processes that occur in the carbon cycle (Figure 9.3). The movement (flux) of carbon between the atmosphere and the land and oceans is naturally and primarily regulated by:

1. the intake of CO_2 during photosynthesis
2. being absorbed by seawater

However, these natural processes can absorb only about half of the 6 billion metric tons of anthropogenic carbon dioxide emissions produced each year.

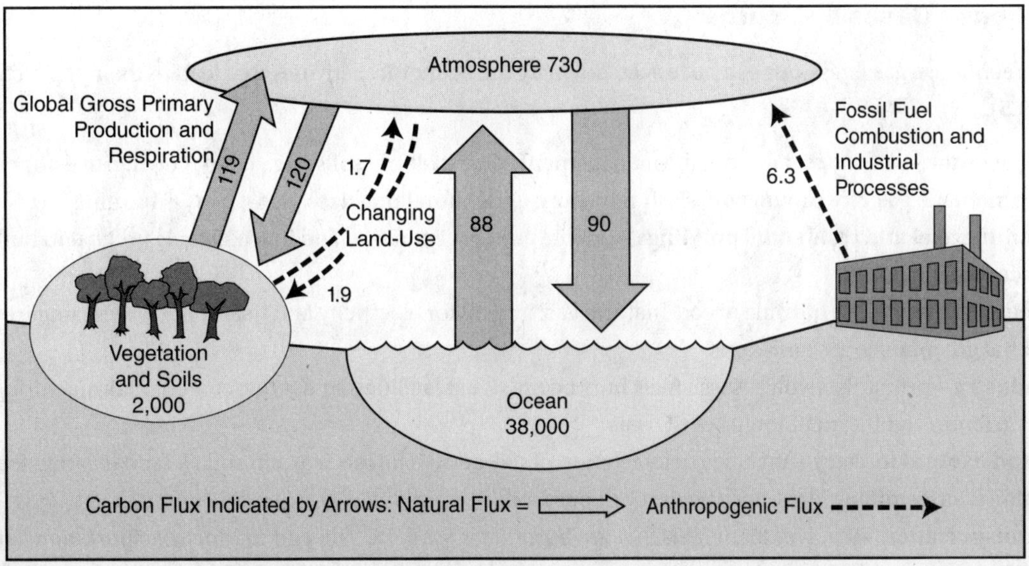

Figure 9.3 Global carbon cycle (billion metric tons carbon)
Source: Intergovernmental Panel on Climate Change

Figure 9.4 shows that for the past 2,000 years, the atmospheric concentrations of CO_2, CH_4, and N_2O—three important, long-lived greenhouse gases—have increased substantially since around the time of the beginning of the Industrial Revolution (approximately 1750), and the rates of increase in the levels of these gases have been dramatic. By analyzing gas bubbles in ice core samples, scientists discovered that CO_2, for instance, never increased more than 30 ppm during any previous 1,000-year period, but has already risen by 30 ppm in the past 20 years.

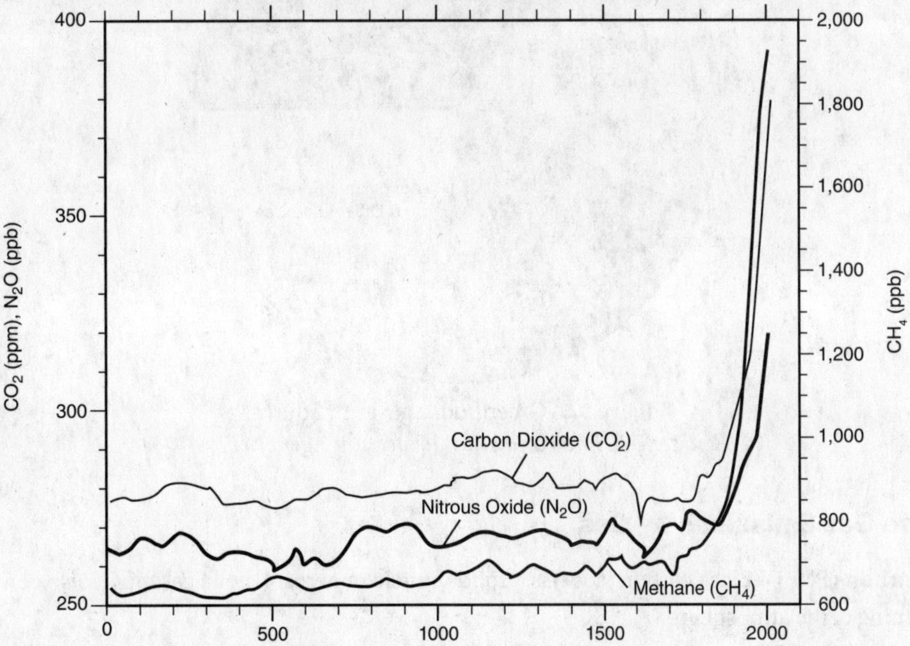

Figure 9.4 Atmospheric CO_2, CH_4, and N_2O concentrations over the last 2,000 years

Increased ocean temperatures decrease the solubility of gases in seawater (i.e., warmer water will release more CO_2 into the atmosphere). Since oceans are one of the most important global carbon sinks, absorbing a quarter of all atmospheric carbon dioxide emissions, as global temperatures increase the oceans thereby become less effective in absorbing carbon dioxide and other greenhouse gases, resulting in a dangerous positive feedback loop.

Greenhouse Gases by Source

Global greenhouse gas emissions can be broken down by the economic activities that lead to their production (Figure 9.5).

- **Agriculture—**mostly comes from the management of agricultural soils (e.g., deep plowing, livestock releasing methane gas, rice production, slash-and-burn agricultural practices, and biomass burning)
- **Commercial and residential buildings**—on-site energy generation and burning fuels for heat in buildings or cooking in homes
- **Energy supply**—The burning of coal, natural gas, and oil for electricity and heat is the largest single source of global greenhouse gas emissions.
- **Industry**—primarily involves fossil fuels burned on-site at facilities for energy; cement manufacturing also contributes significant amounts of CO_2 gas
- **Land use and forestry**—includes deforestation of old-growth forests (carbon sinks), land clearing for agriculture, strip-mining, fires, and the decay of peat soils
- **Transportation**—involves fossil fuels that are burned for road, rail, air, and marine transportation; almost all (95%) of the world's transportation energy comes from petroleum-based fuels, largely gasoline and diesel
- **Waste and wastewater—**landfill and wastewater methane (CH_4), and incineration as a method of waste management

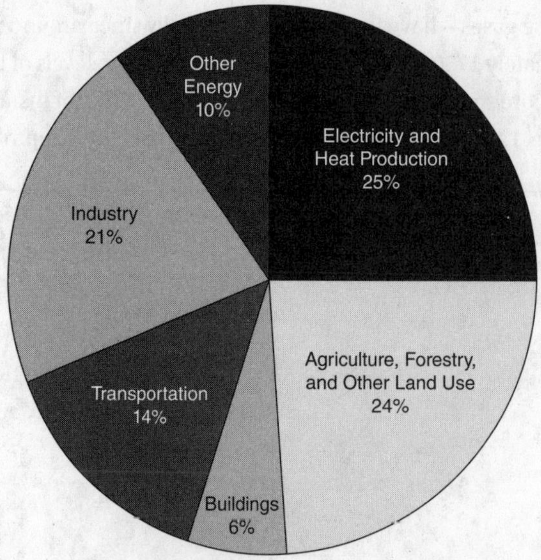

Figure 9.5 Greenhouse gases by source
Source: U.S. Environmental Protection Agency (EPA)

Greenhouse Gas Emissions by Gas

- **Black carbon (PM_x)**—Black carbon (soot) is a solid particle or aerosol, not a gas, but it also contributes to the warming of the atmosphere.
- **Carbon dioxide (CO_2)**—Carbon dioxide (CO_2) is an important heat-trapping (greenhouse) gas, and is released through human activities such as deforestation and burning fossil fuels, as well as natural processes such as respiration and volcanic eruptions. The graph (Figure 9.6) shows atmospheric CO_2 levels measured at Mauna Loa Observatory, Hawaii, in recent years, with average seasonal cycles removed. The circle graph (Figure 9.7) then shows CO_2 levels during the last three glacial cycles, as reconstructed from ice cores.

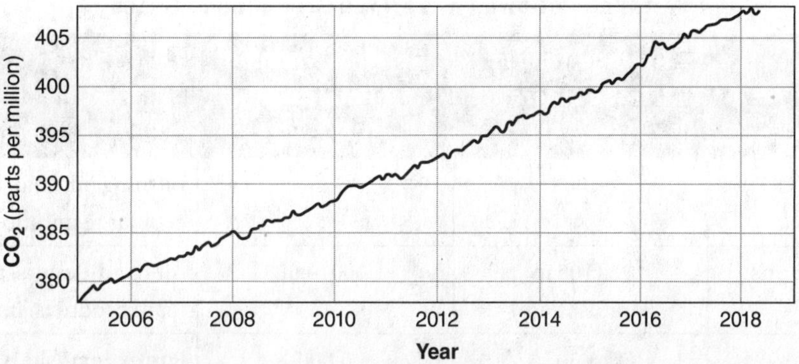

Figure 9.6 Atmospheric CO_2 levels (2006–2018) *(average seasonal cycles removed)*
Source: National Aeronautics and Space Administration (NASA)

- **Fluorinated gases**—Industrial processes, refrigeration, and the use of a variety of consumer products all contribute to emissions of F-gases, which include hydrofluorocarbons (HFCs), perfluorocarbons (PFCs), and sulfur hexafluoride (SF_6).
- **Methane (CH_4)**—Agricultural activities, waste management, and energy use all contribute to CH_4 emissions.
- **Nitrous oxide (N_2O)**—Fertilizer use is the primary source of N_2O emissions.

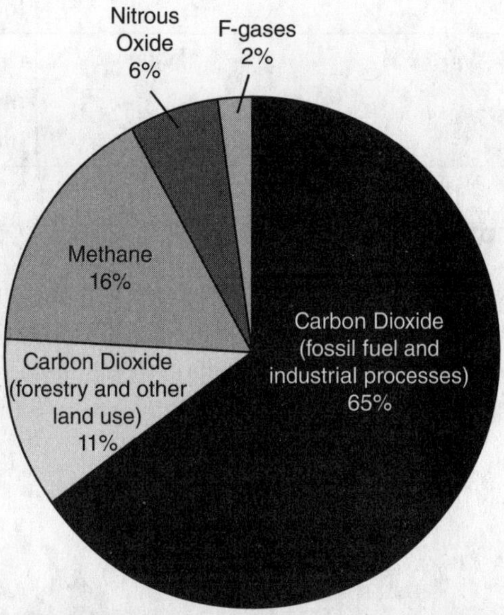

Figure 9.7 Annual greenhouse gas emissions by gas
Source: U.S. Environmental Protection Agency (EPA)

Warming Potential of Various Greenhouse Gases

Greenhouse Gas	Average Time in Troposphere (Years)	Relative Warming Potential ($CO_2 = 1$)	Source
Carbon Dioxide (CO_2)	100	1	Burning oil, coal, deforestation, cellular respiration
Chlorofluorocarbons (CFCs)	15 (100 in stratosphere)	1,000–8,000	Air conditioners, refrigerators, foam products, insulation
Halons	65	6,000	Fire extinguishers
Methane (CH_4)	15	25	Rice cultivation, cattle/sheep raising, landfills, natural gas leaks, coal production
Nitrous Oxide (N_2O)	115	300	Burning fossil fuels, fertilizers, livestock wastes, plastic manufacturing
Sulfur Hexafluoride (SF_6)	3,200	24,000	Electrical industry as a replacement for PCBs
Tropospheric Ozone (O_3)	Varies	3,000	Combustion of fossil fuels, photochemical smog produced by internal-combustion engines

Greenhouse Gases by Country

The top carbon dioxide emitters from fossil fuel combustion, cement manufacturing, and gas flaring are China and the United States (Figure 9.8). Changes in land use are also significant contributors of CO_2, as global estimates indicate that deforestation can account for 5 billion metric tons of CO_2 emissions, or about 16% of emissions from fossil fuel sources. Tropical deforestation in Africa, Asia, and South America is thought to be the largest contributor to emissions.

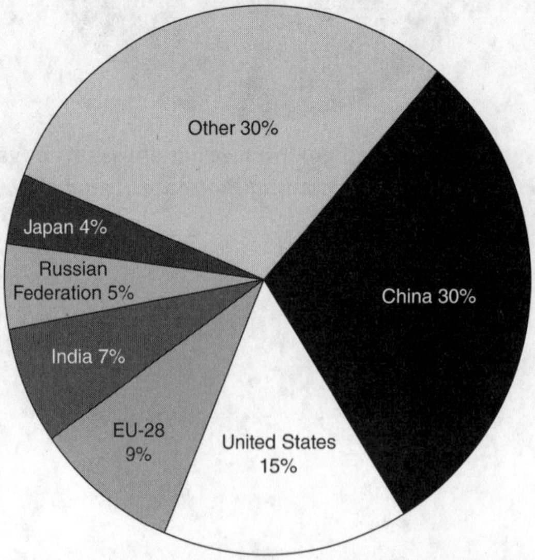

Figure 9.8 Greenhouse gas emissions by country
Source: U.S. Environmental Protection Agency (EPA)

Impacts and Consequences of Global Warming

Global warming affects the weather, the economy, and numerous other aspects of life (Figure 9.9).

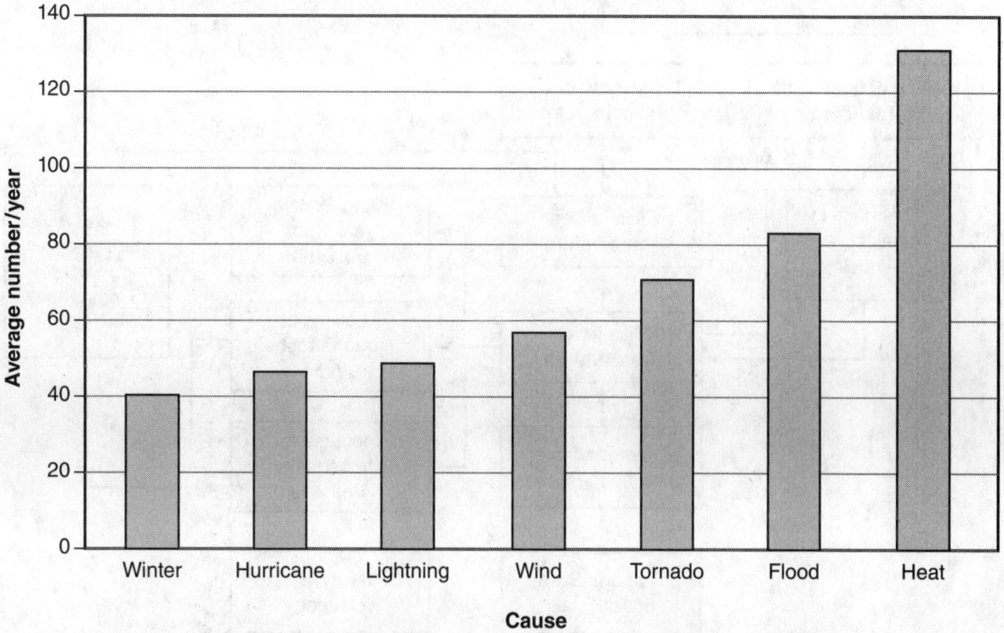

Figure 9.9 U.S. weather fatalities (average per year for 1986–2015)
Source: U.S. National Weather Service

9.5 Global Climate Change

Causes of Long-Term Climate Change

Climate change is nothing new because the Earth's climate has been changing for billions of years. What is different now is the *rate* of the climate change. Below are reasons why the Earth's climate has and is continuing to change.

Carbon Dioxide Content of the Oceans

The world's oceans contain more carbon dioxide than the atmosphere. Furthermore, ocean water only absorbs about a quarter of the CO_2 released into the atmosphere each year. As atmospheric CO_2 levels rise, so do the ocean's levels rise.

Changes at the Poles

The total warming at the poles can be attributed to changes in atmospheric temperatures, cloud cover, surface albedo, and water vapor. However, the north and south poles are warming faster than all other biomes on Earth because of energy in the atmosphere that is carried to the poles through large weather systems (Figure 9.10). Whereas the Earth has warmed ~1.4°F over the last 40 years, the poles have warmed ~3.5°F over the same time period, with most of the warming occurring during that pole's winter.

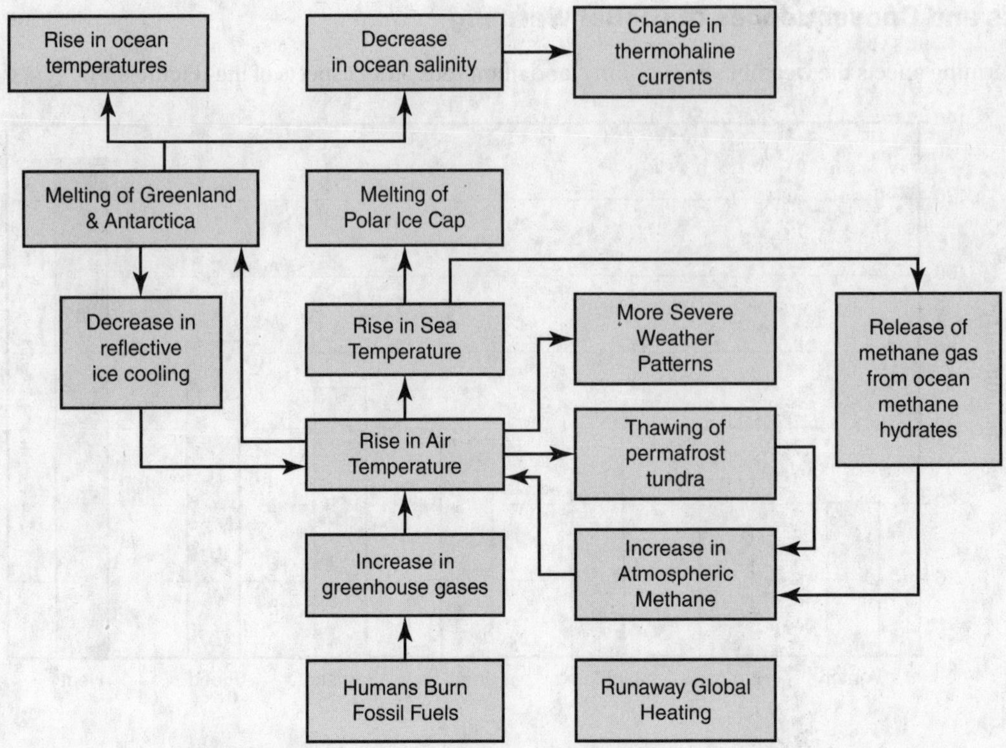

Figure 9.10 Atmospheric warming feedback loop for the poles

Changes in Ocean Currents

Ocean currents carry heat around the Earth. As the oceans absorb more heat from the atmosphere, sea surface temperatures increase and ocean circulation patterns that transport warm and cold water around the globe change. As the oceans store a large amount of heat, even small changes in these currents can have a large and lasting effect on global climate; e.g., warmer ocean water cannot absorb as much carbon dioxide gas as cooler ocean water.

Causes, Changes, and Results of Climate Change

Changes in Tropospheric Weather Patterns

Air temperatures today average 5°F to 9°F (3°C to 5°C) warmer than they were before the Industrial Revolution. Higher average air temperatures may result in higher amounts of rainfall in many areas because of higher rates of evaporation and may also be accompanied by an increase in the frequency or severity of storms, increases in surface water and/or groundwater inputs, increases in sedimentation in bodies of water, increases in flooding and associated water runoff, and increases in the rate of aquifer recharge—all of which have a direct impact on the biodiversity of the biota in an area (Figure 9.11).

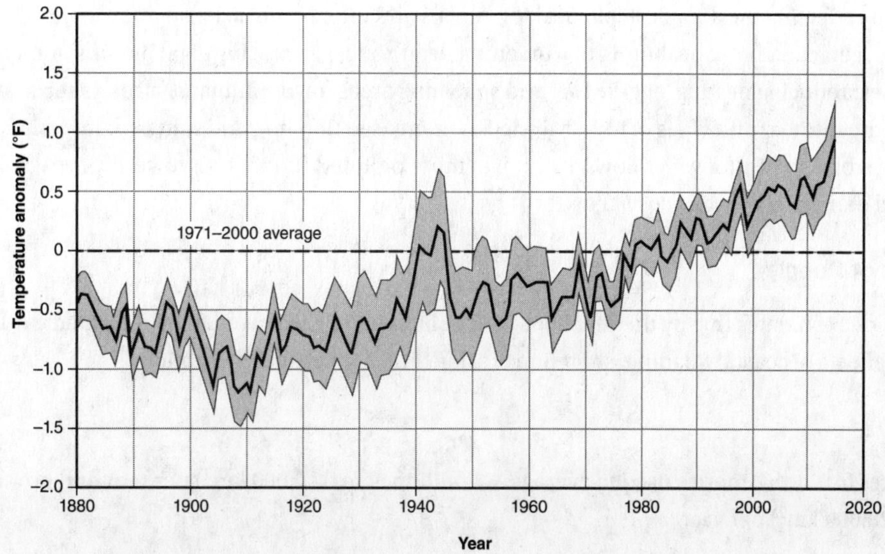

Figure 9.11 Average global sea surface temperature (1880–2020)
Source: U.S. Environmental Protection Agency (EPA)

Worldwide, Category 4 or 5 hurricanes have risen from 20% of all hurricanes in the 1970s to 35% of all hurricanes in the 1990s. Computer simulations and projections for the 21st century show that the number of Category 1, 2, and 3 hurricanes in the Atlantic will diminish, while, during the same period of time, the number of Category 4 and 5 hurricanes will increase dramatically (Figure 9.12).

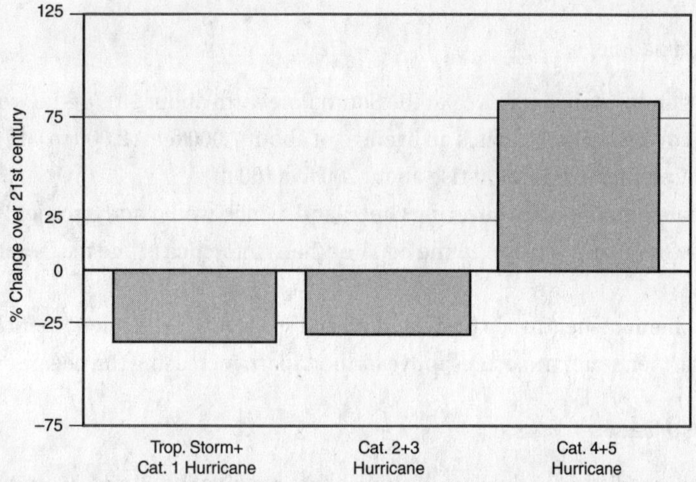

Figure 9.12 Projected changes in Atlantic hurricane frequency (2000–2100)
Source: U.S. National Oceanographic and Atmospheric Administration (NOAA)

Changes in Coastal Ecosystems

Global warming has the potential to completely alter the structure and function of estuaries and coastal wetlands. Sea-level rise threatens to inundate many coastal wetlands, seriously affecting the biota, with little room for them to move inland because of coastal development. In the next 100 years, taking into account the ongoing sinking of land in some coastal areas, net sea-level rise could exceed six feet. Warmer water with potentially higher dissolved carbon dioxide levels, accompanied by a lower pH, will alter the species composition in these extremely sensitive areas by affecting the aquatic ecological tolerances and interactions; altering nutrient cycles; reducing the dissolved oxygen content; contributing to worsening dead zones; increasing algal blooms; increasing the incidence of diseases; reducing nesting, egg-laying, and spawning areas (of amphibians, birds, insects, and reptiles); and increasing the number and types of invasive species. Increased flooding and saltwater intrusion, droughts, hurricanes, and storms will alter water flows, leading to more polluted runoff, thus lowering the water quality and threatening and damaging coastal ecosystems.

Displacement of People

The United Nations estimates that by the year 2050, 150 million people will need to be relocated worldwide because of the effects of coastal flooding, shoreline erosion, and agricultural disruption.

Forest Fires

Boreal forest fires in North America used to average ~2 million acres (10,000 km^2) per year and now average ~7 million acres (28,000 km^2) per year.

Glacier Melting

The total surface area of glaciers worldwide has decreased 50% since the end of the 19th century.

Some causes of glacier melting are as follows:

- Flash floods
- Glacial lake overflow
- Increased variation in the water flows into rivers
- Landslides
- The displacement of people

The main ice-covered landmass is Antarctica at the South Pole, with about 90% of the world's ice and 70% of its freshwater. Antarctica is covered with ice that is an average of about 7,000 feet (2,000 m) thick. If all of the Antarctic ice melted, sea levels around the world would rise about 200 feet (60 m).

There is also a significant amount of ice covering Greenland, which would add another 20 feet (7 m) to the oceans if it melted. Since Greenland is closer to the equator than Antarctica is, the temperatures are higher, so that ice is more likely to melt.

Compounding the problem of melting ice from land raising sea levels is the phenomenon of thermal expansion of water (as water gets warmer, water molecules move farther apart, increasing the oceans' volume).

Increase of Greenhouse Gases

Greenhouse gases trap solar radiation in the Earth's atmosphere, making the climate warmer. Carbon dioxide (CO_2), methane (CH_4), and water vapor are the most abundant greenhouse gases in the atmosphere; however, their residence time in the atmosphere is relatively short. For example, carbon dioxide stays in the atmosphere from years to centuries and contributes to longer periods of warming, whereas methane stays in the atmosphere about nine years and is continually being removed by oxidation (reacting with oxygen gas) into carbon dioxide and water.

Increase in Disease

In areas that experience wetter conditions due to global warming (i.e., an increase in evaporation and atmospheric water content), there are more areas that are suitable for mosquitoes to breed, which will increase the rates of malaria, dengue fever, Zika virus, and yellow fever. Warmer temperatures also allow many insects to:

- have access to more reliable food resources for a longer time period;
- have a longer breeding season; and
- not be subject to winter freezes that naturally tend to thin insect populations.

Warmer water also promotes and increases bacterial activity in water supplies and may promote the spread of amoebic dysentery, cholera, and giardia.

In areas that experience drier conditions due to changes in tropospheric weather patterns as a result of global warming, water stagnation becomes more common in riparian areas and results in higher water temperatures and associated bacterial counts, and is conducive to an increase in mosquito and other disease-carrying vector populations.

Increase in Health and Behavioral Effects

Higher air temperatures have been proven to result in higher incidences of heat-related deaths caused by cardiovascular disease, heat exhaustion, heat stroke, hyperthermia, and diabetes. Furthermore, stress and the resulting rage brought on as by-products of increased air temperatures can dramatically affect human behavior.

Increase in Property Loss

Weather-related disasters have tripled since the 1960s. Insurance payouts have increased fifteen-fold (adjusted for inflation) during this same time period, with much of this being attributed to people's moving to vulnerable coastal areas.

Loss of Biodiversity

Arctic fauna will be the most affected. The food webs of polar bears that depend on ice floes, birds, and marine mammals will be drastically affected. Many species have shifted their ranges toward the poles, averaging 4 miles (6 km) per decade, and bird migrations are averaging over two days earlier per decade. Grasses have become established in Antarctica for the first time. Many species of fish and krill that require cooler waters will be negatively affected, with major repercussions occurring within food webs. Decreased glacier melt, caused by dwindling glaciers, will also negatively affect migratory fish, such as salmon, that need sufficient river flow. The subject of biodiversity will come up many times in this unit as it relates to the topic.

Loss in Economic Development

Money that was earmarked for education, improving healthcare, reducing hunger, and improving sanitation and freshwater supplies will instead be spent on mitigating the effects of global warming.

Natural Drivers and Consequences of Climate Change

Plate Tectonics and Volcanic Eruptions

Over very long periods of time, plate tectonic processes cause the world's continents to move to different locations. The movement of these plates causes volcanoes and mountains to form, which can also contribute to changes in the climate; e.g., large mountain changes affect air circulation patterns around the Earth, which directly influence local climates.

Volcanic gases that reach the stratosphere have a long-term effect on climate. For example, sulfur dioxide released into the stratosphere can cause global cooling, whereas carbon dioxide has the potential to cause global warming. However, the contribution of volcanic gases affecting global climates is (today) very small—about 1% of anthropogenic emissions.

Strength of the Sun

The energy output of the sun is not constant; it varies over time and has an impact on Earth's climate. The fluctuations in the solar cycle impacts Earth's global temperature by $\sim 0.1°$, slightly hotter during solar maximums and slightly cooler during solar minimums. During the last 100 years, Earth's average temperature has increased by approximately 1.1°F (0.6°C) with solar heating accounting for about 25% of this change.

Range Shifts

As temperatures increase, the habitat ranges of many North American species are moving north and to higher elevations. In recent decades, in both land and aquatic environments, plants and animals are being found at higher elevations at a median rate of 36 feet (11 m) per decade and at higher latitudes at a median rate of 11 miles (18 km) per decade. While this means a range expansion for some species, for others it means movement into a less hospitable habitat, increased competition, or range reduction, with some species having nowhere to go because they are already at or near the top of a mountain or at the limit of land suitable for their habitat.

As rivers and streams warm, warm-water fish are expanding into areas previously inhabited by cold-water species. Thus, these cold-water fish, including many highly valued trout and salmon species, are losing their habitat, with projections of 47% habitat loss by 2080. Range shifts disturb the current state of the ecosystem and can limit opportunities for fishing and hunting.

Collectively, the ranges of vegetative biomes are projected to change across 5%–20% of the land in the United States by 2100.

Releases of Methane from Hydrates in Coastal Sediments and Thawing Permafrost

The Arctic region is a large natural source of methane (including methane hydrates, or clathrates, and natural gas deposits), a potent greenhouse gas that is approximately 20 times more effective per molecule in warming potential than carbon dioxide.

Arctic methane release is the release of methane from the seas and soils in permafrost regions of the Arctic due to melting glaciers, and results in a positive feedback loop, as methane is itself a powerful greenhouse gas. Global warming accelerates its release, due to the release of methane from both existing stores and rotting biomass.

Rise in Sea Level

Sea levels have risen 400 feet (120 m) since the peak of the last ice age approximately 18,000 years ago. From about 13,000 years ago to the start of the Industrial Revolution, the rate of sea-level rise averaged 0.1 to 0.2 mm per year; however, since 1900, sea levels have risen about 3 mm per year (a tenfold increase). The IPCC (Intergovernmental Panel on Climate Change) predicts a global rise in sea level between 20 and 40 inches (500–1000 mm) by the year 2100, severely threatening coastal cities (which are usually large population centers) and island nations (Figure 9.13).

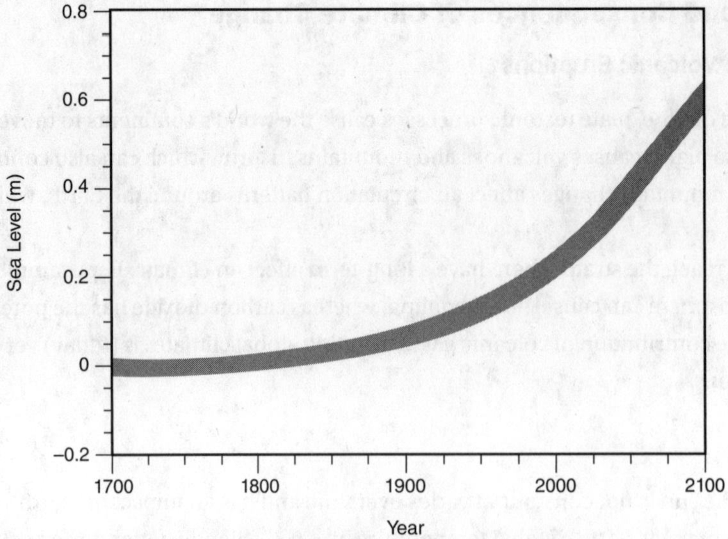

Figure 9.13 Sea level rise, 1700–2100 (projected)
Source: International Panel on Climate Change—Geneva, Switzerland

A significant rise in the sea level results in the following:

- An increase in the number of storm surges, which cause personal property and infrastructure damage—this results in higher insurance premiums, a decrease in property values, and the cost of repair or relocation
- Flooding in wetlands and estuaries, changing the salinity content of the water, the water depth, temperature, and light availability, which can seriously impact various life forms that are unable to adapt
- Increased coastal erosion, which causes increased nutrient loads in the water
- Losses in fish and shellfish catches
- Oil spills that occur at sea, spreading farther inland with serious repercussions to the biota and the economy
- Saltwater intrusion, which causes decreased aquacultural and agricultural production as well as the contamination of local water supplies
- The loss of cultural, recreational, and tourism resources, which causes a loss of livelihood and income
- Waterlogging, which causes a loss of various plant species

Slowing or Shutdown of Thermohaline Circulation

Glacial melting in Greenland would shift the saltwater–freshwater balance in the North Atlantic, resulting in less heavy saltwater sinking than in traditional ocean circulation patterns. Current thermohaline circulation patterns are characterized as follows:

- Between Greenland and Norway, the water cools, sinks into the deep ocean, and begins flowing back to the south.
- Evaporation of ocean water in the North Atlantic increases the salinity of the water as well as cools it, both of which increase the density of the seawater at the surface. Formation of sea ice further increases the salinity of the seawater.
- Global warming, which increases precipitation and glacial melting, could lead to an increased volume of freshwater in the northern oceans that would result in a slowdown or shutdown of thermohaline circulation, causing western Europe to experience a severe ice age.
- Heat coming from the North Atlantic Drift, a branch of the Gulf Stream, moderates temperatures in Western Europe.
- Heat is transported from the equator to the poles, mostly by the atmosphere but also by ocean currents, with warm water occurring near the surface and cold water at deeper levels (Figure 9.14).

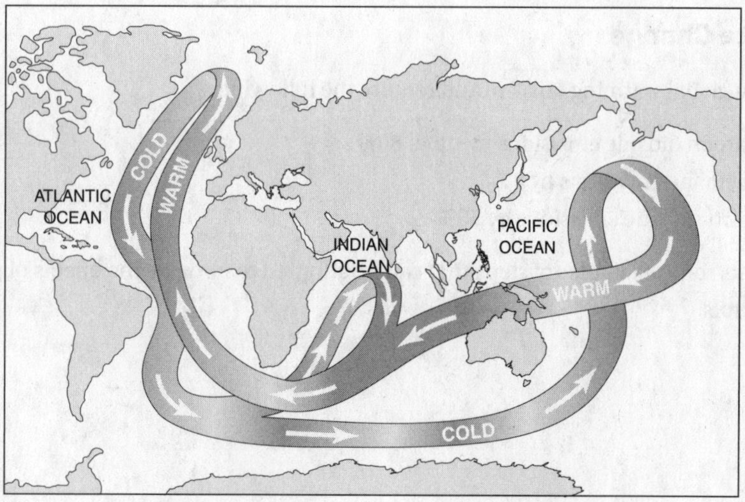

Figure 9.14 Thermohaline circulation pattern
Source: International Panel on Climate Change

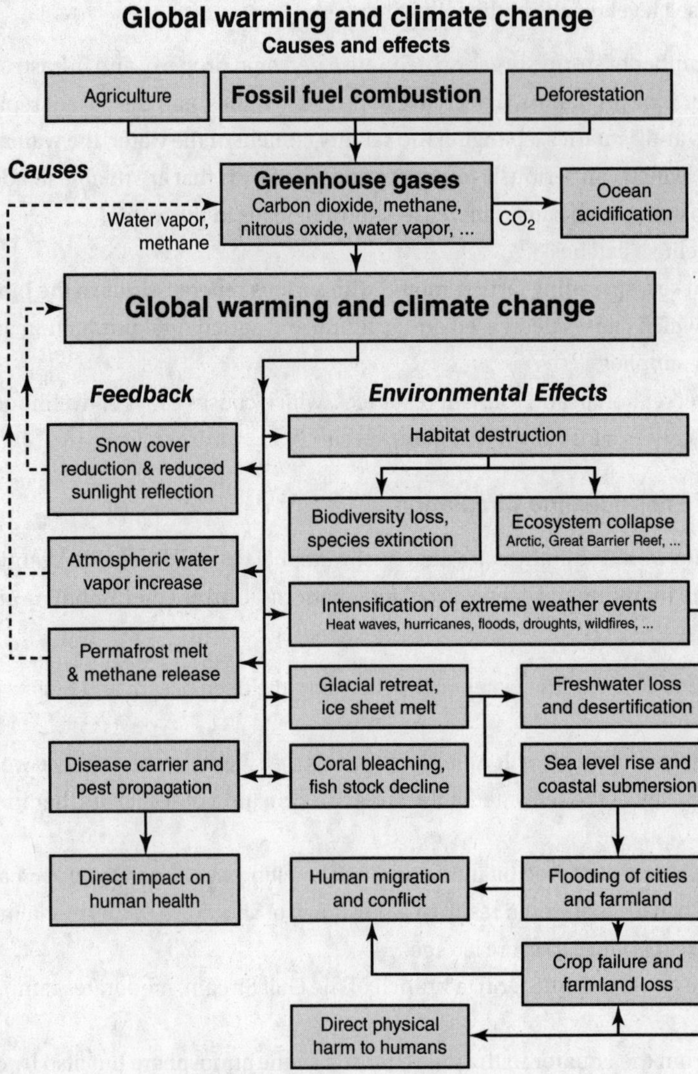

Figure 9.15 Climate change feedback loops

Reducing Climate Change

Stabilizing the current global warming crisis would require the following:

- A decrease in carbon dioxide emissions by up to 80%
- A decrease in methane emissions by 8%
- A decrease in nitrous oxide emissions by 50%

The following table lists some examples of steps that can be adopted to mitigate the effects of global climate change in various sectors.

Steps That Can Be Taken to Mitigate the Effects of Global Climate Change

Sector	Steps That Can Be Taken
Agriculture	▪ Introduce financial incentives (reduced taxes) for large agricultural corporations that institute standardized "good land management policies," such as increasing soil-quality levels, utilizing conservation tillage techniques, reducing the use of water, and reducing the amount of inorganic and nitrogen-based fertilizers.
Buildings	▪ Give subsidies to homeowners and businesses who install photovoltaic solar systems. ▪ Institute a tax on all buildings found not to be in compliance with a universal, standardized level of insulation. ▪ Require all appliances to be Energy Star® certified. ▪ Require all public and private buildings to be insulated to a universal standard.
Energy supply	▪ Introduce a "carbon tax" on nonrenewable energy sources. ▪ Reduce or eliminate government fossil fuel subsidies.
Government	▪ Support treaties and protocols that require reductions in greenhouse gas emissions.
Industry	▪ Institute sliding-scale penalties based on carbon emissions. ▪ Introduce federal tariffs on all imported products that do not meet energy-standard guidelines.
Transport	▪ Increase subsidies for public transportation. ▪ Increase vehicle-miles-per-gallon regulations for both private cars and trucks, with government penalties for miles driven using nonrenewable fuels. ▪ Tax each parking space.
Waste management	▪ Require methane capture at all landfills.

AGREEMENTS AND PROTOCOLS

MONTREAL PROTOCOL (1987)—an international treaty designed to phase out the production of substances that are responsible for ozone depletion. As a result of this international agreement, the ozone hole in Antarctica is slowly recovering.

KYOTO PROTOCOL (2005)—a plan created by the United Nations to reduce the effects of climate change. A total of 192 countries have agreed to reduce their greenhouse gas emissions by 5% to 20%. The main weaknesses of the Kyoto Protocol are that it only requires developed countries to take action and only applies to approximately 14% of the world's emissions. The United States, which produces ~25% of the world's greenhouse gases, initially signed the agreement, but later rejected it.

DOHA AMENDMENT TO THE KYOTO PROTOCOL (2012)—Participating countries have committed to reducing emissions by at least 18% below 1990 levels.

PARIS AGREEMENT (2016)—deals with greenhouse gas emissions and mitigation. The goal is to keep global temperature rise below 2°C above pre-industrial levels while each country determines its own plans to mitigate global warming. In 2017, U.S. President Donald Trump announced his intention to withdraw the United States from the Paris Agreement, leaving the United States as the only country in the world to reject the agreement. However, President Biden, as one of his first acts as president, issued an executive order to rejoin the United States back to the Paris Climate Agreement.

9.6 Ocean Warming

The amount of energy absorbed and stored by the oceans has an important role in the rise of sea level due to thermal expansion. Ocean warming accounts for ~90% of the energy accumulation from global warming, with about one-third of that heat reaching ocean depths as far as ~2,000 feet (~700 m). Ocean warming contributes to global sea-level rise by reducing the density of seawater, thus increasing its volume. Ocean warming also contributes to increased rates of freshwater ice melt of land-based glaciers and ice sheets, which further raises the sea level. At the peak of the most recent ice age, which occurred about 18,000 years ago and is known as the Last Glacial Maximum, sea level was ~300 feet (~100 m) lower than it is today (Figure 9.16).

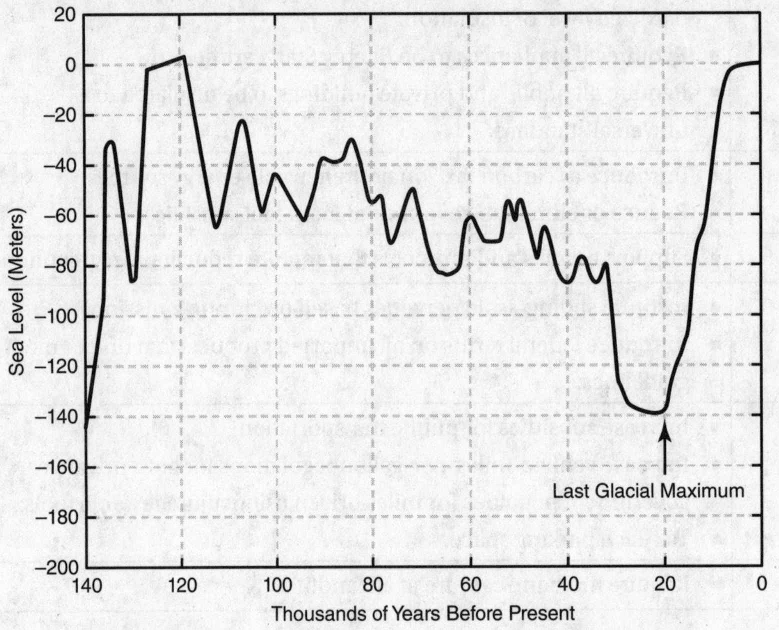

Figure 9.16 Sea-level rise 140,000 years ago to present

9.7 Ocean Acidification

Ocean acidification affects all oceans, including coastal estuaries, deltas, and waterways, as well as people's lives, as many nations rely on food from the ocean as their primary source of protein.

Ocean acidification occurs when atmospheric carbon dioxide (CO_2) reacts with seawater to form carbonic acid (H_2CO_3), which results in a reduction in the pH of ocean water over an extended period of time (oceans absorb ~30% of the CO_2 released into the atmosphere). Currently, over 7 billion tons (6.5 billion m.t.) of carbon dioxide is released into the atmosphere each year from fossil fuel combustion and industrial processes (Figure 9.17).

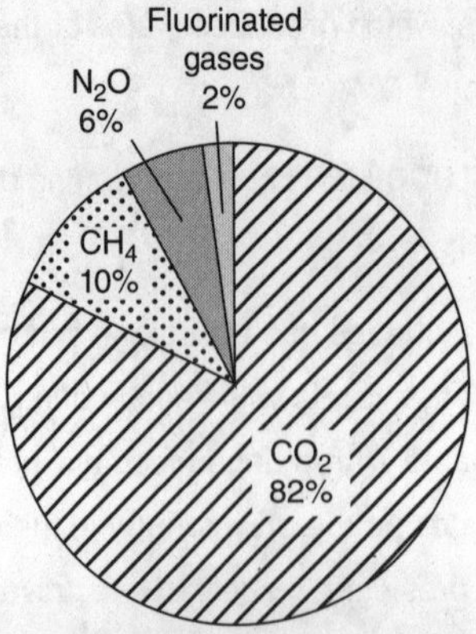

Figure 9.17 U.S. greenhouse gas emissions (2019)
Source: U.S. Environmental Protection Agency

Since the Industrial Revolution (~200 years ago), the concentration of carbon dioxide (CO_2) in the atmosphere has increased due to the burning of fossil fuels and land-use change. For example, the atmospheric concentration of CO_2 prior to the Industrial Revolution was ~280 parts per million (ppm,) and by 2005 it had climbed to ~379 ppm—a nearly 33% increase. As a result, and depending on the extent of future CO_2 emissions, it is predicted that ocean acidity could increase by 150% by 2100.

When carbon dioxide mixes with seawater, it reduces the availability of carbonate ions (CO_3^{2-}), which many marine organisms such as corals, marine plankton, and shellfish need to build their calcium carbonate ($CaCO_3$) shells.

The series of reactions below describes how hydrogen ions are produced when carbon dioxide reacts with seawater, forming bicarbonate ions (HCO_3^-).

$$CO_2 + H_2O \rightarrow H_2CO_3 \rightarrow H^+ + HCO_3^-$$

The next step involves excess hydrogen ions and free carbonate ions forming additional bicarbonate ions, at the expense of lowering the concentration of carbonate ions, which are critical for many shell-forming organisms such as clams, oysters, sea urchins, corals, and calcareous plankton (have shells composed of tiny calcium carbonate plates).

$$H^+ + CO_3^- \Rightarrow HCO_3^-$$

Combining the two reactions

$$\begin{array}{rl} & CO_2 + H_2O \rightarrow H_2CO_3 \rightarrow \cancel{H^+} + HCO_3^- \\ + & \cancel{H^+} + CO_3^- \rightarrow HCO_3^- \\ \hline & CO_2 + H_2O + CO_3^{-2} \rightleftharpoons 2HCO_3^- \end{array}$$

results in one unit of carbonate ion (CO_3^{2-}) being consumed for each unit of carbon dioxide (CO_2) added to seawater, and since the reactions run simultaneously, both pH and the availability of carbonate (CO_3^{2-}) are reduced as the atmospheric concentration of carbon dioxide rises.

Shells also dissolve in environments that are too acidic. In fact, some deep, cold ocean waters are naturally too acidic for marine calcifiers to survive, with these organisms only existing above a certain depth known as the "saturation horizon." As a result of ocean acidification, the saturation horizon is expected to move closer to the surface by 150–700 feet (50–200 m) compared to how it was during the 1800s, and the Southern and Arctic oceans, which are colder and therefore naturally more acidic, may become entirely inhospitable for these organisms. The bottom line is that many species will suffer from the loss of marine calcifiers, which provide essential food and habitats (including coral reefs) for countless marine organisms.

These changes in ocean acidity can also affect the behavior of non-calcifying organisms; e.g., the ability of some fish species to detect predators is decreased in more acidic waters.

9.8 Invasive Species

Invasive or introduced species are animals and plants that are transported to any area where they do not naturally live. The spread of nonnative species has emerged in recent years as one of the most serious threats to biodiversity and has undermined the ecological integrity of many native habitats, pushing some rare species to the edge of extinction. Some introduced species simply out-compete native plants and animals for space, food, or water. Other negative impacts include predation of non-targeted species and the disruption of food webs and/or biogeochemical cycles.

Characteristics of invasive species include the following:*

- Abundant in native range
- Broad diet
- High dispersal rates
- High genetic variability

**May not have all characteristics simultaneously*

- High rates of reproduction
- Living in close association with humans
- Long-lived
- Pioneer species (able to colonize areas after they have been disturbed)
- Short generation times
- Tolerant of a wide range of environmental conditions
- Vegetative or clonal reproduction

About 15% of the estimated 6,000 nonnative plant and animal species in the United States cause severe economic or ecological impacts and have been implicated in the decline of about 40% of the species listed for protection under the federal Endangered Species Act. Examples of introduced species include the following:

- Dutch elm disease (caused by a fungus) is transmitted to elm trees by elm bark beetles. Since 1930, the disease has spread from Ohio through most of the country, killing over half of the elm trees in the northern United States.
- European green crabs found their way into the San Francisco Bay area in 1989. They out-compete native species for food and habitat, and eat huge quantities of native shellfish, threatening commercial fisheries.
- Water hyacinth is an aquatic plant, introduced to the United States from South America. It forms dense mats, reducing sunlight for submerged plants and aquatic organisms, crowding out native aquatic plants, and clogging waterways and intake pipes.
- Zebra mussels first came to the United States from Eurasia in ship ballast water released into the Great Lakes. Since 1988, zebra mussels have spread dramatically, out-competing native species for food and habitat. Zebra mussels can attach to almost any hard surface—clogging water intake and discharge pipes, attaching themselves to boat hulls and docks, and even attaching to native mussels and crayfish.

9.9 Endangered Species

Since 1500 C.E., approximately 816 species that we know about have become extinct, 103 of them since 1800—a rate 50 times greater than the natural background rate (Figure 9.18). In the next 25 years, extinction rates are expected to rise as high as 25%. According to the Nature Conservancy, about one-third of all U.S. plant and animal species are at risk of becoming extinct. Mammals that are listed as "endangered" rose from 484 in 1996 to 520 in 2000, with endangered primates increasing from 13 to 19 different species. Endangered birds increased from 403 species to 503 species. Endangered freshwater fish more than doubled from 10 species to 24 species in four years. One-fourth of all mammals and reptiles, one-fifth of all amphibians, one-eighth of all birds, and one-sixth of all conifers are in some manner endangered to the point of extinction. Factors taken into account for being labeled "endangered" include the following:

- Breeding success rate
- Known threats
- The net increase/decrease in the population over time
- The number of animals remaining in the species

Arguments for protecting endangered species include the following:

- Maintaining genetic diversity
- Maintaining indicator species
- Maintaining keystone species
- Preserving the endangered species' aesthetic, ecological, educational, historical, recreational, and scientific value
- Preserving the yet-to-be-discovered value of certain endangered species (e.g., more than a quarter of all prescriptions written annually in the United States contain chemicals discovered in plants and animals)

The following table provides examples of some characteristics that have contributed to the endangerment or extinction of various species of animals.

Characteristics That Have Contributed to Endangerment or Extinction

Characteristic	Example
Compete for food with humans	African penguins
High infant mortality	Leatherback turtles
Highly sensitive to changes in environmental conditions	Cotton-top tamarins
Hunting for sport	Passenger pigeons, blue whales, Bengal tigers
Introduction of nonnative invasive species	Bandicoots threatened by cats that were introduced by Europeans
Limited environmental tolerance ranges	Frogs, whose eggs are sensitive to water pollution, temperature changes, and the destruction of wetlands
Limited geographic range	Pandas
Long or fixed migration routes	Salmon in the Pacific Northwest that have been driven to extinction because of dam construction, logging, and water diversion
Loss of habitat	Red wolves, whooping cranes
Low reproductive rates or specialized reproductive behavior	Whales, elephants, and orangutans (female orangutans have a maximum of three offspring during their lifetime)
Move slowly	Desert tortoises
No natural predators, which makes them vulnerable as they lack natural defensive behaviors and mechanisms	Dodo birds, Steller's sea cows, sea otters
Not able to adapt quickly	Polar bears
Possess characteristics sought after for commercial purposes	Sharks (fins), elephants (ivory), rhinoceros' horns (perceived aphrodisiac quality), gorillas (meat)
Require large amounts of territory	Tigers
Small numbers of the species, which limits genetic diversity	Tigers
Specialized feeding behaviors and/or diet	Pandas (bamboo)
Spread of disease by humans or livestock	African wild dogs (susceptible to the spread of disease by livestock)
Superstitions	Aye ayes—some people native to Madagascar believe that aye ayes bring bad luck, and therefore kill them

Maintaining Biodiversity

Biodiversity can be protected by doing the following:

- Creating and expanding wildlife sanctuaries
- Establishing breeding programs for endangered or threatened species
- Managing habitats and monitoring land use

- Properly designing and updating laws that legally protect endangered and threatened species (e.g., Endangered Species Act)
- Protecting the habitats of endangered species through private and/or governmental land trusts
- Reducing nonnative and invasive species
- Reintroducing species into suitable habitats
- Restoring compromised ecosystems

RELEVANT LAWS AND TREATIES

ENDANGERED SPECIES ACT (ESA) (1973)—designed to protect critically imperiled species, as well as their habitats, from extinction as a consequence of economic growth and development. All species of plants and animals, except pest insects, are eligible for listing as endangered or threatened. In 2019, the Endangered Species Act was changed to:

(a) allow the federal government to favor economic impacts when considering the ESA;
(b) ignore future impacts to species;
(c) reduce protections to threatened species; and
(d) relax rules for importing threatened and endangered species from trophy hunts from other countries.

CONVENTION ON INTERNATIONAL TRADE IN ENDANGERED SPECIES OF WILD FAUNA AND FLORA (CITES) (1975)—ensures that international trade in specimens of wild animals and plants does not threaten the survival of the species in the wild. It accords varying degrees of protection to more than 34,000 species of animals and plants. The International Union for Conservation of Nature (IUCN) "Red List" is a comprehensive inventory of the global conservation status of biological species requiring some level of protection.

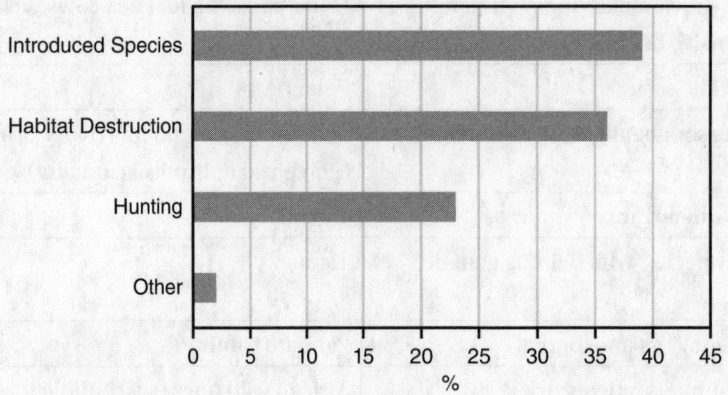

Figure 9.18 Known causes of animal extinctions since 1600 C.E.

9.10 Human Impacts on Biodiversity

Biodiversity is the variety of both animal and plant life found in a given ecosystem. The primary cause of irreversible loss in global biodiversity is the change in land use (primarily for agriculture) which is responsible for a ~30% worldwide decline.

Overfishing, overharvesting, and overhunting for food, medicines, and timber account for an additional ~20% loss, while climate change, pollution, and invasive species together account for ~25%.

The primary human impacts that have, and continue to have, a major impact on the decline of biodiversity include:

- Climate change—primarily brought about by using fossil fuels. Changes in precipitation patterns, extreme weather events, and rising global temperatures, all of which contribute to the displacement or extinction of species.
- Deforestation
- Invasive species
- Loss of habitat, primarily due to urban sprawl and agricultural development
- Overexploitation of natural resources
- Pollution; e.g., overuse of fertilizers and pesticides
- Worldwide spread of both animal and plant diseases

Ways that humans can help to slow down the loss of global biodiversity include:

- Combatting invasive species
- Creating and expanding wildlife sanctuaries
- Establishing breeding programs for endangered or threatened species
- Managing habitats and monitoring land use
- Moving from a current linear economy to a circular economy.
- Promoting sustainable resource use
- Properly designing and updating laws that legally protect endangered and threatened species (e.g., Endangered Species Act)
- Protecting and restoring natural habitats (e.g., coral reefs, forests, wetlands, etc.) through private and/or governmental land trusts
- Reducing greenhouse gasses
- Reducing non native and invasive species
- Reintroducing species into suitable habitats
- Restoring compromised ecosystems

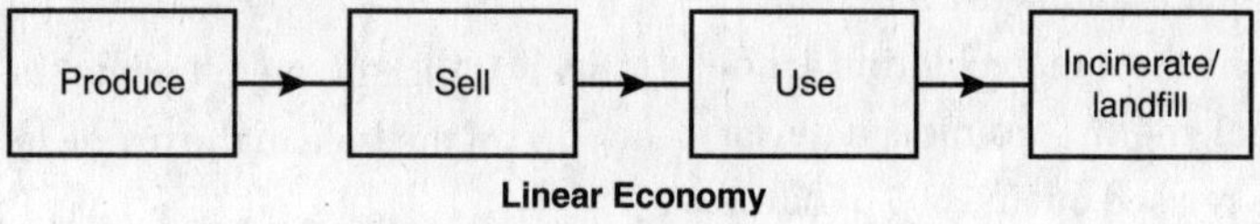

Figure 9.19 Linear Economy

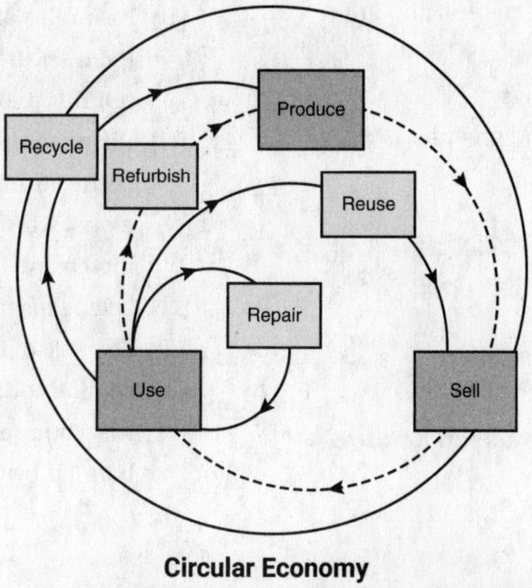

Figure 9.20 Linear vs. circular economy

Practice Questions

1. Which of the following represents the largest contributor of methane emissions to the atmosphere?

   (A) Fossil fuel production
   (B) Growing rice
   (C) Landfills and waste
   (D) Raising livestock

2. Ozone-depletion reactions that occur in the stratosphere are facilitated by

   (A) NH_3
   (B) N_2O
   (C) NO_2^-
   (D) NO_3^-

3. In addition to absorbing harmful solar rays, how do ozone molecules help to stabilize the upper atmosphere?

   (A) They create a buoyant lid on the atmosphere.
   (B) They create a warm layer of atmosphere that keeps the lower atmosphere from mixing with space.
   (C) They release heat to the surroundings.
   (D) All of the above are correct.

4. As a molecule, bromine is about 40 times more damaging to the ozone layer than chlorine. Under the Montreal Protocol, methyl bromide in developing countries was phased out in 2005. What was methyl bromide used for?

   (A) Killing pests on fruits, vegetables, and grain before export
   (B) Killing termites in buildings
   (C) Sterilizing soil in fields and greenhouses
   (D) All of the above

5. Rising sea levels due to climate change are primarily caused by

   (A) melting of glaciers and ice sheets
   (B) melting of sea ice
   (C) thermal expansion of seawater as it warms
   (D) A and C

6. What statements below are true about global warming?

   I. Average carbon dioxide levels in the atmosphere today are the highest they have ever been in Earth's history.
   II. Average global air temperatures today are the highest they have ever been in Earth's history.
   III. The increases in average global air temperatures are equally distributed across Earth's surface.

   (A) I only
   (B) II only
   (C) I and II
   (D) All of these choices are false.

7. One of the reasons why the vortex winds in the Antarctic are important in the formation of the hole in the ozone layer is

   (A) they bring in moist, warm air, which accelerates the ozone-forming process
   (B) they harbor vast quantities of N_2O
   (C) they prevent warm, ozone-rich air from mixing with cold, ozone-depleted air
   (D) they quickly mix ozone-depleted air with ozone-rich air

8. Which of the following statements about the size of the hole in the ozone layer is TRUE?

   (A) Currently, there is no "hole" in the ozone layer over the South Pole.
   (B) Examples of ozone-depleting substances include carbon dioxide, methane, and compounds that contain chlorine and bromine.
   (C) Rising temperatures in the upper layers of the atmosphere in polar regions due to global warming tends to reverse the effect of increasing the size of the "hole" in the ozone layer caused by ozone-depleting substances.
   (D) The size of the hole in the ozone layer is smaller today than it was when chlorofluorocarbons and other halocarbons were first banned by the Montreal Protocol.

9. Over the past 1 million years of Earth's history, the average trend in Earth's temperature would best be described as

(A) a general warming trend
(B) a number of fluctuations varying by a few degrees
(C) very large fluctuations, from periods of extreme heat to periods of extreme cold
(D) very little change, if any; fairly uniform and constant temperature

10. The agency responsible for the identification and listing of endangered species is the

(A) Agriculture Department
(B) Fish and Wildlife Service
(C) Forest Service
(D) National Park Service

11. A treaty that controls international trades in endangered species is known as (the)

(A) CITES
(B) Endangered Species Act
(C) International Treaty on Endangered Species
(D) Lacey Act

12. Managing game species for sustained yields would be consistent with what conservation approach?

(A) Ecosystem approach
(B) Species approach
(C) Sustainable yield approach
(D) Wildlife management approach

13. A certain insect was causing extensive damage to local crops. The farmers, who were environmentally conscious and did not want to use pesticides, decided to introduce another insect into their fields that studies have shown would prey on the pest insect. Before any action is taken, the farmers should consider (the)

(A) Law of Supply and Demand
(B) Murphy's Law
(C) Precautionary Principle
(D) Principle of Natural Balance

14. Which of the following is the biggest threat to wildlife preserves?

(A) Global warming
(B) Hunters and poachers
(C) Invasive species
(D) Tourists

15. A certain species of plant is placed on the threatened species list. Several years later it is placed on the endangered species list. This is an example of

(A) a negative feedback loop
(B) a negative-positive feedback loop
(C) a positive feedback loop
(D) a positive-negative feedback loop

16. The specific form of radiation that is largely responsible for the formation of ozone in the stratosphere is

(A) infrared
(B) UVA
(C) UVB
(D) UVC

17. Which of the following is NOT an effect caused by increased levels of ultraviolet radiation reaching Earth?

(A) Cooling of the stratosphere
(B) Increased genetic damage
(C) Reduction in plant productivity
(D) Warming of the stratosphere

18. Which of the following gases is NOT considered a greenhouse gas, responsible for global warming?

(A) CH_4
(B) H_2O
(C) N_2O
(D) SO_2

19. Which of the following effects would result from the shutdown or slowdown of the thermohaline circulation pattern?

(A) A colder Scandinavia and Great Britain
(B) A saltier ocean
(C) A warmer Scandinavia and Great Britain
(D) Freshwater fish moving into the open ocean

20. The largest number of species are being exterminated per year in
    (A) deserts
    (B) forests
    (C) grasslands
    (D) tropical rainforests

Free-Response Question

By Brian Palm, Catholic Memorial School, West Roxbury, MA
A.B., Dartmouth College, Hanover, NH
MSc, Oxford University, UK

(a) An April 2015 article in *Nature* described the likelihood of the "methane time bomb" that would push our climate situation to a point of irreversible change. This example of a positive feedback loop is one of many that could occur in a "climate-changed" world. Identify AND describe ONE other positive feedback mechanism in an ecological system.

(b) CCS stands for Carbon Capture and Storage, a process that utilizes a number of technologies to capture and then store (hopefully for a long time) the carbon dioxide produced through coal-fired electricity generation. One of the major concerns about CCS is its cost. In 2009, a Harvard University lab estimated that it would cost $150 to capture and sequester one ton of CO_2. If a single 500 MW coal-fired plant generates about 3 million tons of CO_2 each year, calculate the additional annual cost burden for a single coal-fired plant of this size.

(c) IDENTIFY and EXPLAIN one economic consequence of climate change.

(d) Once in the atmosphere, the increased concentration of carbon dioxide forces some of this gas into solution in oceans around the world. Using a chemical equation, IDENTIFY what happens when carbon dioxide goes into solution.

(e) DESCRIBE a protocol for an experiment intended to assess the impact of ocean acidification on shellfish. Assume that you have been given the following materials:
- five test tubes
- distilled water
- four different concentrations of carbonic acid
- five pieces of seashell
- a scale

Answer Explanations

1. **(A)** Methane levels have more than doubled over the last 150 years. There are both natural and human sources of methane emissions. Natural sources, which include wetlands, termite colonies, and the oceans, make up 36% of methane emissions, while human-related sources create the remaining majority (64%) of methane emissions. Since the Industrial Revolution, human sources of methane emissions have been growing. Fossil fuel production and intensive livestock farming have caused the current increase in methane levels. Together, these two sources are responsible for 60% of all human methane emissions. Other human sources of methane include landfills and waste (16%), biomass burning (11%), rice agriculture (9%), and biofuels (4%).

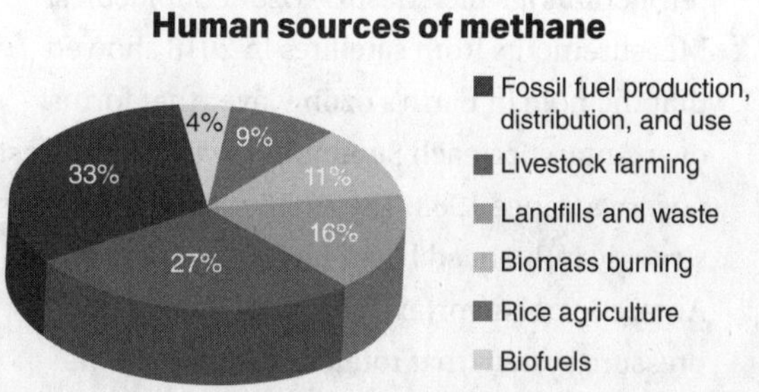

2. **(B)** Nitrous oxide (N_2O) is a greenhouse gas and currently contributes ~4% of the greenhouse effect but is rapidly increasing at about 0.2% per year and is expected to continue to be the largest anthropogenic emission of an O_3-destroying compound in the foreseeable future. In nature, bacteria and fungi in the soil and the oceans break down nitrogen-containing compounds (e.g., fertilizers), releasing nitrous oxide (N_2O), which accounts for ~60% of all naturally produced emissions. Anthropogenic sources of nitrous oxide include the burning of biomass (e.g., deforestation) and fossil fuels. In the stratosphere, nitrous oxide plays a role in the destruction of the ozone layer. The various steps involved in the process are as follows:

   - In the stratosphere, N_2O first reacts with the free oxygen atoms in the presence of UV light to form nitric oxide (NO):

   $$N_2O \text{ (nitrous oxide)} + O \xrightarrow{uv} 2NO \text{ (nitric oxide)}$$

   - Then, nitric oxide reacts with the ozone (decreasing its concentration) to produce nitrogen dioxide and oxygen:

   $$NO + O_3 \rightarrow NO_2 + O_2$$

3. **(D)** Solar energy that is absorbed by ozone molecules and is partly turned into heat creates a warm region in the stratosphere, which creates a stable air mass that resists sinking and mixing with the lower atmosphere, effectively forming a barrier. In the ozone layer, temperature increases with height, creating a stable and buoyant air mass that keeps an effective lid on the lower atmosphere.

4. **(D)** Methyl bromide (CH_3Br), also known as bromomethane, originates from both natural and human sources. In the ocean, marine organisms are estimated to produce approximately 62,000 tons (approximately 56,000 m.t.) annually. Methyl bromide is banned by the Montreal Protocol, but it has continued to receive "critical use" exemptions from the U.S. Environmental Protection Agency to be used as a soil fumigant (primarily in strawberry fields) to prevent spoilage during storage. Both chemical and nonchemical alternatives to methyl bromide exist. For example, steam sterilization of soil is a viable alternative to using chemical fumigants for certain pests and soil types. Other nonchemical alternatives include integrated pest management techniques, pheromones, electrocution, and light traps.

5. **(D)** Global mean sea level has risen about 8 to 9 inches (21–24 centimeters) since 1880, with about a third of that occurring in the last 25 years. The rising seawater level is primarily due to (1) a combination of meltwater from glaciers and ice sheets; and (2) thermal expansion of seawater as it warms. Melting of icebergs and sea ice does not result in a rise in sea level as they are already in the water. However, melting sea ice does contribute to climate change since white sea ice reflects sunlight. As the sea ice melts, the now dark open ocean absorbs sunlight and heats up, raising global temperatures, which in turn cause glaciers and ice sheets on land to melt further. By the end of the 21st century, global mean sea level is likely

to rise at least one foot (0.3 meters) above 2000 levels, even if greenhouse gas emissions are significantly diminished.

6. **(D)** Earth's primitive forests appeared around 300 million years ago during the Carboniferous Period (359 to 299 million years ago during the late Paleozoic Era). Before then, the atmosphere held far more CO_2, but concentrations declined through the Carboniferous Period as plants flourished and absorbed more and more CO_2 from the atmosphere through photosynthesis, creating carbon sinks (coal, oil, etc.). During the Carboniferous Period, the atmosphere became greatly depleted of CO_2 (declining from 2,500 ppm to 350 ppm) so that by the end of the period, the atmosphere was less favorable to plant life, and plant growth slowed. Today, CO_2 concentrations are approximately 380 ppm (0.038% of our atmosphere). The Arctic is feeling the effects of increases in average global air temperatures the most, with average temperatures in Alaska, western Canada, and eastern Russia rising at twice the rate of the global average.

   The Early Eocene Period (54–48 million years ago) is one period in the geologic past that stands out as distinctly warmer than today. The Early Eocene was characterized by atmospheric carbon dioxide levels as high as 2,000 parts per million, which is theorized to be primarily due to increased volcanic activity during this period, compared to today's atmospheric CO_2 concentration of ~415 parts per million. Temperatures during this period can be reconstructed from geochemical measurements of ocean sediments and from vegetation types preserved on land, indicating that the average surface temperature was 9 to 14°C higher than today. The temperature was so warm that Antarctica was ice-free, palm trees grew at high latitudes, and cold-blooded animals, such as crocodiles, lived in the Arctic.

7. **(C)** Ozone depletion follows an annual cycle that corresponds to the amount of light that reaches the Antarctic. The cycle begins every year around June, when vortex winds develop in the Antarctic. Cold temperatures produced by these winds create polar stratospheric clouds that capture floating chlorofluorocarbons (CFCs) and other ozone-depleting compounds. For the next two months, a reaction occurs on the cloud surface that frees the chlorine in the CFCs but keeps it contained within the vortex. In September, sunlight returns to the Antarctic and triggers a chemical reaction, causing chlorine to convert ozone to oxygen gas. November brings a breakdown in the vortex and allows the ozone-rich air to combine with the thinning ozone. Wind currents carry this mixture over the Southern Hemisphere.

8. **(C)** First detected in 1985, the Antarctic ozone (O_3) hole forms during the Southern Hemisphere's late winter as the returning sun's rays catalyze reactions involving man-made, chemically active chemicals containing chlorine and bromine (e.g., chlorofluorocarbons, which were used as refrigerants), which destroy ozone molecules. Measurements from satellites in 2018 showed that the hole in Earth's ozone layer that forms over Antarctica each September was the smallest observed since 1988. The smaller ozone hole was strongly influenced by an unstable and warmer Antarctic polar vortex—the stratospheric low-pressure system that rotates clockwise in the atmosphere above Antarctica and helps minimize polar stratospheric cloud formation in the lower stratosphere. The formation and persistence of these clouds are important first steps leading to the chlorine- and bromine-catalyzed reactions that destroy ozone. The average area of the ozone hole maximums observed since 1991 has been approximately 10 million square miles. Although warmer-than-average stratospheric weather conditions have reduced ozone depletion during the past two years, the current ozone hole is still large because the levels of ozone-depleting substances continue to remain high enough to produce significant ozone loss; however, it is expected that the Antarctic ozone hole will recover back to 1980 levels by around 2070. Carbon dioxide does not directly contribute to ozone depletion.

9. **(A)** Current climatic warming is occurring much more rapidly than past warming events. As Earth moved out of ice ages over the past million years, the global average temperature rose 4–7°C over the last 5,000 years. In the past 100 years, the

temperature has risen 0.7°C, approximately ten times faster than the average rate, with current models predicting that Earth will warm 2–6°C in the next century, a rate approximately 20 times faster than normal.

10. **(B)** The Fish and Wildlife Service, in the Department of the Interior, and the National Oceanic and Atmospheric Administration (NOAA), in the Department of Commerce, share responsibility for the administration of the Endangered Species Act. An endangered species is one that is in danger of extinction throughout all or a significant portion of its range and is likely to become extinct in the foreseeable future.

11. **(A)** CITES (the Convention on International Trade in Endangered Species) is an international agreement between governments. Its aim is to ensure that international trade in specimens of wild animals and plants does not threaten their survival.

12. **(D)** The strength of the traditional wildlife management approach is that it explicitly uses and enhances natural processes to perpetuate populations.

13. **(C)** The Precautionary Principle states that if the consequences of an action are unknown but are judged to have some potential for major or irreversible negative consequences, that action should be avoided. The concept includes risk prevention, cost effectiveness, ethical responsibilities toward maintaining the integrity of natural biological systems, and risk assessment. It is not the risk that must be avoided, but the potential risk that must be prevented.

14. **(A)** Global warming is an issue that universally affects all ecosystems and needs to be addressed on a global scale, which requires universal and international cooperation (something not always easy to achieve). All choices but (C) can be dealt with on a local level through specific enforcement and management practices.

15. **(C)** A positive feedback loop occurs in a situation in which a change in a certain direction (toward extinction in this case) causes the system to change in the same direction.

16. **(D)** Ozone is produced by oxygen and sunlight in the UVC wavelength range (< 240 nm). In the atmosphere, this reaction works only at higher altitudes where there is adequate high-energy UV penetration. The atomic oxygen so formed can combine with oxygen gas (O_2) to form ozone (O_3).

17. **(D)** The stratosphere has been cooling over the past three decades. The stratosphere contains the ozone layer, which absorbs sunlight and heats the stratosphere. This long-term cooling trend is generally accepted to result from the loss of the ozone layer caused by human-made influences. However, the cooling trend is not uniform like ozone loss but, rather, is broken into a series of jumps or discontinuities most likely associated with major volcanic eruptions that inject aerosols into the stratosphere. The aerosols also absorb sunlight and heat from the stratosphere, thus temporarily offsetting the cooling trend from ozone loss.

18. **(D)** Sulfur dioxide (SO_2) contributes to acid rain, not global warming.

19. **(A)** There is speculation that global warming could melt glaciers in Greenland, increasing the amount of freshwater in the North Sea. This disruption in balance between saltwater and freshwater could theoretically slow down or shut down the thermohaline circulation that is responsible for the North Atlantic Drift (a section of the Gulf Stream), which currently stabilizes temperatures in Great Britain and Scandinavia.

20. **(D)** The tropics contain the highest biodiversity found anywhere on Earth. Deforestation in the tropics, in order to convert the land for agricultural purposes and cattle grazing, is occurring at unprecedented levels.

Free-Response

10 Total Points Possible

**Note: Even though it is possible to earn 12 points, no more than 10 points are assigned.*

(a) *Maximum 2 points total: 1 point for identifying a viable positive feedback mechanism and 1 point for correctly describing the feedback loop. Note that the explanation must describe/correlate to the feedback loop that was identified.*

An April 2015 article in Nature *described the likelihood of the "methane time bomb" that would push our climate situation to a point of irreversible change. This example of a positive feedback loop is one of many that could occur in a "climate-changed" world. Identify AND describe ONE other positive feedback mechanism in an ecological system.*

A positive feedback loop/mechanism refers to changes within a system in which subsequent events are enhanced, amplified, or made larger. The result of such a mechanism is the movement of that system away from an equilibrium, or steady, state, making it less stable.

Students must link their identified positive feedback loop with an explanation that refers correctly to that mechanism.

Positive Feedback Loop Example	Associated Explanation of the Mechanism
Warmer temperatures = more evaporation = more water vapor = warmer temperatures	As the climate system warms, evaporation rates (oceans/lakes/soil/rivers) will increase. An increase in evaporation causes greater amounts of water vapor. Water vapor is a greenhouse gas that will then cause more warming. The output of more water vapor causes more warming, which causes more water vapor.
Warmer temperatures = loss of sea/land ice (Arctic and Antarctica) = lower albedo = less reflected light (greater conversion of light energy into infrared heat) = warmer temperatures	As the climate system warms, ice will melt (polar and other). As the ice melts, Earth's overall average albedo gets lower, as the surface color is converted from areas that were once white (ice) to areas that are darker (soil/vegetation/water). These areas will absorb light energy, converting it into IR (infrared) instead of reflecting that light energy. The increase in IR will cause further warming, which will melt more ice.
Warmer temperatures = melting of permafrost = release of methane trapped in permafrost = warming due to more methane (methane is a greenhouse gas)	As the climate system warms, permafrost melts in the higher latitudes (near the poles/Canada/Siberia/Alaska). When permafrost melts, the methane gas bubbles that were frozen in the permafrost will melt. The release of methane causes further warming because methane is a potent greenhouse gas.

(b) *Maximum 2 points total: 1 point for the correct calculation setup and 1 point for the correct answer.*

CCS stands for Carbon Capture and Storage, a process that utilizes a number of technologies to capture and then store (hopefully for a long time) the carbon dioxide produced through coal-fired electricity generation.

One of the major concerns about CCS is its cost. In 2009, a Harvard University lab estimated that it would cost \$150 to capture and sequester one ton of CO_2. If a single 500 MW coal-fired plant generates about 3 million tons of CO_2 each year, calculate the additional annual cost burden for a single coal-fired plant of this size.

Responses must include a setup that is similar to the one below.

While units are not required in this setup, the numbers should be consistent, and they should demonstrate the correct mathematical operation (in this case, the fact that \$150 is multiplied by 3,000,000 [or 3×10^6]. The correct answer is \$450,000,000. This can also be displayed as $\$4.5 \times 10^6$ or as \$450 million. Full credit is given for any variations, including $\$45 \times 10^7$ or 450×10^6, though one of the first two formats is preferred.

$$\frac{\$150.00}{\text{\sout{tons of CO}}_2} \times \frac{\text{\sout{3,000,000 tons of CO}}_2}{\text{year}} = \text{\sout{\$450,000,000/year}}$$

$$= \$4.5 \times 10^8/\text{year}$$

(c) *Maximum 2 points total: 1 point for IDENTIFYING and 1 point for EXPLAINING one <u>economic</u> consequence.*

IDENTIFY and EXPLAIN one economic consequence of climate change.

Examples and explanations will vary for each of these consequences, recognizing that consequences do not have to be necessarily positive or negative, but they must correctly describe an ECONOMIC impact (i.e., the economic consequence must be related to relevant topics like jobs or money spent/earned).

<u>Economic consequences:</u>

- Due to sea-level rise resulting from climate change and loss of land ice/thermal expansion, coastal residents and businesses may need to redirect funds to pay for the construction of ocean defense mechanisms, including barriers, new roads, and other modifications to infrastructure.
- Due to increases in drought frequency, yields from agriculture will decrease, putting farmers at risk financially.
- Losses in fishing jobs and wages will result from reduction in species habitats/range.

(d) *Once in the atmosphere, the increased concentration of carbon dioxide forces some of this gas into solution in oceans around the world. Using a chemical equation, IDENTIFY what happens when carbon dioxide goes into solution.*

Maximum 3 points total: 1 point for identifying the reactants as CO_2 and H_2O. 1 point for identifying a form of bicarbonate OR carbonic acid. 1 BONUS point can be earned for correctly writing the following balanced equation, showing both the bicarbonate ion and the hydrogen or hydronium ion.

$$CO_2 + H_2O \rightarrow H^+ + HCO_3^-$$

(e) *Maximum 3 points total: 1 point for describing a setup with varying acid levels AND a control (distilled water). 1 point for describing a reasonable means to quantify changes in the shells after exposure to the varied solutions. 1 BONUS point may be earned for recognizing the need to control other variables.*

DESCRIBE a protocol for an experiment intended to assess the impact of ocean acidification on shellfish. Assume that you have been given the following materials:

- *5 test tubes*
- *distilled water*
- *4 different concentrations of carbonic acid*
- *5 pieces of seashell*
- *a scale*

While responses may vary, a typical experimental setup would include five test tubes, each containing the same amount of solution. (Note: the effort to describe a consistent volume of solution demonstrates appreciation of controlling variables.)

<u>Sample Setup</u>:

Test Tube A might have 20 ml of distilled water (acting as the control).
Test Tube B would have 20 ml of the weakest acid (highest pH).
Test Tube C would have 20 ml of the next most concentrated acid.
Test Tube D would have 20 ml of the next most concentrated acid.
Test Tube E would have the most concentrated acid.

Each piece of shell should be massed PRIOR to being exposed to the solutions. A reasonable amount of time should pass with the shells in each of the solutions and the shells should be taken out, dried, and massed for a second time *(recognition that they should be exposed for the same amount of time demonstrates appreciation of controlling variables)*. Lost mass could be calculated to determine if different acid concentrations weaken or deteriorate shell structure.

Responses might also include a less quantitative measure, such as a visual scale determined by how transparent or brittle the shell was at the end of the exposure compared to its starting condition.

Practice Tests

ANSWER SHEET
Practice Test 1

1. Ⓐ Ⓑ Ⓒ Ⓓ	21. Ⓐ Ⓑ Ⓒ Ⓓ	41. Ⓐ Ⓑ Ⓒ Ⓓ	61. Ⓐ Ⓑ Ⓒ Ⓓ
2. Ⓐ Ⓑ Ⓒ Ⓓ	22. Ⓐ Ⓑ Ⓒ Ⓓ	42. Ⓐ Ⓑ Ⓒ Ⓓ	62. Ⓐ Ⓑ Ⓒ Ⓓ
3. Ⓐ Ⓑ Ⓒ Ⓓ	23. Ⓐ Ⓑ Ⓒ Ⓓ	43. Ⓐ Ⓑ Ⓒ Ⓓ	63. Ⓐ Ⓑ Ⓒ Ⓓ
4. Ⓐ Ⓑ Ⓒ Ⓓ	24. Ⓐ Ⓑ Ⓒ Ⓓ	44. Ⓐ Ⓑ Ⓒ Ⓓ	64. Ⓐ Ⓑ Ⓒ Ⓓ
5. Ⓐ Ⓑ Ⓒ Ⓓ	25. Ⓐ Ⓑ Ⓒ Ⓓ	45. Ⓐ Ⓑ Ⓒ Ⓓ	65. Ⓐ Ⓑ Ⓒ Ⓓ
6. Ⓐ Ⓑ Ⓒ Ⓓ	26. Ⓐ Ⓑ Ⓒ Ⓓ	46. Ⓐ Ⓑ Ⓒ Ⓓ	66. Ⓐ Ⓑ Ⓒ Ⓓ
7. Ⓐ Ⓑ Ⓒ Ⓓ	27. Ⓐ Ⓑ Ⓒ Ⓓ	47. Ⓐ Ⓑ Ⓒ Ⓓ	67. Ⓐ Ⓑ Ⓒ Ⓓ
8. Ⓐ Ⓑ Ⓒ Ⓓ	28. Ⓐ Ⓑ Ⓒ Ⓓ	48. Ⓐ Ⓑ Ⓒ Ⓓ	68. Ⓐ Ⓑ Ⓒ Ⓓ
9. Ⓐ Ⓑ Ⓒ Ⓓ	29. Ⓐ Ⓑ Ⓒ Ⓓ	49. Ⓐ Ⓑ Ⓒ Ⓓ	69. Ⓐ Ⓑ Ⓒ Ⓓ
10. Ⓐ Ⓑ Ⓒ Ⓓ	30. Ⓐ Ⓑ Ⓒ Ⓓ	50. Ⓐ Ⓑ Ⓒ Ⓓ	70. Ⓐ Ⓑ Ⓒ Ⓓ
11. Ⓐ Ⓑ Ⓒ Ⓓ	31. Ⓐ Ⓑ Ⓒ Ⓓ	51. Ⓐ Ⓑ Ⓒ Ⓓ	71. Ⓐ Ⓑ Ⓒ Ⓓ
12. Ⓐ Ⓑ Ⓒ Ⓓ	32. Ⓐ Ⓑ Ⓒ Ⓓ	52. Ⓐ Ⓑ Ⓒ Ⓓ	72. Ⓐ Ⓑ Ⓒ Ⓓ
13. Ⓐ Ⓑ Ⓒ Ⓓ	33. Ⓐ Ⓑ Ⓒ Ⓓ	53. Ⓐ Ⓑ Ⓒ Ⓓ	73. Ⓐ Ⓑ Ⓒ Ⓓ
14. Ⓐ Ⓑ Ⓒ Ⓓ	34. Ⓐ Ⓑ Ⓒ Ⓓ	54. Ⓐ Ⓑ Ⓒ Ⓓ	74. Ⓐ Ⓑ Ⓒ Ⓓ
15. Ⓐ Ⓑ Ⓒ Ⓓ	35. Ⓐ Ⓑ Ⓒ Ⓓ	55. Ⓐ Ⓑ Ⓒ Ⓓ	75. Ⓐ Ⓑ Ⓒ Ⓓ
16. Ⓐ Ⓑ Ⓒ Ⓓ	36. Ⓐ Ⓑ Ⓒ Ⓓ	56. Ⓐ Ⓑ Ⓒ Ⓓ	76. Ⓐ Ⓑ Ⓒ Ⓓ
17. Ⓐ Ⓑ Ⓒ Ⓓ	37. Ⓐ Ⓑ Ⓒ Ⓓ	57. Ⓐ Ⓑ Ⓒ Ⓓ	77. Ⓐ Ⓑ Ⓒ Ⓓ
18. Ⓐ Ⓑ Ⓒ Ⓓ	38. Ⓐ Ⓑ Ⓒ Ⓓ	58. Ⓐ Ⓑ Ⓒ Ⓓ	78. Ⓐ Ⓑ Ⓒ Ⓓ
19. Ⓐ Ⓑ Ⓒ Ⓓ	39. Ⓐ Ⓑ Ⓒ Ⓓ	59. Ⓐ Ⓑ Ⓒ Ⓓ	79. Ⓐ Ⓑ Ⓒ Ⓓ
20. Ⓐ Ⓑ Ⓒ Ⓓ	40. Ⓐ Ⓑ Ⓒ Ⓓ	60. Ⓐ Ⓑ Ⓒ Ⓓ	80. Ⓐ Ⓑ Ⓒ Ⓓ

Practice Test 1

Section I (Multiple-Choice Questions)

TIME: 90 MINUTES
80 QUESTIONS
60% OF TOTAL GRADE

DIRECTIONS: For each question or statement, select the one lettered choice that is the best answer, and fill in the corresponding circle on the answer sheet.

1. An APES class went on a field trip and took a soil sample. The sample was analyzed back at school and the results presented below.

Poor nutrient-holding capacity
Good water infiltration capacity
Poor water-holding capacity
Good aeration properties

The site that they most likely visited was

(A) a coniferous forest
(B) a sandy beach
(C) a temperate grassland
(D) a tropical rainforest

Refer to the following diagram to answer Question 2.

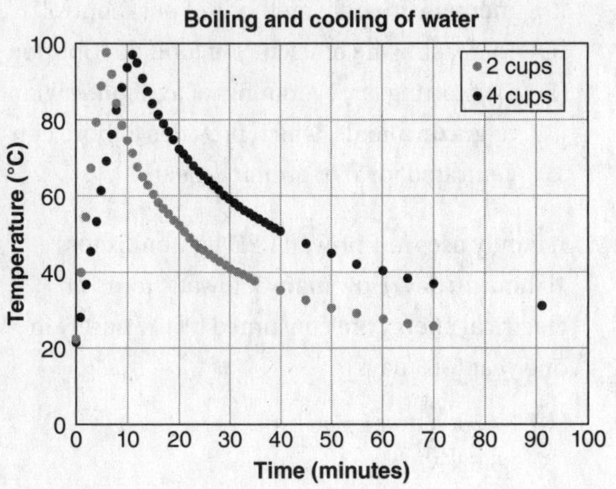

2. An APES class was studying how the proximity to oceans affects climate. They began their experiment by filling a beaker with two cups of room-temperature water, placing it on a hot plate set at medium heat, and taking temperature readings every minute. When the water began to boil, they removed the beaker and continued taking temperature readings until the temperature returned to its original temperature. They repeated the experiment with four cups of water, keeping everything else the same. According to the results above, which of the following statements is FALSE?

(A) Boiling water has the same specific heat capacity as cold water.
(B) It takes the same amount of time to boil two cups of water as it does four cups of water.
(C) The rate of cooling for two cups of water is the same as the rate of cooling for four cups of water.
(D) It takes twice as long to heat four cups of water to the boiling point as it takes for two cups of water.

3. Which of the following descriptions of air temperature from the ground up best depicts the existence of an inversion layer?

(A) Cool, cooler, cold
(B) Cool, warm, cool
(C) Hot, warm, cool
(D) Warm, cool, cooler

Refer to the diagram of the water cycle below to answer Question 4.

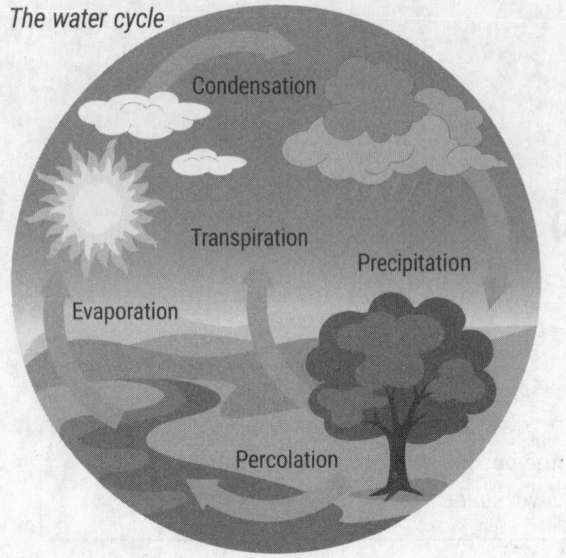

4. Which of the following relationships is correct?

(A) Precipitation = (evaporation – runoff) – change in water storage in soil or bedrock
(B) Precipitation = (runoff – evaporation) + change in water storage in soil or bedrock
(C) Precipitation = (evaporation + runoff) + change in water storage in soil or bedrock
(D) Precipitation = (evaporation × runoff) + change in water storage in soil or bedrock

Refer to the following report of an APES field trip in order to answer Question 5.

An APES class was doing a field study of a lake. The class determined the amount of dissolved oxygen was 8 mg O_2 per liter in the lake at a depth of two feet through a titration technique. Then the class filled and sealed two clear glass bottles with lake water from the same location and depth, labeling one bottle "light" and the other bottle "dark." The "dark" bottle was wrapped completely with several layers of foil. Both bottles were then lowered three feet into the lake at the same location where the initial sample was taken. After one hour, the bottles were retrieved and, again through titration, the amount of oxygen in the water in the "light" bottle was determined to be 10 mg O_2 per liter, while the amount of oxygen in the water in the "dark" bottle was determined to be 5 mg O_2 per liter.

5. Calculate the respiration rate (RR), gross primary productivity (GPP) and net primary productivity (NPP).

(A) RR = 2 mg/L/hr; GPP = 3 mg/L/hr; NPP = 5 mg/L/hr
(B) RR = 3 mg/L/hr; GPP = 5 mg/L/hr; NPP = 2 mg/L/hr
(C) RR = 2 mg/L/hr; GPP = 5 mg/L/hr; NPP = 3 mg/L/hr
(D) RR = 5 mg/L/hr; GPP = 3 mg/L/hr; NPP = 2 mg/L/hr

Refer to the following diagram to answer Question 6.

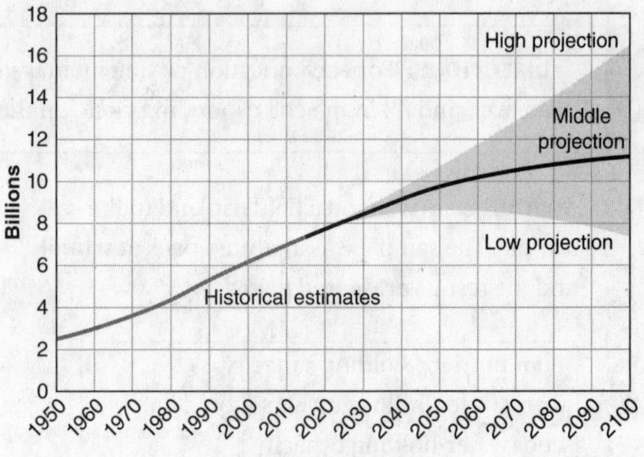

Source: United Nations

6. The world population is expected to grow from 7.6 billion in 2018 to 12 billion by 2100. Which of the following would NOT be a viable strategy for ensuring adequate nutrition for 12 billion people?

(A) Developing new hybrid crops that are more drought- and pest-resistant
(B) Expanding the use of non-toxic and environmentally friendly methods of pest control
(C) Increasing the efficiency of food distribution
(D) Converting large amounts of agricultural land to grazing lands, which provides a more concentrated form of protein (meat)

7. A family uses ten 10-watt LED lightbulbs for 10 hours a day. How many kilowatt-hours of electrical energy are consumed by the family in one year (365 days)?

(A) 3.65 kWh/yr
(B) 36.5 kWh/yr
(C) 365 kWh/yr
(D) 3,650 kWh/yr

Refer to the following diagram to answer Question 8.

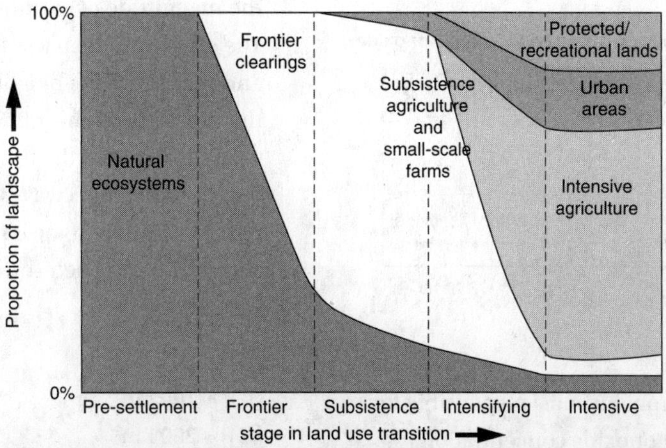

Source: Land-use transitions through history (Foley et al. 2005)

8. Which of the following choices lists land use in order from the smallest collective area on Earth to the largest?

    (A) Agricultural < cities < forests
    (B) Agricultural < forests < cities
    (C) Cities < forests < agricultural
    (D) Forests < agricultural < cities

9. The major sink for phosphorus is

    (A) atmospheric gases
    (B) marine sediments
    (C) plants
    (D) seawater

Refer to the following newspaper article to answer Question 10.

Barronsville Gazette

FOUR LOCAL BUSINESSMEN INDICTED ON FEDERAL CHARGES OF OPERATING AN ILLEGAL WASTE DISPOSAL FACILITY

Monday, October 5: Four local residents were named in a federal indictment today at the Federal Courthouse in Dallas, Texas. The four, John, Steve, Dan, and Linda Smith, were arrested on charges of allowing their customers to illegally dispose of solid hazardous wastes on their unlicensed property. Investigators who had been working on this case over the last two months discovered piles of open waste scattered across their farm, including exposed containers of asbestos and mercury—deemed "hazardous" by the Environmental Protection Agency. In a statement to the Gazette, Dan Smith replied to the charges, "We've done nothing wrong—it's just garbage that nobody wants and we're helping them to get rid of it and besides, we're making good money by doing it." The four defendants posted bail and are awaiting trial.

10. Which federal act listed below were the Smiths arrested on?

    (A) Clean Water Act (CWA)
    (B) Comprehensive Environmental Response, Compensation, and Liability Act (CERCLA)
    (C) Resource Conservation and Recovery Act (RCRA)
    (D) Toxic Substances Control Act (TOSCA)

Refer to the following graph to answer Question 11.

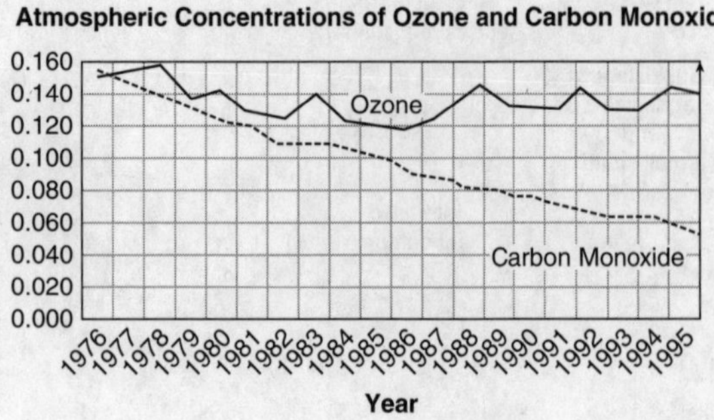

11. The factor that was probably most responsible for the trend in carbon monoxide concentration in the atmosphere from 1976 to 1995 was

(A) catalytic converters
(B) clean coal technology
(C) more efficient mufflers
(D) scrubbers on smokestacks

Refer to the seismogram below to answer Question 12.

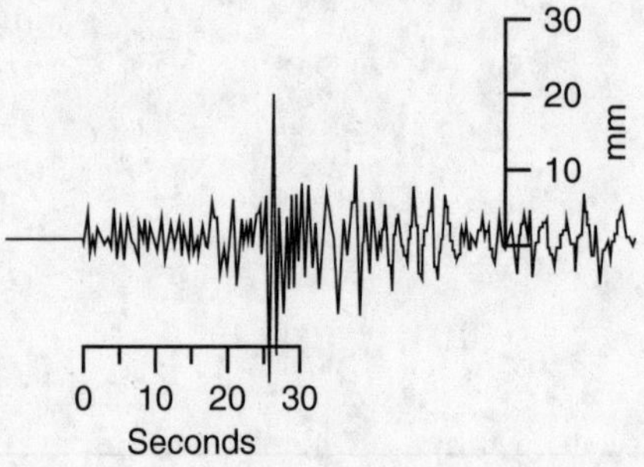

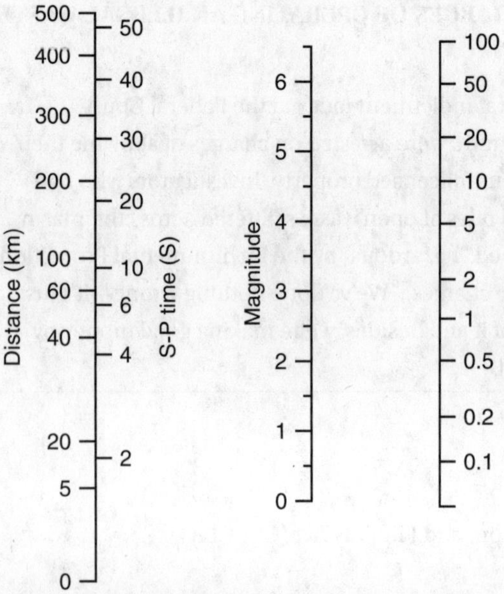

12. An APES class spent a class period determining the magnitude of earthquakes from seismograms. The class was given the seismogram above and measured the height from the baseline to the top of the highest P wave as 20 mm. They then determined the time interval between the beginning of the S waves and the beginning of the first P waves to be 25 seconds. Using the chart above, the class determined the distance to the epicenter to be

(A) ~20 km
(B) ~100 km
(C) ~200 km
(D) ~400 km

13. Which of the following are contributing factors in determining a lake's natural pH and its ability to resist the effects of acid deposition?

I. The types of soil, trees, and decaying leaves that exist in the watershed that drains into the lake.
II. The types of rocks and soil that exist in the watershed that drains into the lake.
III. The types of fish and plants that live in the water of the lake.

(A) I
(B) II
(C) III
(D) I and II

Refer to the graph below to answer Question 14.

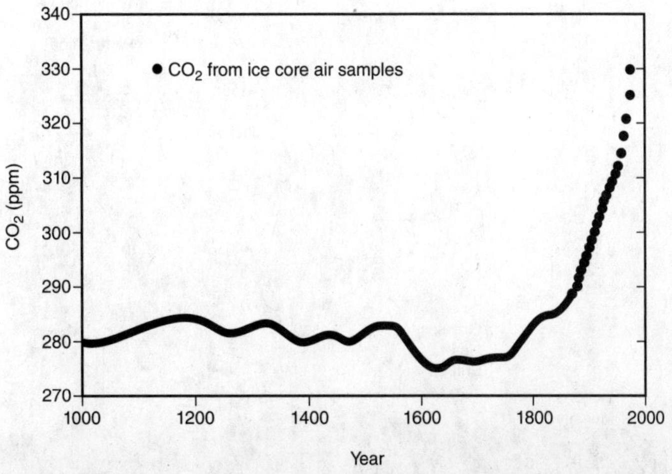

14. A scientific team took ice core samples to measure the concentration of carbon dioxide that was present in the atmosphere for the last 1,000 years. Their findings are shown above. What was the approximate net change in CO_2 concentration that occurred from the year 1000 to the year 2000?

(A) −50 ppm
(B) 50 ppm
(C) 50 million ppm
(D) 330 ppm

Refer to the following diagram to help answer Question 15.

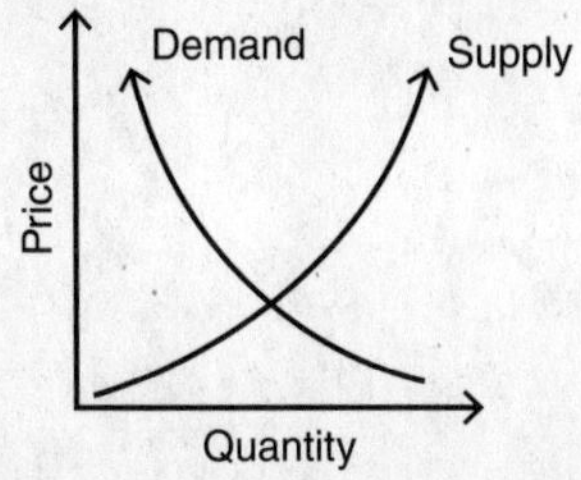

15. Which of the following does NOT affect the price of crude oil?

(A) The demand for oil
(B) The supply of oil
(C) The worldwide refining capacity
(D) All of the above affect the price of crude oil

Refer to the following diagram to answer Question 16.

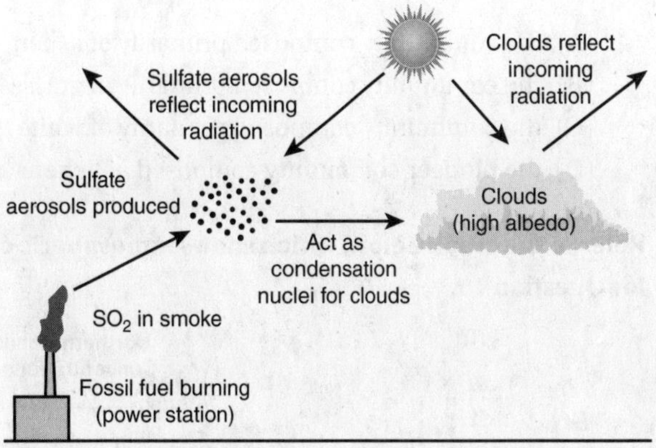

16. The largest natural source of sulfur in the atmosphere is:

(A) carbon disulfide (CS_2)
(B) sulfur dioxide (SO_2)
(C) hydrogen sulfide (H_2S)
(D) sulfate aerosols (SO_4^{2-})

Refer to the following diagram to answer Question 17.

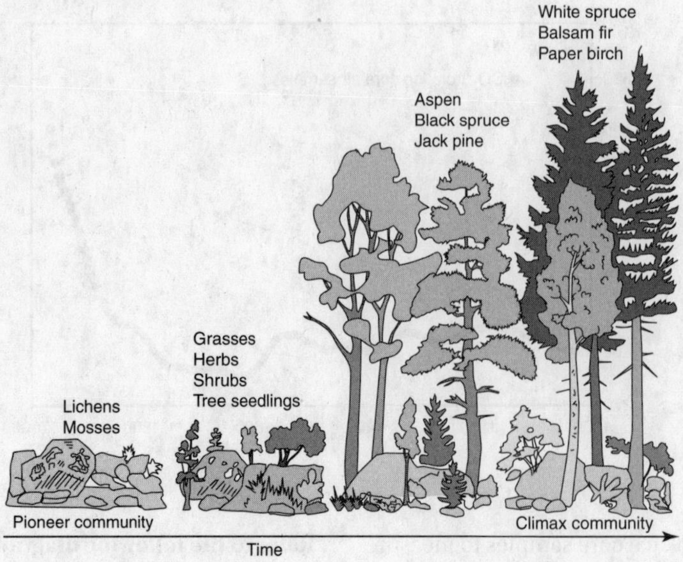

17. The greatest species diversity would be found in

(A) the community composed primarily of aspen, black spruce, and jack pine
(B) the community composed primarily of grasses, herbs, shrubs, and tree seedlings
(C) the community composed primarily of white spruce, balsam fir, and paper birch
(D) the pioneer community composed of lichens and mosses

Refer to the graph below, which shows atmospheric concentrations of CFCs, CCl_4, and SF_6 from 1940 to 2020, for Question 18.

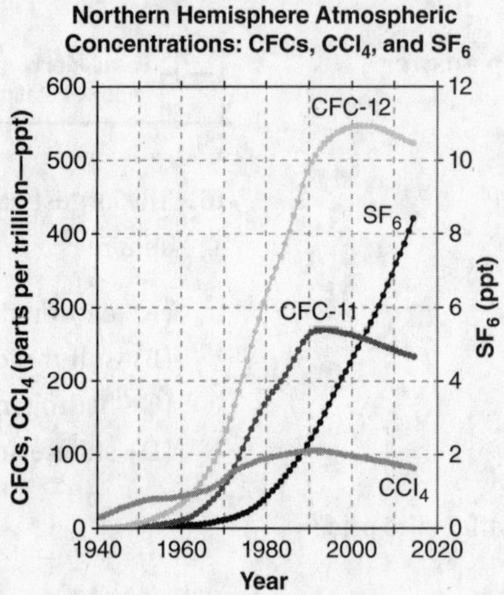

Source: National Oceanic and Atmospheric Administration United States Department of Commerce

18. According to the graph above, how many more parts per trillion (ppt) were there of atmospheric CFC-12 compared to SF_6 in 1990?

(A) ~8 ppt
(B) ~400 ppt
(C) ~500 ppt
(D) ~700 ppt

19. Which of the following choices lists soil particles in order from the largest to the smallest?

(A) sand → silt → clay → gravel
(B) clay → sand → silt → gravel
(C) gravel → sand → silt → clay
(D) silt → clay → sand → gravel

Refer to the diagram below to answer Question 20.

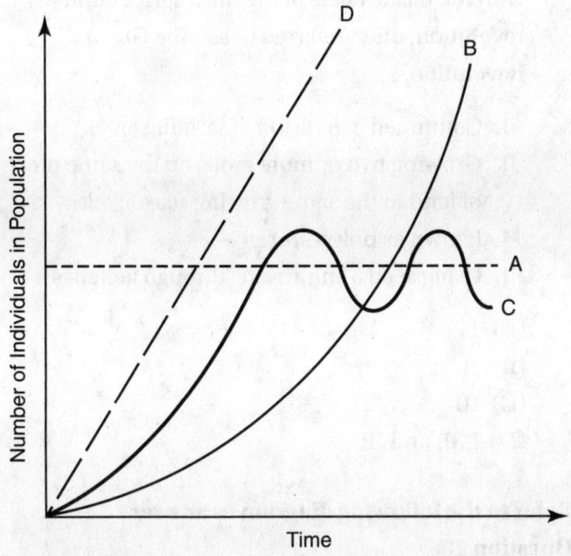

20. Which growth curve shown above best represents the effects of environmental resistance acting on a sustainable population?

(A) A
(B) B
(C) C
(D) D

Consider the following facts for country XYZ in order to answer Question 21.

- All cars in XYZ run on diesel fuel.
- The average car in XYZ can travel 5 km on one liter of diesel fuel.
- The average car in XYZ is driven 20,000 km/year.
- XYZ has 200,000 cars.
- Fifty percent of the cars in XYZ are expected to run on hydrogen fuel cells by the year 2030.

21. What is the potential reduction in diesel fuel that could be achieved by 2030 assuming the facts above remain constant?

(A) 1.0×10^5 liters
(B) 2.0×10^6 liters
(C) 4.0×10^7 liters
(D) 4.0×10^8 liters

22. Which international agreement has been helpful in protecting endangered animals and plants by listing those species and products whose international trade is controlled or banned?

(A) Convention on International Trade in Endangered Species of Wild Fauna and Flora
(B) Endangered Species Act
(C) International Treaty on Endangered and Threatened Species
(D) World Environmental Policy Act

Refer to the following diagram to answer Question 23.

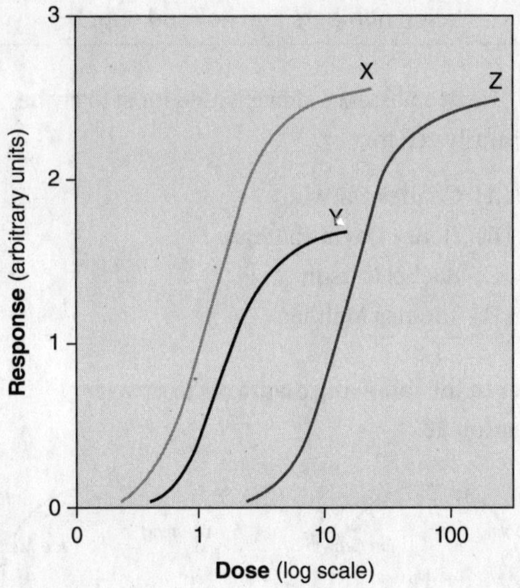

23. Which drug(s) shown above has/have the greatest therapeutic effect (greatest response for the least amount of drug)?

(A) X
(B) Y
(C) Z
(D) X and Z

24. Which of the following is/are NOT a current threat to the decrease in biodiversity over time?

    I. Alteration and loss of plant habitats
    II. Introduction of invasive species and genetically modified organisms
    III. Overexploitation of resources

    (A) I
    (B) II
    (C) III
    (D) All are current threats to the decrease in biodiversity

Refer to the following two passages to answer Question 25:

> "Populations multiply geometrically while food grows arithmetically; therefore, whenever the food supply increases, populations will rapidly grow to eliminate the abundance."

> "Environmental disasters serve to maintain population numbers and its food supply."

25. The two passages above would most likely be attributed to:

    (A) Charles Darwin
    (B) Henry David Thoreau
    (C) Rachel Carson
    (D) Thomas Malthus

Refer to the following diagrams to answer Question 26.

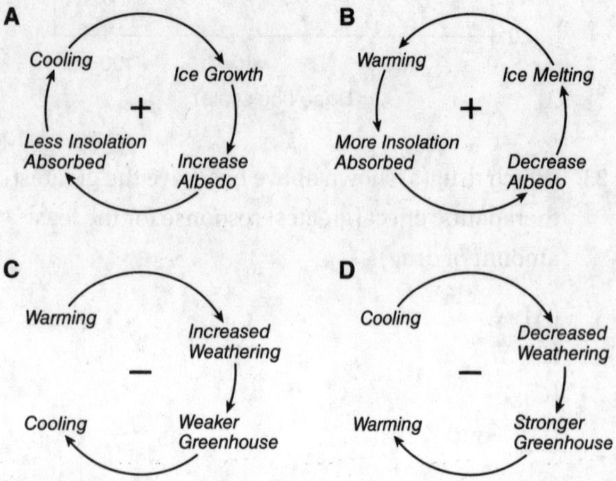

26. Which of the feedback mechanisms shown above is INCORRECT?

    (A) A
    (B) B
    (C) C
    (D) D

27. Which of the statement(s) below is/are consistent with the basic tenets of the third agricultural revolution, often referred to as "The Green Revolution"?

    I. Continued expansion of farming areas
    II. Growing two or more crops on the same piece of land in the same growing season, also known as polyculture
    III. Using seeds "improved" through genetics

    (A) I
    (B) II
    (C) III
    (D) I, II, and III

Refer to the following diagram to answer Question 28.

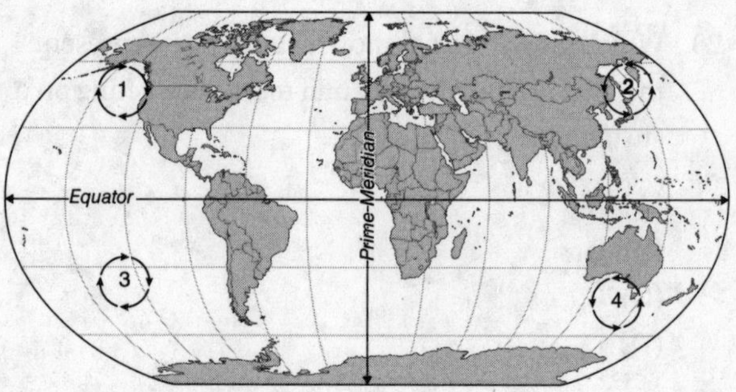

28. In which directions do high pressure air masses spin?

    (A) 1 and 2
    (B) 1 and 3
    (C) 1 and 4
    (D) 2 and 3

29. Which of the following factors reduce atmospheric carbon dioxide levels?

(A) Glacial periods
(B) Interglacial periods
(C) Summer in the Northern Hemisphere as compared to the winter in the Northern Hemisphere
(D) A, B, and C

Refer to the graph below to answer Question 30.

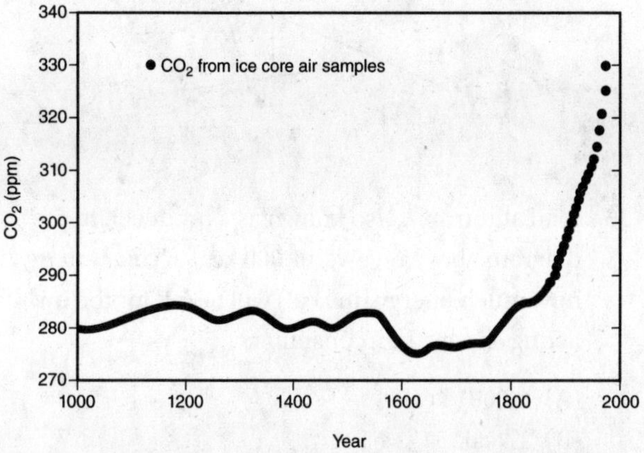

30. A scientific study took ice core samples to measure the concentration of carbon dioxide that was present in the atmosphere for the last 1,000 years. Their findings are shown above. What was the approximate percentage increase in carbon dioxide concentration over the last 1,000 years? (Use the year 1000 as the base year for comparison.)

(A) ~0.18%
(B) ~1.8%
(C) ~18%
(D) ~180%

31. Which energy source listed below is the most efficient in producing electricity; for example, which has the highest energy return on energy invested?

(A) Hydroelectric
(B) Nuclear
(C) Photovoltaic (solar)
(D) Wind

Examine the two age-structure pyramids shown below to answer Question 32.

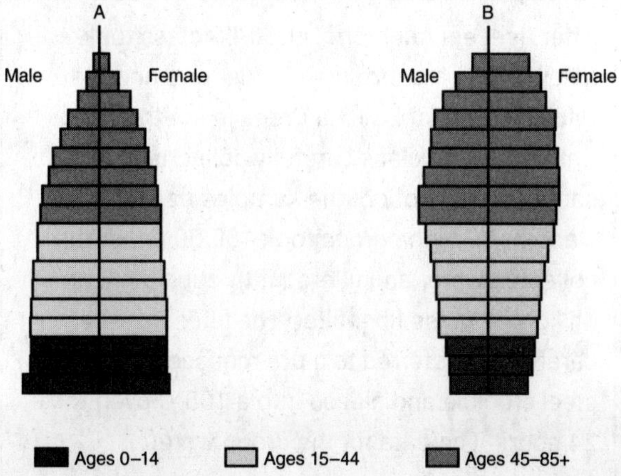

32. Which of the following statements would be TRUE?

(A) Diagram A shows negative population growth, while Diagram B shows zero population growth.
(B) Diagram A shows positive population growth, while Diagram B shows zero population growth.
(C) Diagram A shows zero population growth, while Diagram B shows negative population growth.
(D) Diagram A shows zero population growth, while Diagram B shows positive population growth.

33. Which of the following air pollutants is classified as a "secondary" air pollutant?

(A) Lead
(B) NO_x
(C) Tropospheric ozone (O_3)
(D) Volatile organic compounds (VOCs)

Refer to the following in order to answer Question 34.

After a recent rainstorm, an APES class took a field trip to a storm drain outlet entering the Pacific Ocean at Ballona Creek in southern California. The class carefully collected water samples and brought the samples back to their classroom. Alpha group took 100.00 mL of the collected water and filtered it through a 1.2 µm Millipore™ glass fiber filter. The filter was then carefully transferred to a pre-massed stainless steel crucible and placed into a 105°C oven for 24 hours. The data for the group were:

Weight of crucible + filter + residue after heating: 100.000 g

Weight of crucible + filter = 80.000 g

34. Which setup below would be used to determine the total suspended solid amount in terms of $mg \cdot L^{-1}$?

    (A) (80.000 / 100.00) × 100%

    (B) $\frac{(100.000 - 80.000) \times 1{,}000 \times 1{,}000}{100.00}$

    (C) $\frac{(100.000 - 80.000) \times 1{,}000}{100}$

    (D) $\frac{(100.000 - 80.000) \times 1{,}000}{1{,}000}$

35. Which of the following actions would lead to an increase in biodiversity?

    I. Reducing or eliminating invasive species
    II. Restoring connectivity between otherwise fragmented protected areas and the reintroduction of local predators
    III. Restricting the amount and frequency of human contact or interference in remote, wilderness areas

    (A) I
    (B) II
    (C) III
    (D) I, II, and III

Refer to the diagram below to answer Question 36.

36. If an autotroph, also known as a producer, in the diagram above received 1,000 kcal of solar energy, how much energy (in kcal) will be left for the owl acting as a tertiary consumer?

    (A) 0.0001 kcal
    (B) 1 kcal
    (C) 1,000 kcal (energy can neither be created nor destroyed)
    (D) 10,000 kcal (the owl requires more energy than the mouse)

37. Which of the following is NOT a characteristic of an invasive species?

    (A) Fast growth
    (B) High dispersal ability
    (C) High fertility rate
    (D) *K*-selected species

Examine the graph below to answer Question 38.

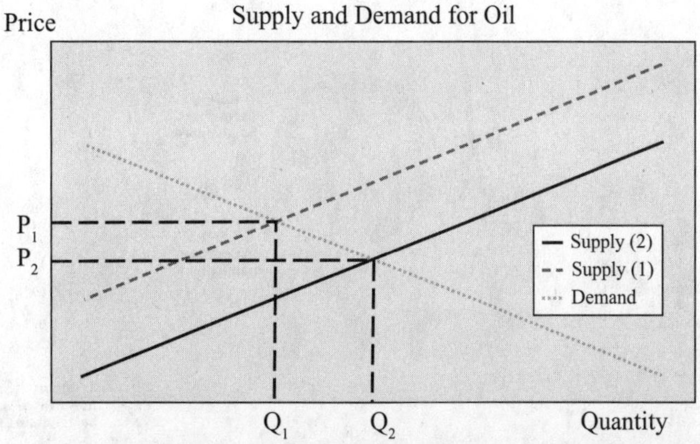

38. If the price of gasoline were to suddenly and dramatically decrease, it would

I. hinder the expansion and development of alternative sources of energy, e.g., hydrogen, methane clathrates, solar, wind, etc.

II. make it more economically feasible to explore for new sources of oil, e.g., offshore oil rigs, shale and sandstone formations, etc.

III. make it too expensive to explore and drill for new sources of oil

(A) I
(B) II
(C) III
(D) I and III

39. An electrical power plant that uses X joules of energy derived from natural gas can generate a maximum of Y joules of electrical energy from that amount of natural gas. Which of the following is always TRUE?

(A) $X = Y$ due to the law of conservation of energy
(B) $X < Y$ due to the First Law of Thermodynamics
(C) $X > Y$ due to the Second Law of Thermodynamics
(D) None of the above

Refer to the following diagram to answer Question 40.

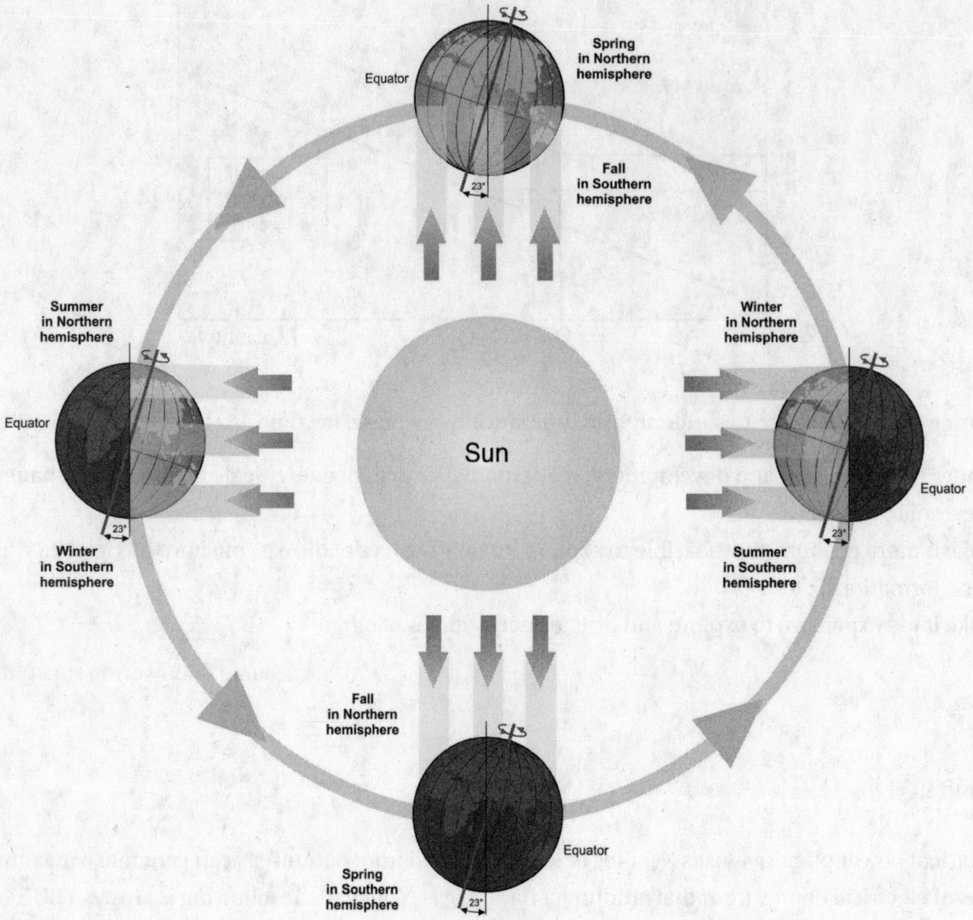

40. Earth's climate is primarily determined by

(A) Earth's longitude
(B) the amount of cloud cover
(C) the distance between Earth and the sun
(D) the tilt of Earth's axis

41. Which of the following would NOT be an example of a point source of pollution?

(A) Noise pollution from a jet engine
(B) Oil, grease, and toxic chemicals found inside a storm drain in a neighborhood
(C) Thermal pollution from a nuclear power plant
(D) Water pollution from an oil refinery wastewater discharge outlet

Examine the maps below to answer Question 42:

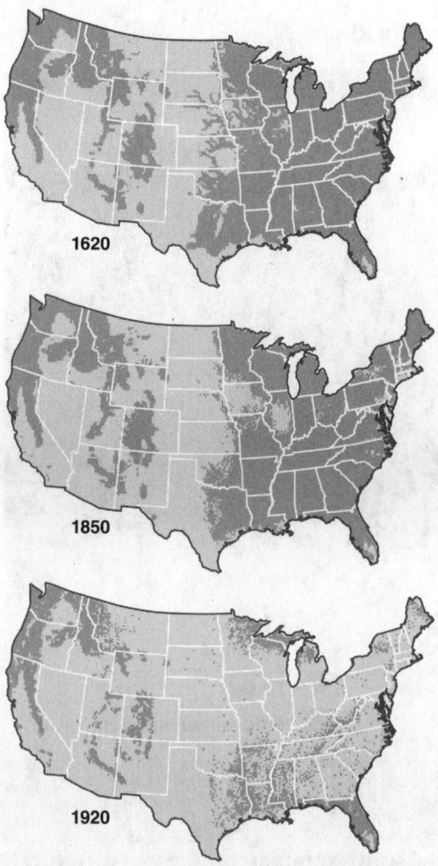

42. The maps above show the extent of harvesting trees in the United States from 1620 to 1920 (dark areas represent original forests). The type of harvesting done from 1620 to 1920 was mostly likely

(A) Clear-cutting
(B) Seed-tree harvesting
(C) Selective cutting
(D) Shelterwood cutting

43. Starting with an increase in atmospheric CO_2 concentration, which of the choices below would be the next most logical sequence of events?

(A) Less CO_2 uptake by the oceans → warmer oceans → warmer atmosphere
(B) Warmer atmosphere → warmer oceans → less CO_2 uptake by oceans
(C) Warmer atmosphere → less CO_2 uptake by oceans → warmer oceans
(D) Less CO_2 uptake by the oceans → warmer atmosphere → warmer oceans

Refer to the following diagram to answer Question 44.

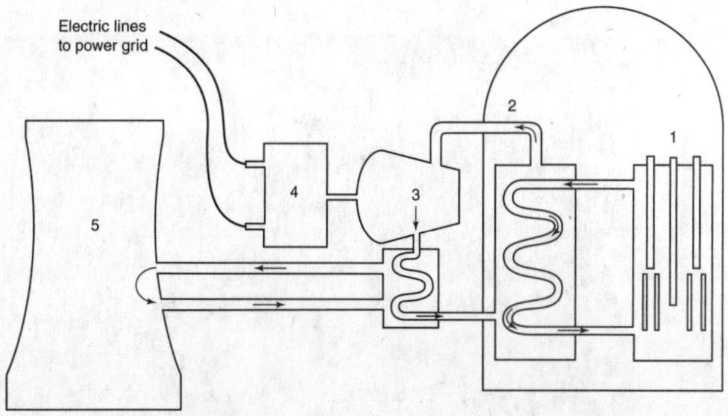

44. What is the name and function of the part labeled "5"?

(A) Cooling tower—cools and recycles water and releases unwanted heat to the environment.
(B) Heat exchanger—transfers heat from the fission reactor to water, which then turns to steam capable of powering steam turbines.
(C) Moderator—the container holding a material (e.g., graphite or water) that is used for slowing down neutrons so that they have a higher probability of inducing nuclear fission.
(D) Turbine—steam under very high pressure and temperature strikes blades mounted on a wheel, enabling the large wheel to turn.

45. Suppose that the ecological efficiency at each trophic level of a particular ecosystem is 20%. If the green plants of the ecosystem capture 100 units of energy, about _____ units of energy will be available to support herbivores and about _____ units of energy will be available to support primary consumers.

(A) 2, 20
(B) 4, 20
(C) 20, 2
(D) 20, 4

Refer to the map of world biomes shown below to answer Question 46.

The Main Biomes in the World

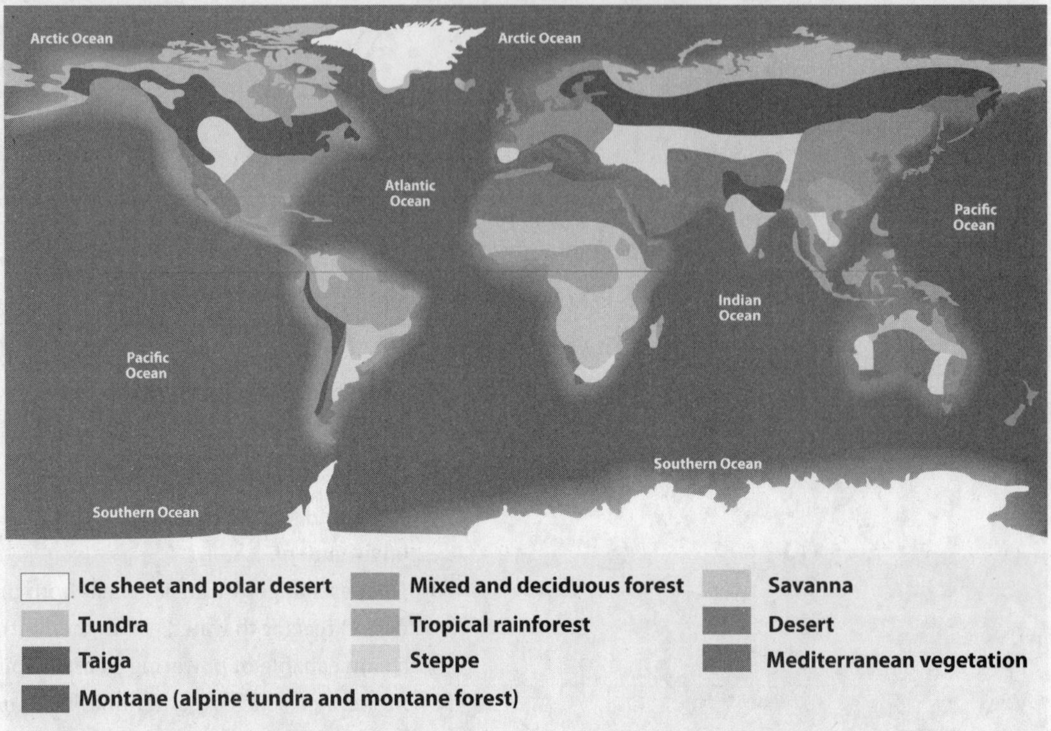

46. Which biome consists of dry, grassy plains with cold winters and warm summers?

(A) Montane
(B) Savanna
(C) Steppe
(D) Taiga

47. Which of the following choices listed below represent(s) an example of "The Tragedy of the Commons"?

(A) Population growth above a 2.1 TFR
(B) The Great Pacific Garbage Patch
(C) Traffic congestion on a public highway
(D) All choices above are examples of "The Tragedy of the Commons"

Refer to the graph below to answer Question 48.

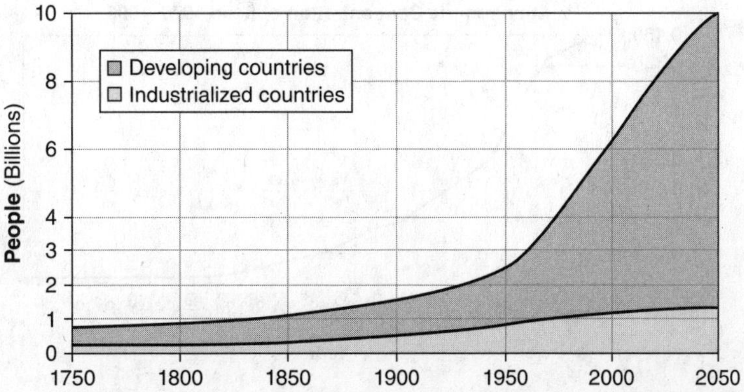

48. The graph above shows the difference in world population growth between developing and industrialized countries between 1750 and 2050 (projected). Which of the following would NOT be a factor in explaining the difference between the population growth patterns of developing and industrialized countries?

(A) In developing countries compared to industrialized countries, larger families ensure that aging parents will be taken care of as the parents become older.
(B) In developing countries compared to industrialized countries, more children from larger families can assist in farm labor chores or other labor activities that aid the family's financial situation.
(C) In industrialized countries, empowerment of women through higher education, which creates greater social equality and more accessibility to birth control, results in smaller average family sizes.
(D) Overall, death rates are lower in developing countries than industrialized countries.

49. Three friends decided to go to a rock music festival featuring a new group called "The APES." Before they left for the festival, one of their parents reminded the group to bring ear protection. Which of the following responses listed below was correct?

(A) Abbey: "Heck, loud sound isn't dangerous as long as you don't feel any pain in your ears."
(B) Ben: "Relax, hearing loss is only temporary—we'll be OK tomorrow."
(C) Caren: "I've heard that all sounds permanently cause hearing loss—so it doesn't matter."
(D) None of their responses were correct.

Refer to the following graph to answer Question 50.

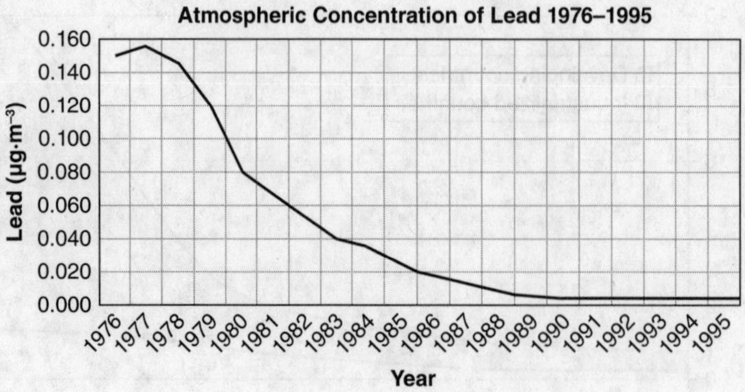

50. The atmospheric concentration of lead decreased significantly from 1976 to 1995, as seen in the graph above. The factor that was most responsible for the decrease in atmospheric lead levels was

(A) mandating scrubbers be used in power plants that use coal or oil
(B) outlawing lead in paints
(C) phasing out lead in gasoline
(D) the increase in diesel cars and trucks that do not use leaded gasoline

51. Groundwater contamination is addressed in all of the following acts EXCEPT the

(A) Clean Water Act
(B) Resource Conservation and Recovery Act
(C) Safe Drinking Water Act
(D) Superfund

Refer to the following diagram to answer Question 52.

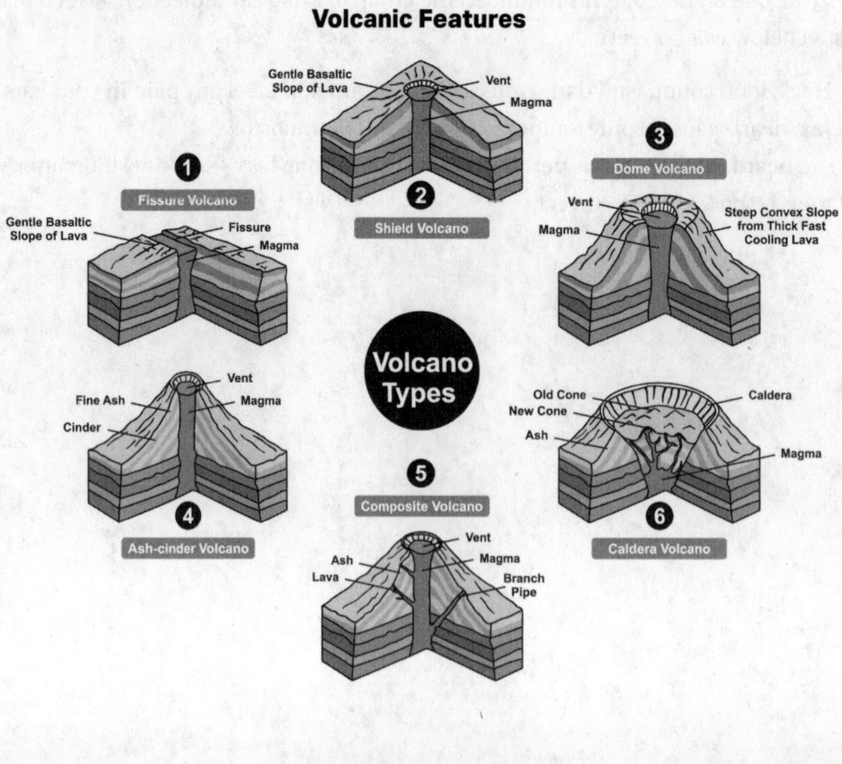

52. Some volcanoes consist almost entirely of lava that flows out from the crater and, over time, forms a broad, gently sloping dome-shaped cone. The build-up of these highly fluid lava flows eventually spreads out over great distances. These types of volcanoes are known as

(A) cinder cones
(B) composite volcanoes
(C) shield volcanoes
(D) stratovolcanoes

53. Which action listed below would most likely be the most effective method in reducing the emissions of carbon dioxide gas in the United States?

(A) Charging a large tax on the usage of electricity produced from burning coal and oil
(B) Planting millions of trees
(C) Switching from burning coal and oil to producing electricity by using renewable resources such as solar, wind, etc.
(D) Switching from gasoline and diesel engines to battery-powered vehicles

Examine the map below to answer Question 54.

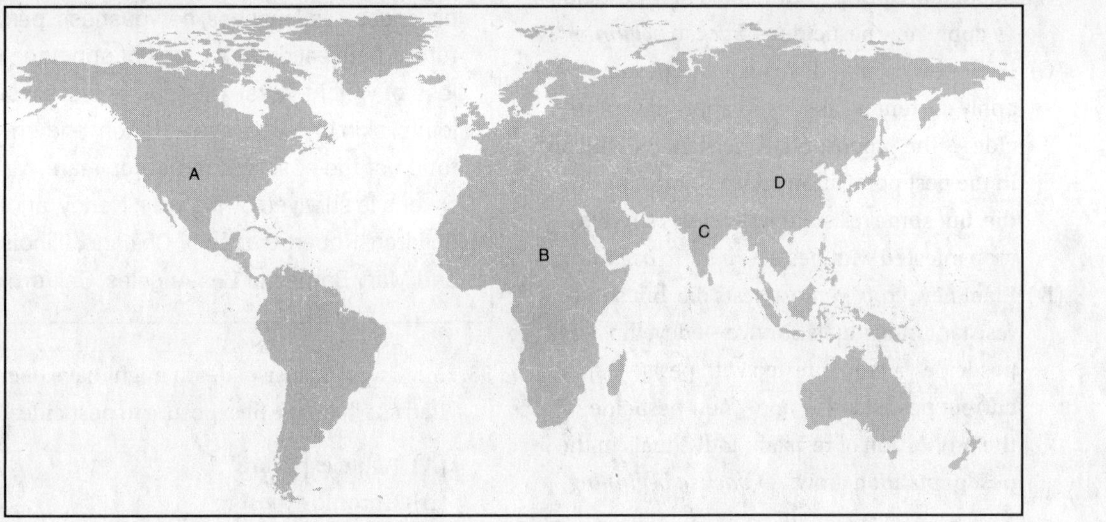

54. Referring to the map above, which location represents an area with the highest human fertility rate?

(A) A
(B) B
(C) C
(D) D

55. "We can burn coal to produce electricity to operate a refrigerator" is an example of the _____, and "if we burn coal to produce electricity to operate a refrigerator, we lose a great deal of energy in the form of heat" is an example of the _____.

(A) First Law of Thermodynamics, Second Law of Thermodynamics
(B) First Law of Thermodynamics, Third Law of Thermodynamics
(C) Second Law of Thermodynamics, First Law of Thermodynamics
(D) Second Law of Thermodynamics, Third Law of Thermodynamics

56. Which of the following choices correctly shows the stages involved in the "pesticide treadmill" starting with "Plant new crop"?

(A) Plant new crop → apply current pesticides → develop a new pesticide → the proportion of resistant individuals in the pest population grows → most pests die, but some resistant individuals survive → apply new pesticide → crop infested with pests → *back to beginning*
(B) Plant new crop → crop infested with pests → apply current pesticides → most pests die, but some resistant individuals survive → the proportion of resistant individuals in the pest population grows → develop a new pesticide → apply new pesticide → *back to beginning*
(C) Plant new crop → develop a new pesticide → apply current pesticides → apply new pesticide → the proportion of resistant individuals in the pest population grows → most pests die, but some resistant individuals survive → crop infested with pests → *back to beginning*
(D) Plant new crop → most pests die, but some resistant individuals survive → develop a new pesticide → crop infested with pests → apply current pesticides → apply new pesticide → the proportion of resistant individuals in the pest population grows → *back to beginning*

57. Currently (after the implementation and enforcement of the Montreal Protocol), the gas contributing most to stratospheric ozone breakdown is

(A) carbon dioxide (CO_2)
(B) methane (CH_4)
(C) nitrogen dioxide (NO_2)
(D) nitrous oxide (N_2O)

Refer to the following obituary to answer Question 58.

FUNERAL SERVICES FOR BILL OWENS THIS SATURDAY, JUNE 6 AT WOODLAWN CEMETERY—2:00 PM

Bill Owens, 53, passed away last week due to medical complications that he had been suffering from for many years. Mr. Owens, owner of Owens Agricultural Pest Control Service, a company that has been doing business in the area for over 30 years, had been complaining of headache, fatigue, weakness, dizziness, restlessness, nervousness, perspiration, nausea, diarrhea, loss of appetite, loss of weight, thirst, moodiness, soreness in joints, skin irritation, eye irritation, and irritation of the nose and throat for years. Mr. Owens is survived by his wife, Nancy, and his children, Robert Owens of Chicago, Illinois, and Mary Barnes of Los Angeles, California.

58. Mr. Owens' cause of death might have been classified as what type of exposure to pesticides?

(A) Acute exposure
(B) Chronic exposure
(C) Hazardous exposure
(D) Sustained exposure

59. Which of the following statements regarding coral reefs is FALSE?

(A) Coral reefs are among the most biologically diverse ecosystems on Earth.
(B) Coral reefs are among the most endangered ecosystems on Earth.
(C) Coral reefs capture about half of all the calcium flowing into the ocean every year, fixing it into calcium carbonate rock.
(D) Coral reefs store large amounts of organic carbon and are very effective sinks for carbon dioxide.

Refer to the following diagram below to answer Question 60.

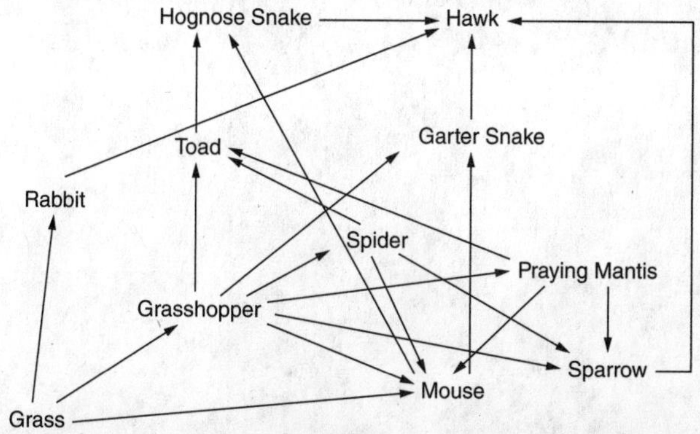

60. Which organism in the food web above would accumulate the highest concentration of a fat-soluble pollutant?

 (A) Grass
 (B) Grasshopper
 (C) Hawk
 (D) Sparrow

61. Which of the following choices represents the correct order of events that might occur in the process of primary succession?

 (A) Bare rock → grasses → mosses → small bushes → conifers → short-lived hardwood trees → long-lived hardwood trees
 (B) Bare rock → mosses → grasses → small herbaceous plants → small bushes → conifers → short-lived hardwood trees → long-lived hardwood trees
 (C) Grasses → small bushes → small herbaceous plants → lichen → conifers → long-lived hardwood trees → short-lived hardwood trees
 (D) Long-lived hardwood trees → short-lived hardwood trees → conifers → small bushes → small herbaceous plants → grasses → bare rock

Examine the following illustration to answer Question 62.

62. The type of fishing represented above is known as

 (A) dragline
 (B) drift net
 (C) longline
 (D) purse seine

63. A wind turbine was rated as being able to produce 15 million kWh of electrical energy per year working at 60% capacity (taking into account factors such as presence of wind, average wind speed, offline for maintenance, etc.). The average U.S. household consumes ~10,000 kWh of electricity per year. If a company planned to install 50 of these wind turbines, working at 60% capacity, how many households could the 50 windmills support?

 (A) 5,000
 (B) 7,500
 (C) 50,000
 (D) 75,000

Refer to the following map to answer Question 64.

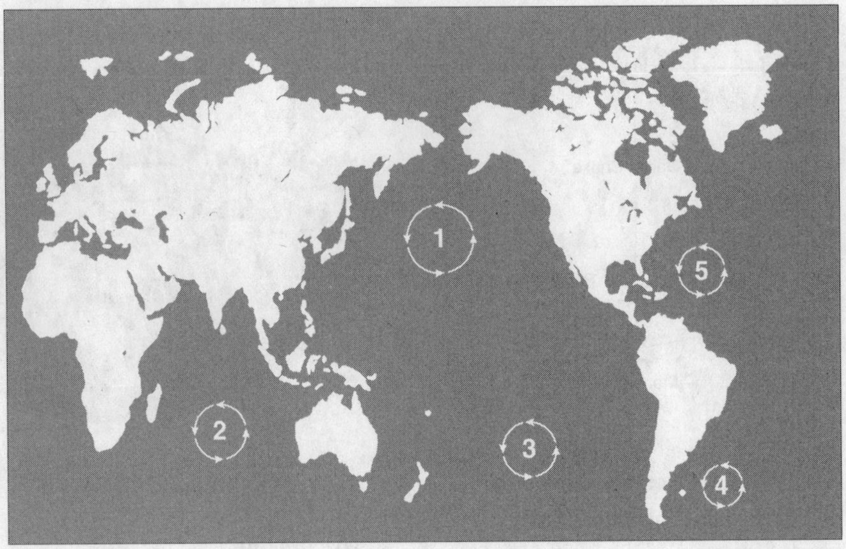

64. The Great Pacific Garbage Patch (#1 above) is the largest of the five offshore plastic accumulation zones in the world's oceans, and is located halfway between Hawaii and California. Which of the following statements about the Great Pacific Garbage Patch is FALSE?

    (A) Ten percent of the material in the Great Pacific Garbage Patch is composed of nurdles, small pieces of plastic formed through photodegradation (breakdown by sunlight).
    (B) Approximately 80% of the plastic originates on land, floating in rivers to the ocean or being blown by the wind into the ocean, with the remaining 20% originating from oil platforms and ships.
    (C) In the Great Pacific Ocean Gyre there is six times more plastic than plankton, which is the main food source for many marine animals.
    (D) Within a few years, the plastic will decompose and sink to the bottom of the ocean.

65. The two greatest threats to a species' survival are generally considered to be

    (A) lack of food and pollution
    (B) loss of habitat and invasive species
    (C) loss of habitat and pollution
    (D) pollution and invasive species

Refer to the following newspaper article to answer Question 66.

County Supervisors to Vote on Proposed Water Project

In Barronsville, N.Y., where the quality of drinking water has fallen below standards required by the U.S. Environmental Protection Agency (EPA), county authorities are studying the possibility of restoring the polluted Catskill Watershed.

Dr. Thorpe, environmental consultant for the area, informed the Board of Supervisors, "Once the input of sewage and pesticides to the watershed is reduced through natural abiotic processes such as soil adsorption and filtration of chemicals, together with biotic recycling via root systems and soil microorganisms, the water quality should improve to levels that meet government standards."

The cost of this investment in natural capital is estimated to be approximately $2 billion compared to the estimated construction cost of a water filtration plant of $8 billion with an annual operating cost of $300 million.

The Board of Supervisors will be meeting on September 17 at 6:00 PM and welcomes public comments.

66. The article above demonstrates the value and cost-effectiveness of

(A) economic services
(B) ecosystem services
(C) natural resource services
(D) social services

67. Almost one-third of the anthropogenic (of, relating to, or resulting from the influence of human beings on nature) carbon dioxide emitted into the atmosphere

(A) is absorbed by the oceans
(B) is taken in by plants and converted to oxygen through photosynthesis
(C) naturally decomposes in the presence of ultraviolet light
(D) remains in the atmosphere

Examine the phosphorus cycle below to help answer Question 68.

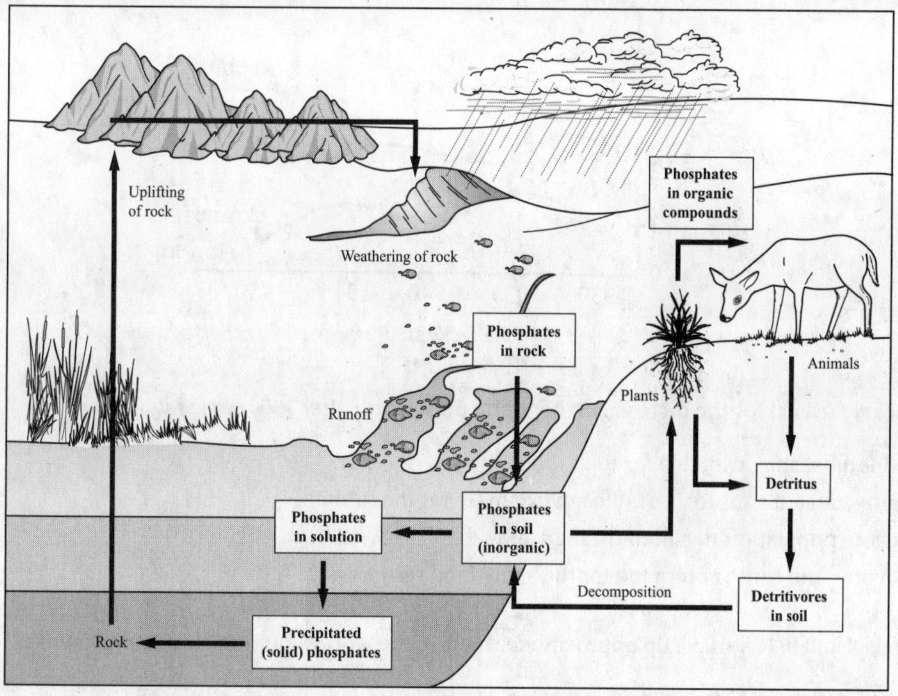

68. Which of the following statements is/are FALSE?

(A) Phosphates found in rain are responsible for adding nutrients to water supplies and result in explosive plant growth. This plant growth is often followed by eutrophication.
(B) The phosphorus cycle includes both biological and geological processes.
(C) The phosphorus cycle moves in only one direction—from soil to plant to animal.
(D) Choices A and C

69. As the populations of developing countries undergo demographic transition, which of the following is/are most likely to occur?
    I. Food supply shortages and famine will decrease as urban populations increase and the efficiency of food production and distribution improves—Law of Supply.
    II. Rural populations will increase as larger populations will require more food—Law of Demand.
    III. Urban populations will increase as people migrate to cities to find employment.

    (A) I
    (B) II
    (C) III
    (D) I and III

Refer to the following passage and graph to answer Question 70:

In 1940 ranchers introduced cattle in an area. The graph below shows the effect of cattle ranching on the populations of two animals (rabbits and coyotes) present in the area before the introduction of cattle.

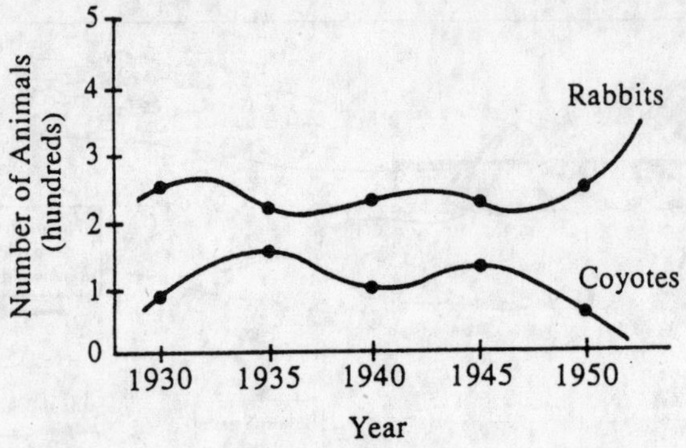

70. The most likely reason for the increase in the rabbit population after 1950 was

    (A) more food became available for the rabbits
    (B) the cattle created a more favorable environment for the rabbits
    (C) the coyote population declined dramatically
    (D) the coyotes and cattle competed for the same food resource

71. Which foods listed below make up approximately 60% of the world's human food energy intake?

    (A) Beans, rice, corn
    (B) Meat, fish, and milk and milk products
    (C) Rice and fish
    (D) Rice, corn, and wheat

Refer to the following diagram to help answer Question 72.

The El Niño Phenomenon

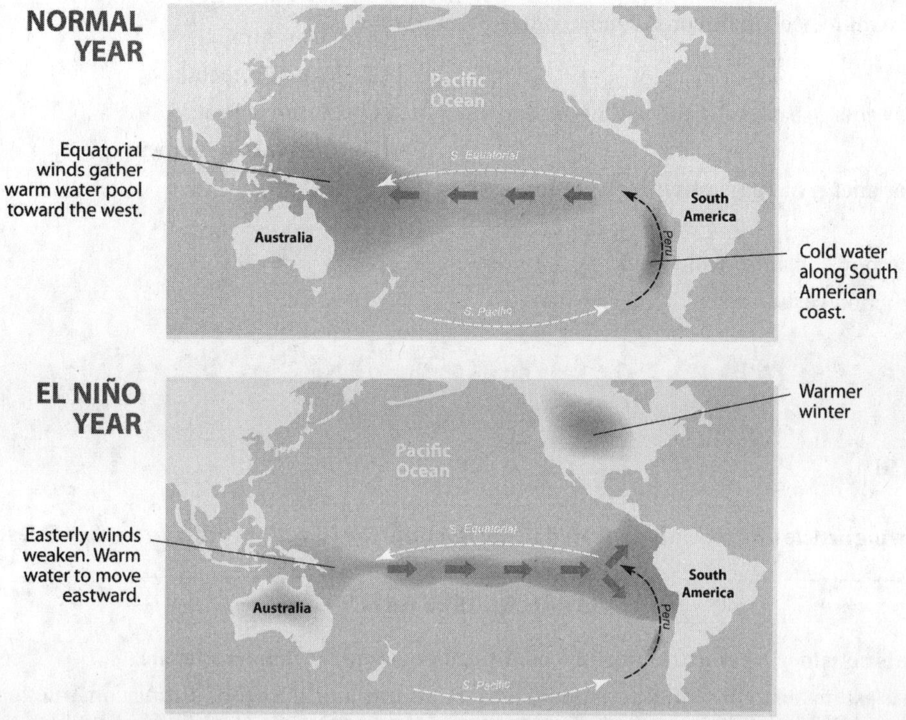

72. Which of the following events are NOT typical of or true during an El Niño event?

I. Usually the wind blows strongly from east to west along the equator in the Pacific and water piles up in the western part of the Pacific. In the eastern part of the Pacific, deeper, colder water gets pulled up from below to replace the water pushed west. Therefore, the normal situation is warm water in the western Pacific and colder water in the eastern Pacific.

II. In an El Niño, the winds pushing the water get weaker. As a result, some of the warm water piled up in the west flows back down to the east, and not as much cold water gets pulled up from below, which tends to make the water in the eastern Pacific warmer.

III. The warmer ocean makes the winds weaker and begins a cycle known as a positive feedback loop, which creates a stronger El Niño effect and is often associated with wet winters over the southeastern United States and droughts in Indonesia and Australia.

(A) I
(B) II
(C) III
(D) All listed events are TRUE and occur during an El Niño

73. A small country that had depended on burning coal to generate power has recently decided to switch to nuclear power. Which of the following statements would need to be considered in order to switch to nuclear power?

    I. Nuclear energy has a relatively low net-energy yield.
    II. Nuclear energy has a relatively high net-energy yield.
    III. Nuclear power plants involve high construction costs.

    (A) I
    (B) II
    (C) III
    (D) I and III

74. Which of the following was instrumental in significantly reducing the worldwide production of chlorofluorocarbons?

    (A) Clean Air Act
    (B) Kyoto Protocol
    (C) Montreal Protocol
    (D) The Paris Agreement

Read the following article that recently appeared in the *Barronsville Gazette* in order to answer Question 75.

> **BARRONSVILLE GAZETTE**
>
> **ABC Energy is considering construction of a coal-fired power plant in Barronsville.**
>
> The plant is estimated to bring 150 permanent jobs to the area and 2,000 jobs during construction. At an estimated cost of $1.5 billion, county officials are looking at potential tax revenue from the plant that would be used to fund improvements to roads and schools. Local businesses would also benefit from the influx of new residents.
>
> But the promise of jobs to Barronsville has not clouded everyone's vision of the downside of the plan—air pollution.
>
> At a town meeting last evening, some Barronsville residents had strong feelings about preserving Barronsville's environment and are not willing to sacrifice health concerns for economic prosperity. Others were skeptical about the promised jobs being filled by Barronsville residents.
>
> ABC Energy has hired the firm of Anderson and Fletcher and their team of accountants and economists to analyze the project with another town meeting being scheduled for next month.

75. What type of study would the firm of Anderson and Fletcher most likely undertake that would monetarily sum up the benefits of building the coal plant and then subtract the costs associated with it?

    (A) A cost-benefit analysis
    (B) A site assessment plan
    (C) An audit
    (D) An environmental impact report

76. Which of the following types of public lands would mining NOT be allowed on?

    I. National Forests
    II. National Resource Lands
    III. National Wildlife Refuges

    (A) I
    (B) II
    (C) III
    (D) Mining would be allowed on all types of public lands listed above

77. A new road connecting a planned community to an existing urban area was needed due to high traffic volume on the existing roadway. The least expensive option was to build the road through a forested area owned by the county. An environmental impact study was conducted and the developers agreed to the following conditions in exchange for the contract:

- preserve as much existing vegetation as possible during the construction of the new highway
- replace or restore forested areas that would be affected during construction of the new highway
- the developer will purchase a similar amount of privately held land and place it in conservation easements
- use selective cutting and clearing practices

This trade-off approach used to address the environmental impact of building the new road is known as

(A) mitigation
(B) preservation
(C) restoration
(D) sustainability

78. Orchids are found growing in the branches of a tree. The orchid benefits by having a solid base on which to grow and by living in an elevated spot that gives better access to sunlight. Its position in the tree also allows it to absorb nutrients falling from the tree's upper leaves and limbs and from the nutrient particles that enter rainwater. The orchid has no negative effect(s) on the tree and the tree neither benefits nor is harmed by the orchid. Which of the following choices below describes this relationship?

(A) Amensalism
(B) Commensalism
(C) Mutualism
(D) Parasitism

Use the diagram below to help answer Question 79.

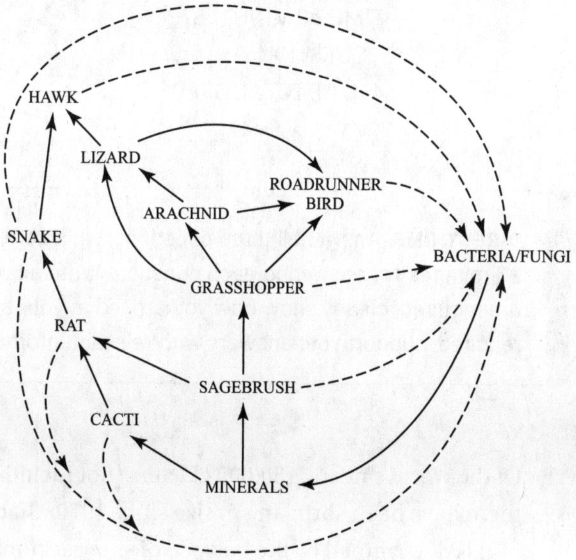

79. The snake in the food web above is acting as

(A) an autotroph
(B) a primary consumer
(C) a secondary consumer
(D) a tertiary (third order) consumer

80. Nutrient density is defined as

(A) the amount of kilocalories in a food divided by the amount of kilocalories needed in a day
(B) the amount of a nutrient in a serving of food divided by the amount of the nutrient needed for that day
(C) the amount of a particular nutrient in a serving of food divided by the number of kilocalories in that serving
(D) the amount of a particular nutrient in a serving of food divided by the number of grams of protein

Section II (Free-Response Questions)

TIME: 90 MINUTES
3 QUESTIONS
40% OF TOTAL GRADE

DIRECTIONS: Answer all three questions, which are weighted equally. The suggested time is about 23 minutes for answering each question. Write all your answers on scrap paper. Where calculations are required, clearly show how you arrived at your answer. Where an explanation or discussion is required, support your answers with relevant information and/or specific examples.

1. Of the world's nearly 150,000 glaciers (not including the ice sheets of Antartica and Greenland), 90% of them have been shrinking in size since 1970. Glacial volume (and whether the glacier is growing or receding) is determined by the balance of snow accumulation and snow loss on different parts of the glacier. Refer to the figure below, which describes the ice balance of a glacial system, to answer the following questions.

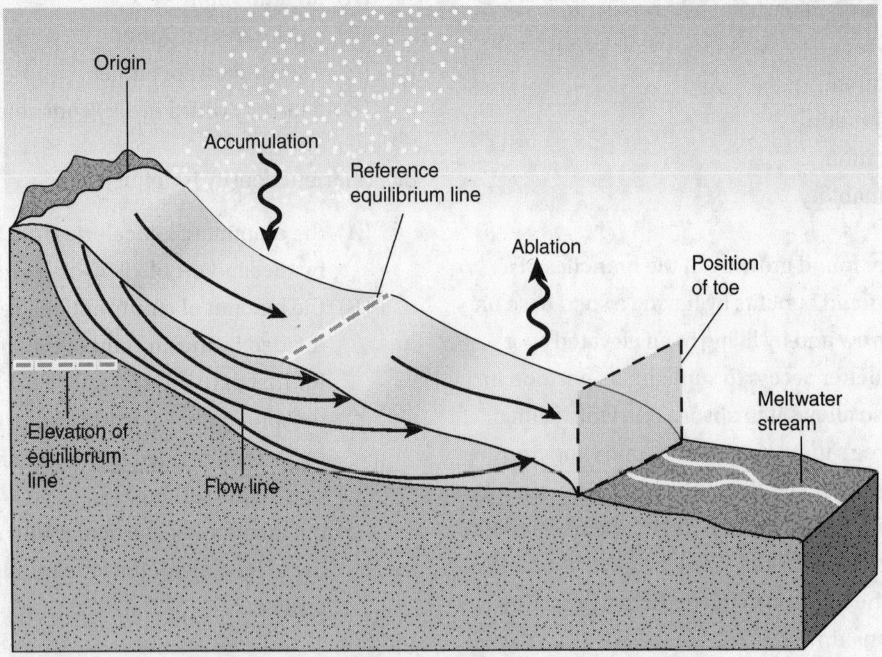

(a) IDENTIFY an input (gain) and output (loss) in the system above, and DESCRIBE conditions when the glacial system would be growing AND when the glacial system would be shrinking. Reference the "equilibrium line" in your response.

(b) Glaciers are often an importance source of year-round freshwater for regions whose rivers are fed by glacial ice melt. As glaciers recede, these water systems are changing. DESCRIBE how ONE abiotic condition of that river might change as the glacier that feeds it melts.

(c) EXPLAIN how that change in abiotic conditions might affect one of the biotic components in that river ecosystem.

(d) IDENTIFY and DESCRIBE ONE societal OR environmental impact that might result from global, glacial ice loss.

(e) Propose one way to mitigate or reduce the impact described in part (d) above.

2. (a) Describe what the Coriolis effect is and how it affects global atmospheric circulation patterns.

(b) Complete the diagram below by drawing in the arrows shown in the legend to tell how the Coriolis effect changes global wind patterns.

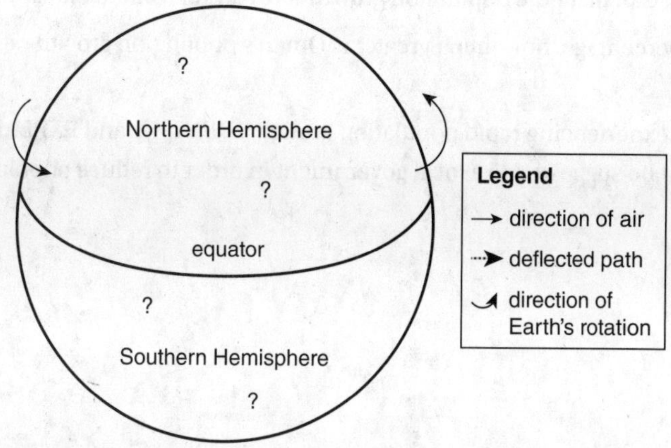

(c) Discuss how global surface ocean current circulation patterns are affected by global atmospheric circulation patterns and the Coriolis effect.

(d) Using the map shown below, draw arrows showing the path of the "Great Ocean Conveyor Belt."

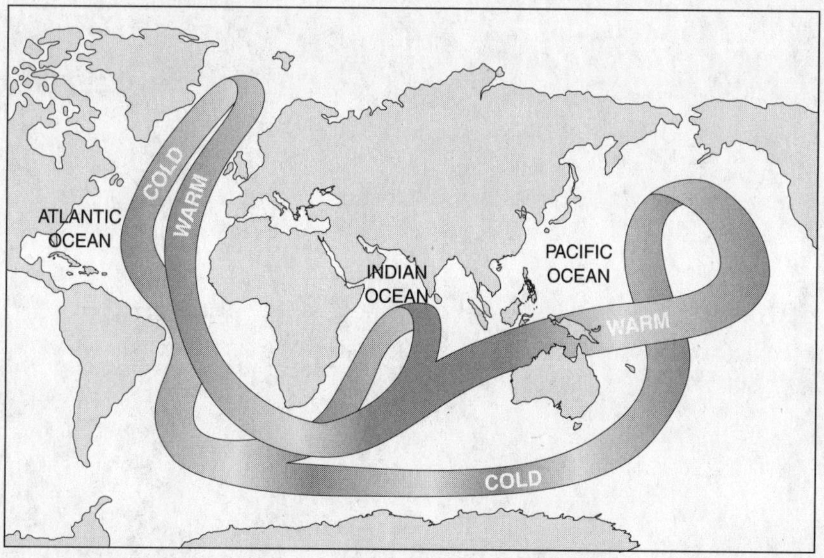

(e) Discuss how global climate patterns are affected by the relative amount of land mass in each hemisphere.

(f) How does the current distribution of plants and animals across Earth reflect the historical changes that have occurred in global climatic conditions and the historical changes that have occurred in the location of land masses?

3. (a) In 2018, the total fertility rate (TFR) in the United States was 1,728 births per 1,000 women, a record low. DESCRIBE the term "total fertility rate."

   (b) Replacement level fertility describes the point at which a population is neither growing nor declining. Using ONLY the United States' TFR data from 2018, EXPLAIN whether the population of the United States is projected to expand or contract.

   (c) In 2017, the crude birth rate (CBR) in the United States was 12.5 and the crude death rate (CDR) was 8.2. That same year, the CBR in one of the world's fastest growing countries, Oman, was 24 and the CDR was 2.5. Calculate the population growth rates for the United States and Oman in 2017.

   (d) Expressed as a percentage, how much greater is Oman's population growth rate than that of the United States?

   (e) For any country experiencing rapid population growth, IDENTIFY and EXPLAIN one strategy or policy that might be suggested to a local government in order to reduce population growth.

Multiple-Choice Explanations

1. **(B)** The most common component of sand is silicon dioxide (SiO_2) in the form of quartz, which is hard and insoluble in water, and doesn't decompose easily into smaller grains from weathering processes such as wind, rain, and freezing/thawing cycles. Streams, rivers, and wind transport quartz particles to the seashore, where the quartz accumulates as beach sand. Soil texture refers to the relative amounts of sand, silt, and clay in the soil and refers only to particle size, not to the type of mineral that comprises them. Sand is the largest soil size and provides excellent aeration and water drainage; however, it erodes easily, and has a low capacity for holding water and nutrients. *Topic 4.3*

Properties	Sand	Silt	Clay
Cohesion between particles	Very low	Medium to high	Very high
Drainage rate	High	Low to medium	Very low
Organic material content	Low	Medium to high	Medium to high
Shrink and swell potential	Low	Medium to high	High
Water-holding capacity	Low	Medium to high	High

2. **(B)** It took approximately five minutes to boil two cups of water and approximately 11 minutes to boil four cups of water. Water has a high specific heat due to hydrogen bonding between the hydrogen atom of one water molecule and an oxygen atom of another, which means it absorbs more energy to raise its temperature than air does.

   The ocean serves as a heat sink, moderating air temperatures nearby, and during the summer, the ocean absorbs a large amount of heat from the sun. It retains that heat as temperatures fall during the winter; in fact, the ocean stores more heat in the uppermost 10 feet (3 meters) of water than Earth's atmosphere.

   When air passes over the warm ocean, its temperature increases and moderates inland temperatures during colder months; e.g., Los Angeles experiences milder temperature swings from summer to winter than Chicago. Ocean currents can also transfer heat between regions; e.g., the Gulf Stream transfers warmth from the equator to Northern Europe. *Topic 4.8*

3. **(B)** On most days, the temperature of the air in the atmosphere is cooler the higher up in altitude you go since most of the sun's energy is converted to heat at the ground, which in turn warms the air at the surface. The warm air then rises in the atmosphere, where it expands and cools. Sometimes, however, the temperature of air actually increases with height. The situation of having warm air on top of cooler air is referred to as a temperature inversion, because the temperature profile of the atmosphere is "inverted" from its usual state.

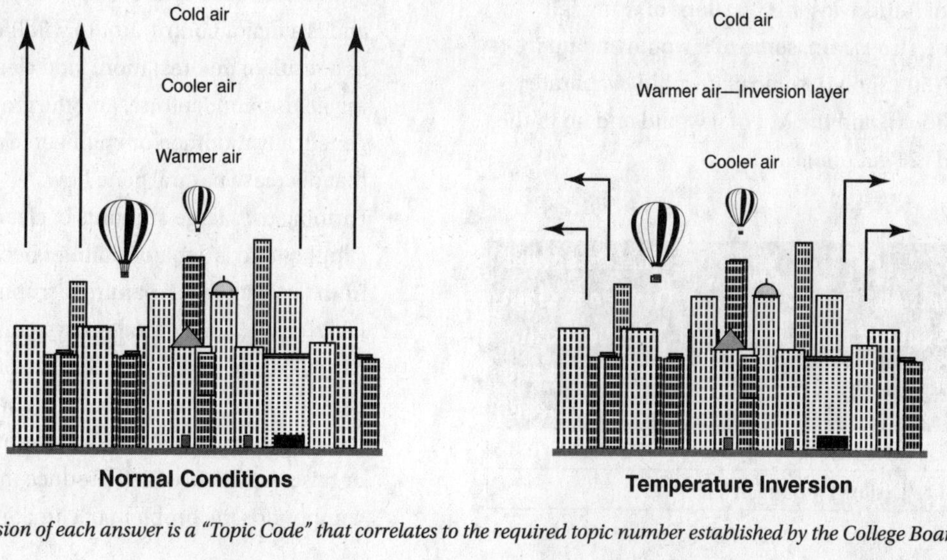

Normal Conditions **Temperature Inversion**

**At the conclusion of each answer is a "Topic Code" that correlates to the required topic number established by the College Board.*

The most common manner in which surface inversions form is through the cooling of the air near the ground at night. Once the sun goes down, the ground loses heat very quickly, which cools the air that is in contact with the ground. However, since air is a very poor conductor of heat, the air just above the surface remains warm. Conditions that favor the development of a strong surface inversion are calm winds, clear skies, and long nights. Calm winds prevent warmer air above the surface from mixing down to the ground, and clear skies increase the rate of cooling at the Earth's surface. Long nights allow for the cooling of the ground to continue over a longer period of time, resulting in a greater temperature decrease at the surface. Since nights in the wintertime are much longer than nights during the summertime, surface inversions are usually stronger and more common during the winter months. During the daylight hours, surface inversions normally weaken and disappear as the sun warms Earth's surface. However, under certain meteorological conditions, such as strong high pressure over an area, these inversions can persist as long as several days, even accumulating and increasing air pollution. In addition, local topographical features can enhance the formation of inversions, especially in valley locations, e.g., Los Angeles. *Topic 7.3*

4. **(C)** The general water balance equation is: $P = Q + E + \Delta S$, where P stands for precipitation, Q is runoff, E is evaporation, and ΔS is the change in water storage (in soil or bedrock). Think of it this way: if 100 gallons of water fell from the sky as rain, some of it would end up in runoff (50 gallons), some of it would evaporate (25 gallons), and the rest of it would end up in the ground (25 gallons). *Topic 1.7*

5. **(B)**

Initial Dissolved Oxygen (ppm)	Final Dissolved Oxygen in Dark Bottle	Final Dissolved Oxygen in Light Bottle (both respiration and photosynthesis)
8 mg O_2/L (light)	–	10 mg O_2/L
8 mg O_2/L (dark)	5 mg O_2/L	–

Respiration Rate

Only respiration (the decrease in dissolved oxygen over time) not photosynthesis can occur in a dark bottle. To determine the respiration rate, subtract the (dark) dissolved oxygen value from the initial (dark) dissolved oxygen value, then divide by the time.

$$\textbf{Respiration rate} = \frac{\text{initial DO (mg/L)} - \text{dark DO (mg/L)}}{\text{time}} = \frac{8\text{ mg/L } O_2 - 5\text{ mg/L } O_2}{1\text{ hr}} = \mathbf{3\ mg/L/hr}$$

Gross Primary Productivity (GPP)

Gross primary productivity (GPP) is the total amount of carbon that was fixed by organisms over time.

$$\textbf{GPP} = \frac{\text{dissolved oxygen (light)} - \text{dissolved oxygen (dark)}}{\text{time (hr)}} = \frac{10\text{ mg/L} - 5\text{ mg/L}}{1\text{ hr}} = \mathbf{5\ mg/L/hr}$$

Net Primary Productivity (NPP)

Net primary productivity is the total amount of carbon that was fixed (converted) by organisms into living tissue minus that used over time for respiration.

$$\text{NPP} = \text{GPP} - \text{R} = 5\text{ mg/L/hr} - 3\text{ mg/L/hr} = \mathbf{2\ mg/L/hr}$$

Topic 8.1

6. **(D)** It takes much more grain, land, energy, and water to raise an animal to produce a pound of meat than it does to produce the same number of calories by growing crops. However, there are environmental issues to consider. Examples include:
   - Agriculture accelerates the loss of biodiversity and is a major contributor to wildlife extinction as a result of deforestation, land clearing, burning fields, monoculture, and the growing of genetically modified or engineered organisms that decrease natural gene flow.
   - Farming uses large amounts of clean water supplies and is a major polluter because runoff from fertilizers and manure disrupts fragile lakes, rivers, and coastal ecosystems.
   - Nitrous oxide (N_2O) from fertilized fields and carbon dioxide from the cutting down of old-growth forests (carbon sinks) to grow crops or raise livestock (which produce methane) compounds the problem of climate change.

- Raising cattle and growing rice releases more greenhouse gases than all cars, trucks, trains, and airplanes combined.

Topic 3.5

7. **(C)**

$$\frac{10\ \cancel{\text{LEDs}}}{1} \times \frac{10\ \cancel{\text{watts}}}{1\ \cancel{\text{LED}}} \times \frac{1\ \text{Kw}}{1{,}000\ \cancel{\text{watts}}}$$

$$\times \frac{10\ \text{hours}}{1\ \cancel{\text{day}}} \times \frac{365\ \cancel{\text{days}}}{1\ \text{year}} = 182.5\ \text{kWh/yr}$$

Topic 6.2

8. **(C)** The primary change in natural land cover on Earth since the mid-18th century has been the deforestation of temperate regions and the conversion of these regions to areas of intensive agriculture and urbanization, with environmental consequences of desertification, salinization, and soil erosion. *Topic 5.4*

9. **(B)** Phosphorus normally occurs in nature as a phosphate ion, PO_4^{3-}. Most phosphates are found as salts, e.g., in ocean sediments and rocks as calcium phosphate, $Ca_3(PO_4)_2$. Over time, geologic processes can bring ocean sediments to land, and weathering will carry the terrestrial phosphates back to the ocean. *Topic 1.6*

10. **(C)** The Resource Conservation and Recovery Act (RCRA) is the public law that creates the framework for the proper management of hazardous and nonhazardous solid waste and its disposal.

The Toxic Substances Control Act (TOSCA) does not address dumping of hazardous wastes; rather, it regulates the introduction of new or already existing chemicals. Its three main objectives are (1) to assess and regulate new commercial chemicals before they enter the market, (2) to regulate chemicals already on the list that pose an "unreasonable risk to health or to the environment" (e.g., PCBs, lead, mercury, and radon), and (3) to regulate the listed chemicals' distribution and use. *Topic 8.11*

11. **(A)** Three-way catalytic converters were developed in the late 1970s and are able to catalyze the conversion of the three main air pollutants (carbon monoxide, nitrogen oxides, and unburned hydrocarbons) simultaneously. In a three-way catalytic converter, uncombusted fuel residues are oxidized (burned in oxygen) to produce carbon dioxide (also a greenhouse gas) and water while, simultaneously, nitrogen oxides are converted to nitrogen and carbon monoxide is oxidized to carbon dioxide. *Topic 7.1*

12. **(C)** Determine the time interval between the first P wave and the first S wave. In this case, the time is ~25 seconds (see chart below). Now, find 25 seconds on the S-P timeline below and read directly across to find the distance (~215 km). *Topic 4.1*

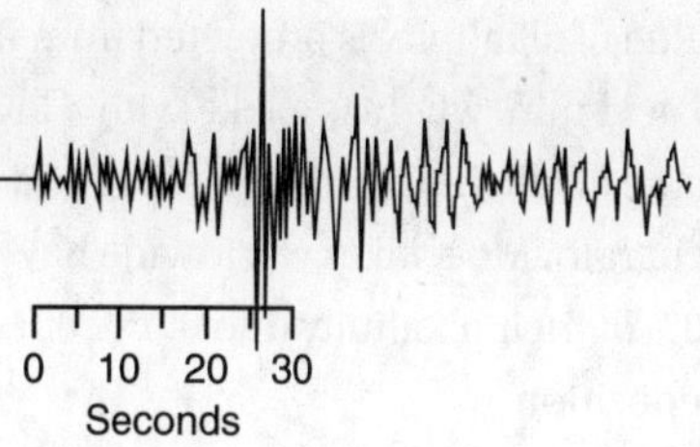

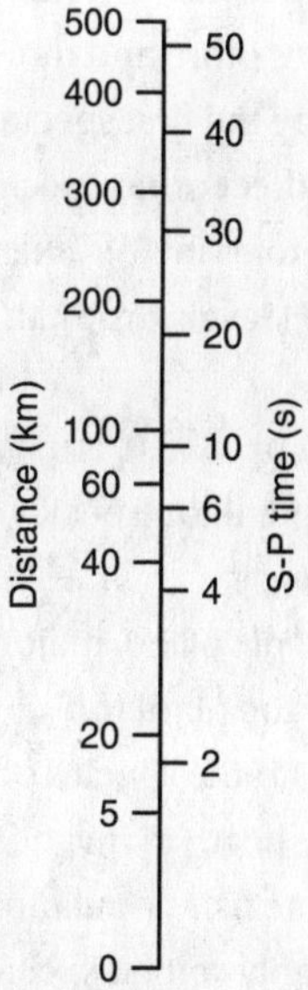

13. **(D)** Most healthy lakes and streams have a pH between 6 and 8, while acid rain has a pH less than 5. As the rain falls or particulates are deposited, many lake ecosystems become less able to buffer this increased acid. Natural sources of acids include sulfur oxides, sulfuric acid from volcanoes, forest fires, decomposition of plants and animals, carbon dioxide in air forming carbonic acid, and lightning that converts nitrogen molecules in the air to nitrogen oxides and then to nitric acid.

The types of soil, trees, and decaying leaves that surround a lake have the strongest influence on whether a lake is acidic or basic. Water from rain and snow comes into contact with materials surrounding the lake before it drains from the watershed into the lake. The layers of decayed leaves and pine needles surrounding a lake are rich in organic matter and produce acids. Rocks that contain limestone contain bases that can neutralize acids present from the rain, snow, or soil. Alkalinity is a measurement of the lake's ability to buffer or "absorb" acidity. A lake with a high value of alkalinity is protected from the effects of acid rain, whereas a lake with a low or zero value of alkalinity will most likely be affected by acid rain. Most lakes worldwide have lost up to 40% of their alkalinity due to the effects of acid deposition.

Acidic lakes can be harmful to fish, with some species being more tolerant to acidic conditions than others. As a lake becomes more and more acidic, clams and snails will be the first species to disappear. This, in turn, will affect other species, including those that are acid tolerant. As lakes become even more acidic (pH levels drop), the following changes are seen:

- As the water approaches a pH of 6.0, crustaceans, insects, and some plankton species begin to disappear.
- As the water approaches a pH of 5.0, major changes in the makeup of the plankton community occur, less desirable species of mosses and plankton may begin to invade, and the progressive loss of some fish populations is likely, with the more highly valued species being generally the least tolerant of acidity.
- As the water approaches a pH less than 5.0, the water is largely devoid of fish, the bottom is covered with undecayed material, and the near shore areas may be dominated by mosses. *Topic 2.5*

14. **(B)** 1,000 years ago: $[CO_2] \sim 280$ ppm; Present: $[CO_2] \sim 330$ ppm
Net change = 330 ppm − 280 ppm = 50 ppm.
Since CO_2 concentration increased, the sign is positive. *Topic 9.4*

15. **(D)** With all other factors being equal, as the price of a good or service increases, the quantity of goods or services that suppliers offer will increase, and vice versa. The Law of Supply says that as the price of an item goes up, suppliers will attempt to maximize their profits by increasing the quantity offered for sale; e.g., when consumers start paying more for cupcakes than for donuts, bakeries will increase their output of cupcakes and reduce their output of donuts in order to increase their profits.

The Law of Demand states that quantity purchased varies inversely with price; i.e., the higher the price, the lower the quantity demanded. As long as nothing else changes, people will buy less of something when its price rises and they'll buy more when its price falls; e.g., when the local coffee shop raises the price from \$1.75 to \$2.50, the demand for coffee will decrease; people will make their own. *Topic 6.2*

16. **(D)** Major natural contributors of sulfate to the environment are sulfur released from erosion and sulfide-containing rocks and minerals as well as volcanoes, bacterial action in soils, and chemical reactions occurring in coastal wetlands. Sulfate is taken up by plants and microorganisms, and is later consumed by animals, thereby moving sulfur through the food chain. One-third of the sulfur reaching the environment is generated from industrial activities such as mining and mineral processing, agriculture, paper and pulp manufacturing, and the combustion of fossil fuels and wastes. Sulfate in mine discharge waters is a major issue around the world where acidic tailings are periodically affected by drought and rain, and in mines where sulfuric acid is used to process the ore and mineral concentrates. *Topic 7.1*

17. **(C)** Climax communities are characterized by stability, high species diversity, low competitive interactions, limited niche overlap, large body size, few offspring per year, one reproductive cycle per year, and *K*-selective species. *Topic 2.1*

18. **(C)** In 1990, the atmospheric concentration of SF_6 (sulfur hexafluoride) was 2 ppt (parts per trillion). In that same year, the atmospheric concentration of CFC-12 was 500 ppt, a difference of 498 parts per trillion. *Topic 9.3*

19. **(C)** Soil texture is determined by particle size. Gravel is the largest. Sand, silt, and clay are the smallest. Soil texture has agricultural applications, such as determining crop suitability and predicting the response of the soil to environmental and management conditions, such as drought. *Topic 4.3*

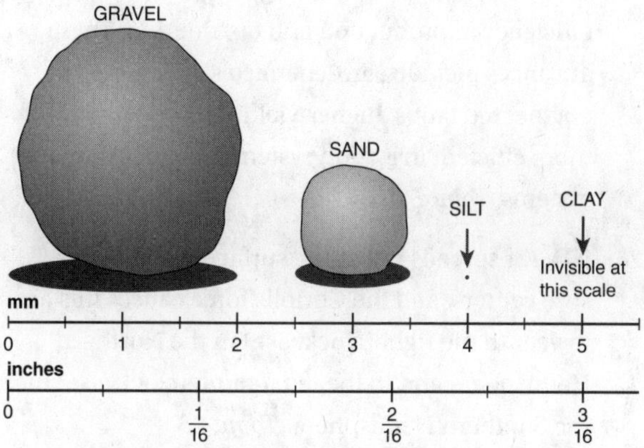

20. **(C)** Environmental resistance, or limiting, factors consist of all factors in an environment that limit a population's ability to increase in number, and work to hold population size in check. Environmental resistance can be either density-dependent or density-independent. Limiting factors that are density-dependent are stronger when a population has a higher density (is more crowded), and include predation, parasitism, disease, and competition for food or space, whereas density-independent (usually abiotic) limiting factors harm organisms whether conditions are crowded or not, and include floods, storms, earthquakes, and fire. Biotic potential and environmental resistance work together to stabilize the numbers that can be supported by the supply of natural resources. When environmental resistance acts on J-curve population growth, the curve eventually stabilizes into what is known as an S-curve and reflects a balanced community. *Topic 3.5*

21. **(D)**

$$\frac{2 \times 10^5 \text{ \sout{cars}}}{1} \times \frac{2 \times 10^4 \text{ \sout{km}}}{\text{\sout{car}}} \times \frac{1 \text{ L}}{5 \text{ \sout{km}}}$$

$$\times (0.50) = 4.0 \times 10^8 \text{ L}$$

Topic 6.11

22. **(A)** The Convention on International Trade in Endangered Species of Wild Fauna and Flora (CITES) is an international treaty to prevent species from becoming endangered or extinct because of international trade. Under this treaty, countries work together to regulate the international trade of animal and plant species and ensure that this trade is not detrimental to the survival of wild populations. Any trade in protected plant and animal species should be sustainable, based on sound biological understanding and principles.

Today, 182 countries and the European Union implement CITES, which accords varying degrees of protection to over 35,000 species of animals and plants. *Topic 9.9*

23. **(A)** Drug X has greater biologic activity per dose and is thus more potent than drugs Y and Z. In other words, drug X required the least dose but got the highest response. *Topic 8.15*

24. **(D)**

I. Alteration and loss of plant habitats: the transformation of natural areas determines not only the loss of the plant species, but also results in a decrease in the animal species associated with and dependent on them. There are no known examples of natural ecosystems that contain *only* plants.

II. Introduction of invasive species and genetically modified organisms: species originating from a particular area that are introduced into new natural environments can lead to imbalance in ecological equilibrium (balance). One example is the introduction of the Burmese python to southern Florida. Due to their large size, Burmese pythons have few predators, feed on native mammals and birds, and compete with threatened native species.

III. Overexploitation of resources: when the activities connected with capturing and harvesting (farming, fishing, or hunting) a renewable natural resource in a particular area are intense, the resource itself may become exhausted. For example, the passenger, or wild, pigeon had populations in North America in the 19th century estimated at five billion. They became extinct due to hunting, and the last passenger pigeon, Martha, died in the Cincinnati Zoo in 1914. *Topic 2.1*

25. **(D)** Thomas Malthus was an 18th-century British philosopher and economist famous for his ideas about population growth. Malthus' population theories were outlined in his book, *An Essay on the Principle of Population*, first published in 1798. In it, he theorized that populations will continue to grow until growth is stopped or reversed by disease, famine, war, or calamity. He developed what is now referred to as the Malthusian growth model, an exponential formula used to forecast population growth. *Topic 3.4*

26. **(B)** Positive feedback is a process in which the end products of an action cause more of that action to occur and amplifies the original action. For choice (B) to be correct, all arrows would need to be reversed in order to illustrate positive feedback. Other examples of positive feedback mechanisms in climatology include:
    - global warming affects the cloud distribution. Clouds at higher altitudes result in more infrared radiation being reflected back at the Earth's surface than immediately reflected back into space.
    - methane hydrates can be unstable, so that a warming ocean could release more methane, which is also a greenhouse gas.
    - peat, occurring naturally in peat bogs, contains carbon. When peat dries, it decomposes, and may additionally burn. Peat also releases nitrous oxide.

    Negative feedback occurs when the end results of an action slow down or inhibit that action from continuing to occur. *Topic 9.5*

27. **(D)** The first agricultural revolution began around 10,000 B.C.E. and involved subsistence and sustainable farming through the innovation of cultivation. The second agricultural revolution coincided with the Industrial Revolution (1800–1900), when farming became mechanized and commercial with the development of new inventions and technology (e.g., combine, seed drill, and tractor). This is consistent with the second stage of demographic transition as cities expanded and urban societies grew. The third agricultural revolution, began in the mid-1960s and continues today. It involves advanced technology to increase farming efficiency in production and distribution. These advances include agro-chemicals, genetically engineered crops, higher-yielding crop varieties, more efficient irrigation systems and distribution systems. *Topic 5.3*

28. **(C)** Air spreads out at the surface in high-pressure centers, and the Coriolis force causes this air to veer to the right (clockwise) in the Northern Hemisphere and to the left (counterclockwise) in the Southern Hemisphere. *Topic 4.5*

29. **(D)** Glacial periods were times of colder ocean temperatures—colder water absorbs and retains more carbon dioxide gas than warmer water. An interglacial period is a geological interval of warmer global average temperature, lasting thousands of years, that separates consecutive glacial periods within an ice age. Warmer air temperature is conducive to higher photosynthetic output—more plants absorbing more CO_2. Summer in the Northern Hemisphere also results in greater photosynthetic activity. *Topic 9.4*

30. **(C)**

    $$\frac{330 \text{ ppm} - 280 \text{ ppm}}{28 \text{ ppm}} \times 100\% = \sim 18\%$$

    Topic 9.4

31. **(A)** Much of the energy content from the available sources of power generation is wasted by inefficiencies in energy conversion and distribution. For example, to generate electricity based on steam turbines, 65% of all energy produced is wasted as heat, and further energy is lost due to energy conversion at the point of the end-user; e.g., only 2% of the electrical energy used to power an older incandescent lightbulb is converted to light energy, with most of it being lost through transmission line loss and heat generated by the bulb.

The chart below shows the theoretical efficiencies of various electricity generation sources. *Topic 6.9*

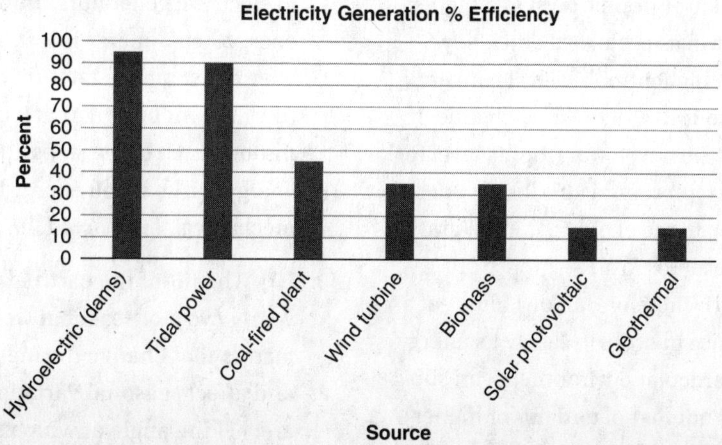

32. **(C)** Zero population growth (ZPG) is a state of demographic balance where the number of people in a specified population neither grows nor declines, considered by many as a social goal. In the long term, zero population growth can be achieved when the birth rate of a population equals the death rate; i.e., the replacement level is met and the growth rate is stable. According to some, zero population growth is the ideal toward which countries and the world should aspire in the interests of accomplishing long-term environmental sustainability. Achieving zero population growth is difficult because a country's population growth is often determined by economic factors, such as incidence of poverty, natural disasters, disease, etc. *Topic 3.6*

33. **(C)** Secondary air pollutants are not emitted directly into the air; rather, they form in the air when primary pollutants react or interact. Tropospheric ozone is a greenhouse gas and forms when nitrogen oxides (NO_x), carbon monoxide (CO), and volatile organic compounds (VOCs) produced by the incomplete combustion of fossil fuels react in the presence of sunlight in the lower atmosphere. *Topic 7.1*

34. **(B)** Let A = weight of crucible + filter + residue after 24 hours at 105°C

Let B = weight of crucible + filter

$$\text{Total suspended solids} = \frac{(\text{A} - \text{B}) \times 1{,}000\ \text{mL/L}}{\text{sample volume (mL)}} \times \frac{1{,}000\ \text{mg}}{\text{g}}$$

$$= \frac{(100.000\ \cancel{\text{g}} - 80.00\ \cancel{\text{g}}) \times 1{,}000\ \cancel{\text{mL}}/\text{L}}{100.00\ \cancel{\text{mL}}} \times \frac{1{,}000\ \text{mg}}{\cancel{\text{g}}}$$

Topic 8.2

35. **(D)** Invasive species are nonnative plants or animals that are introduced accidentally or deliberately into an ecosystem where they are not normally found, often with serious negative consequences. Common characteristics of invasive species include high dispersal ability, rapid reproduction and growth, the ability to quickly adapt to new conditions, and the ability to survive on various types of food.

Invasive species are among the leading threats to native wildlife. Choice II is known as "rewilding" and involves restoring connectivity between otherwise fragmented protected areas and the reintroduction of local predators and keystone species to create a self-regulatory and self-sustaining ecosystem.

Humans are often responsible for the spread of invasive species by, for example, releasing nonnative animals (e.g., Burmese pythons into the Florida Everglades) or accidentally introducing animals like quagga and zebra mussels—into aquatic systems. *Topic 9.10*

36. **(B)** The amount of energy with which energy is transferred from a lower trophic level to a higher trophic level is known as ecological efficiency. Consumers at each higher trophic level convert on average only about 10% of the total chemical energy available at the next lower trophic level for their needs, which represents a 90% loss of energy for each higher step in the energy pyramid: grass (1,000 kcal) → insect (100 kcal) → mouse (10 kcal) → owl (1 kcal). Therefore, an owl must eat a large number of mice in order to satisfy its energy requirements. A particular environment can support only a limited number of tertiary (or higher) consumers. The owl is considered an apex predator, as it is not preyed upon as a healthy adult in the wild. Quaternary consumers are very rare, as there are generally not enough calories available for their survival. The diagram contains a hawk (an apex predator) serving the role of a quaternary consumer. *Topic 1.11*

37. **(D)** Invasive, introduced, or exotic species are any species that are not native to that ecosystem and whose introduction is likely to cause economic or environmental harm to species already living in that ecosystem. Invasive species are typically *r*-selected species that have the ability to disperse offspring widely, retain an early maturity onset, have a high fecundity (capacity, especially in female animals, of producing young in great numbers), a short generation time, and small body size. *Topic 9.8*

38. **(D)** A free market economy determines the price of oil. In this graph, the original price of oil (P_1) allows for production of a certain quantity that is profitable (Q_1). But when the price drops (P_2), it results in more production and a higher quantity (Q_2); for example, when the price of gasoline at the pump drops, people drive more, resulting in a larger demand for gasoline. This increases competition, and lowers prices. *Topic 6.2*

39. **(C)** The chemical energy stored in chemical bonds in fossil fuels (e.g., coal) is released as thermal (heat) energy when the coal is burned in the presence of oxygen, as mechanical energy when the heat is used to boil water and the steam powers spinning turbine blades, and finally as electrical energy as the turbines are connected to electrical generators. Through each of these energy steps, energy is lost. The Second Law of Thermodynamics states that the quality of energy (the amount of work that it can do) decreases through each of the steps and that this loss in energy is true for all energy forms—chemical, mechanical, etc. *Topic 1.10*

40. **(D)** The tilt of the earth is responsible for the yearly cycle of seasonal weather changes. Two factors that change during the course of a year and affect seasonal variations in temperatures are: (1) the angle at which sunlight enters the atmosphere and strikes the ground; and (2) the number of daylight hours. The sun appears lower in the sky during winter, when the hemisphere affected is tilted away from it, which decreases the number of hours for the sun to heat the ground (or ocean), resulting in lower temperatures. *Topic 4.7*

41. **(B)** Point sources of pollution refer to easily identified locations that are discharging pollutants, whereas non-point sources of pollution are often termed "diffuse" pollution and refer to those inputs and impacts that occur over a wide area and are not easily attributed to a single source. Non-point sources are often associated with particular land uses, as opposed to individual point source discharges. *Topic 8.1*

42. **(A)** Clear-cutting is a forestry/logging practice in which most or all trees in an area are uniformly cut down. Along with shelterwood and seed tree harvests, it is used by foresters to promote select species that require an abundance of sunlight or grow in large, even-age stands. Clear-cutting is the most common and economically profitable method of logging. However, it also creates detrimental side effects, such as the loss of topsoil and destruction of native habitat, and is a major contributor to climate change. In addition to being used to harvest wood, clear-cutting is also used to create land for farming and urban expansion. The human demand for wood and arable land through unsustainable logging has led to the loss of over half of the world's rainforests. *Topic 5.2*

43. **(B)** The oceans are an important sink for CO_2. As atmospheric CO_2 levels increase, the warming of the atmosphere increases, which ultimately increases the temperature of the oceans. However, based on the principle of solubility, the ability of the oceans to remove CO_2 from the atmosphere decreases. *Topic 3.4*

44. **(A)** For every three units of energy produced by the reactor core of a nuclear power plant, two units are discharged to the environment as heat energy. Nuclear power plants are often built on the shores of lakes, rivers, and oceans because of the large quantities of water needed to handle the amount of waste heat. *Topic 6.6*

45. **(D)** According to the question, 100 units of energy are found in green plants (producers); therefore, 100×0.20 or 20 units of energy are available to the herbivores (e.g., rabbits) and only 20×0.20 or 4 units of energy are available to the primary carnivores (coyotes). *Topic 1.8*

46. **(C)** A steppe is a dry, grassy plain and found in temperate climates. They lie between the tropics and polar regions. These temperate regions have distinct seasonal temperature changes, with cold winters and warm summers, and are semiarid, with 10–20 inches (25–50 cm) of rain each year—enough rain to support short grasses but not enough for tall grasses or trees to grow. The largest temperate grassland in the world is the Eurasian steppe, which reaches almost one-fifth of the way around Earth. The dry, shortgrass prairie of the North American Great Plains is also a steppe. Many of the world's steppes have been converted to cropland and pasture, with the short grasses that grow naturally on steppes providing grazing for animals such as camels, cattle, goats, horses, and sheep. When the short grasses of the steppe are plowed under for agriculture, the soil can erode very quickly, and important nutrients that were anchored in the soil by grasses are blown or washed away. Agricultural development can also degrade the soil with fertilizer and other chemicals. *Topic 1.2*

47. **(D)** Growth of the human population on a planet with finite, shared resources is an example of "The Tragedy of the Commons." In this case, each person uses the same air, food, land, mineral, and water resources, causing a depleting the supply for the surrounding environment.

Throughout the world's oceans, garbage (especially plastic) accumulates in the center of circular ocean currents known as "gyres" and is likely to affect all humans and wildlife, as these pollutants cycle through the food chain.

Traffic congestion on public roads is also an example of common property shared by many people. As the roads become more and more congested, air pollution for all increases, which increases the risk of health issues, e.g., asthma, along with increasing greenhouse gases, which affect all life on Earth. *Topic 5.1*

48. **(D)** Developing countries have higher death rates compared to industrialized countries due to contaminated water supplies, diets that are low in calories and protein, endemic disease in some countries, poor access to medical services, and poor housing conditions. *Topic 3.9*

49. **(D)** Repetitive exposure to prolonged or excessive noise has been shown to cause many health problems ranging from stress, poor concentration, productivity losses in the workplace, communication difficulties, and fatigue from lack of sleep to more serious issues, such as cardiovascular disease, cognitive impairment, and permanent hearing loss.

Noise can also have a detrimental effect on animals, increasing the risk of death by changing the delicate balance in predator-prey detection and avoidance, and interfering with the use of sound in communication, especially in relation to reproduction and navigation. *Topic 7.8*

50. **(C)** In 1970, the Clean Air Act required the refining industry to stop using tetra-ethyl lead to increase octane levels, and switch to unleaded fuels that did not affect the performance of catalytic converters and exhaust emission control devices. These are the devices that convert toxic gases and pollutants in exhaust gas from internal combustion engines into less-toxic pollutants. *Topic 7.6*

51. **(A)** The Clean Water Act (1972) was enacted to reduce point source pollution and protect surface water, but did not address groundwater quality. *Topic 8.1*

PRACTICE TEST 1

52. **(C)** Shield volcanoes are found in the Hawaiian Island chain, the Galapagos Islands, and Iceland. They are wide structures rather than tall ones, built by one eruption of lava over another over vast periods of time. *Topic 4.1*

53. **(C)** Carbon dioxide emissions in the United States account for approximately 77% of all U.S. anthropogenic (caused by humans) greenhouse gas emissions, primarily from burning fossil fuels for energy, industrial processes, and transportation.

Charging large taxes on vital commodities (e.g., gasoline, heating oil) is a discriminatory tax, as it allows those that can afford it to heat their homes, while those who cannot are forced to live in unheated homes.

Planting millions of trees is not a bad choice, as trees can reduce the amount of CO_2, but it does not eliminate the "root" cause of the problem.

Switching to battery-powered vehicles is also a step in the right direction, but again, could be viewed as discriminatory in cases where people would not have enough money to purchase these types of vehicles at this time, as their prices are higher than the more mass-produced gasoline-powered vehicles. Furthermore, if the electricity required to recharge the batteries was produced from nonrenewable energy resources, there would be no net decrease in carbon dioxide emissions. *Topic 7.4*

54. **(B)** The vast majority of the countries in the world with the highest fertility rates are in Africa, with Niger topping the list at 7.153 children per woman. The current worldwide average total fertility rate (TFR) is 2.4. *Topic 3.7*

55. **(A)** The First Law of Thermodynamics states that energy can be changed from one form to another (heat → electricity), but energy cannot be created nor destroyed. This means that the total amount of energy (and matter) in the universe remains constant. It merely changes from one form to another. The Second Law of Thermodynamics states that as energy is transferred or transformed, more and more of it is "wasted" (usually as heat energy) and that there is a natural tendency to move in a direction of disorder (entropy). *Topic 6.5*

56. **(B)** The future effectiveness of pesticides is threatened by the evolution of resistant pests, resulting in a cycle commonly referred to as "the pesticide treadmill." Most pesticides used today are synthetic compounds, where target species are often able to develop resistance to newly introduced compounds. This pesticide resistance allows for the rapid evolution under strong selective pressures, similar to that of human resistance resistance with antibiotics. The figure below shows two ways pesticide resistance can develop. *Topic 5.6*

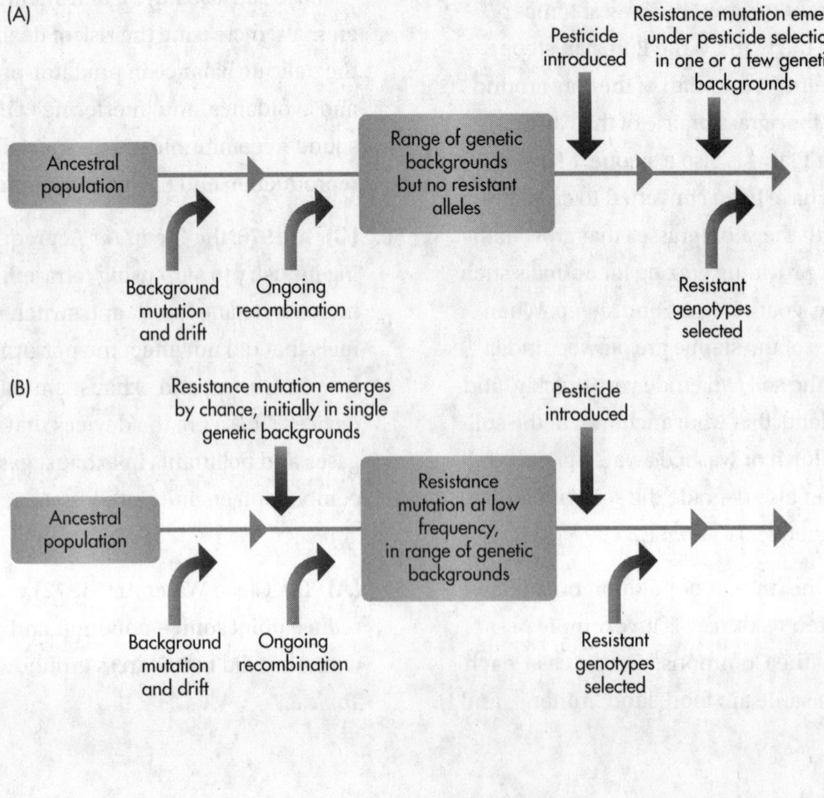

57. **(D)** Currently, nitrous oxide emissions from human activities are more than twice as high as the next leading ozone-depleting gas. Nitrous oxide is emitted from natural sources as a by-product of agricultural fertilization, livestock manure, sewage treatment, and industrial processes. With chlorofluorocarbons (CFCs) and other ozone-destroying compounds being addressed and phased out by the 1987 Montreal Protocol, nitrous oxide has become the largest contributor to ozone depletion in the stratosphere. It is also a greenhouse gas. *Topic 9.3*

N_2O Emissions	Thousands of Metric Tons
Agricultural soils	8,005
Other agricultural activities	885
Manure management	728
Fossil fuel combustion	703
Production of nitric acid	531
Biomass combustion	108
Other nonagricultural activities	60

Source: U.S. Environmental Protection Agency

58. **(B)** Toxicity is the degree to which a chemical substance or a particular mixture of substances can damage an organism. The more toxic something is, the less of it is necessary to produce a harmful effect(s) or the less time it takes for the effect(s) to become apparent. In chronic exposures, the dose of the toxin is delivered over a fairly long period of time (e.g., radon gas in a home or, as in this case, occupational contact with pesticides), whereas with acute exposures (e.g., exposure to a nerve agent), the exposure happens in a single dose and absorption is rapid; i.e., chronic exposures happen at low concentrations over a long time period, while acute exposures happen at high concentrations over a short time period. *Topic 8.12*

59. **(D)** Because the precipitation of calcium carbonate results in the sequestering of carbon, it frequently has been thought that coral reefs function as sinks of global atmospheric CO_2. However, the precipitation of calcium carbonate is accompanied by a shift of pH that results in the release of CO_2. This release of CO_2 is less in seawater than freshwater systems; nevertheless, coral reefs are sources, not sinks, of atmospheric carbon. Coral reef carbonate production is approximately 0.4% to 1.4% of the current anthropogenic CO_2 production due to fossil fuel combustion. *Topic 9.7*

60. **(C)** Bioaccumulation refers to how pollutants enter a food chain and represents the increase in concentration of a pollutant from the environment to the first organism in a food chain (in this case, grass). Biomagnification refers to the tendency of pollutants to concentrate as they move from lower trophic levels to higher trophic levels, and describes the increase in concentration of a pollutant from one link in a food chain to the next. In order for biomagnification to occur, the pollutant must be long-lived and fat-soluble. *Topic 8.8*

61. **(B)** In primary succession, which occurs in a virtually lifeless area (e.g., an area below a retreating glacier), pioneer species like lichen, algae, and/or fungus, as well as abiotic factors (e.g., water, wind), start to change or create conditions better suited for vascular plant growth. These changes include accumulation of organic matter in the litter or humus layer, alteration of soil nutrients, and changes in the pH or water content of the soil.

These pioneer plants are dominated and often replaced by plants better adapted to less austere conditions (grasses and shrubs) and are able to live in thin, mineral-based soils. The structure of the plants themselves can also alter the community. For example, when larger species like trees mature, they produce shade on the developing forest floor, excluding the light-requiring species, allowing more shade-tolerant species to invade and live in the area. *Topic 2.7*

62. **(D)** Purse seine fishing is a method of fishing that employs a fishing net called a seine that hangs vertically in the water with its bottom edge held down by weights and its top edge buoyed by floats. Seine fishing is the preferred fishing technique for capturing fish species that school near the surface, e.g., anchovies, herring, mackerel, sardines, and certain species of tuna and salmon. *Topic 5.8*

PRACTICE TEST 1

63. **(D)**

$$\frac{50 \text{ wind turbines}}{1} \times \frac{1.5 \times 10^7 \text{ kWh}}{1 \text{ wind turbine}} \times \frac{1 \text{ home}}{10{,}000 \text{ kWh}} = 75{,}000 \text{ homes}$$

Topic 6.12

64. **(D)** Despite their durability, many plastics do deteriorate over long periods of time into extremely small fragments (<5 mm) known as microplastics. This is often a result of exposure to ultraviolet radiation and mechanical stress. Plastics can last in the marine environment for decades or even hundreds of years when in surface waters, and far longer when in the deep sea. They accumulate in microorganisms; invertebrates; microbial communities and other filter feeders, such as clams, krill, sponges, and baleen whales; and many fish, such as sharks. Most synthetic polymers are buoyant in waters, while other plastics may sink. A "nurdle" is a small piece of plastic formed by photodegradation. *Topic 8.2*

65. **(C)** Habitat loss poses one of the greatest threats to species survival. The world's wetlands, forests, deserts, and grasslands continue to disappear at record levels as they are deforested or converted for agriculture. Industrialization, and pollution have become one of the greatest threats to biodiversity. Air and water pollution do not recognize borders, so they travel worldwide, entering every food chain on the planet. *Topic 2.5*

66. **(B)** Humankind benefits from a multitude of resources and processes supplied by natural ecosystems that are collectively known as ecosystem services, for example, clean drinking water, decomposition of biodegradable wastes, and the natural pollination of crops and other plants. Ecosystem services are grouped into four broad categories:

I. Provisioning—the production of food and water
II. Regulating—the control of climate and disease
III. Supporting—nutrient cycles and oxygen production
IV. Cultural—spiritual and recreational benefits

Topic 2.2

67. **(A)** The world's oceans are the largest natural reservoir of carbon dioxide, absorbing approximately one-third of the carbon dioxide added to the atmosphere by human activities each year. It is believed that dissolved CO_2 in the ocean will double over its pre-industrial level by 2050, with accompanying ocean acidity that results in:

- decreased rate of skeletal growth in reef-building corals
- excessive CO_2 levels in the blood of fish
- impaired developmental stages of invertebrates
- impaired production (calcification) of the shells of clams, oysters, mussels, microscopic algae, and other animals that form the base of marine food webs
- increase in corals' susceptibility to bleaching and disease
- reduced rate of calcium carbonate production in marine algae
- reduced survival of larval marine species, including commercial fish and shellfish

Topic 9.3

68. **(D)** Phosphates are not found in rain. However, rain does cause natural weathering of rocks (part of the geological cycle), which allows various phosphate compounds to enter the water cycle. Usually the very large concentrations of phosphorus in water supplies are due to human causes; for example, partially treated and untreated sewage, runoff from agricultural sites, and application of some lawn fertilizers.

As with other cycles on Earth, the phosphorus cycle does not have a single direction of movement, for example, when phosphorus moves from the animal to the soil. As seen in the figure, phosphates are created through both biological processes (plants converting inorganic phosphates into various organic compounds) and geological processes, such as weathering of rock. *Topic 1.5*

69. **(C)** The human population has lived a rural lifestyle throughout most of history. However, today, the world's population is becoming more and more urbanized as people migrate to cities seeking jobs and better educational opportunities, and as family-owned farms are being replaced by large corporate farms. In 1950, less than 30% of

the world's population lived in cities; by 2030 that percentage is expected to jump to 60%. *Topic 3.9*

70. **(C)** The relationship indicated by the graph is a predator-prey relationship. Prior to the introduction of cattle, the rabbit and coyote populations were in balance. However, after the introduction of cattle, the coyote population decreased dramatically, most likely due to ranchers eliminating the coyotes to save their cattle. *Topic 3.5*

71. **(D)** Of more than 50,000 edible plant species in the world, only a few hundred contribute significantly to human food supplies. Just 15 crop plants provide 90% of the world's energy intake, with three (corn [maize], rice, and wheat) making up approximately two-thirds of it. *Topic 5.15*

72. **(D)** El Niño is the warm phase of the El Niño-Southern Oscillation (ENSO) and is associated with a band of warm ocean water that develops in the central and east-central equatorial Pacific, including the area off the Pacific coast of South America. The ENSO is the cycle of warm and cold sea surface temperatures of the tropical central and eastern Pacific Ocean. El Niño is accompanied by high air pressure in the western Pacific and low air pressure in the eastern Pacific. During the development of El Niño, rainfalls usually develop between September and November. The cool phase of ENSO is La Niña, with sea surface temperatures in the eastern Pacific below average and air pressure high in the eastern and low in the western Pacific. Both El Niño and La Niña cause global changes in temperature and rainfall. *Topic 4.9*

73. **(D)** The usable amount of energy available from a given quantity of energy resource is called the net energy. Net energy is the total amount of energy *left over* after the amount of energy necessary to find, extract, process, and deliver the energy to the consumer has been subtracted. Nuclear power plants are expensive to build, involve costly regulation, are costly to run, have a limited lifetime, and also have radioactive disposal issues. *Topic 6.6*

74. **(C)** The 1987 Montreal Protocol on Substances that Deplete the Ozone Layer is an international treaty designed to protect the ozone layer by phasing out the production of numerous substances that are responsible for ozone depletion; specifically, chemicals known as CFCs (chlorofluorocarbons). These are chemicals that were used as refrigerants and solvents and found in fire extinguishers, and replaced by HCFCs (hydrochlorofluorocarbons). HCFCs are able to break down in the lower atmosphere. As a result of the international agreement, the ozone hole in Antarctica is slowly recovering. Climate projections indicate that the ozone layer will return to 1980 levels between 2050 and 2070. The Montreal Protocol is also expected to have effects on human health by reducing the amount of harmful ultraviolet radiation that reaches Earth, thus preventing over 280 million cases of skin cancer, 1.5 million skin cancer deaths, and 45 million cataracts in the United States. HFCs (hydrofluorocarbons—no chlorine) are beginning to replace HCFCs and are currently being used in developed countries. The downside of HFCs is that they have a high global warming potential (GWP). HFCs are 10,000 times more potent than carbon dioxide as greenhouse gases. *Topic 9.2*

75. **(A)** Cost-benefit analysis is a systematic process for calculating and comparing the benefits and costs of a decision, policy, or project by determining if an investment or decision is sound. Do the benefits outweigh the costs, and by how much? Using cost-benefit analysis, benefits and costs are expressed in monetary terms. The primary costs of the plant are the "external costs" of pollution (health effects, impact on agriculture, tourism, etc.). The benefits of the plant include the profit earned by the plant. Both of these numbers can be monetized using various economic methods. The following are some of the steps that are involved in cost-benefit analysis:
    - Adopt a plan that has the most benefit with the least cost.
    - Come up with alternatives to the project.
    - Define the goals and objectives of the project/activities.
    - List all stakeholders.
    - Predict and quantify all costs to build the project (e.g., planning, construction, and legal) and quantify all benefits (e.g., income from selling power and tax breaks from the local government).

    Topic 7.6

76. **(D)** In the United States, cities, counties, states, and the federal government all manage land that is referred to as public lands, or public domain, held in trust for the American people by the federal government and managed by the Bureau of Land Management, the United States National Park Service, the Bureau of Reclamation, the Fish and Wildlife Service, the Forest Service, or the U.S. Army Corps of Engineers. In 2017, President Donald Trump signed an executive order titled "Presidential Executive Order on Promoting Energy Independence and Economic Growth" that overturned several environmental regulations implemented during the Obama administration, such as a three-year moratorium on the leasing of federal lands for coal mining. *Topic 5.9*

77. **(A)** Environmental mitigation is used to describe projects or programs intended to offset known environmental impacts to an existing natural resource, such as an endangered species, forest, stream, or wetland. Actions taken to avoid or minimize environmental damage are considered the most effective method of mitigation. *Topic 5.10*

78. **(B)** Commensalism is an association between two organisms in which one benefits and the other one neither benefits nor is harmed. *Topic 2.6*

79. **(C)**

cacti (producer) → rat (primary consumer) → snake (secondary consumer)

Within an ecological food chain, consumers are categorized as primary consumers, secondary consumers, or tertiary consumers. Primary consumers are herbivores, feeding on plants. Secondary consumers are carnivores, and prey on other animals. Omnivores, which feed on both plants and animals, can also be considered secondary consumers. Tertiary consumers, sometimes also known as apex predators, are usually at the top of food chains, capable of feeding on secondary consumers and primary consumers. Tertiary consumers can be either fully carnivorous or omnivorous, e.g., humans. *Topic 1.2*

80. **(C)** Nutrient density is defined as the amount of nutrients contained in a food relative to the number of calories in that food. Calculating nutrient density allows a ranking of food based on its nutrient richness. Nutrient-dense foods such as fruits and vegetables are often the opposite of energy-dense food (also called "empty calorie" food), such as alcohol, foods high in added sugar, and processed cereals. Nutrient density can be correlated with soil quality and mineralization levels of the soil. The top three foods with the highest nutrient density are: #1 liver; #2 cacao; #3 eggs. *Topic 3.5*

Free-Response Explanations with Scoring Rubric

Question 1

Though 11 potential points are available, students may earn a maximum score of 10.

(a) IDENTIFY an input (gain) and output (loss) in the system above, and DESCRIBE conditions when the glacial system would be growing AND when the glacial system would be shrinking. Reference the "equilibrium line" in your response.

Maximum 3 points total: 2 points for accurately identifying the inputs and the outputs (one of each must be mentioned in order to earn both points). 1 point for describing that the system is growing when the accumulation rates are higher than the losses. Such a situation would push the equilibrium line farther down the mountain slope. If the rates of loss (melt) are higher than accumulation rates, the equilibrium line will move farther up the slope. The equilibrium line must be included in the response to earn this point.

The input (gain) for this system is snowfall. The output (loss) is melting (loss of solid ice turning to water) or ablation (loss of ice turning to gas). The most important accumulation area is above the equilibrium line, but snowfall anywhere in this system would be considered a gain. Melting would most likely take place below the equilibrium line, but melting or ablation anywhere in the system would be considered a gain.

The glacier is growing when accumulation rates are greater than rates of loss (snow > melting/ablation). When these conditions are present, the equilibrium line will move farther down the mountain (likely pushing the snout of the glacier farther into the valley below).

The glacier is receding when rates of loss are greater than rates of accumulation (melting/ablation > snow). When these conditions are present, the equilibrium line will move back up toward the origin and the snout of the glacier will move back up the valley.

(b) Glaciers are often an importance source of year-round freshwater for regions where rivers are fed by glacial ice melt. As glaciers recede, these water systems are changing. DESCRIBE how ONE abiotic condition of a river might change as the glacier that feeds it melts.

Maximum 2 points total: 1 point is earned for referencing an abiotic factor that might change as a glacier recedes. 1 point is earned for describing the way in which the abiotic factor changes when the glacier recedes. Students must link their identified abiotic factor with a description that correctly shows how that abiotic factor will change when the glacier recedes. Note: As long as the description is well reasoned, credit will be given. For example, one might argue that turbidity will increase OR decrease, as long as the conclusion is supported by a reasonable explanation. Examples AND their explanations will vary, but may include the following:

Viable Abiotic Factor	Description of the Change as the Glacier Recedes
Temperature	As the glacier recedes, the amount of meltwater will decrease, which will cause the temperature of the water remaining in the river system to rise.
Turbidity	As the glacier recedes, the river system will transport the fine sediment that was in the ice. Additionally, the initial increase in meltwater will pick up more sediment, causing turbidity to increase.
Velocity	As the melt rate increases with warming, the flow rate of the river will initially increase. As the volume of ice in the glacier decreases or disappears, the flow rate of the river will go down.

(c) EXPLAIN how that change in abiotic conditions might affect one of the biotic components in that river ecosystem.

Maximum 2 points total: 1 point is earned for referencing a biotic factor that might change as a glacier recedes/an abiotic factor in the river changes. 1 point is earned for describing the way in which the biotic factor changes when the glacier recedes/the abiotic factor changes. Students must link their identified biotic factor with a description that correctly shows how that biotic factor will change when the glacier recedes and the abiotic factor changes. Note: As long as the description is well reasoned, credit will be given. Examples AND their explanations will vary, but may include the following:

Viable Biotic Factor	Description of the Change as the Glacier Recedes
Fish species, potentially an indicator species such as trout or salmon	Fish species that require cold water temperatures would die off/be unable to inhabit the river system as water temperatures increase.
Predators that rely on vision for feeding	As turbidity increases, predators that feed visually become impaired and less effective. These predators would be at a disadvantage, likely decreasing in number.
Algae and aquatic plants	As turbidity increases (or decreases), the amount of light penetrating the water column will change. Since plants demand light for food production, populations of plants will be affected by changes in turbidity.

(d) IDENTIFY and DESCRIBE ONE societal OR economic impact that might result from global, glacial ice loss.

Maximum 2 points total: 1 point is earned for identifying a viable societal or environmental impact. 1 point is earned for describing how that environmental impact is connected to global ice loss.

Societal impacts are related to effects on HUMAN populations such as government, social structure/stability, livelihood, or housing. Economic impacts are related to effects on financial institutions and systems (at the individual or regional level).

Students must link their identified impact with a description that correctly connects to loss of ice. Examples AND their explanations will vary, but may include the following:

Societal or Economic Impact	Connection to Glacial Ice Loss
The need to move and resettle cities, towns, or villages.	■ As glaciers melt, freshwater supplies that have traditionally "fed" downstream populations will become unreliable, and settlements will need to move to find other reliable freshwater. ■ As land ice (glaciers and ice sheets) melt, sea levels will rise. Populations in low-lying areas will be at risk for flooding and/or become flooded. ■ Business, individuals, and governments will bear the cost of this relocation. Subsidizing this move will be a significant burden for governments. Most at risk will be the low income/impoverished segments of society. ■ As populations flee from the rising waters and dangerous areas, climate refugees will place new social, economic, and political stressors on adjacent cities and countries.
Loss of coastal wetland habitat and economic/societal impacts of fisheries (or other maritime-based industries).	Inundated coastal wetlands and estuaries will have a significant impact on fisheries as well as shellfish and other organisms that rely on these areas for nurseries/habitat. The economic consequences can be measured in loss of jobs. Societal consequences can include loss of traditional ways of life and disruption of social structure, as cultures change the way they get food and traditional food sources.

Societal or Economic Impact	Connection to Glacial Ice Loss
Loss of sea ice for indigenous populations will equate to losses in traditional hunting and transportation methods.	Indigenous populations that rely on stable, long-lasting sea ice for hunting and transportation will need to develop other means to compensate. This will likely have economic impacts as they rely on new ways to meet caloric requirements, and will have societal impacts as traditions and cultural practices are lost.

(e) Propose one way to mitigate or reduce the likelihood of increased rates of global ice loss.

Maximum 2 points total: 1 point is earned for identifying a viable method for reducing the probability of global ice loss. A second point is earned for recognizing that increasing global temperatures are the cause for this ice loss; therefore, students must describe a viable way to reduce warming potential.

Examples AND their explanations will vary but may include the following:

Method for Reducing the Likelihood of Ice Loss	Connection to Climate Change and Reducing Probability of Climate Change
Reducing use of fossil fuels	Fossil fuel combustion generates greenhouse gases, which cause an increase in global temperatures and, therefore, losses of all forms of ice.
Reforestation, alternative agriculture that decreases carbon dioxide or methane production	By planting more trees, the biosphere becomes more capable of absorbing carbon dioxide. When this greenhouse gas is removed from the environment, temperatures will not rise as much. Avoiding the release of methane or carbon dioxide by techniques like no-till farming or alternative strategies for producing protein will reduce the number of greenhouse gases and the associated global warming potential.
Increased reliance on low-carbon forms of energy (renewables or nuclear)	Increased use of renewable or nuclear power would mean a reduction in fossil fuel use. These low-carbon energy sources avoid the production of greenhouse gases, which reduces global warming potential and ice loss.

Topic 6.1; Topic 9.3; Topic 9.4; Topic 9.5

Question 2

(a) *Maximum 2 points: 1 point for correctly describing what the Coriolis effect is and 1 point for correctly describing how it affects global atmospheric circulation patterns.*

Radiation from the sun that reaches Earth is the primary factor that determines Earth's climate patterns, which in turn is the primary factor that determines the location and types of biomes found on Earth. These biomes range from extremely cold and dry polar regions to hot and humid tropical regions.

Due to more sunlight (measured in both duration and intensity) reaching the tropical areas (between 30°N and 30°S latitudes), in those regions air is able to hold more water vapor. As this moist air warms, it becomes less dense and rises until it reaches the upper levels of the atmosphere, where it begins to cool. This rising warm air also pushes the air mass away from the equatorial regions toward the polar regions. As the air cools, the water vapor condenses, forming clouds that eventually will produce rain. These events occur in atmospheric areas of Earth known as Hadley cells.

Starting at around 30°N and 30°S latitudes, this air begins to descend. As it descends, the air expands, gets farther from the dew point (becomes drier), and draws moisture out of the soil, resulting in arid deserts.

If the Earth did not spin on its axis, this cycle of evaporation, condensation, and precipitation would move air and its water vapor along a north-south axis from the equator to the poles, but this does not happen due to the Coriolis effect. Due to the Earth's rotation, air returning to Earth's surface is deflected by the Coriolis force, which shifts the flow of air to the right in the Northern Hemisphere and to the left in the Southern Hemisphere. Winds blowing toward the equator are deflected to the west, creating the easterly trade winds (easterly winds blow from east to west). In the temperate zones, where the winds blow toward the poles, the Coriolis force deflects them toward the east, with prevailing westerlies (blowing from west to east).

(b) *Maximum 4 points: 1 point for each arrow that is correctly placed into the diagram showing how the Coriolis effect affects global wind patterns.*

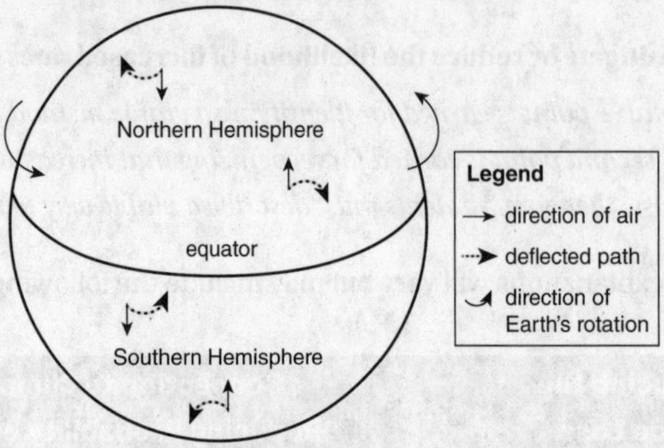

(c) *Maximum 1 point for correctly describing how global ocean surface current circulation patterns are affected by global atmospheric patterns and the Coriolis effect.*

Earth's rotation affects ocean currents in a manner similar to the way it affects atmospheric air currents. Ocean currents are driven by surface winds, Earth's rotation, and differences in salinity.

Ocean surface waters near the equator flow from east to west, due in part to trade winds. And, as in the atmosphere, the Coriolis force causes ocean currents to be deflected away from the equator (or northward and clockwise in the Northern Hemisphere and southward and counterclockwise in the Southern Hemisphere). As these waters approach their respective poles, the surface waters become colder and colder, with some of the water freezing; however, during the freezing process, salt does not remain in the ice, but rather remains in the ocean water, which ultimately results in a higher seawater salt content in the colder water. With this higher ocean salinity and colder water temperature occurring near the poles, the surface waters sink and set up a large "conveyor belt" of seawater that travels along the ocean floors, eventually reaching the Atlantic Ocean.

(d) *Maximum 1 point for correctly showing with arrows the path of the "Great Ocean Conveyor Belt."*

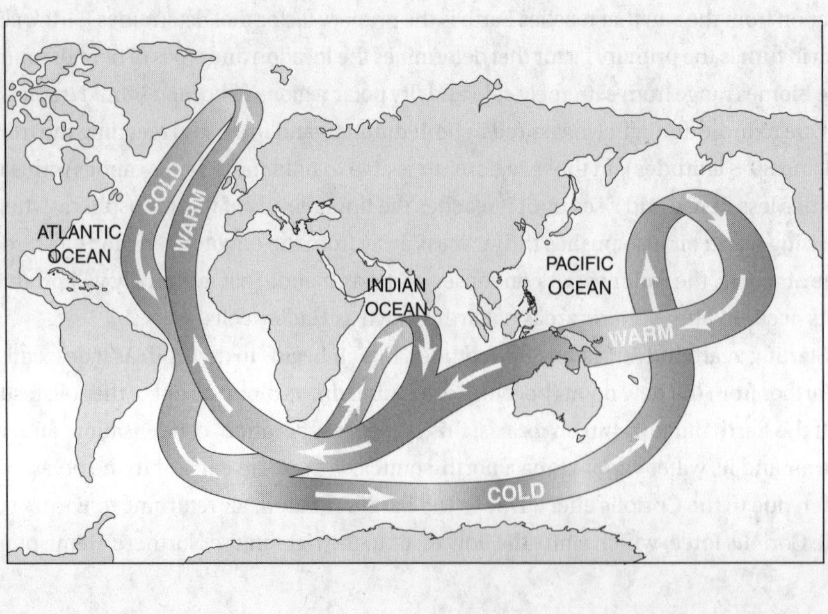

(e) *Maximum 1 point for correctly describing how global climate patterns are affected by the relative amount of land mass in each of Earth's hemispheres.*

The combined effects of oceanic and atmospheric circulation redistribute the Earth's heat and moisture across the planet. Furthermore, as the relative amount of land mass in the Northern Hemisphere is more concentrated than in the Southern Hemisphere (e.g., the Southern Hemisphere contains more islands that are spread out), the Earth's seasons in the Northern Hemisphere have a greater temperature range than in the Southern Hemisphere, primarily because of the difference in the specific heat capacity of land vs. water. In the Southern Hemisphere, however, where the larger land masses are located closer to the equator and the majority of the surface area is covered by water, climates are influenced more by the presence and absence of precipitation than they are by changes in air temperature.

(f) *Maximum 1 point for correctly explaining how the current distribution of plants and animals across Earth reflects the historical changes that have occurred in global climatic conditions and the historical changes that have occurred in the location of land masses.*

Biogeography is the study of the distribution of species and ecosystems in various habitats over long periods of time. Island biogeography is a field within biogeography that examines factors that affect the species richness and diversification of islands and other isolated natural communities.

Organisms and biological communities are distributed across Earth according to a wide variety of factors, including competition for resources, continental drift, glaciation, migration, resource availability, speciation, tolerance to changing environmental conditions, and wind pattern seed dispersal.

Observing the differences in plants and animals in various areas of Earth is easier when studying isolated islands, as they are more "condensed" than larger ecosystems found on the continents and can likewise range widely in diversity, e.g., from arctic to tropical climates. Studying smaller islands allows scientists to get a clearer picture of the dynamics that have occurred (and continue to occur) in larger ecosystems.

Topic 4.5

Question 3

(a) In 2018, the total fertility rate (TFR) in the United States was 1,728 births per 1,000 women, a record low. DESCRIBE the term "total fertility rate."

Maximum 1 point total: 1 point for accurately describing the term "total fertility rate."

Total fertility rate (TFR) is a measure of the number of children born to a woman over the course of her lifetime. The inclusion of the term "lifetime" or some similar expression is required in order to earn this point. Note: Though typically described as the number of births per 1,000 women, it may also be represented as the average number of births for an average female (in this case, 1.728). The data are calculated at a single moment in time by gathering birth rates from all the women of childbearing age.

(b) Replacement level fertility describes the point at which a population is neither growing nor declining. Using ONLY the United States' TFR data from 2018, EXPLAIN whether the population of the United States is projected to expand or contract.

Maximum 2 points total: 1 point for referencing the idea that, typically, replacement level fertility is reached when the total fertility rate (TFR) is around 2.1. 1 point is earned for recognizing/stating that the total fertility rate in the United States was under 2.1 in 2018 (it was 1.728) AND connecting this data point to the idea that since it is below 2.1, the population is declining.

Though responses might vary, the above conclusions should be reached when students consider ONLY the component of total fertility rate. Students should understand that replacement fertility is slightly greater than 2.0, though they do not need to use 2.1 as a specific number. Connecting the data point to the idea that a shrinking population would be predicted for values less than 2.1 (or 2.0) is critical for earning this point.

Note: Since students were asked to consider ONLY the TFR in their answer, they should not describe the population as growing (although this may indeed be true for some populations, depending on the age distribution and the number of individuals younger than childbearing age, at childbearing age, and beyond childbearing age).

(c) In 2017, the crude birth rate (CBR) in the United States was 12.5 and the crude death rate (CDR) was 8.2. That same year, the CBR in one of the world's fastest growing countries, Oman, was 24 and the CDR was 2.5. Calculate the population growth rate of the United States and Oman in 2017.

Maximum 4 points total: 2 points for the correct calculation setups for the United States AND Oman and 2 points for the correct answers for the United States and Oman.

Responses must include a setup similar to below. While units are not required in the setup, the numbers should be consistent, and students should apply the correct mathematical operation. The setups should include subtracting the CDR value from the CBR value and EITHER dividing that number by 1,000 (since both values are "per 1,000") and then multiplying by 100 (to describe answers as a percentage) OR simply dividing the difference by 10 (this method is acceptable since many students learn the shortcut: dividing by 1,000 and multiplying by 100 is the same as dividing by 10). The correct answer for the growth rate in the U.S. is 0.43% and the correct answer for the growth rate in Oman is 2.15%.

United States calculation and correct answer:

$$\frac{12.5 - 8.2}{1000} = \frac{4.3}{1000} \times 100 = 0.43\% \quad or \quad 12.5 - 8.2 = \frac{4.3}{10} = 0.43\%$$

Oman calculation and correct answer:

$$\frac{24 - 2.5}{1000} = \frac{21.5}{1000} \times 100 = 2.15\% \quad or \quad 24 - 2.5 = \frac{21.5}{10} = 2.15\%$$

(d) Expressed as a percentage, how much greater is Oman's population growth rate than that of the United States?

Maximum 2 points total: 1 point for the correct calculation setup and 1 point for the correct answer.

The general formula and setup of: $\frac{v_1 - v_2}{v_2} \times 100 = \%$ change should be used to make this calculation. In this case, it makes more sense to have v_1 be Oman's value (2.15) and the v_2 value be the U.S. value (0.43). This will make the value a positive number, showing that Oman's growth rate is larger than the growth rate in the U.S. The setup and correct answer are shown below:

$$\frac{2.15 - 0.43}{0.43} \times 100 = \frac{1.72}{0.43} \times 100 = 4 \times 100 = 400\%$$

Oman's population growth rate is 400% greater than the growth rate in the U.S.

Students who understand that 1.72 is 3 times the value of 0.43 might respond that it is 300% greater; however, this is an incorrect answer. Instead, a correct answer could also be described in the following manner: 2.15 is 300% of 0.43. To earn credit for this statement, students would need to include the specific language "300% of."

(e) For any country experiencing rapid population growth, IDENTIFY and EXPLAIN one strategy or policy that might be suggested to a local government in order to reduce population growth.

Maximum 2 points total: 1 point for IDENTIFYING a viable strategy or policy and 1 point for correctly EXPLAINING how that strategy slows population growth.

Students must link their identified strategy/policy with an explanation that refers correctly to that mechanism. Examples AND their explanations will vary but may include the following:

Mechanism/Strategy/Policy to Slow or Discourage Population Growth	Associated Explanation of the Mechanism
Education of women/girls	As females gain access to basic education and/or higher levels of education, total fertility rates go down and populations grow less quickly. The reasons for this effect are varied, but include higher age of marriage for educated women, higher age of first child for educated women, and greater involvement with and control of household finances, which often results in lower birth rates.
Apply a one child per couple policy	Combinations of incentives or penalties can be deployed to implement a one-child policy, resulting in families with only one child. This would cause the TFR to move close to/equal to 1. Population growth slows and may even turn negative if the policy is maintained for long enough.
Provide sex education, access to contraceptives, and/or family planning services	Increased access to any of these services will encourage smaller families and reduce birth rates.

Topic 3.7

ANSWER SHEET
Practice Test 2

1. Ⓐ Ⓑ Ⓒ Ⓓ	21. Ⓐ Ⓑ Ⓒ Ⓓ	41. Ⓐ Ⓑ Ⓒ Ⓓ	61. Ⓐ Ⓑ Ⓒ Ⓓ
2. Ⓐ Ⓑ Ⓒ Ⓓ	22. Ⓐ Ⓑ Ⓒ Ⓓ	42. Ⓐ Ⓑ Ⓒ Ⓓ	62. Ⓐ Ⓑ Ⓒ Ⓓ
3. Ⓐ Ⓑ Ⓒ Ⓓ	23. Ⓐ Ⓑ Ⓒ Ⓓ	43. Ⓐ Ⓑ Ⓒ Ⓓ	63. Ⓐ Ⓑ Ⓒ Ⓓ
4. Ⓐ Ⓑ Ⓒ Ⓓ	24. Ⓐ Ⓑ Ⓒ Ⓓ	44. Ⓐ Ⓑ Ⓒ Ⓓ	64. Ⓐ Ⓑ Ⓒ Ⓓ
5. Ⓐ Ⓑ Ⓒ Ⓓ	25. Ⓐ Ⓑ Ⓒ Ⓓ	45. Ⓐ Ⓑ Ⓒ Ⓓ	65. Ⓐ Ⓑ Ⓒ Ⓓ
6. Ⓐ Ⓑ Ⓒ Ⓓ	26. Ⓐ Ⓑ Ⓒ Ⓓ	46. Ⓐ Ⓑ Ⓒ Ⓓ	66. Ⓐ Ⓑ Ⓒ Ⓓ
7. Ⓐ Ⓑ Ⓒ Ⓓ	27. Ⓐ Ⓑ Ⓒ Ⓓ	47. Ⓐ Ⓑ Ⓒ Ⓓ	67. Ⓐ Ⓑ Ⓒ Ⓓ
8. Ⓐ Ⓑ Ⓒ Ⓓ	28. Ⓐ Ⓑ Ⓒ Ⓓ	48. Ⓐ Ⓑ Ⓒ Ⓓ	68. Ⓐ Ⓑ Ⓒ Ⓓ
9. Ⓐ Ⓑ Ⓒ Ⓓ	29. Ⓐ Ⓑ Ⓒ Ⓓ	49. Ⓐ Ⓑ Ⓒ Ⓓ	69. Ⓐ Ⓑ Ⓒ Ⓓ
10. Ⓐ Ⓑ Ⓒ Ⓓ	30. Ⓐ Ⓑ Ⓒ Ⓓ	50. Ⓐ Ⓑ Ⓒ Ⓓ	70. Ⓐ Ⓑ Ⓒ Ⓓ
11. Ⓐ Ⓑ Ⓒ Ⓓ	31. Ⓐ Ⓑ Ⓒ Ⓓ	51. Ⓐ Ⓑ Ⓒ Ⓓ	71. Ⓐ Ⓑ Ⓒ Ⓓ
12. Ⓐ Ⓑ Ⓒ Ⓓ	32. Ⓐ Ⓑ Ⓒ Ⓓ	52. Ⓐ Ⓑ Ⓒ Ⓓ	72. Ⓐ Ⓑ Ⓒ Ⓓ
13. Ⓐ Ⓑ Ⓒ Ⓓ	33. Ⓐ Ⓑ Ⓒ Ⓓ	53. Ⓐ Ⓑ Ⓒ Ⓓ	73. Ⓐ Ⓑ Ⓒ Ⓓ
14. Ⓐ Ⓑ Ⓒ Ⓓ	34. Ⓐ Ⓑ Ⓒ Ⓓ	54. Ⓐ Ⓑ Ⓒ Ⓓ	74. Ⓐ Ⓑ Ⓒ Ⓓ
15. Ⓐ Ⓑ Ⓒ Ⓓ	35. Ⓐ Ⓑ Ⓒ Ⓓ	55. Ⓐ Ⓑ Ⓒ Ⓓ	75. Ⓐ Ⓑ Ⓒ Ⓓ
16. Ⓐ Ⓑ Ⓒ Ⓓ	36. Ⓐ Ⓑ Ⓒ Ⓓ	56. Ⓐ Ⓑ Ⓒ Ⓓ	76. Ⓐ Ⓑ Ⓒ Ⓓ
17. Ⓐ Ⓑ Ⓒ Ⓓ	37. Ⓐ Ⓑ Ⓒ Ⓓ	57. Ⓐ Ⓑ Ⓒ Ⓓ	77. Ⓐ Ⓑ Ⓒ Ⓓ
18. Ⓐ Ⓑ Ⓒ Ⓓ	38. Ⓐ Ⓑ Ⓒ Ⓓ	58. Ⓐ Ⓑ Ⓒ Ⓓ	78. Ⓐ Ⓑ Ⓒ Ⓓ
19. Ⓐ Ⓑ Ⓒ Ⓓ	39. Ⓐ Ⓑ Ⓒ Ⓓ	59. Ⓐ Ⓑ Ⓒ Ⓓ	79. Ⓐ Ⓑ Ⓒ Ⓓ
20. Ⓐ Ⓑ Ⓒ Ⓓ	40. Ⓐ Ⓑ Ⓒ Ⓓ	60. Ⓐ Ⓑ Ⓒ Ⓓ	80. Ⓐ Ⓑ Ⓒ Ⓓ

Practice Test 2

Section I (Multiple-Choice Questions)

TIME: 90 MINUTES
80 QUESTIONS
60% OF TOTAL GRADE

DIRECTIONS: For each question or statement, select the one lettered choice that is the best answer, and fill in the corresponding circle on the answer sheet.

1. Your APES class was on a field trip and discovered the following fossil:

What type of rock was this fossil most likely found in?

(A) Igneous
(B) Lava
(C) Metamorphic
(D) Sedimentary

2. What was Earth's original source of free atmospheric oxygen (O_2)?

(A) Breakdown of stratospheric ozone by ultraviolet light
(B) Chemosynthesis by primitive bacteria
(C) Oxygen produced from photosynthetic marine algae
(D) Photosynthesis by cyanobacteria (blue-green algae)

3. Not all water taken from a source (e.g., aquifer, river, well, etc.) reaches the root zone of plants, as some of it is lost through inefficiency (e.g., percolation through soil layers below the root zone, seepage from irrigation systems, surface runoff, etc.).

Field application efficiency (e_a) is the percentage of irrigation water that is used efficiently. Conveyance efficiency (e_c), also expressed as a percentage, is a value that numerically expresses the efficiency of an irrigation method (e.g., drip, flooding, etc.).

You are in charge of a project to design a long canal form of field irrigation. The project requires a long canal to be constructed of clay liners, and the irrigation method uses furrows dug into the field.

Using the tables below, what is the project's irrigation efficiency (e) where $e = [e_i \cdot e_c] / 100\%$?

Conveyance Efficiency (e_c)

	Earthen Canals			Lined Canals
	Sand	Loam	Clay	
Canal Length				
Long (>2000 m)	60%	70%	80%	95%
Medium (200–2000 m)	70%	75%	85%	95%
Short (<200 m)	80%	85%	90%	95%

Field Application Efficiency (e_a)

Irrigation Method	e_a
Surface (basin flood, furrow, etc.)	60%
Sprinkler	75%
Drip	90%

(A) 14%
(B) 24%
(C) 48%
(D) 62%

4. Which of the following choices would be most likely to increase competition among members of a squirrel population in a given area?

(A) An epidemic of rabies within the squirrel population
(B) An increase in the food supply
(C) An increase in the number of hawks in the area
(D) An increase in the number of squirrels in the area

PRACTICE TEST 2

Examine the nitrogen cycle diagram below to answer Question 5.

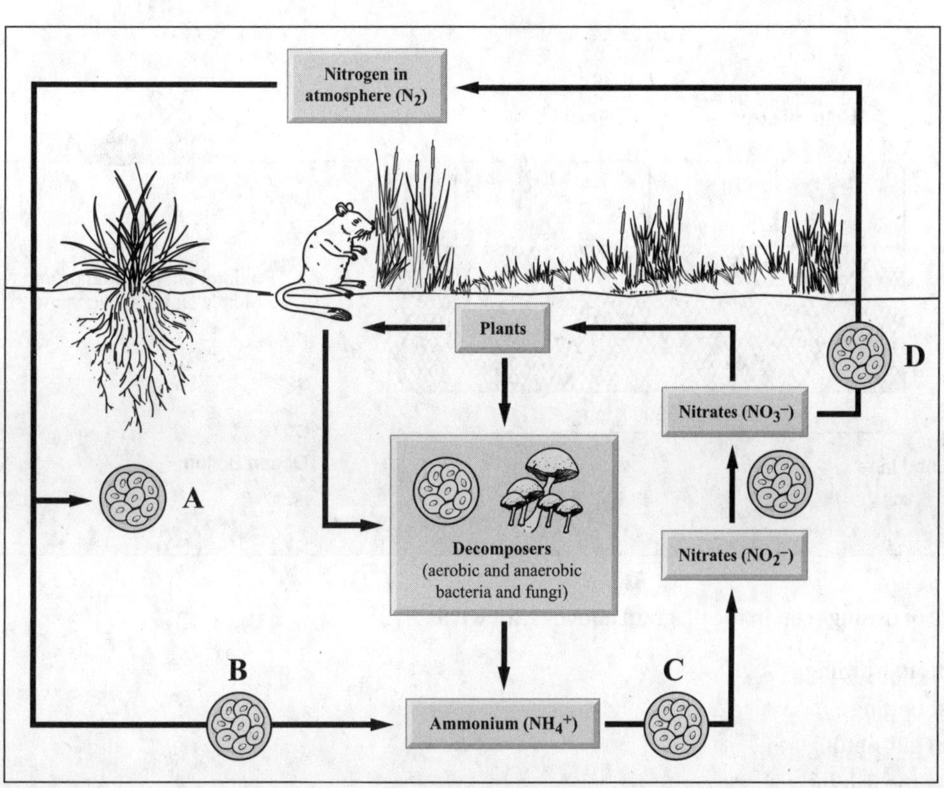

5. Nitrogen-fixing bacteria are primarily found at location(s)

    (A) A
    (B) B
    (C) C
    (D) A and B

6. An APES class was doing a field study of effluent water entering a river that passed through a mangrove forest. Upstream from the mangrove forest, a large factory was dumping material from its production processes through large drainage pipes. Which of the following choices would be a logical conclusion based upon their observations?

    (A) The students detected high concentrations of metal ions in the river bottom upstream of the factory; found nothing but anaerobic bacteria, fungi, and sludge worms living in the mud; and concluded that the factory was producing inorganic fertilizer and releasing some of the fertilizer into the water.
    (B) The students detected measurable amounts of coliform bacteria in the river and low dissolved oxygen content in the water downstream of the drainage pipe and concluded that the factory could be a meat-packing plant.
    (C) The students found the turbidity of the water coming out of the drainage pipe to be extremely high and concluded that the factory was producing pharmaceuticals.
    (D) The students observed a large number of insect larvae in the mangrove forest one mile downstream from the drainage pipe and concluded that the water was polluted by the factory releasing food wastes into the water.

Refer to the diagram below to answer Question 7.

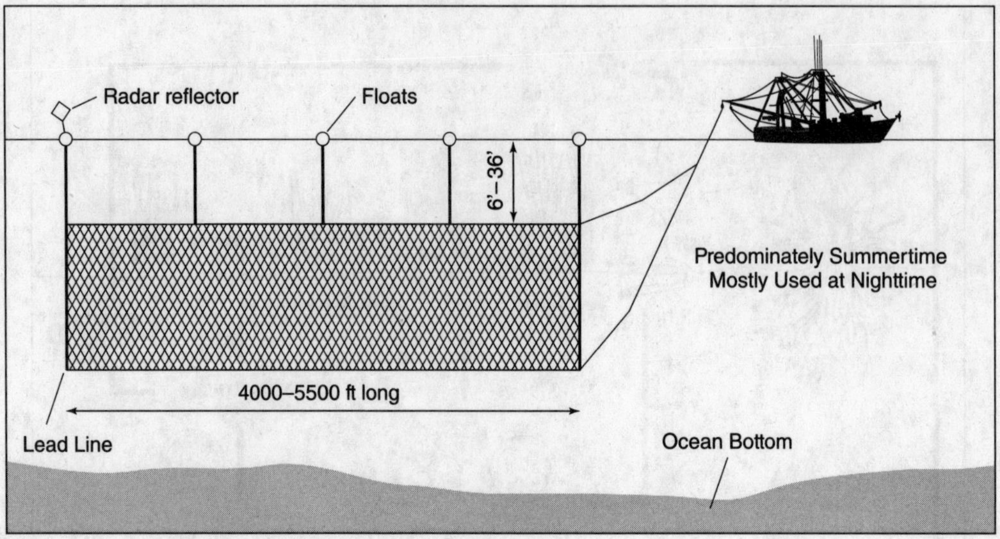

7. The type of fishing seen in the diagram above is known as

(A) bottom trawling
(B) cast netting
(C) drift gill-netting
(D) longline fishing

8. According to Thomas Malthus, which of the following would be true?

(A) Humans will be able to exceed and maintain the natural carrying capacity of Earth and thereby increase the standard of living as a result of technological advances in agriculture, medicine, and sanitation.
(B) The human population will eventually exceed its carrying capacity, which will result in mass starvation.
(C) The human population will eventually reach its carrying capacity and maintain it.
(D) The human population will never reach its carrying capacity.

9. The graph below shows the relationship between photosynthetic rate and temperature for plant species A–D. Based on these results, which species is *best* adapted to Arctic conditions, where the mean temperature does not exceed 8°C during the growing season?

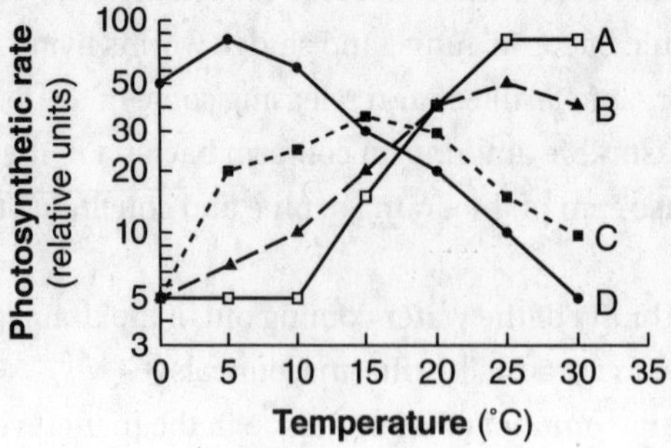

(A) A
(B) B
(C) C
(D) D

10. An APES class measured the biological oxygen demand along eight miles of a stream located in a wetland. Their results are presented below:

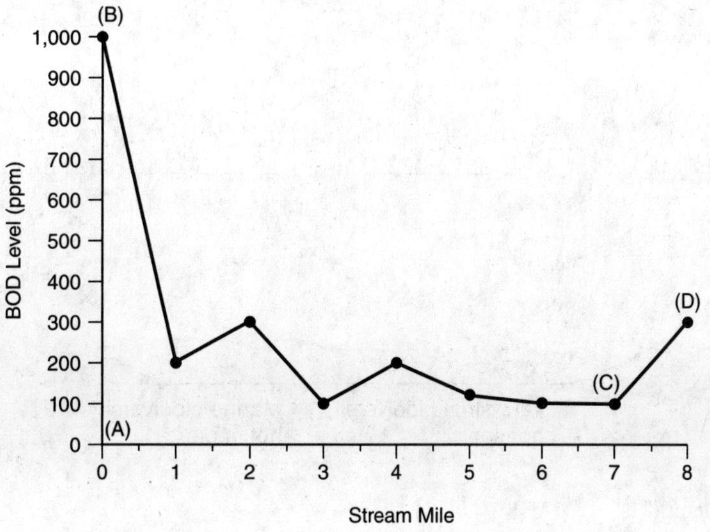

At which point on the graph would the highest concentration of anaerobic (living with little or no oxygen) bacteria be found?

(A) A
(B) B
(C) C
(D) D

11. Which of the following atmospheric conditions would contain the most amount of water vapor in a given volume of air?

(A) 10°F at 10% relative humidity
(B) 10°F at 80% relative humidity
(C) 50°F at 80% relative humidity
(D) 80°F at 10% relative humidity

12. Which of the following statements about the role of carbon dioxide (CO_2) in the carbon cycle is/are TRUE?

I. Carbon dioxide concentration in the atmosphere decreases when trees are cut down and trees decay.
II. Carbon dioxide is produced during photosynthesis.
III. The primary source of carbon dioxide entering Earth's atmosphere NOT caused by humans is outgassing from Earth's interior.

(A) I
(B) II
(C) III
(D) I and III

Refer to the following diagram to answer Question 13.

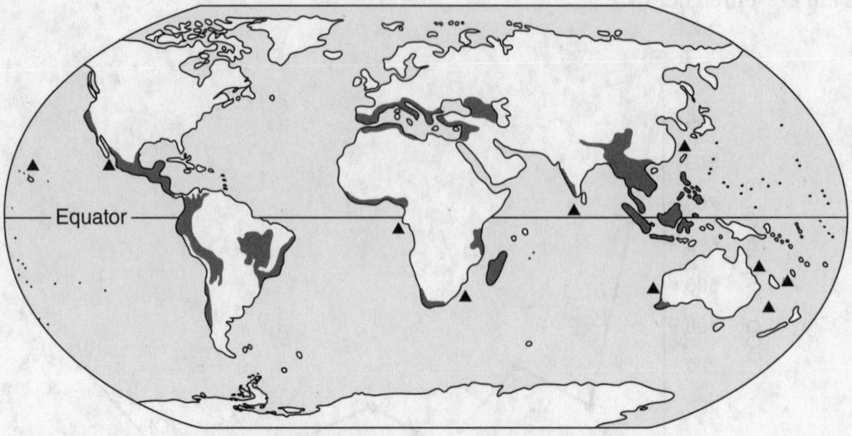

13. A biodiversity "hotspot" is

(A) a region with high biodiversity that is under threat from humans
(B) a region with high biodiversity that is not under threat from humans
(C) an area that has been pinpointed as being the origin of an outbreak of a contagious disease
(D) an area that is experiencing a rapid birth rate

14. Which of the following choices is NOT a benefit associated with integrated pest management (IPM)?

(A) Decreasing the chance of pests developing resistance to a pesticide
(B) A more reliable and effective method of pest control
(C) Reducing the amount of the most hazardous types of pesticide needed
(D) Total elimination of pest species

Refer to the diagram below to answer Question 15.

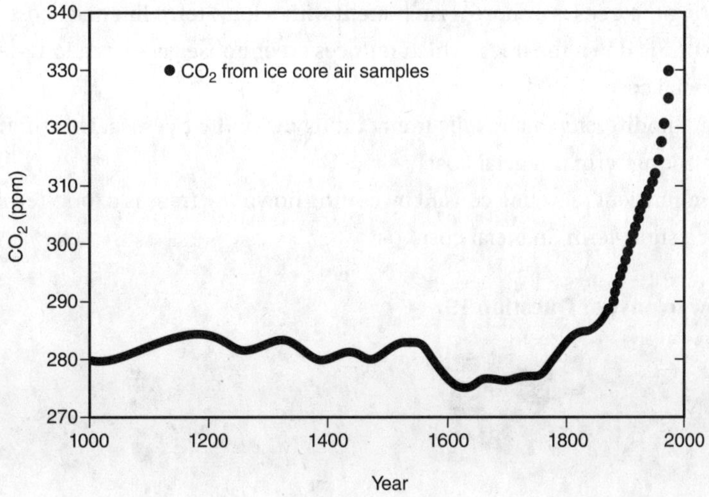

15. A scientific study took ice core samples to measure the concentration of carbon dioxide that was present in the atmosphere for the last 1,000 years. What was the approximate net change in CO_2 concentration that occurred from the year 1000 to the year 2000?

(A) 10 ppm
(B) 50 ppm
(C) 200 ppm
(D) 330 ppm

16. The overall threat of a greenhouse gas depends on its

I. global warming potential
II. lifetime in the atmosphere
III. relative or potential abundance in the atmosphere

(A) I
(B) II
(C) III
(D) I, II, and III

17. Polar bears rely heavily on their sea ice environment for traveling, hunting, mating, resting, and, in some areas, maternal dens. In particular, they depend heavily on sea ice–dependent prey, such as ringed and bearded seals. Additionally, their long generation time and low reproductive rate may limit their ability to adapt to rapid changes in the environment. Which of the following statements below would be TRUE if a warming polar climate trend continues?

(A) The carrying capacity would decrease, as would the polar bear population.
(B) The carrying capacity would decrease, but the polar bear population would remain the same.
(C) The carrying capacity would remain the same, but the population of polar bears would decrease.
(D) The number of polar bear deaths would increase but the number of births would also increase, so the population size would remain the same.

18. You raise cattle and are thinking of converting a section of wooded land that you own into grazing pasture for cattle. Which of the following choices listed below would be TRUE?

(A) The cost of cutting the trees is a short-term benefit with a long-term financial cost.
(B) The impact of cutting down the trees, which reduces greenhouse gases, is a long-term benefit with a long-term financial cost.
(C) The reduction in biodiversity that results from cutting down the trees is a short-term environmental cost combined with a long-term financial cost.
(D) The reduction in nutrient recycling caused by cutting down the trees is a long-term environmental cost combined with a short-term financial cost.

Examine the map below to answer Question 19.

19. Which areas listed below would be found in the regions of Earth primarily influenced by Hadley cells?

(A) 1
(B) 2
(C) 3
(D) 4

20. Which statement(s) below concerning ozone (O_3) is/are TRUE?

I. Ozone is considered an air pollutant in the troposphere or lower atmosphere but is beneficial in the stratosphere because it protects Earth and all life forms from excessive ultraviolet radiation.
II. Ozone is considered beneficial in the troposphere, as it protects Earth from harmful ultraviolet radiation, but is considered an air pollutant in the stratosphere.
III. Ozone is considered beneficial in both the troposphere and the stratosphere because it protects Earth and all life forms from excessive ultraviolet radiation.

(A) I
(B) II
(C) III
(D) I and II

Refer to the diagram below to answer Question 21.

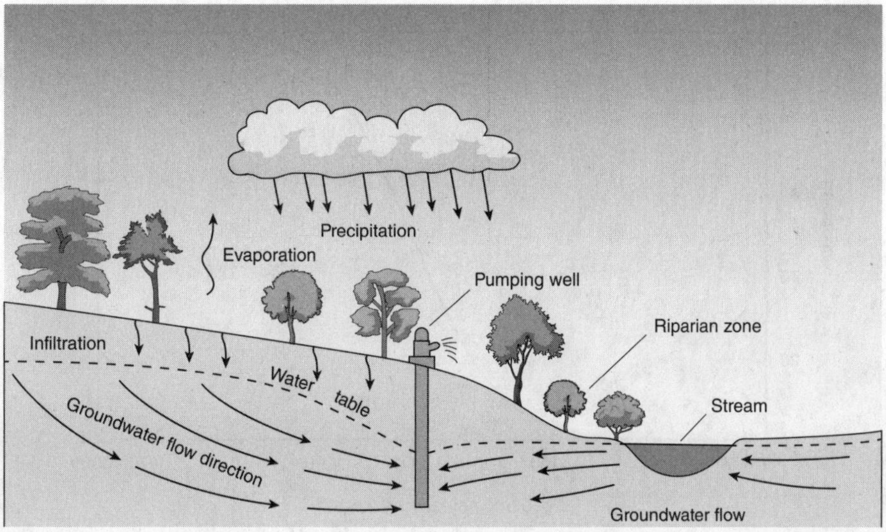

Source: U.S. Geological Survey

21. Groundwater flow

(A) can be replenished indefinitely
(B) has a faster flow rate than surface waters
(C) is not affected by the permeability of soil or rock
(D) moves downward by the force of gravity in whichever direction the water table slopes

22. The distribution of biomes around the world is in large part determined by long-term seasonal weather patterns, which themselves are primarily determined by

(A) ocean currents
(B) the amount of solar radiation that is released during certain times of the year
(C) the angle at which solar radiation reaches Earth
(D) the distance from Earth to the sun during a particular time of year

Read the following passage to answer Question 23.

An APES class was doing a field study of a lake. The class determined the amount of dissolved oxygen in the lake at a depth of two feet to be 8 mg of O_2 per liter. Then the class filled and sealed two clear glass bottles with lake water from the same location and depth, labeling one bottle "light" and the other bottle "dark." The "dark" bottle was wrapped completely with several layers of foil. Both bottles were then lowered into the lake at the same location and depth where the initial sample was taken. After one hour, the bottles were retrieved and the amount of oxygen in the "light" bottle was determined to be 10 mg O_2 per liter, while the amount of oxygen in the "dark" bottle was determined to be 5 mg O_2 per liter.

23. The net primary productivity of the lake water sample was

(A) 2 mg O_2/L/hr
(B) 3 mg O_2/L/hr
(C) 4 mg O_2/L/hr
(D) 5 mg O_2/L/hr

Refer to the graph below to answer Question 24.

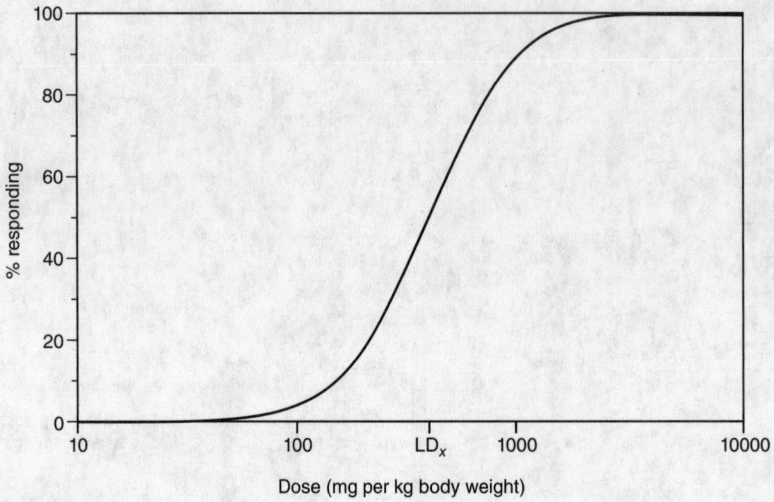

24. A large agricultural company that manufactures pesticides had developed a new rodenticide to kill a certain species of rat that was damaging citrus crops in a developing country. After repeated testing of various doses of this new rodenticide using laboratory rats, the results are presented above. Which of the following statements is the most accurate regarding this rodenticide?

(A) Any rat exposed to 400 mg of this rodenticide will die.
(B) For every 100 rats exposed to 400 mg of this rodenticide, 50 will die.
(C) Out of 100 people exposed to 400 mg of this rodenticide, 50 will become acutely ill and die.
(D) Out of 100 rats each receiving a dose of 400 mg of this rodenticide per kilogram of body weight, 50 will die.

25. Mitigation techniques that can be used for overgrazing include all of the following EXCEPT

(A) planting warm-season perennial grasses, e.g., switchgrass
(B) selecting proper breeds of animals that work best with the natural land ecology and the resources that are available
(C) using rotational grazing schedules
(D) All of the choices above are methods that can be used to mitigate overgrazing.

26. Refer to the following graph, which illustrates survival rates for five animal populations and shows the relationship of the number of individuals in a population to units of life span.

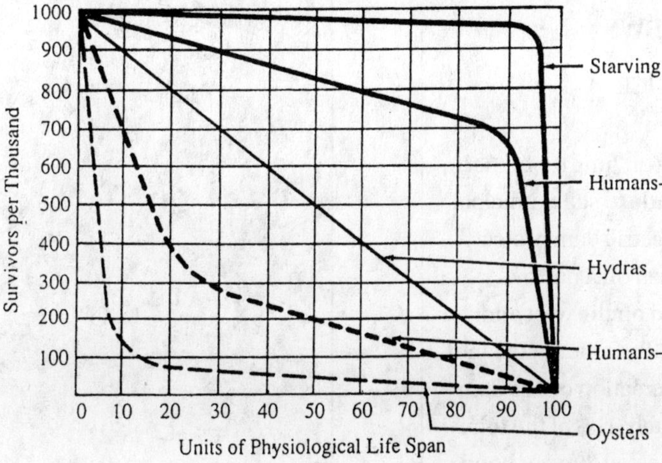

When survival curves are calculated, the following assumptions are made:

I. All individuals of a given population are the same age.
II. No new individuals enter the population.
III. No individuals leave the population.

According to the graph, it can be assumed that

(A) oysters outlive fruit flies
(B) the life span of humans is the same as oysters
(C) the population of hydras is remaining stable
(D) there is a high mortality rate among young oysters

27. Kerosene, gasoline, motor oil, and diesel fuel all come from crude oil. Refineries take advantage of what physical property in order to separate these components from the original crude oil?

(A) Boiling point
(B) Density
(C) Solubility
(D) Viscosity

28. Which of the following pollutants contributes to the formation of both acid rain AND photochemical smog?

(A) NO_2
(B) O_3
(C) Particulates
(D) SO_2

Refer to the following newspaper article to answer Question 29.

> **SEATTLE NEWS**
>
> PUBLIC NOTICE
>
> Seattle voters will be deciding on June 6 whether or not to allow the City of Seattle to sell municipal bonds to build a new hydroelectric dam project on the Duwamish River, located on a narrow isthmus between Puget Sound on the west and Lake Washington on the east. The amount of land that would be flooded by the creation of this dam would be equal to approximately 41% of the total area of the city of Seattle. Projected 20-year bond yields are expected to be 4.2%. A public meeting will be held February 2 at 7:00 PM at Seattle City Hall.

29. Which of the following choices would be (a) possible environmental consequence(s) of this hydroelectric project?

    I. Increased sediment downstream of the dam that would clog fish gills and need to be periodically dredged
    II. Altered water temperatures affecting fish and wildlife
    III. Reduced oxygen content (hypoxia) of the water released by the dam

    (A) I
    (B) II
    (C) III
    (D) II and III

30. Which two greenhouse gases listed below contribute most to climate change?

    (A) Carbon dioxide and methane
    (B) Carbon dioxide and nitrous oxide
    (C) Carbon dioxide and water vapor
    (D) Carbon dioxide and sulfur hexafluoride

Refer to the diagram below to answer Question 31.

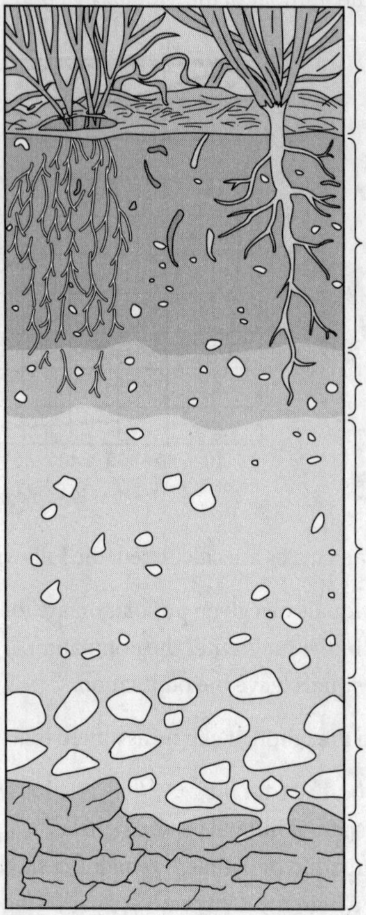

31. According to the diagram above, which soil layer consists of clay, iron oxides, and other components that came from the zone of leaching?

    (A) A
    (B) B
    (C) C
    (D) O

32. Ecosystem services are the direct and indirect contributions of ecosystems to human well-being, and support directly or indirectly human survival and quality of life. Which of the following choices below is/are NOT examples of key services provided by ecosystem services?

    (A) Climate regulation
    (B) Water purification
    (C) Both A and B
    (D) Neither A nor B

Refer to the following diagram to answer Question 33.

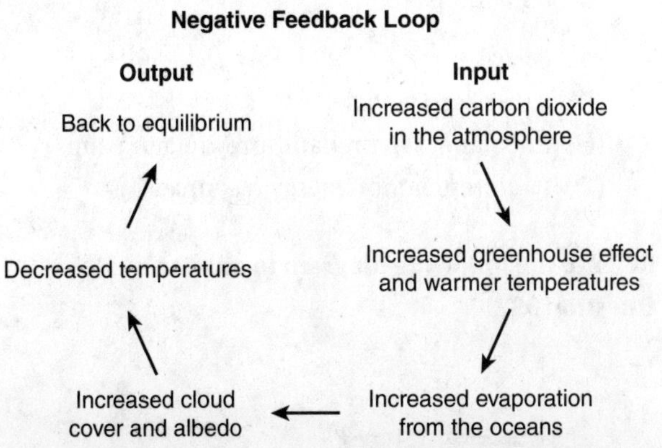

33. Which of the examples described below is/are consistent with a negative feedback loop?

I. Atmospheric cooling tends to increase ice cover, hence increasing the albedo, which thereby reduces the amount of solar energy absorbed by Earth and leads to more cooling.

II. Global climatic warming is expected to increase rates of land degradation and desertification, which in turn results in the emission of more disturbed windblown dust. These particles then act to cool Earth's surface by increasing atmospheric reflectivity, thereby decreasing the amount of solar radiation absorbed by the land surface.

III. Global warming would decrease soil moisture in some areas and may bring about an increased frequency of natural fires. The burning vegetation would pump even more CO_2 into the atmosphere.

(A) I
(B) II
(C) III
(D) I and III

Read the following passage in order to answer Question 34.

In 1990, fewer than fifty green sea turtles were documented nesting at the Archie Carr National Wildlife Refuge on Florida's east coast. This 20-mile stretch of beach hosted more than 10,000 green sea turtle nests in 2013, making this one of the greatest conservation success stories of our time.

34. The success of the green sea turtles' comeback from the brink of extinction is primarily due to the

(A) Convention on International Trade in Endangered Species
(B) Endangered Species Act
(C) International Treaty for Protecting Endangered Animals and Plants
(D) All of the above

Refer to the following diagram to answer Question 35.

35. The type of lake shown above is known as a(n)

(A) eutrophic lake
(B) mesotrophic lake
(C) oligotrophic lake
(D) oxbow lake

Refer to the following diagram to answer Question 36.

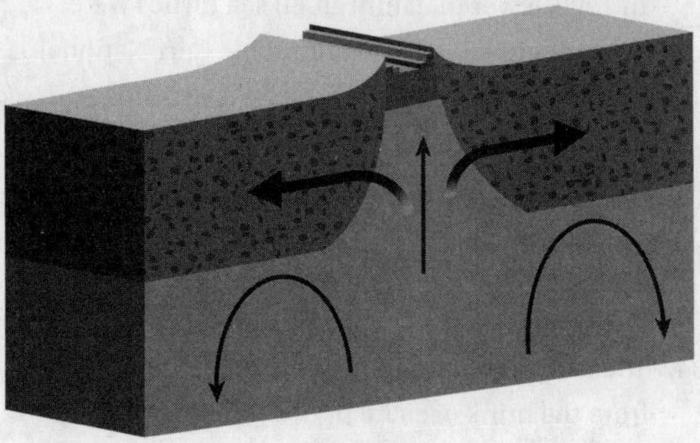

36. The locations where two tectonic plates slide apart from each other, with the space that was created being filled with molten magma from below (e.g., East African Great Rift Valley, Mid-Atlantic Ridge), as seen in the diagram above are known as

(A) convergent boundaries
(B) divergent boundaries
(C) tectonic boundaries
(D) transform boundaries

Refer to the following diagram to answer Question 37.

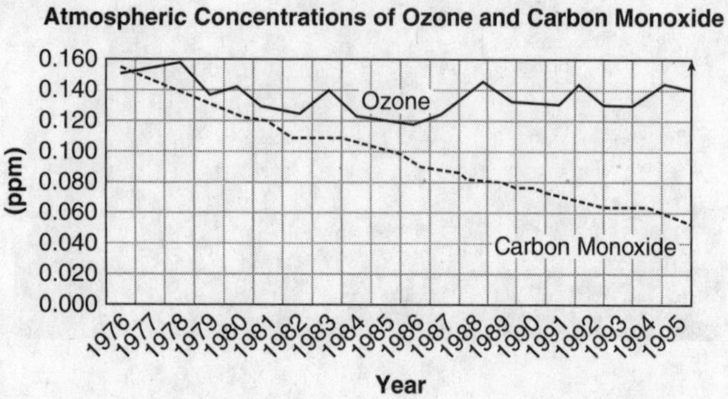

37. The diagram above shows the change in atmospheric concentrations of carbon monoxide and ozone between 1976 and 1995. The percent change in atmospheric carbon monoxide concentration was

(A) ~ −10%
(B) ~ −63%
(C) ~ 0%
(D) ~ 63%

38. Which of the following is LEAST likely to result in density-dependent effects on the growth of natural populations?

(A) Disease
(B) Increased rainfall
(C) Imbalances in predator-prey relationships
(D) Requirement for energy resources

Refer to the following diagram to answer Question 39.

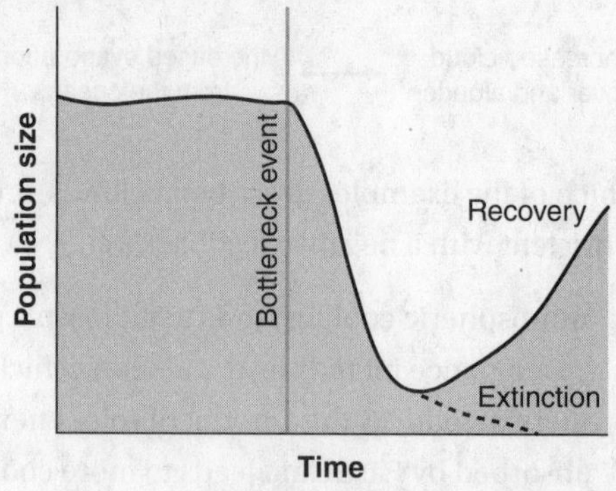

39. The diagram above shows a "bottleneck" event occurring over time. Which of the following choices below is NOT consistent with a bottleneck event?

(A) The northern elephant seal population fell to ~30 during the 1890s. Although it has recovered and now numbers in the hundreds of thousands, only the dominant male bulls generally mate, which results in limited genetic diversity, making the species more vulnerable to diseases and genetic mutations.
(B) The number of greater prairie chickens in Illinois plummeted from ~100 million in 1900 to ~50 in 1990 and was the result of hunting and habitat destruction, resulting in a drastic reduction of species diversity.
(C) The population of American bison drastically declined around the turn of the 20th century due to overhunting, which cleared land for agriculture and ultimately reduced genetic variation in the species and nearly led to extinction, though the American bison has since begun to recover.
(D) All choices are examples of population (or genetic) bottleneck.

PRACTICE TEST 2

40. Which of the following agricultural practices has the LEAST impact on controlling soil erosion?

(A) Drip irrigation
(B) No-till planting
(C) Surface irrigation
(D) Terracing

Refer to the seismogram below to answer Question 41.

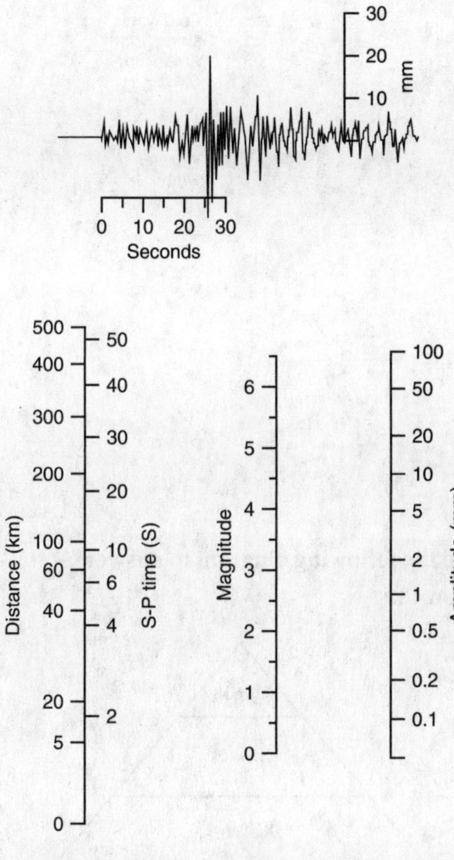

41. An APES class spent a class period determining the magnitude of earthquakes from seismograms. The class was given the seismogram above, and measured the height from the baseline to the top of the highest P wave as 20 mm. They then determined the time interval between the beginning of the S waves and the beginning of the first P waves to be 25 seconds. Using the seismogram above, the class determined the Richter magnitude of the earthquake to be

(A) 2.0
(B) 3.0
(C) 4.0
(D) 5.0

42. Most of the energy used by the world today comes from (in descending order)

(A) natural gas > oil > coal > nuclear
(B) coal > oil > nuclear > natural gas
(C) oil > coal > natural gas > hydro
(D) oil > natural gas > coal > hydro

Refer to the following diagram to answer Question 43.

The El Niño Phenomenon

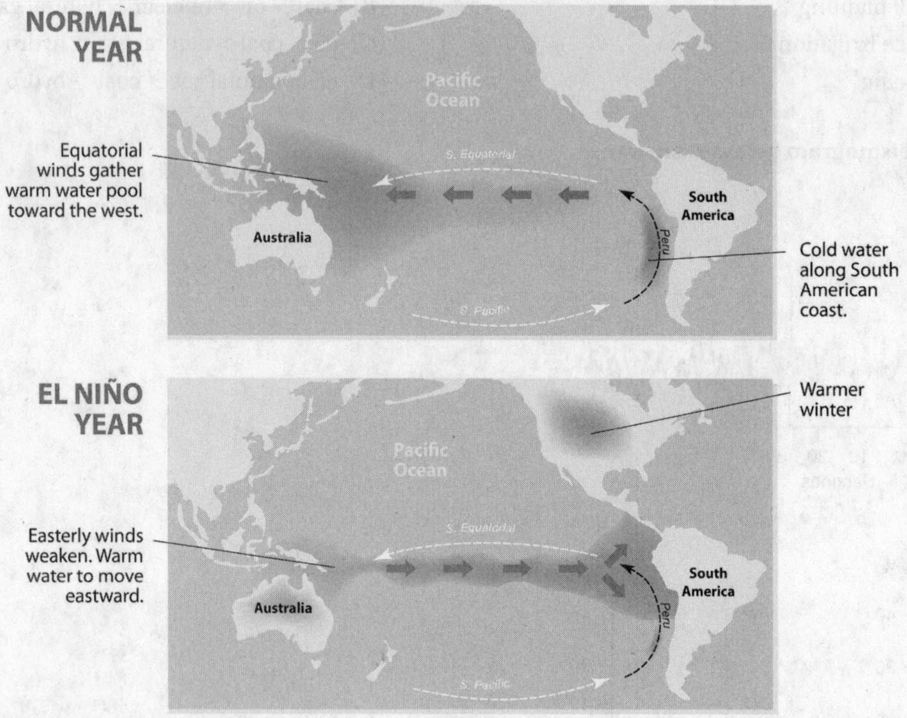

43. Which of the following best represents El Niño conditions in the Pacific Ocean near Peru?

	Water Temperature	Rainfall
(A)	Low	Low
(B)	Low	High
(C)	High	Low
(D)	High	High

44. The concept of net primary productivity

(A) is the rate at which producers manufacture chemical energy through photosynthesis
(B) is the rate at which producers use chemical energy through respiration
(C) is the rate of photosynthesis plus the rate of respiration
(D) is usually reported as the energy output of an area of producers over a given time period

Refer to the following diagram to answer Question 45.

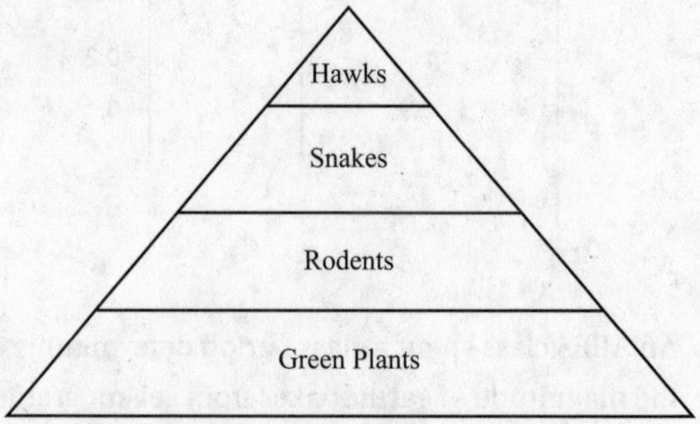

45. The food pyramid above implies that in order to live and grow, 1,000 kilograms of snakes would require

(A) less than 1,000 kilograms of green plants
(B) less than 1,000 kilograms of hawks
(C) more than 1,000 kilograms of rodents
(D) ~1,000 kilograms of rodents

46. If a city with a population of 100,000 experiences 4,000 births, 3,000 deaths, 500 immigrants, and 200 emigrants within the course of one year, what is the net annual percentage growth rate?

(A) 0.13%
(B) 1.3%
(C) 13%
(D) 130%

Refer to the diagram below for help in answering Question 47.

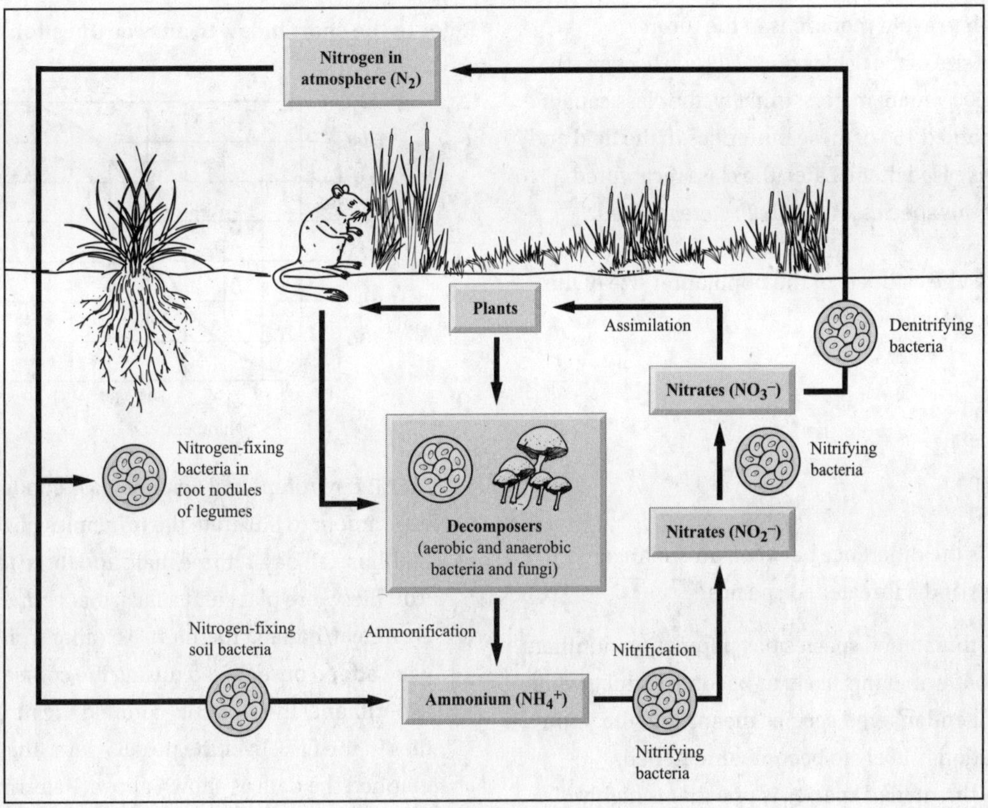

47. In which stage of the nitrogen cycle do soil bacteria convert ammonium ions (NH_4^+) into nitrate ions (NO_3^-)—a form of nitrogen that can be used by plants?

(A) Ammonification
(B) Assimilation
(C) Nitrification
(D) Nitrogen fixation

48. Which of the following proposals would NOT increase the sustainability of ocean fisheries?

(A) Establish fishing quotas based on past harvests.
(B) Set quotas for fisheries well below their estimated maximum sustainable yields.
(C) Sharply reduce fishing subsidies.
(D) Shift the burden of proof to the fishing industry for them to show that their operations are sustainable.

Refer to the following information to answer Question 49.

An APES class was doing field research using a mark-recapture technique to estimate the population size of a population of monarch butterflies, which have a specific dietary resource requirement (milkweed) that is their only food source in the caterpillar stage. An animal that eats a monarch butterfly usually doesn't die, but feels sick enough to avoid monarchs in the future.

The size of the field was exactly two hectares (ha), or 20,000 square meters. Initially, the class caught and marked 150 of these butterflies in the field and then released them. Later, the class recaptured 100 of this species, of which 20 were marked.

49. What is the estimate of the population size of this species of butterfly?

(A) 75
(B) 750
(C) 1,500
(D) 3,000

50. What is the difference between an endangered species and a threatened species?

(A) A threatened species has population numbers so low that it is likely to become extinct, while an endangered species means that the population is likely to become threatened.
(B) A threatened species is one that is highly endangered. An endangered species is one that is highly threatened.
(C) A threatened species means that the population is likely to become endangered. An endangered species has population numbers so low that it is likely to become extinct.
(D) A threatened species has a population size that is not capable of sustaining itself. An endangered species means that the population's numbers have decreased greatly over the last 10 years.

51. A store uses one hundred 200-watt lightbulbs for 10 hours per day. How many kilowatt-hours of electrical energy are used in one year if the store is open 50 weeks per year, 6 days per week?

(A) 1,500
(B) 3,000
(C) 6,000
(D) 60,000

Refer to the chart below to answer Question 52.

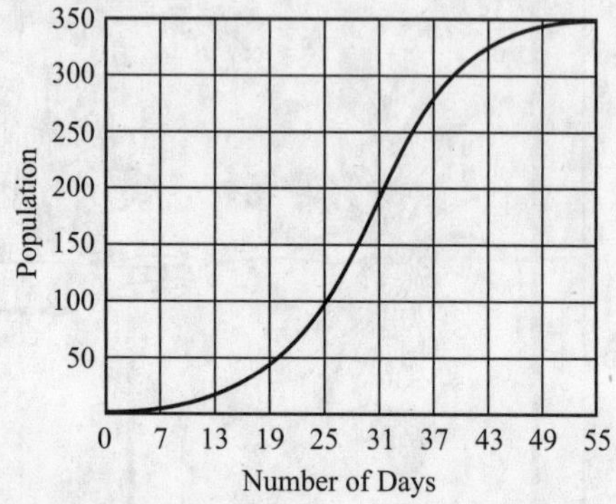

52. An AP Environmental Science class conducted an experiment to illustrate the principles of Thomas Malthus. On day 1, three male and three female fruit flies were placed in a flask that contained a cornmeal/banana medium. No other fruit flies were added or removed during the course of the experiment. The students counted the number of flies in the flask over the next 55 days. They then graphed the data as shown above. Referring to the students' graph, the rate of reproduction is equal to the rate of death on day/days

(A) 1
(B) 25
(C) 55
(D) 1 and 55

Refer to the following newspaper article to answer Question 53.

> **BARRONSVILLE GAZETTE**
>
> AIR POLLUTION TO REACH DANGEROUS LEVELS
>
> Weather experts are expecting air pollution levels to be extremely high in Barronsville from July 1–July 7.
>
> Temperatures in Barronsville next week may actually be higher in the local mountains than they will be at lower elevations.
>
> Driving visibility may also be affected due to current dangerous levels of air pollution, and fog may also be present. Schools will make an announcement in the next few days as to whether they will be open during this period of excessive air pollution. Residents are urged to keep windows and doors closed and refrain from physical exertion.

53. Which of the following conditions is consistent with the newspaper article above?

(A) Cold air that is heavier lies above a mass of warmer air that wants to rise, trapping particulates close to the ground.
(B) Stable, warm air lies above a mass of colder air, trapping air particulates close to the ground.
(C) There is no cold air mass during the summer to provide a difference in air temperatures; therefore, the air stalls and traps particulates close to the ground.
(D) Warm air rises while colder air sinks. Particulates within the cold air are found closer to the ground and their concentration increases, producing high levels of smog.

Refer to the following graphs, which show possible relationships between the amount of habitat area (m^2) and the possible number of species that are able to survive in that habitat, to answer Question 54.

A

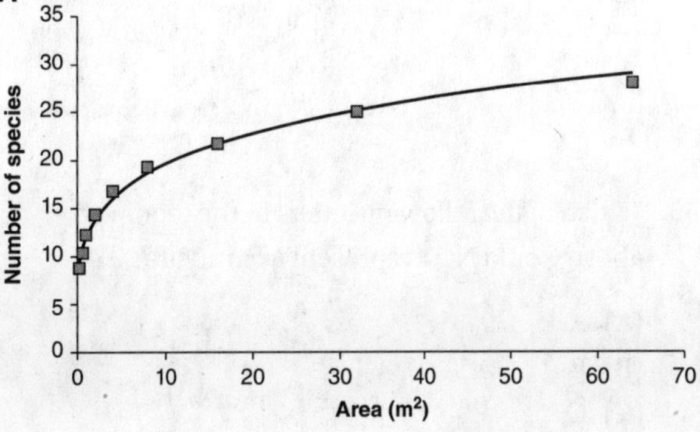

B

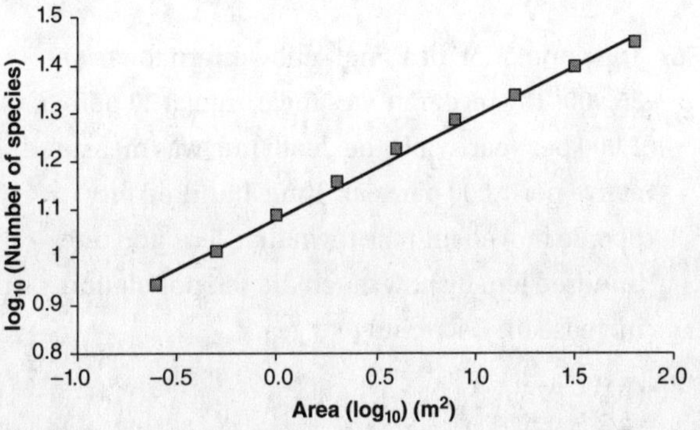

C

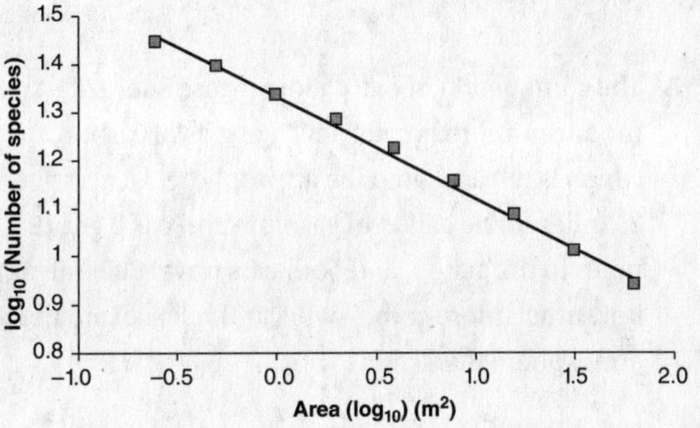

54. Which graph(s) above show correctly the relationship between the number of species in an area as a function of the size of that area?

(A) A
(B) B
(C) C
(D) A and B

Use the following diagram to help answer Question 55.

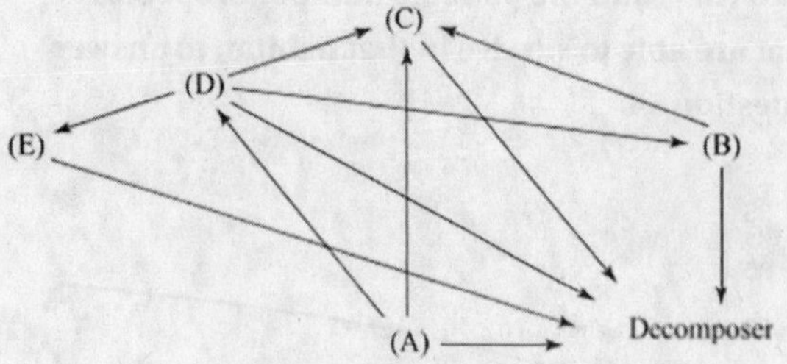

55. Which of the following letters in the food web above would NOT represent a carnivore?

(A) A
(B) B
(C) C
(D) A, C, and D

56. The population of a small midwestern town was 35,000. The birth rate was measured at 20 per 1,000 per year, while the death rate was measured at five per 1,000 per year. Three hundred fifty people moved into the town that year, and one hundred left. By how much did the population increase (or decrease) that year?

(A) −775
(B) +775
(C) +1,200
(D) +1,775

57. The combustion of one gallon of gasoline produces approximately five pounds of carbon, which is released into the atmosphere. If car A gets 25 miles to the gallon of gasoline and car B get 15 miles to the gallon, and both cars travel 600 miles, how much more carbon will car B release into the atmosphere than car A?

(A) 4 pounds
(B) 24 pounds
(C) 40 pounds
(D) 80 pounds

Refer to the following diagram to answer Question 58.

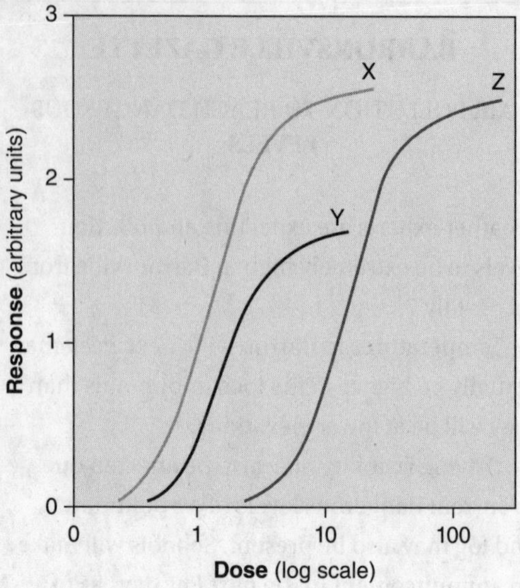

58. Which drug(s) has/have the least potency for the greatest therapeutic effect?

(A) X
(B) Y
(C) Z
(D) X and Z

59. The primary cause(s) of worldwide tropical deforestation is/are

(A) agricultural expansion (crops and cattle)
(B) clearing of land for cities
(C) commercial logging
(D) all of the above

Examine the two islands below to answer Question 60. The number of arrows reaching an island are indicators of "more."

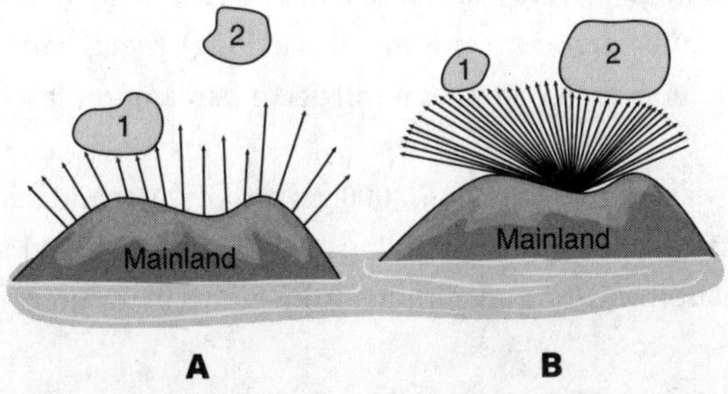

60. Which statement below is FALSE?
    (A) In diagram A, Island 1 receives more random dispersion of organisms.
    (B) In diagram A, Island 2 receives less random dispersion of organisms.
    (C) In diagram B, Island 1 receives less random dispersion of organisms.
    (D) In diagram B, Island 2 receives less random dispersion of organisms.

61. Place the following steps involved in the formation of photochemical (brown) smog in chronological order.
    I. Concentrations of nitrogen oxides and VOCs increase.
    II. Production of ozone stops.
    III. Nitrogen dioxide is broken down and forms ozone, nitric acid, and nitric oxide. NO_2 also reacts with volatile organic compounds to produce PANs (peroxyacyl nitrates).
    IV. Nitrogen oxides and VOCs begin to react, forming nitrogen dioxide.

    (A) I, II, III, IV
    (B) I, IV, III, II
    (C) II, IV, I, II
    (D) III, I, II, IV

Refer to the following diagram to answer Question 62.

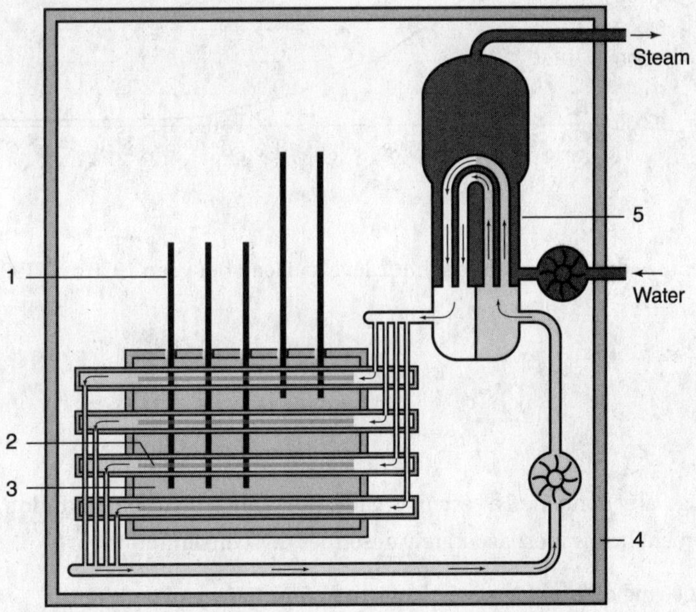

62. In the diagram above of a heavy-water moderated nuclear reactor, #1 represents the
    (A) control rods
    (B) fuel rods
    (C) heat exchanger
    (D) moderator

63. Mrs. Saunders' AP Environmental Science class was studying watersheds. She handed the class a field study report conducted by the U.S. Forest Service on determining the potential for increasing water yield in a local watershed by manipulating ponderosa pine and pinyon-juniper vegetation types. Mrs. Saunders explained that water runoff (R) or streamflow is a function of precipitation (P) and evapotranspiration (E_t)—water loss by evaporation or transpired from plants—and to have perennial flow (streamflow all year long), precipitation must exceed E_t. If precipitation exceeds E_t only part of the time, then ephemeral will occur (streamflow lasting for only a short period of time).

Data for that year showed that runoff in the study area was estimated to be 70,000 gallons of water and that the rainfall amount in the watershed for that year was estimated to be 100,000 gallons. Mrs. Saunders asked the class to determine the amount of water lost by evaporation and/or transpiration from plants in the area of study.

The correct amount of water lost due to evaporation and/or transpiration was estimated to be:

(A) 30,000 gallons
(B) 40,000 gallons
(C) 100,000 gallons
(D) 170,000 gallons

Use the graph below, which shows the atmospheric concentrations of lead from 1976 to 1995, to answer Question 64.

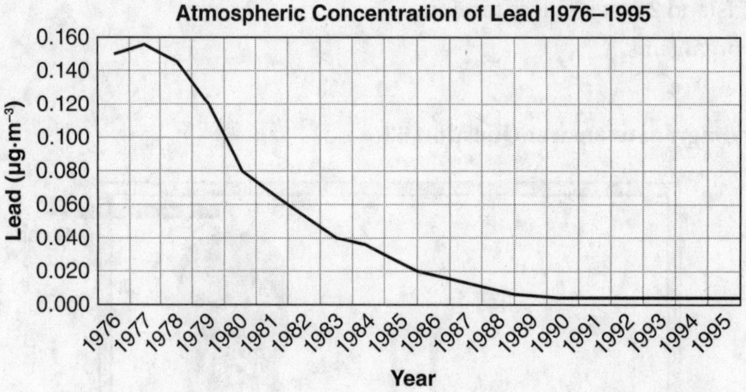

64. What was the percentage decrease in atmospheric levels of lead between 1976 and 1995?

(A) 200%
(B) −100%
(C) 15%
(D) 1.5%

65. A sample of drinking water contained a certain organic pollutant with a concentration of 100 parts per million (ppm). Which statement below most accurately describes this concentration?

(A) In every 1,000,000 mg of drinking water, you would find 100 mg of pollutant.
(B) In every 1,000,100 mg of drinking water, you would find 100 mg of pollutant.
(C) There are 100 mg of pollutant for every 999,900 mg of water in the sample.
(D) Both A and C

66. A company that manufactures electronic parts containing high levels of lead and mercury planned on building a factory to produce these products in a large metropolitan city. The county supervisors required the company to submit plans for how they would dispose of the wastes. The company submitted the plans shown below:

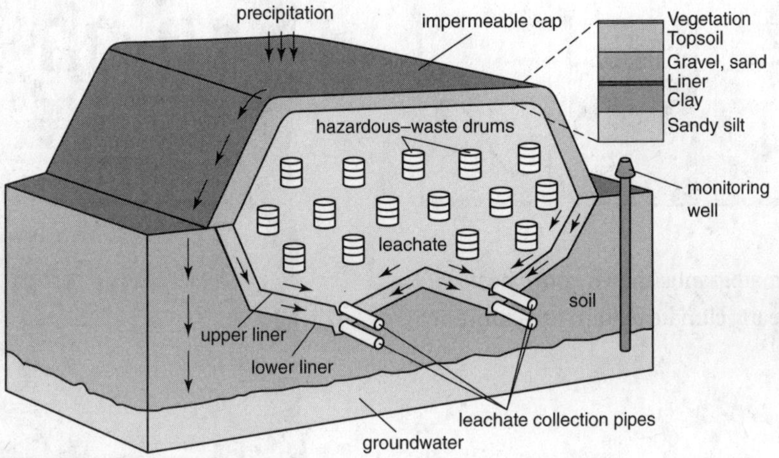

The plans call for constructing a

(A) hazardous waste landfill
(B) sanitary landfill
(C) sewage treatment facility
(D) surface impoundment facility

67. The population of a country in 1994 was 200 million, and its rate of growth was measured at 1.2%. Assuming that the rate of growth remains unchanged and all other factors (e.g., available resources) remain constant, in what year will the population of the country reach 400 million?

(A) 2024
(B) 2034
(C) 2044
(D) 2054

For Question 68, use the following climatographs, which show a year's temperature and rainfall amounts for four different biomes.

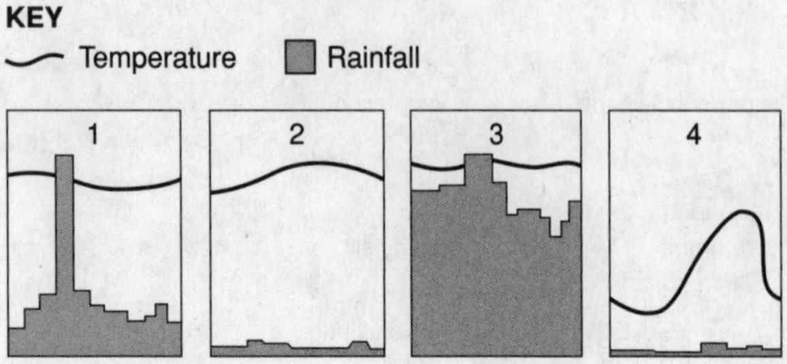

68. Which of the climatographs shown above best represents the yearly climate pattern for a tundra?

(A) 1
(B) 2
(C) 3
(D) 4

Refer to the following list of pollutants to answer Question 69.

I. Asbestos
II. Chlorofluorocarbons (CFCs)
III. Peroxyacyl nitrates (PANs)
IV. Radon
V. Volatile organic compounds (VOCs)

69. Which of the pollutants listed above are NOT naturally occurring indoor air pollutants?

(A) I and II
(B) II, III, and IV
(C) III and V
(D) V

Refer to the following illustration to answer Question 70.

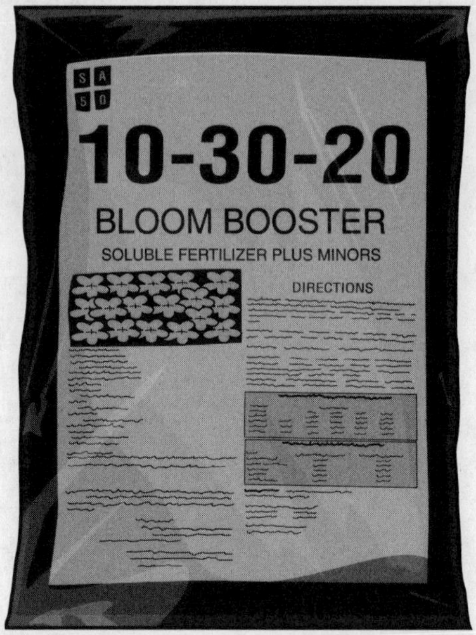

70. Many fertilizer labels have three bold numbers as seen above. The first number (10) shown on the bag represents the relative amount (%) of nitrogen (N) in the bag, the second number (30) is the relative amount (%) of phosphate in the bag, and the third number (20) is the relative amount (%) of potassium in the bag. For example, a bag of 10-10-10 fertilizer would contain 10% (by weight) nitrogen, 10% (by weight) phosphate, and 10% (by weight) potassium.

Suppose the bag shown above weighs 25 kilograms and a soil analysis of your 1,000 square foot lawn reported that you needed 1 kg of potassium for every 200 square feet of lawn. How many bags of this fertilizer should you buy?

(A) 0.5 bag
(B) 1 bag
(C) 2 bags
(D) 3 bags

71. Which of the following factors are NOT primary causes for current and projected rise in sea levels across the world?

I. Melting of land-based ice
II. Melting of sea ice
III. More frequent rain and flooding near the equator
IV. Thermal expansion of the oceans

(A) I and II
(B) I and IV
(C) II and III
(D) All of the choices above are primary causes for the current and projected rise in sea levels across the world.

Examine the graphs below to answer Question 72.

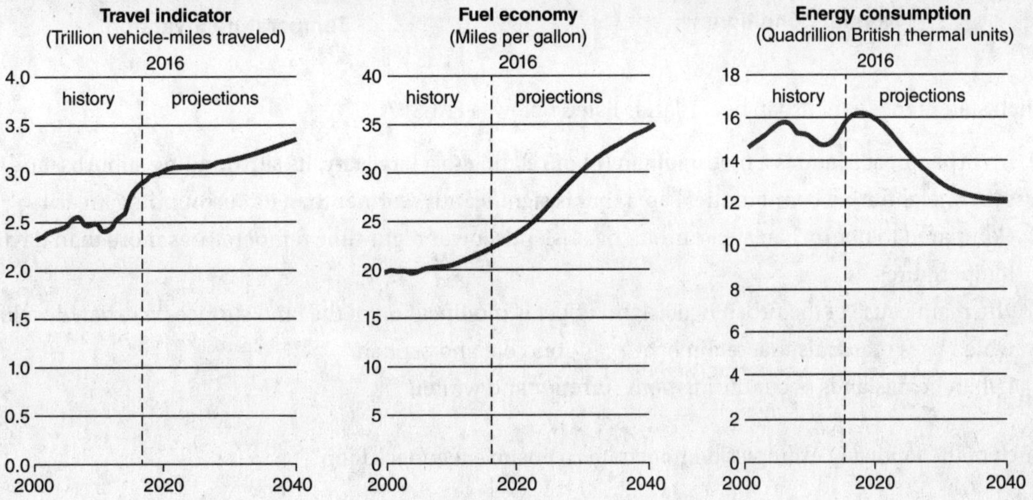

Source: U.S. Energy Information Agency

72. What is the projected percentage increase in fuel economy between 2000 and 2040?

(A) 25%
(B) 50%
(C) 75%
(D) 175%

73. Which of the following characteristics would generally put a species at risk for extinction?

I. High reproductive rate
II. Feeding at a low trophic level
III. Fixed migratory pattern

(A) I
(B) II
(C) III
(D) I, II, and III

Refer to the following diagrams to answer Question 74.

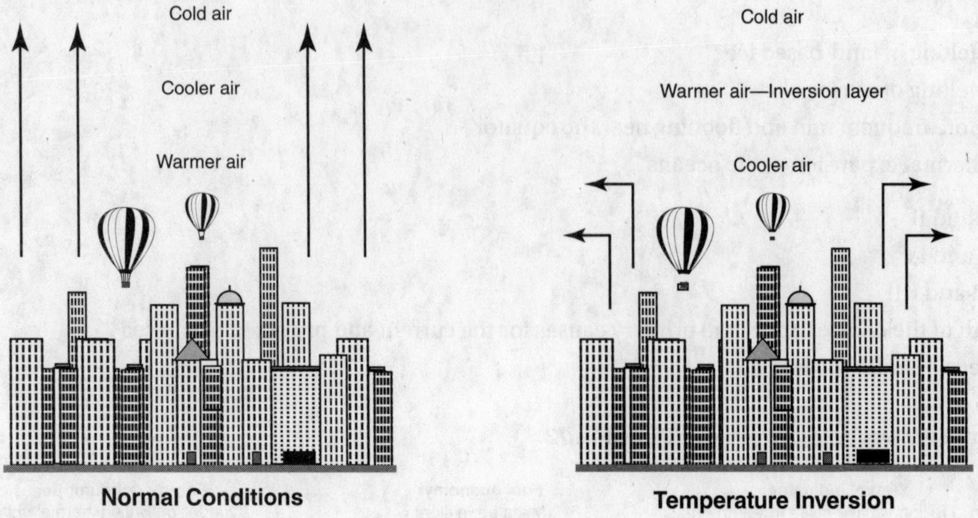

74. Which statement about urban heat islands listed below is FALSE?

(A) An urban heat island is a metropolitan (of or relating to a large city, its surrounding suburbs, and/or other neighboring communities) area that is significantly warmer than its surrounding areas.
(B) Compared to non-urban areas, urban heat islands lower night-time temperatures more than daytime temperatures.
(C) The main cause of the urban heat island effect is modification of the land surface by urban development, which uses materials that retain heat, e.g., concrete and asphalt.
(D) Urban heat islands occur during both summer and winter.

75. Which of the following examples demonstrates a positive feedback loop?

(A) Due to a warmer atmosphere evaporation will increase, condensation will decrease, and the atmosphere will contain more water vapor, which is a greenhouse gas that will warm the atmosphere further.
(B) A warmer atmosphere will melt ice, which changes the albedo, which further warms the atmosphere.
(C) A colder climate will cause ice caps and glaciers to grow, changing the albedo and further cooling the atmosphere.
(D) All of the above

76. All of the following are realistic and legal environmental consequences of sanitary landfills in the United States EXCEPT

(A) off-gassing of methane generated by decaying organic matter
(B) pollution of the local environment, such as contamination of groundwater and/or aquifers by leachate leakage
(C) release of toxic emissions into the atmosphere and groundwater
(D) residual soil contamination during landfill usage as well as after landfill closure

Use the following information to answer Question 77.

- An office building has 10,000 square feet of space.
- 100,000 BTUs of heat per square foot are required to heat the office space during the winter months.
- One cubic foot of propane supplies 1,000 BTUs of heat energy.
- Propane is used to heat the office space, and costs $1 per thousand cubic feet.
- The heating system is only 50% efficient.

77. How much will it cost to heat the office building for one winter?

(A) $100
(B) $1,000
(C) $2,000
(D) $3,000

Read the following passage and then answer Question 78.

Originally from Eastern Europe, zebra mussels were picked up in the ballast water of ships and brought to the Great Lakes in the 1980s. They spread dramatically, outcompeting native species for food and habitat. A single zebra mussel will produce about five million eggs during its lifetime, with about 100,000 reaching adulthood. Today, there are an estimated 10 trillion zebra mussels in the Great Lakes.

Zebra mussels blanket the bottom of the Great Lakes and filter water as they eat plankton and have succeeded in doubling lake water clarity during the past decade; however, fewer plankton floating in the water means less food for native fish. Clearer water also allows more sunlight to penetrate through the water, creating ideal conditions for algae to grow, which promotes the growth and spread of deadly algae blooms.

Zebra mussels harm native fish populations by stripping the food web of plankton, which has a ripple effect throughout the ecosystem, causing native populations of fish and other species to plummet.

78. What effect will an invasive species such as the zebra mussel have on the carrying capacity of the ecosystem?

(A) The carrying capacity for all species will increase.
(B) The carrying capacity for all species will decrease.
(C) The carrying capacity for all species will stay the same
(D) The carrying capacity will decrease for those in competition for the same resource.

Refer to the following four graphs in order to answer Question 79.

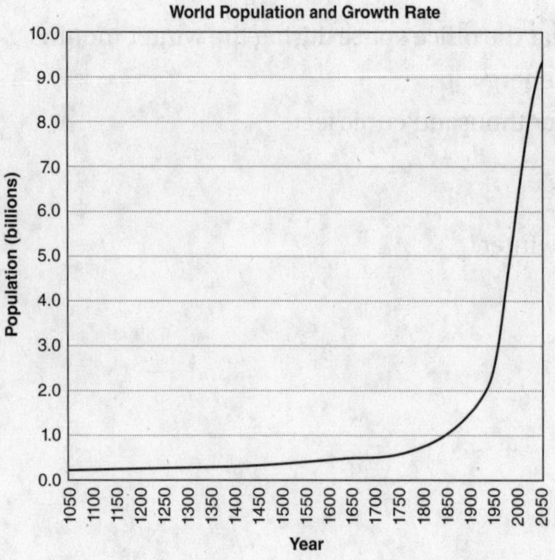

(A)

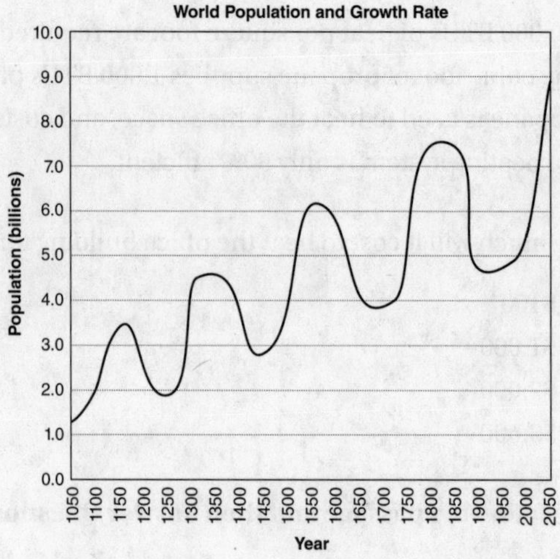

(C)

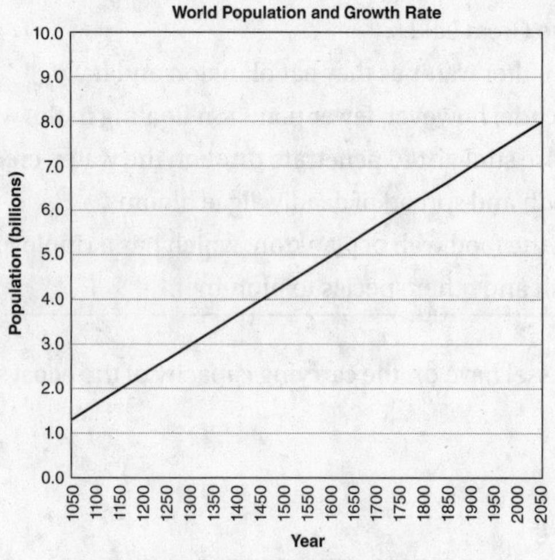

(B)

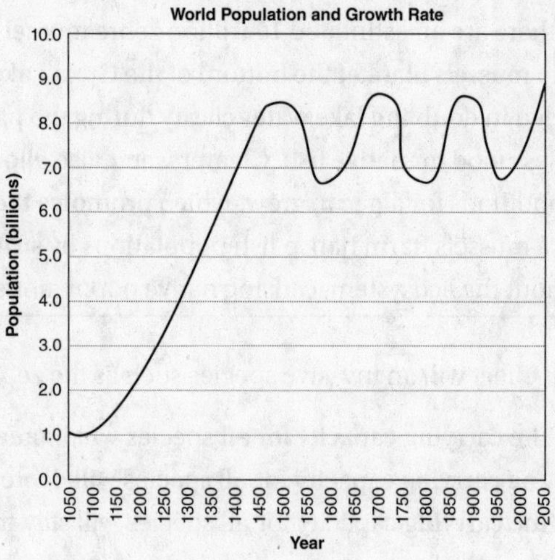

(D)

79. Which of the above graphs shows human population growth over the last 1,000 years?

(A) A
(B) B
(C) C
(D) D

Refer to the table below to answer Question 80.

	LEDs	Incandescent
# of lights required	20	20
Lifespan	20,000 hours	1,000 hours
Watts per bulb (equiv. to 50 watts)	10	50
Cost per bulb	$5	$1
kWh of electricity used over 25,000 hours	200	1,000
Cost of electricity @ $0.10 per kWh	$20	$100
Bulbs needed for 20,000 hours of use	1	25
Equivalent 20,000 hours bulb expense	$5	$25
Total cost for 20,000 hours of lighting (1 bulb)	$25	$125
Total cost for 20,000 hours of lighting using 20 bulbs for a home	$500	$2,500
Total savings to household for 20,000 hours of lighting using 20 bulbs	?	0

80. A homeowner was considering whether or not to replace all 20 incandescent lights in her home with LEDs. She compiled data on her decision (see above). What would be her total savings for 20,000 hours of lighting using LEDs?
    (A) $125
    (B) $1,250
    (C) $2,000
    (D) $2,500

Section II (Free-Response Questions)

TIME: 90 MINUTES
3 QUESTIONS
40% OF TOTAL GRADE

DIRECTIONS: Answer all three questions, which are weighted equally. The suggested time is about 23 minutes for answering each question. Write all your answers on scrap paper. Where calculations are required, clearly show how you arrived at your answer. Where an explanation or discussion is required, support your answers with relevant information and/or specific examples.

1. You loved your high school APES class and went on to major in environmental science in college. During your junior year in college, you received an internship offer to spend your upcoming summer break helping a research team in Panama investigate the Lake Gatún region.

   Lake Gatún is an artificial lake created for the operation of the ship locks in the Panama Canal. A significant environmental problem in the area is the drainage of up to 52 million gallons (197 million liters) of freshwater leaving the lake each time a ship passes through the canal. Although there is sufficient annual rainfall to replenish the water used by the canal operations, the seasonal nature of the rainfall requires the storage of this water from one season to the next.

   The rainforest surrounding the lake traditionally played a major role by absorbing rainwater and releasing it slowly; however, due to deforestation throughout the area of the lake, rain flows rapidly down through the deforested slopes, carrying silt into the lake, with the excess silted water flowing through estuaries and into the ocean, resulting in water shortages during the dry season and silt buildup in the lake and estuaries, which must periodically be dredged.

   (a) You agreed to accept the internship and are beginning to pack your bags. Describe the climate (air temperature and rainfall) you might expect to find during your stay in Panama in June and July.

   (b) Draw a sketch showing how sunlight reaches the equator in terms of its angle compared to how sunlight reaches the poles.

   (c) Sketch a climatograph of yearly temperature and precipitation patterns for that area (you do not need to include actual temperature or precipitation amounts; instead, you may use the labels "cold," "cool," "warm," and "hot" for the temperature axis and the terms "low," "moderate," and "high" for the precipitation axis).

   (d) Describe the type of global wind circulation cell that exists over the area, and include a sketch showing the global wind circulation pattern that exists in the area.

   (e) Describe the type of vegetation that you would find in the tropical forests near Lake Gatún, Panama.

   (f) Describe the type of soil, and its characteristics, that you might find near your location in Panama.

   (g) One of your first projects when you arrived in the Lake Gatún area was to join a team investigating the water quality in Lake Gatún. Describe ONE water quality test that you might do and what you might expect from the results. The water quality test that you choose MUST be related to the environmental issue(s) described in the initial background information for the question.

(h) You then joined a team that was investigating the effects of deforestation in the area. Define deforestation and describe the effects of deforestation on the hydrologic cycle and water quality in the area.

2.

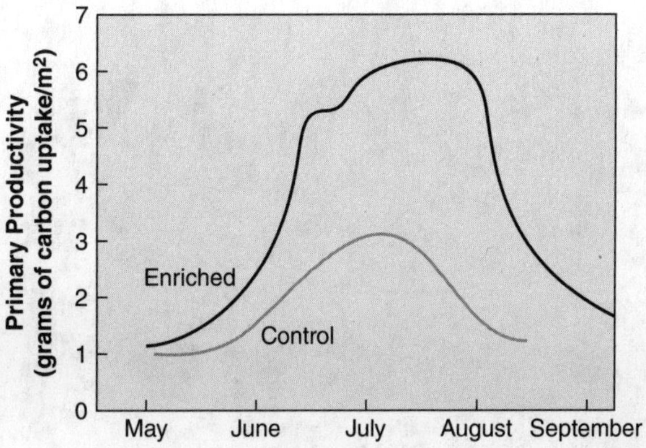

Figure 1

(a) Figure 1 displays the results of an experiment measuring plant growth (primary productivity) in two ponds from May through September in the Northern Hemisphere. The enriched pond is located in a heavily fertilized urban area, while the control pond is found in a heavily forested, rural setting nearby. Using the graph above, DESCRIBE the trends in algae growth in both ponds.

(b) Using your understanding of the process of eutrophication, provide a hypothesis to EXPLAIN the difference between the control and enriched conditions in the experiment shown in Figure 1.

(c) The scientists performing this experiment also noted elevated rates of growth in decomposer populations in late June for the enriched pond. EXPLAIN how the presence of this decomposer population impacts ONE abiotic factor in the pond.

(d) In the Northern Hemisphere, fish kills are most frequently observed in late August and early September. Using your understanding of the connection between dissolved oxygen, nitrate, and phosphate levels, DESCRIBE the sequence of events that would result in fish kills in the enriched pond.

(e) IDENTIFY ONE recommendation that you would make to local landowners within the drainage area of the enriched pond to improve water quality by reducing the likelihood of eutrophication. JUSTIFY your recommendation's impact on water quality.

3. Refer to the following food web to answer this Free-Response Question.

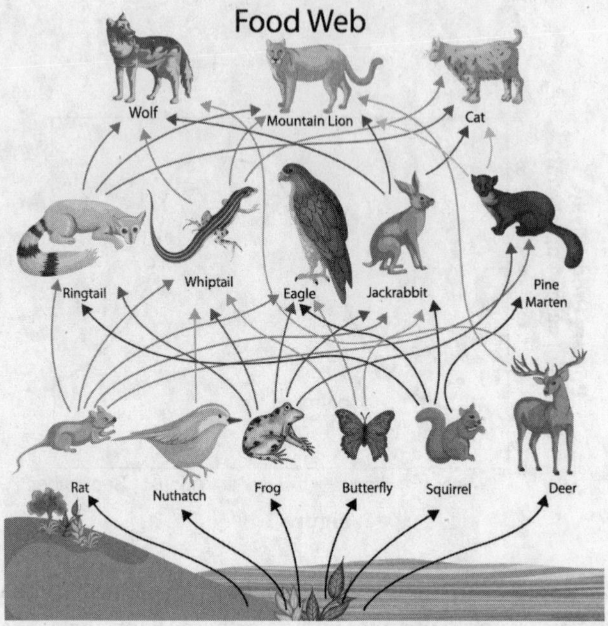

(a) Examine the food web above. Identify a biome that would be consistent with the organisms shown in the food web and describe three characteristics of that biome.

(b) In terrestrial and near-surface marine communities, energy flows from the sun to producers in the lowest trophic levels and then upward to higher trophic levels. Describe how energy decreases as it flows through ecosystems.

(c) Suppose 1,000,000 joules of radiant energy from the sun reaches this biome in a defined time period. Draw a model (known as an energy pyramid) of how the energy from the sun will ultimately be available to the wolf. Be sure to show how many joules of energy are available at each trophic level in your model.

(d) Explain how energy loss through higher and higher trophic levels is consistent with the first and second laws of thermodynamics.

(e) The 10% Rule is an approximation of the energy efficiency between trophic levels. Net productivity, or the rate of energy storage as biomass, by the various trophic levels in the food web diagram above is as follows:

- Primary producers: 7,500 kcal/m^2/yr
- Primary consumers: 1,000 kcal/m^2/yr
- Secondary consumers: 100 kcal/m^2/yr
- Tertiary consumers: 5 kcal/m^2/yr

Using the data above:

1. Calculate the transfer efficiency between primary producers and tertiary consumers (be sure to include the setup in your answer).
2. Calculate the percentage difference between the theoretical energy loss and the actual energy loss.

Topic 1.10-ENG-1.C.1

Multiple-Choice Explanations

1. **(D)** After an organism's soft tissues decay in sediment, the hard parts (e.g., bones) are left behind. Water and its minerals seep into the spaces within the remains, where they form crystals. These crystallized minerals cause the remains to harden along with the encasing sedimentary rock.

   In another fossilization process, called replacement, the minerals in groundwater replace the minerals that make up the bodily remains after the water completely dissolves the original hard parts of the organism through a process called water erosion. *Topic 4.2*

2. **(D)** Oxygen has been present in the Earth's atmosphere for about 2.5 billion years. The earliest cells utilized the process of anaerobic fermentation (no oxygen) to break down more complex compounds into less complex compounds and used the energy that was released to grow and reproduce. The evolutionary process that resulted in utilizing photosynthesis made it possible for these primitive organisms to manufacture their own food. Photosynthesis captures the energy of sunlight and stores it in energy-rich molecules (adenosine triphosphate [ATP]) that can then be used to make sugars such as fructose, glucose, and sucrose, which are used in plant growth and development. *Topic 4.4*

3. **(C)** Conveyance efficiency: $e_c = 80\%$ (from top chart)

   Field application efficiency: $e_a = 60\%$ (from bottom chart)

   $e = [e_c \times e_a] / 100 = [80\% \times 60\%] / 100 = 48\%$

   Topic 5.5

4. **(D)** Scramble (or finite) competition refers to a situation in which a resource is accessible to all competitors, so that the resource per individual declines with increasing population density (density dependent). Since the resource is usually finite, scramble competition may lead to decreased survival rates for all competitors if the resource is used to its carrying capacity; e.g., the more time that a squirrel spends time seeking food, the less energy the squirrel has to defend itself against predators, to maintain body temperature, etc. *Topic 3.5*

5. **(D)** Atmospheric nitrogen (N_2) is relatively inert; i.e., it does not easily react with other chemicals to form new compounds. The fixation process frees nitrogen atoms from its strongly bonded diatomic form, N≡N, so that it can be converted to useful compounds, such as amino acids that build proteins or ammonia (NH_3), and for the synthesis of DNA and RNA. Nitrogen fixation is carried out in the soil by a wide range of nitrogen-fixing bacteria, which have symbiotic relationships with plants such as legumes (e.g., alfalfa, clover, peanuts, and soybeans) that contain symbiotic bacteria in their root systems. When these types of plants die, the "fixed" nitrogen is then released, making it available to other plants as fertilizer for the soil. Nitrogen fixation can also occur when lightning flashes react with nitrogen (N_2) and oxygen (O_2) gases in the atmosphere, producing nitrogen oxides (NO and NO_2). *Topic 1.5*

6. **(B)** Large amounts of organic wastes at the site of the drainage pipe would supply vital nutrients for bacteria that use up large amounts of dissolved oxygen, depriving other organisms of that oxygen and eventually leading to cultural eutrophication.

   In choice (A), high concentrations of metal ions *upstream* of the factory generally rule out the factory. Also, metal ions would not be a waste product from a fertilizer plant.

   In choice (C), turbidity measures the amount of suspended material in the water, and would not likely be a waste product of manufacturing pharmaceuticals.

   In choice (D), large numbers of insect larvae, which may be indicator species, indicate that the water is probably very clean. Many insect larvae are not tolerant to low levels of dissolved oxygen, high water temperature, large amounts of sediment in the water, nutrient enrichment, and/or toxic chemicals and heavy metals. *Topic 8.1*

**At the conclusion of each answer is a "Topic Code" that correlates to the required topic number established by the College Board.*

7. **(C)** Drift gill-netting is a fishing technique in which nets called drift nets are allowed to drift free in a sea or lake. With drift gill nets, fish try to swim through certain-sized mesh openings but are unable to swim forward, as their gills become entangled in the nets. Nets of up to 30 miles are used in this process and are sometimes lost or discarded in the sea, entangling other marine life. More than half of the marine life caught by drifting gill nets is discarded—usually dead or dying. *Topic 5.8*

8. **(B)** Thomas Malthus argued that the human population would increase faster than the available food supply. The population would eventually reach its carrying capacity, and any further increases in population would result in a population crash caused by famine, disease, and war. *Topic 3.4*

9. **(D)** The growing season in the Arctic regions occurs during the summer, the warmest time of year. The question states how temperatures do not exceed 8°C during this time of year; therefore, species D, which has the highest photosynthetic rate closest to 8°C rapidly declines above that temperature, is most adapted to Arctic conditions. *Topic 2.6*

10. **(B)** The area of the graph where oxygen in the water is lowest is the area where the biological oxygen demand (BOD) is the highest—in this case, location B. Anaerobic bacteria live in areas devoid of oxygen. The bacteria use the oxygen attached to the nitrate ion (NO_3^-)—through a process known as anaerobic respiration. This frees up nitrogen gas that then can escape into the atmosphere along with carbon dioxide, hydrogen sulfide (H_2S), and methane.

    The most common areas within a stream ecosystem where anaerobic bacteria may be found are the lower layer of sand and deep gravel beds. *Topic 8.5*

11. **(C)** Relative humidity is the amount of water vapor in a mixture of air. With all factors held constant, warmer air can hold more water vapor than cooler air. That is why water droplets condense on a cold piece of glass—the air cannot hold the moisture. *Topic 4.4*

12. **(C)** The primary source of carbon dioxide entering the Earth's atmosphere is outgassing from Earth's interior at mid-ocean ridges, hot-spot volcanoes, and volcanic arcs. Much of the CO_2 released at subduction zones is derived from carbonate-containing rocks and their becoming harder and more completely crystalline as they subduct into the ocean's crust. Much of the outgassing of CO_2, especially at mid-ocean ridges and hot-spot volcanoes, was originally stored in the earth's mantle. Of this outgassing CO_2, some becomes dissolved in the oceans, some is temporarily stored in living and dead biomass, and some is bound into carbonate rocks. *Topic 1.4*

13. **(A)** Biodiversity hotspots are regions that are both rich in their distribution of plants and animals and highly threatened. They constantly face destruction and degradation due to development, illegal logging, pollution, and deforestation. There are 35 identified hotspots around the world today and represent only about 2% of Earth's land surface but support more than half of the world's endemic plant specie. This is nearly half of all endemic bird, mammal, reptile, and amphibian species. *Topic 2.1*

14. **(D)** Integrated pest management (IPM) is an ecosystem-based strategy that focuses on long-term prevention of pests and their damage through a combination of techniques such as biological control, habitat manipulation, modification of cultural practices, and use of resistant crop varieties. Pesticides are used only after monitoring indicates they are needed according to established guidelines, and treatments are made with the goal of removing only the target organism. Pest control materials are selected and applied in a manner that minimizes risks to human health, and the environment. IPM does not attempt to eliminate all pest species, but is used to prevent infestation, observe patterns of infestation when they occur, and intervene without pesticides when necessary. *Topic 5.14*

15. **(B)** 1,000 years ago: $[CO_2] \sim 280$ ppm

    Present $[CO_2]$: ~ 330 ppm

    $\Delta = 330 \text{ ppm} - 280 \text{ ppm} = 50 \text{ ppm}$. Since CO_2 concentration increased, the sign is positive. *Topic 9.4*

16. **(D)** A greenhouse gas is a gas in an atmosphere that absorbs and emits radiation within the thermal infrared range and is the fundamental cause of the greenhouse effect. The primary greenhouse

gases in Earth's atmosphere are water vapor, carbon dioxide, methane, nitrous oxide, and ozone. However, even though water vapor is a greenhouse gas, it does not contribute significantly to global climate change because it has a short residence time in the atmosphere. Clouds are the major non-gas contributor to Earth's greenhouse effect, as they absorb and emit infrared radiation. Global warming potential (GWP) depends on both the efficiency of the molecule as a greenhouse gas and its atmospheric lifetime. GWP is measured relative to the same mass of CO_2. *Topic 9.3*

17. **(A)** Carrying capacity is defined as the maximum number of organisms of a particular species that can be supported indefinitely in a given environment. When a population is below its carrying capacity, the population tends to increase due to the availability of resources. When the population exceeds the carrying capacity of its environment, the population will decrease; e.g. starvation. The carrying capacity is different for each species in a habitat due to that species' unique environmental requirements.

At the current rate of sea ice melting, it is estimated that two-thirds of the current polar bear population will be gone by the 2050s, and extinction of the species could occur by the end of this century. *Topic 3.5*

18. **(D)** Reducing nutrient recycling by reformatting the land from a woodland to a pasture has the following long-term environmental issues:

    1. The input of nitrogen by biological nitrogen fixation is reduced.

    2. The capture and bringing up of nutrients from below the rooting zone is reduced.

    3. There would be an increase in nutrient losses through higher levels of erosion and leaching.

    Short-term financial costs refer to financing the needs for a small period of time, normally less than a year, whereas long-term financial costs involve paying for something over a much longer period of time, e.g., financing a new car over a number of years. In this question, the cost of cutting down the trees in order to convert the land to pasture would take a relatively short period of time at a relatively low cost. *Topic 5.4*

19. **(B)** The Hadley cells are a global scale tropical atmospheric circulation that features air rising near the equator flowing poleward at a height of 6 to 9 miles (10 to 15 km) above Earth's surface, descending in the subtropics, and then returning toward the equator near the surface. Refer to the following diagram to locate Hadley cells. *Topic 4.5*

GLOBAL ATMOSPHERIC CIRCULATION

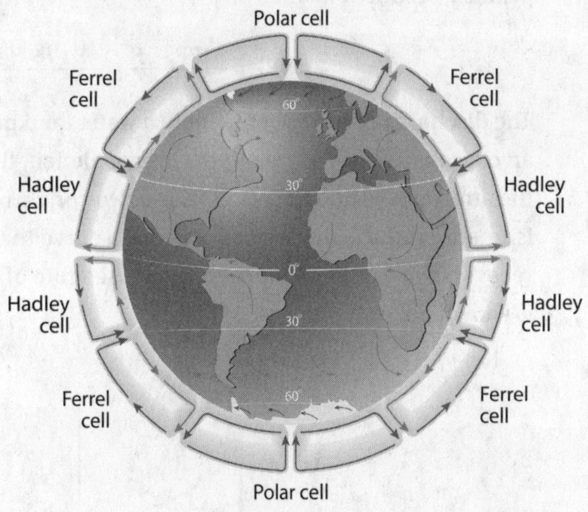

20. **(A)** Ozone (O_3) occurs in two layers of the atmosphere: (a) in the troposphere—the lowest layer of Earth's atmosphere—where nearly all weather conditions take place and where approximately 75% of the atmosphere's mass and 99% of the total mass of water vapor, is and (b) in the stratosphere, which is located just above the troposphere. The temperature of the stratosphere increases with distance from Earth, while cooler layers are closer to Earth, as a result of the absorption of the sun's ultraviolet radiation by the ozone layer. *Topic 9.1*

21. **(D)** Groundwater is water located beneath the ground surface in soil pore spaces and in the fractures of rock formations. The depth at which soil pore spaces and spaces in rock become completely saturated with water is called the water table. An aquifer is an underground layer of water-bearing permeable rock, gravel, sand, or silt from which groundwater can be usefully extracted using a well. Groundwater flow is the movement of groundwater through the earth. Like surface water, groundwater flow moves downward in whichever direction the water table slopes with its flow rate being much slower than that of surface water. Groundwater can take over 10 years to move one mile.

    Approximately two billion people worldwide depend on groundwater, with about 25% of all water used in the United States coming from groundwater. *Topic 1.7*

22. **(C)** World biomes are primarily determined by climate, which is long term, vs. weather, which is short term and is currently occurring. Two important factors that determine an area's long-term climate are air temperature and precipitation.

    Solar radiation reaching Earth is at its strongest between 23° North and 23° South latitudes and results in the warmest areas on the planet. The closer an area is to the North or South Pole, the smaller the angle of radiation, and therefore the less intense it is, which results in cooler air temperatures (cool air cannot hold as much moisture as warm air). The reason it rains a lot in tropical rainforests near the equator is that very moist air near the ground in this region rises and therefore cools, with the moisture returning to Earth as rain. *Topic 4.7*

23. **(A)** Net primary productivity is the amount of energy going into new plant growth. During daylight hours, plants take in carbon dioxide and release oxygen through photosynthesis, and at night about half that carbon is released through cellular respiration; however, plants still remain a net carbon sink. The difference between a plant's photosynthesis, or gross primary production (GPP), and its cellular respiration (R) is known as net primary production, or NPP.

$$\text{NPP} = \frac{\text{GPP} - \text{R}}{\text{Time}} = \frac{10 \text{ mg } O_2/\text{L} - 8 \text{ mg } O_2/\text{L}}{1 \text{ hr}} = 2 \text{ mg } O_2/\text{L}/\text{ hr}$$

The discharge of sewage effluent in a watershed produces a biochemical oxygen demand (BOD) that causes an oxygen deficit; the greater the oxygen deficit, the greater the rate of natural oxygen replenishment from the atmosphere into the water. These two concurrent processes of oxygen consumption and oxygen replenishment produce an oxygen sag curve, a curve that sags initially, due to the increased oxygen demand, and recovers downstream due to the increased rate of oxygen replenishment. An example of an oxygen sag curve is shown below:

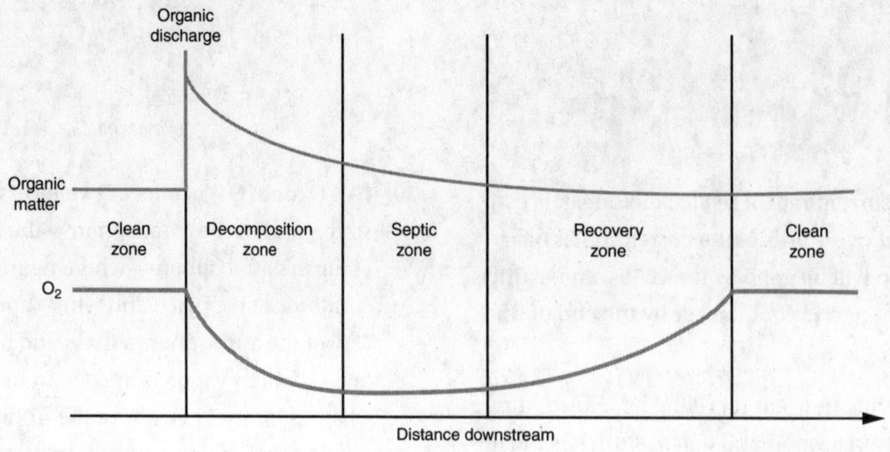

Topic 1.10-STB-3.B.6

24. **(D)** The median lethal dose (LD_{50}) of a toxin, radiation, or pathogen is the dose required to kill half the members of a tested population after a specified test duration. LD_{50} figures have frequently been used as a general indicator of a substance's acute toxicity; however, new and more humane methods of determining toxicity and lethal dose are being developed and adopted. *Topic 8.13*

25. **(D)** Overgrazing occurs when wildlife or livestock excessively feed on pasture before it has recovered from prior grazing and arises as a result of having too many animals grazing on a single piece of land. Planting switchgrass on croplands dramatically reduces soil erosion. The perennial (year-round) nature of switchgrass provides continual protection of the soil surface against erosive forces as a result of overgrazing. *Topic 5.4*

26. **(D)** Note that only 10% of the young oysters reach the 20th unit of life span. *Topic 3.3*

27. **(A)** The boiling point of a crude oil component or fraction is the temperature at that the component evaporates, and is dependent upon the length of the carbon chain in the molecule. The fractions with shorter chains (gasoline) evaporate or boil off faster than the with ones with the longer chains (crude oil). *Topic 6.3*

28. **(A)** Nitrogen dioxide (NO_2) reacts with water vapor (H_2O) to form nitrous acid (HNO_2) and nitric acid (HNO_3), which are both components found in acid deposition (rain):

$$2NO_2 + H_2O \rightarrow HNO_2 + HNO_3$$

Nitrogen dioxide also absorbs light energy and breaks down to form nitric oxide (NO) and atomic oxygen (O):

$$NO_2 \rightarrow NO + O$$

Nitric oxide can then combine with ozone (O_3) to form nitrogen dioxide (NO_2) and oxygen (O_2):

$$NO + O_3 \rightarrow NO_2 + O_2$$

If the concentration of nitrogen dioxide is greater than the concentration of nitric oxide, ozone is formed and can reach dangerous levels. However, if the concentration of nitrogen dioxide is less than the concentration of nitric acid, the ozone is destroyed almost instantly, keeping the concentration of ozone stable and below a harmful threshold. *Topic 7.2 and Topic 7.7*

29. **(D)** Dams slow the flow of water. As water backs up behind a dam, more sediment and nutrients also back up a consequence known as silting. By slowing water flow, most dams increase water temperatures, which often negatively impacts native fish populations. If water is released from the top of the reservoir behind the dam, the temperature is usually warmer, as it gives the water more time to warm up. If the water is released near the bottom of the dam, the water is usually cooler but lower in dissolved oxygen, because of the decomposition of organic matter by bacteria. *Topic 6.9*

30. **(C)** Global warming potential (GWP) is a relative measure of how much heat a greenhouse gas traps in the atmosphere and compares the amount of heat trapped by a gas to the amount of heat trapped by a similar mass of carbon dioxide, whose GWP is 1. Although some gases have extremely high GWP, the amount of the gas may be significantly lower than the volume of CO_2 in the atmosphere.

Gas	Formula	Lifetime	20 year GWP
Carbon dioxide	CO_2	Variable	1
Methane	CH_4	12	56
Nitrous oxide	N_2O	120	280
Sulfur hexafluoride	SF_6	3,200	16,300

Water vapor is the largest contributor to Earth's greenhouse effect and accounts for around 60% of the global warming effect, a necessary requirement that allows life as we know it. If there was no water vapor in the

PRACTICE TEST 2

Earth's atmosphere, the average air temperature on Earth would be about 35°F (~2°C). Furthermore, water vapor does not break down in the environment, as other greenhouse gases do, and there is the possibility that additional water vapor in the atmosphere could produce a negative feedback loop, as more water vapor leads to more cloud formation, which would reflect more sunlight, which would reduce the amount of solar energy that reaches Earth's surface. However, additional cloud cover alone does not necessarily mean cooler air temperature; all factors being equal, it is often warmer on a cloudy winter day than on a clear one. *Topic 9.3*

31. **(B)** The B horizon is commonly referred to as "subsoil" and consists of mineral layers that may contain concentrations of clay or minerals (such as iron or aluminum oxides) or organic material that arrives in the layer through leaching. This layer is also known as the "zone of accumulation." Plant roots penetrate through this layer, but there is very little humus and the soil is usually brownish or red because of the clay and iron oxide washed down from the A horizon. *Topic 4.2*

32. **(D)** Ecosystem services can be divided into four categories:

   1. Cultural services—include non-material benefits that people obtain from ecosystems, e.g., aesthetic values, intellectual development, and recreation and/or spiritual enrichment
   2. Habitat services—highlight the importance of ecosystems in providing habitat for migratory species and maintaining the viability and diversity of gene pools
   3. Provisioning services—the products obtained from ecosystems, e.g., food, fresh water, and medicines
   4. Regulating services—the benefits obtained from the regulation of ecosystem processes, e.g., climate regulation, pest control, waste management, and water purification

Climate regulation is one of the most important ecosystem services globally. For example, peat soils, a heterogeneous mixture of decomposed plant (humus) material that has accumulated in a water-saturated environment in the absence of oxygen, contain the largest single sink (store) of carbon. However, the climate-regulating function of peatlands depends on land use and intensification, e.g., drainage and conversion to agriculture, and is likely to have profound impacts on the capacity of soil to store carbon and on carbon emissions. *Topic 2.2*

Hint: be careful of questions with double negatives (NOT and Neither), which turn into a positive making the question "Which of the following choices below ARE. . ."

33. **(B)** The term "feedback" refers to the effect that occurs when the output of a system becomes an input. Feedback loops may be positive or negative. Positive feedback occurs when the effects of an original change are amplified or accelerated. Negative feedback occurs when the effects of an initial change are decreased by subsequent changes, with the result that the system reverts to its original condition. Another example of a negative feedback loop involving climatic warming is an increase in air temperature and how it increases the amount of cloud cover, reducing incoming solar radiation and warming. An example of a positive feedback loop involving global warming is the fact that oceans are an important sink for CO_2 by absorbing CO_2 into the water's surface; i.e., as atmospheric CO_2 increases, it increases air temperature and if air temperatures become warmer, that area of the ocean also becomes warmer, releasing even more CO_2 into the atmosphere. *Topic 9.4*

34. **(B)** The 1973 Endangered Species Act provides for the conservation of ecosystems upon which threatened and endangered species of fish, wildlife, and plants depend and is one of America's most effective and important environmental laws. *Topic 9.9*

35. **(D)** As a mature river begins to meander and change course, it cuts and erodes into the outside of the curve of the bend in the river and deposits sediment on the inside of the curve, as the water velocity is greater on the outside of the curve. As the erosion and deposition continue over time, the curve becomes larger and more circular and

eventually the river begins to cut the "loop" off, forming what is known as an oxbow lake. *Topic 1.3*

36. **(B)** In plate tectonics, a divergent (plate) boundary exists between two tectonic plates that are moving away from each other. Divergent boundaries within continents initially produce rifts, which eventually become rift valleys. Most active divergent plate boundaries occur between oceanic plates and exist as mid-oceanic ridges. Divergent boundaries also form volcanic islands, which occur when the plates move apart to produce gaps that molten lava rises to fill. *Topic 4.1*

37. **(B)** $([y_2 - y_1] / y_1) \times 100\% =$ percentage change

$y_1 = 0.160 \qquad y_2 = 0.060$

$([0.060 - 0.160]/0.160) \times 100\% = \sim -3\%$

Topic 9.1

38. **(B)** A density-dependent factor is a limiting factor that depends on the population size. Examples of density-dependent factors include the following below.

- Diseases, such as the flu and certain water-borne diseases, that require organisms to living closely to one another for disease to spread.
- Imbalances in predator-prey relationships. For example, a reduction in the number of jackrabbits in one area could result in less available food for the local coyote population, demanding an adjustment—whether the coyotes starve or migrate elsewhere.
- Parasites that can limit the growth of a population, taking nourishment from their hosts.
- The demand for energy sources affects populations in a way that is proportional to their density. For example, if only one locust were to inhabit an area, chances are that food demand would not be impacted. However, locusts live in swarms and can deplete an entire area of food before they move to a new area.

A density-independent limiting factor affects all members of a population in similar ways, regardless of the population size. Increased rainfall, air temperature, floods, etc., affect all members equally. *Topic 3.5*

39. **(D)** A population bottleneck, or genetic bottleneck, is a sharp reduction in the size of a population due to major environmental events, e.g., famines, earthquakes, floods, fires, disease, or droughts, and often reduces the diversity in the gene pool, resulting in a reduction in the robustness of the population and in its ability to adapt to and survive environmental changes. These may include climate change, pest control, overhunting or overfishing, selective agricultural and seed production. *Topic 2.1*

40. **(C)** Surface, or flood, irrigation is a technique used to distribute and apply irrigation water over the entire soil surface through gravity, and is the most common form of irrigation in the world. However, it is associated with environmental impacts, such as over-irrigating, salinization, and waterlogging. *Topic 5.5*

41. **(D)**

1. Measure the amplitude of the strongest wave. The amplitude is the height of the strongest wave. On this seismogram, the amplitude is 20 millimeters. Find 20 millimeters on the right side of the diagram and mark that point.

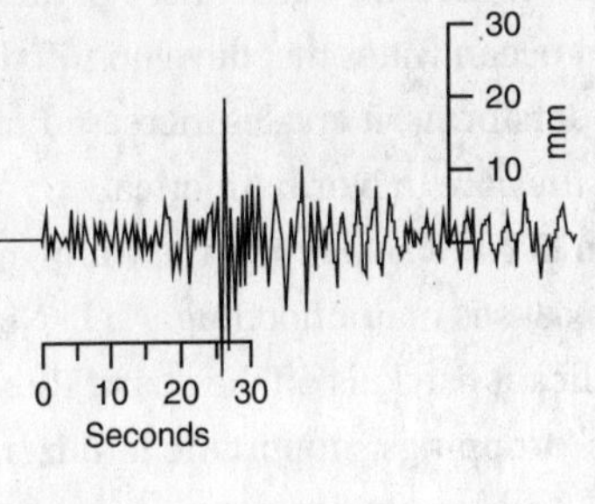

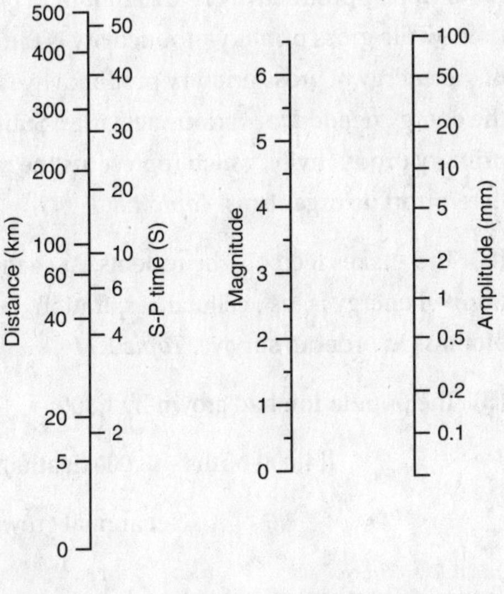

2. Place a ruler (or the straight edge of a piece of paper) on the diagram and connect the points you marked for time (25 seconds) and the amplitude (20 mm). The point where your straight edge crosses the middle line on the chart marks the magnitude (strength) of the earthquake; in this case, the earthquake had a magnitude of 5.0.

Topic 4.1

42. **(C)**

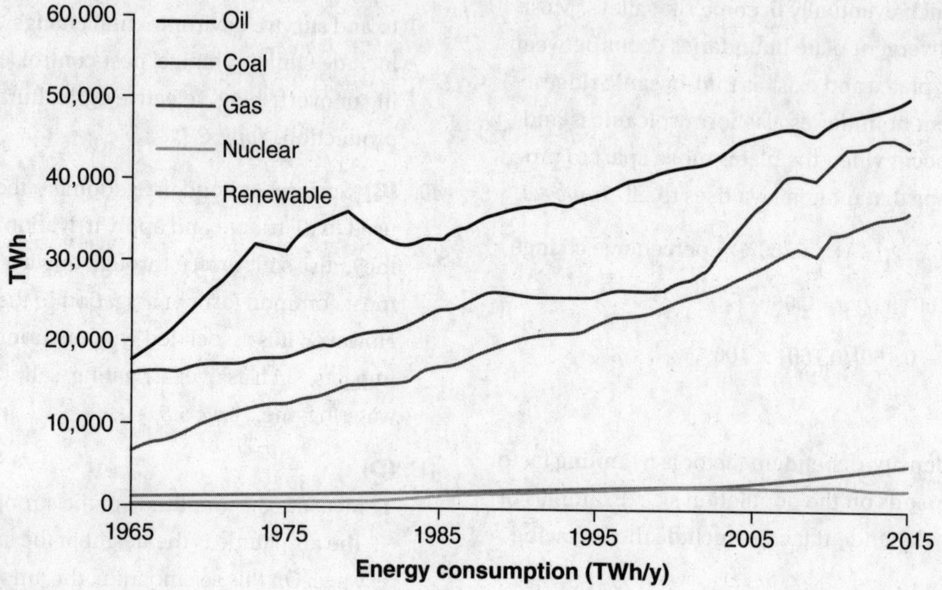

Topic 6.2

43. **(D)** El Niño is the warm phase of the El Niño–Southern Oscillation (ENSO) and is associated with a band of warm ocean water that develops off the Pacific coast of South America. Because El Niño's warm pool feeds thunderstorms, it creates increased rainfall off the South American west coast, with effects that are stronger than they are in North America.

An El Niño also brings the warm and wet weather months in April–October along the coast of Peru, and often causes major flooding. An El Niño event reduces upwellings of cold, nutrient-rich waters, leading to significant fish kills off the coast. These negatively impact seabirds that depend on the fish for food, while the birds' droppings support the fertilizer (guano) industry in South America. *Topic 4.9*

44. **(D)** Primary productivity is the amount of biomass produced through photosynthesis per unit area per unit time, while gross primary productivity is the total energy fixed by plants through photosynthesis. A portion of the energy of gross primary productivity is used by plants for respiration. Respiration provides a plant with the energy needed for various activities. Subtracting respiration from gross primary productivity gives net primary productivity, which represents the rate of production of biomass that is available for consumption by heterotrophic organisms. *Topic 1.8*

45. **(C)** The snakes feed on the rodents. At each trophic level, energy is lost by various means (organisms die and trapped energy is lost, cellular respiration, heat, etc.). Therefore, the snakes must consume more than their biomass in order to survive. *Topic 1.11*

46. **(B)** The population had grown by 1,300:

$$[(4{,}000 \text{ births} - 3{,}000 \text{ deaths}) + (500 \text{ immigrants} - 200 \text{ emigrants})] = 1{,}300$$

$$\text{Net annual growth rate} = \frac{1{,}300}{100{,}000} \times 100\% = 1.3\%$$

Topic 3.5 (mathematical routines)

47. **(C)** Ammonia is produced by the breakdown of organic sources of nitrogen when organisms die. Nitrification is the process by which this ammonia is converted to nitrites (NO_2^-) by bacteria of the genus *Nitrosomonas* and then to nitrates (NO_3^-) by bacteria of the genus *Nitrobacter. Topic 1.5*

48. **(A)** Demand for seafood and advances in technology have led to fishing practices that are depleting fish and shellfish populations around the world. Commercial fishing removes more than 170 billion pounds (~77 billion kilograms) of wildlife from the sea each year, and continuing to fish at this rate may soon result in a collapse of the world's fisheries. In order to continue relying on the ocean as an important food source, sustainable fishing practices such as the following need to be employed:

- managing wild fisheries and monitoring populations accurately
- mitigating bycatch and reducing the practice of dredging and other damaging fishing practices
- national and international enforcement of fishing regulations that help to maintain healthy fish populations
- sustainable aquaculture operations that minimize pollution, disease, damage to native ecosystems, and the practice of using wild-caught fish as feed for aquaculture stocks
- targeting only plentiful species and those species smaller and lower on the food chain to allow for quick reproduction and replacement
- the use of exclusionary devices to avoid catching non-target species such as sea turtles, dolphins, and sharks. *Topic 5.8*

49. **(B)** To determine the estimate of the population size *N*, multiply the number marked in the first catch, M_1, by the total number caught in the second catch, *C*, and divide that number by the number of marked recaptures in the second catch, M_2:

$$N = \frac{(M_1 \times C)}{M_2} = \frac{(150 \times 100)}{20} = 750$$

Topic 3.5

50. **(C)** Endangered species are species that are on the verge of extinction and require prompt action to protect them from complete destruction. Threatened species refers to a group of species that could become endangered in the near future. Thus, threatened species is a larger term, which includes endangered species. There are approximately 1,500 endangered and threatened animals currently listed. *Topic 9.9*

51. **(D)**

$$\frac{100\ \cancel{\text{lightbulbs}}}{1} \times \frac{200\ \cancel{\text{watts}}}{1\ \cancel{\text{lightbulb}}} \times \frac{10\ \text{hours}}{1\ \cancel{\text{day}}} \times \frac{6\ \cancel{\text{days}}}{1\ \cancel{\text{week}}} \times \frac{50\ \cancel{\text{weeks}}}{1\ \text{year}} \times \frac{1\ \text{kilowatt}}{1{,}000\ \cancel{\text{watts}}}$$

$$= 60{,}000\ \text{kilowatt-hours/year}$$

Topic 6.2-6.C (mathematical routines)

52. **(D)** The graph represents what is known as logistic growth, whereby a resource important to a species' survival can act as a limit to unlimited growth—i.e., a limiting factor. In the students' experiment, water availability, temperature, food supply, and/or space to grow would have been intraspecific environmental limiting factors. Intraspecific competition for resources may not affect populations that are well below their carrying capacity when resources are plentiful and all individuals can obtain what they need; however, as population size increases, the competition for these limited resources intensifies. In addition, the accumulation of waste products within the flask would also reduce the carrying capacity.

Wherever on the graph the slope of the line is horizontal (or 0), there is no population growth; i.e., the rate of reproduction is equal to the rate of death. *Topic 3.8*

53. **(B)** A temperature inversion is a reversal of the normal behavior of temperature in the lower atmosphere (troposphere) in which a layer of cool air at the surface is below a layer of warmer air above. Under normal conditions air temperature usually decreases with height.

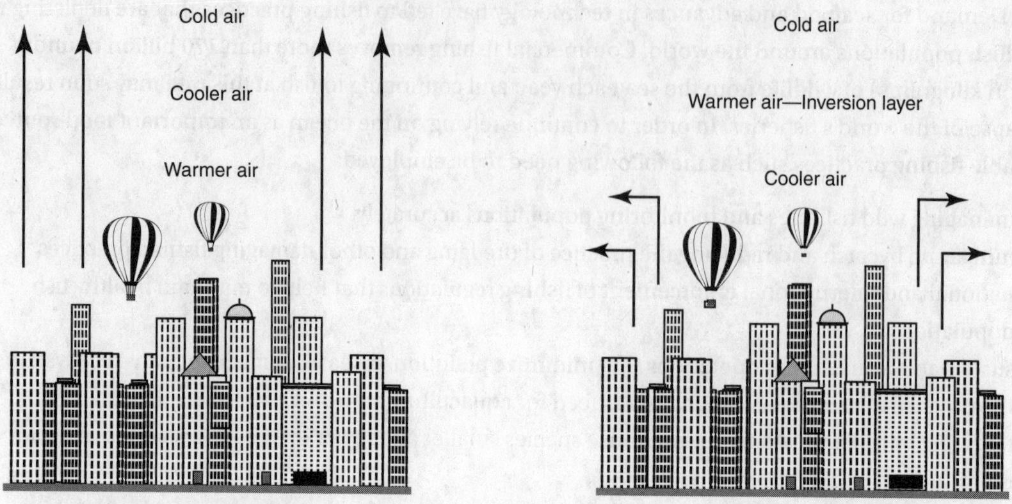

Normal Conditions **Temperature Inversion**

Inversions play an important role in determining how clouds form, are part of precipitation patterns, and visibility as they act as a cap on the upward movement of air from the layers below.

During an inversion, dust, smoke, and other air pollutants are trapped near the surface, reducing visibility and causing respiratory issues. Convective clouds cannot grow high enough to produce rain and because air near the base of an inversion tends to be cool, fog may also be present.

Inversions also affect diurnal (daily) variations in air temperature. The principal heating of air during the day is produced by its contact with a land surface that has been heated by the sun's radiation, as heat from the ground warms air through conduction and convection; but, since a temperature inversion will usually control the upper level to where heat is carried by convection, only a shallow layer of air will be heated if the inversion is low and large, and the rise in temperature will be significant. *Topic 7.3*

54. **(D)** Graphs A and B are plotting the same data—A is plotted arithmetically and B (using the same data) is plotted as a log function. Graph C is the inverse of Graph B. The type of graphs shown in the question are known as species-area curves and describe the relationship between the area of a habitat and the number of species found within that area; bottom line—larger areas tend to contain larger numbers of species. The species-area relationship is usually constructed for a single type of organism at a specified site, such as all species at a specific trophic level (frogs) within a particular site (pond). Factors that determine the shape and slope of the species-area relationship depend upon the relative balance between immigration and extinction, the rate and the magnitude of disturbance on small vs. large areas, predator-prey relationships, and clustering patterns of individuals of the same species. The species-area relationship can also be viewed as a function of the Second Law of Thermodynamics where entropy is a thermodynamic driving force that propels natural selection. This is the mechanism of evolution, since a characteristic of all living organisms is that they are open systems that maintain greater order than their surroundings by consuming relatively ordered free energy (e.g., concentrated nutrients) and exporting relatively disordered heat and waste products. *Topic 2.1*

55. **(D)** A carnivore derives its energy and nutrient requirements from a diet consisting mainly or exclusively of animal tissue, whether through predation or scavenging. Animals that depend solely on animal flesh for their nutrient requirements are called obligate carnivores (cats), while those that also consume non-animal food are known as facultative carnivores (dogs), and carnivores that sit at the top of the food chain are termed apex

predators, (lions, saltwater crocodiles). Some plants (Venus flytraps) and fungi are carnivores (more precisely "insectivores") and prey on amoebae, flatworms, or insects.

In the food web shown in the question, "A" is a producer. Producers, also known as autotrophs, occupy the lowest position in a food web and are typically photosynthetic plants, which use sunlight as their energy source, while "C" and "D" are both omnivores—animals that eat both plants and animals as primary food sources. *Topic 1.11*

56. **(B)**

$$\frac{35{,}000\ \cancel{\text{population}}}{1} \times \frac{20\ \text{births}}{1{,}000\ \cancel{\text{population}}} = 700\ \text{births} + 350\ \text{immigrants}$$

$$= 1{,}050\ \text{new people that year}$$

$$\frac{35{,}000\ \cancel{\text{population}}}{1} \times \frac{5\ \text{births}}{1{,}000\ \cancel{\text{population}}} = 175\ \text{deaths} + 100\ \text{emigrants}$$

$$= 275\ \text{less people that year}$$

$$= 1{,}050\ \text{additional} - 275\ \text{less} = \mathbf{775\ people\ added}$$

Topic 3.8

57. **(D)**

$$\text{Car A: } \frac{600\ \cancel{\text{miles}}}{1} \times \frac{1\ \cancel{\text{gallon gasoline}}}{25\ \cancel{\text{miles}}} \times \frac{5\ \text{lbs carbon}}{1\ \cancel{\text{gallon gasoline}}} = 120\ \text{lbs C}$$

$$\text{Car B: } \frac{600\ \cancel{\text{miles}}}{1} \times \frac{1\ \cancel{\text{gallon gasoline}}}{15\ \cancel{\text{miles}}} \times \frac{5\ \text{lbs carbon}}{1\ \cancel{\text{gallon gasoline}}} = 200\ \text{lbs C}$$

200 lbs − 120 lbs = 80 pounds of carbon

Topic 6.2-6.C

58. **(C)** Drugs X and Z have equal efficacy (their therapeutic effects are equal); but drug Z requires much more drug (x-axis is a log scale), which means it is by far the least potent of the three drugs. *Topic 8.13*

59. **(D)** "Slash and burn" is a method sometimes used by farmers to create short-term yields from marginal soils. When practiced repeatedly, or without intervening fallow periods, the nutrient-poor soils that are used for agricultural purposes (crops and livestock) may be exhausted or eroded to an unproductive state.

Large industrial farms have taken over rural areas and expanded farther into remaining forests, as efforts are made to supply both domestic urban populations and a growing international agricultural market. Because of often poor nutrient levels in tropical soils (which are bound in the biomass), frequent applications of usually inorganic fertilizers are required. Other environmental impacts of deforestation include the release of carbon dioxide, increases in soil and stream temperatures, and flooding. *Topic 5.2*

60. **(D)** Island biogeography examines the factors that affect the species richness (the number of different species) and diversification rate (the rates at which new species form) of isolated natural communities. The term "island" refers to any area of habitat suitable for a specific ecosystem that is surrounded by an expanse of unsuitable habitat, for example, mountain peaks, isolated lakes, or a grassland surrounded by housing developments. The basic premise of island biogeography is that the number of species found on an "island" is determined by immigration and extinction. It is important to note that island size, as shown in diagram B, has a direct correlation with the variety of habitats found on the island. On smaller islands there is not as much area available for variations in habitat; consequently, there is an increased probability of extinction. Over large time spans, the opposing forces of extinction vs. immigration tend to balance out, resulting in an equilibrium that remains until it is disturbed. *Topic 2.3*

61. **(B)** Photochemical smog is a type of smog produced when ultraviolet light from the sun reacts with nitrogen oxides in the atmosphere. It is visible as a brown haze, and is most prominent during the morning and afternoon, especially in densely populated, warm cities such as Beijing, Los Angeles, and Mexico City.

The largest contributors to photochemical smog are automobiles and coal-fired power plants that produce photochemical smog. *Topic 7.2*

62. **(A)** Control rods regulate the number of neutrons in the core of a nuclear reactor and control the rate of the reaction. These tubes contain material that absorb neutrons. *Topic 6.6*

63. **(A)** Runoff (R) or streamflow is a function of the difference between precipitation (P), which includes the inputs of rain and snow melt, and evapotranspiration (E_t), which includes the outputs of water lost by evaporation and transpiration through plants.

Therefore, the amount of water lost through evapotranspiration (E_t) must be equal to the total amount of precipitation (P) minus the amount of runoff or streamflow (R): $E_t = P - R$ or 100,000 gallons of precipitation (P)—70,000 gallons of runoff (R) = 30,000 gallons of water lost through evapotranspiration (E_t). *Topic 4.6*

64. **(B)** $[(y_2 - y_1) / y_1] \times 100 = \%$ change

$[(0 - 0.150) / 0.150] \times 100 = -100\%$

Topic 7.6-7.D

65. **(D)** A solution is made up of two components—the solute, or the thing in the smaller amount (in this case the pollutant), and the solvent, or the thing doing the dissolving and usually in the larger quantity (in this case, pure water). Together the solute and the solvent make the solution; or the drinking water. (A) is correct because it is saying there are 100 mg of pollutant (solute) in every 1,000,000 mg of the drinking water sample (solution). (C) is correct because there would be 100 mg of pollutant in 999,900 mg of water (remember, the total solution is the pollutant + the water, or 100 parts of pollutant + 999,900 parts water = 1,000,000 parts drinking water).

(B) is incorrect because it is saying there are 100 mg of pollutant per 1,000,100 mg of drinking water. *Topic 8.14*

66. **(A)** Hazardous waste is waste that has substantial or potential threats to public health or the environment and exhibits one or more of the following properties: corrosive, ignitable, highly reactive, or toxic. The treatment, storage, and disposal of hazardous waste is regulated in the United States under the Resource Conservation and Recovery Act.

Hazardous wastes must be deposited in secure landfills that provide at least 10 feet (3 m) of separation between the bottom of the landfill and the underlying bedrock or groundwater table. It must have two impermeable liners and a network of perforated pipes, which helps to prevent the accumulation of trapped leachate. Finally, a cover is placed over the landfill to reduce groundwater leaching. Regular groundwater and air sampling are also required to detect any leaks. *Topic 8.10*

67. **(D)** To find the doubling time of a population, divide 72 by the annual growth rate:

$72 \div 1.2 = 60$ years $\quad 1994 + 60 = 2054$

Topic 3.5

68. **(D)** Tree growth in the tundra is scattered and hindered by low temperatures and short growing seasons, with ground vegetation being composed of dwarf shrubs, grasses, mosses, and lichens. The ecotone (or ecological boundary) between the tundra and the forest is known as the tree line, or timberline. There are three types of tundra: Arctic tundra, alpine tundra, and Antarctic tundra. Tundra climates have at least one month per year with an average temperature high enough to melt snow, but no month generally exceeds ~60°F (~10°C). Rainfall and snowfall are generally slight, due to the low vapor pressure of water in the cold atmosphere, and there are many soggy swamps and bogs. *Topic 1.2*

69. **(D)** "Naturally occurring" means occurring in nature in a free state. A trick to answering these types of questions is to write the words "are" and "are not" in that order next to the choices—in this case we're looking for "are not" followed by "are."

Both radon and asbestos *are* naturally occurring and *are* sometimes found in indoor air samples. Chlorofluorocarbons (CFCs) such as Freon™, commonly used as a refrigerant for air conditioning systems, are *not* naturally occurring and are *not* considered to be indoor air pollutants. Volatile organic compounds (e.g., formaldehyde) *are not* naturally occurring but *are* common indoor air pollutants found in carpeting, paneling, furniture, paints, plastics, and cleaning fluids. *Topic 7.5*

70. **(B)**

$$\frac{1\text{ bag fertilizer}}{25\ \cancel{\text{kg fertilizer}}} \times \frac{1\ \cancel{\text{kg fertilizer}}}{1{,}000\ \cancel{\text{g fertilizer}}} \times \frac{100\ \cancel{\text{g fertilizer}}}{20\ \cancel{\text{g potassium}}} \times \frac{1{,}000\ \cancel{\text{g potassium}}}{200\ \cancel{\text{feet}^2\text{ lawn}}} \times \frac{1{,}000\ \cancel{\text{feet}^2\text{ lawn}}}{1} = \mathbf{1\ bag}$$

Topic 5.3

71. **(C)** The three main reasons atmospheric warming causes the sea level to rise are: ice sheets losing ice faster than it forms from snowfall, glaciers melting, and oceans expanding as they get warmer. Sea ice melt contributes only slightly to global sea level rise. If the melt water from ice floating in the sea was exactly the same as sea water, then, according to Archimedes' principle, no rise would occur. However, melted sea ice contains less dissolved salt than sea water and is less dense. Although the melted sea ice weighs the same as the sea water it was displacing, when it was ice, its volume is still slightly greater. Therefore, if *all* floating ice shelves and icebergs were to melt, sea level would only rise by about 1.6 inches (~4 cm).

More rain and flooding near the equator would also not have a significant impact on sea level rise, as that cycle does not include ice. Ocean water evaporates from warm equatorial waters and returns to Earth as rain within the Hadley cell. *Topic 9.5*

72. **(C)** $[(y_2 - y_1) / y_1] \times 100\% = \%$ change $=$ $[(35 - 20) / 20] \times 100\% = 75\%$

Topic 6.13

73. **(C)** Migratory animals (e.g., humpback whales, monarch butterflies, and wildebeests) are especially vulnerable to the threat of extinction by virtue of the long and fixed distances they travel. Their populations can be harmed not only by the loss of their breeding habitats, but also by changes in their wintering grounds and stopover sites. Furthermore, many migratory animals aggregate at specific places during certain times of the year, a habit that makes them vulnerable to exploitation. Climate change also has a high potential to disrupt or drastically change migratory routes. *Topic 9.9*

74. **(B)** Though the warmer air temperature within the urban heat island is generally most noticeable at night, the air temperature *difference* between the urban heat island and the surrounding environment is *greater* at night and *less* during the day. During the day, particularly when the skies are free of clouds, urban surfaces are warmed by the sun and tend to heat up at a faster rate than surfaces in surrounding rural areas, as urban surfaces act as giant reservoirs of stored heat energy. Concrete and asphalt have much higher heat capacities than soil or grass, and concrete can hold ~2,000 times more heat than an equivalent volume of air. This warming can also generate winds. At night, the situation reverses. In the absence of solar heating, the atmospheric convection decreases, causing the urban boundary layer to stabilize. If enough stabilization occurs, an inversion layer is formed that traps warm urban air close to the ground, keeping surface air warm from the still-warm urban surfaces. *Topic 4.5*

75. **(D)** "Positive feedback" refers to a situation in which some effect causes more of itself; i.e., "A produces more of B, which in turn produces more of A." A system undergoing positive feedback is unstable—that is, it will tend to spiral out of control as the effect amplifies itself. *Topic 9.4*

76. **(C)** Typically, in nonhazardous, sanitary waste landfills, techniques are applied by which the wastes are confined to as small an area as possible, compacted to reduce their volume, and covered (usually daily) with layers of soil. Under federal law, sanitary landfills designed to accept

and store municipal solid wastes may not under any circumstances accept any form of toxic wastes; however, this does not mean that toxic materials do not enter these facilities illegally.

Facilities that transport, store, and dispose of hazardous and toxic substances are governed by the Resource, Conservation, and Recovery Act (RCRA), which provides "cradle-to-grave" regulation of toxic and hazardous substances based partly on their persistence, degradability, corrosiveness, and flammability. The Comprehensive Environmental Response, Compensation, and Liability Act (CERCLA) directs the Environmental Protection Agency (EPA) to identify sites at which hazardous or toxic substances may have been released and ascertain the parties potentially responsible for cleaning up these sites. These parties include the owners and operators of sites where hazardous materials have been discharged, as well as the people responsible for discharging the hazardous material. *Topic 8.9*

77. **(C)**

$$\frac{2 \times 10^6 \text{ ft}^3}{1} \times \frac{\$1}{1000 \text{ ft}^3} = \$2{,}000$$

Topic 6.3

78. **(D)** An invasive, introduced, or exotic species is any species, including its seeds, eggs, spores, or other biological material capable of propagating that species, that is not native to an ecosystem, and whose introduction does or is likely to cause economic or environmental harm to species already living in that ecosystem. Invasive species typically are *r*-selected species with the following characteristics: high fecundity, small body size, early maturity onset, short generation time, and the ability to disperse offspring widely. *Topic 9.8*

79. **(A)** The human population has dramatically increased in the last ~100 years. Two thousand years ago, Earth had ~300 million humans—less than the population of the United States today (~329 million). Today, the world population is ~7.5 billion: a ~2,300% increase over the last 2,000 years. Even though birth rates have fallen, population momentum will continue with ~1 billion people added to the planet about every 15 years or so. Only after TFR (total fertility rates) have remained close to or below 2.1 for a generation or two, will population sizes begin to level off or decrease. Explosive growth over the last 100 years or so is due primarily to death rates falling faster than birth rates, primarily as a result of antibiotics, immunizations, clean water (though not in all areas), increased food production and distribution, and major declines in infant and child mortality. *Topic 3.8*

80. **(C)** Her total cost for 20,000 hours of lighting her home that required 20 incandescent light-bulbs (but which needed to be replaced periodically due to a limited life span) was $2,500. However, using LEDs, which did not need to be replaced and required much less electricity, would cost only $500, a difference of $2,000. *Topic 6.13*

Free-Response Explanations with Scoring Rubric

Question 1

(a) *1 point maximum: 1 point for correctly describing the climate in Panama.*

- Air temperature ranges from warm to hot and varies little throughout the year.
- Precipitation is evenly distributed throughout the year. In other climates, evaporated water vapor is often transported away and condenses in far-off areas (e.g., water vapor from the Pacific Ocean condensing and falling as rain in the Midwest.)
- The intense sunlight due to the angle that the sunlight reaches the equatorial area and the amount of sunlight available during the day (~12 hours) warms the land and sea, evaporating large quantities of water, which creates very humid conditions. As the air rises, it cools and is not able to hold as much water vapor. The water vapor then condenses and falls as rain frequently.
- Warmer air can hold more water vapor than colder air and, as a result, the air is very humid.

(b) *1 point maximum: 1 point for correctly drawing a sketch showing the difference in the sun's angle of sunlight reaching the equator vs. reaching the poles.*

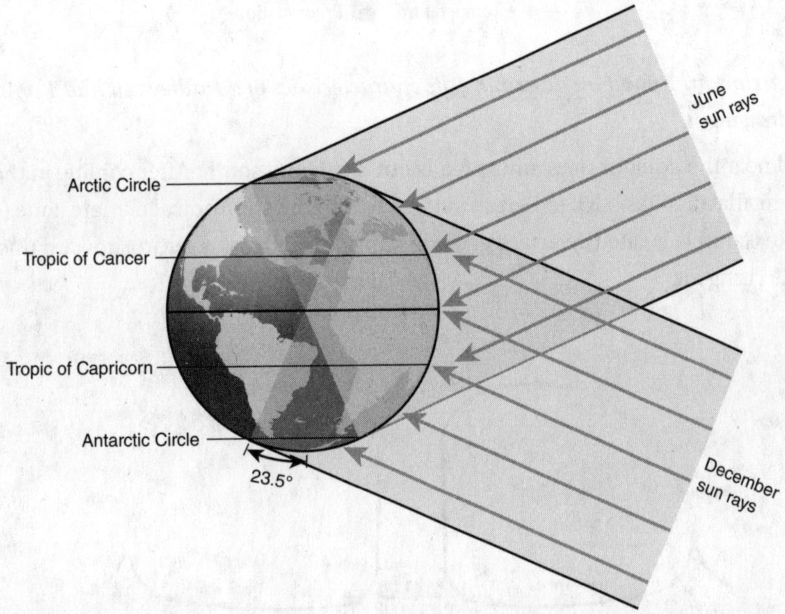

(c) *1 point maximum: 1 point for correctly sketching a climatograph for the area around Lake Gatún showing both relative temperature and precipitation patterns over a one-year time scale.*

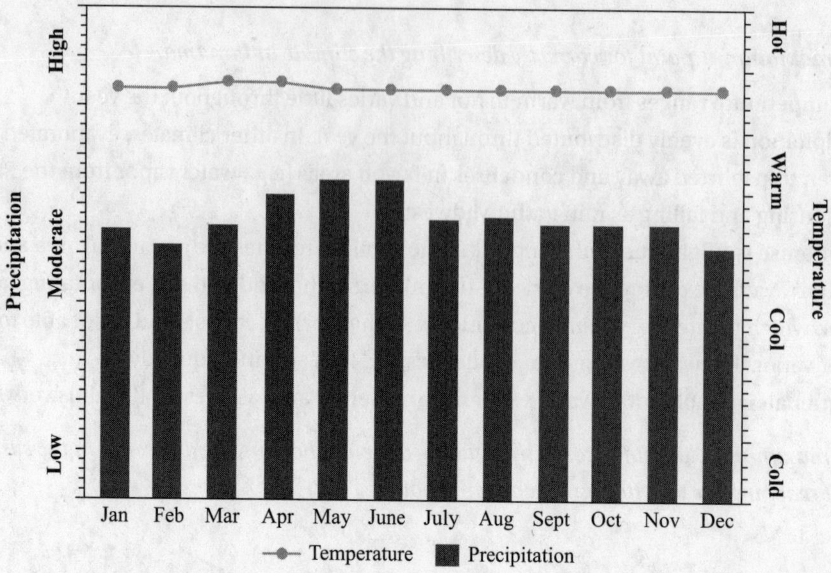

(d) *2 points maximum: 1 point for describing the characteristics of a Hadley cell and 1 point for correctly drawing it.*

Air heated near the equator rises and spreads out north and south. After cooling in the upper atmosphere, the air sinks back to Earth's surface within the subtropical climate zone (the area where Lake Gatún is located). Surface air from subtropical regions returns toward this region to replace the rising air.

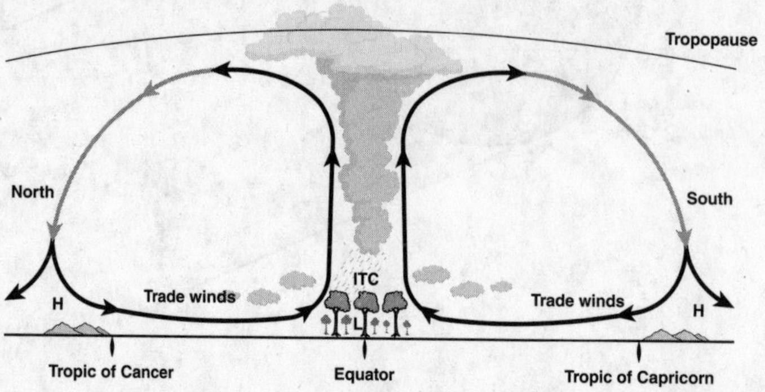

(e) *1 point maximum: 1 point for correctly describing the type of vegetation that you might expect to find in the area.*

- Biotic diversity is high.
- Many leaves in the multilayered canopy block out the sun from the lower layers. The canopy contains over 50% of the rainforest wildlife. Lianas (vines) climb toward the canopy to reach sunlight.
- Many trees have wide buttress roots to support them, as they grow very tall due to competition for sunlight.
- The forest floor is usually dark and damp and contains a layer of rotting leaves called "litter."
- The shrub layer has the densest plant growth and consists of plants (e.g., shrubs and ferns) that need less light.
- The undercanopy contains many bare tree trunks and lianas.

(f) *1 point maximum: 1 point for correctly describing the type of soil and its characteristics that you might expect to find in the area.*

- Forest floor litter (e.g., dead plant materials such as leaves, bark, etc.) found in the O soil horizon decomposes rapidly to form a thin humus, which is rich in nutrients but quickly absorbed by the flora, resulting in poor soil quality.
- Topsoil holds very large amounts of carbon, known to have a major potential influence on atmospheric CO_2 levels and hence a major potential influence on climate change.
- Tropical soils are often several years (meters) deep, but the soils are often strongly leached, with large amounts of nutrients and minerals being removed from the subsoils. Over many millions of years, this leaching has left most of the soils lacking many of the fundamental nutrients needed by vegetation above the ground.

(g) *1 point maximum: 1 point for correctly choosing a water quality test related to the environmental issue(s) described in the initial background information for the question.*

I would choose to measure water turbidity, as it is a measure of the degree to which the water loses its transparency in the presence of suspended particles—in this case, the silt entering the water, which is an environmental consequence of deforestation. As the rate of deforestation increases, the turbidity of the water likewise increases as more topsoil enters the watershed. One of the environmental consequences of increased turbidity in watersheds is the decreased light levels available for aquatic plants, which die off and consequently release less oxygen into the water, which begins to affect all living organisms higher in the food chain.

(h) *2 points maximum: 1 point for a correct definition of deforestation and 1 point for a correct description of the effects of deforestation on the hydrologic cycle and water quality in the area.*

Deforestation is the permanent destruction of forests in order to make the land available for other uses. Deforestation affects the amount of water in the soil and groundwater, thus affecting the recharge of aquifers and the moisture content in the atmosphere. Deforestation also increases surface water runoff, which can contribute to flash or extreme flooding. Increased deforestation can significantly reduce atmospheric water vapor, resulting in water and food shortages elsewhere.

Question 2

(a) Using Figure 1, DESCRIBE the trends in algae growth in both ponds.

Maximum 2 points total: 1 point is earned for accurately describing EACH of the curves through the growing season. Responses must represent the relative difference between each curve, noting that the ponds start out the growing season with similar primary productivity levels. The enriched pond shows higher primary productivity during ALL parts of the season. While answers do not need to describe this level of detail, the following observations are valid and students must have at least one observation for each pond in order to earn both points:

- Both ponds show similar rates of primary productivity early in the growing season.
- Primary productivity decreases to minimal levels by September. The enriched pond demonstrates steeper increases in primary productivity earlier in the season than the control pond. Primary productivity climbs to levels that are more than twice the productivity rates of the control pond earlier in the season (late June), and they stay at that elevated level for longer.
- Primary productivity decreases in mid/late August, but returns to a level that is higher than the control pond.
- The control pond experiences a gradual increase in productivity until early/mid July, when primary productivity starts to decrease.

(b) Using your understanding of the process of eutrophication, provide a hypothesis that EXPLAINS the difference between the control and enriched conditions in the experiment in Figure 1.

Maximum 2 points total: 2 points are earned connecting the fact that nutrient-rich water (water that has higher levels of nitrogen and/or phosphorus) encourages higher rates of plant growth (measured as primary productivity). While answers do not need to include all of the following observations, responses receiving 2 points will contain at least two of the following valid statements in order to earn both points:

- Plant growth can be measured or identified by rates of primary productivity.
- Similar to adding fertilizer to a garden, the addition of nitrogen and phosphorus to aquatic ecosystems will increase all types of plant growth.
- The main growing season in the Northern Hemisphere is in the summer months (identified in the graph) and both systems will experience increased plant growth, but the pond with more nitrogen and phosphorus will experience a faster growth rate.

(c) The scientists performing this experiment also noted elevated rates of growth in decomposer populations in late June for the enriched pond. EXPLAIN how the presence of these decomposer populations impacts ONE abiotic factor in the pond.

Maximum 2 points total: 1 point is earned for identifying an abiotic factor that might change as decomposer populations increase. 1 point is earned for describing the way in which the abiotic factor changes when decomposer numbers increase.

Students DO NOT need to describe the reason for the growth in decomposer populations. Instead, they must link their identified abiotic factor with a description that correctly shows how that abiotic factor would be impacted by the presence of decomposer populations. Examples AND their explanations will vary, but may include the following:

Viable Abiotic Factor	Description of Environmental Change Due to Population Increase
Dissolved oxygen	As decomposer populations increase, these organisms require oxygen (they are consumers) and therefore they will *draw down (lower)* the levels of dissolved oxygen.
Biological oxygen demand (BOD)	As decomposer populations increase, these organisms require oxygen (they are consumers) and as a result the biological oxygen demand of the system will *increase.*
Nitrogen/phosphorus/ available nutrients	As the decomposer population increases, the rate of nutrient cycling will also *increase,* which may cause an *increase* in the amount of biologically available nitrogen or phosphorus.

(d) In the Northern Hemisphere, fish kills are most frequently observed in late August and early September. Using your understanding of the connection between dissolved oxygen, nitrate, and phosphate levels, DESCRIBE the sequence of events that would result in fish kills in the enriched pond.

Maximum 3 points total: 2 points may be earned for describing any two of the following steps in the sequence that results in a fish kill. An additional point may be earned for students who also recognize that DO levels decrease as water temperatures increase in late summer.

While answers do not need to include all of the following observations, responses receiving 2 points will contain at least two of the following valid statements in order to earn both points (the sequence below is ordered to show which steps occur early and which occur later in the process):

- After an influx of nutrients (nitrates and phosphates in particular) into the water system, plant growth/primary productivity will increase (often rapidly), causing an algal bloom.
- Additional material from dying fish populations becomes available for decomposers, causing a further decrease in DO.
- As decomposer populations arrive, dissolved oxygen levels decrease (BOD increases with this new/larger consumer population).
- Available dissolved oxygen decreases. Fish species that require high levels of DO will be the first affected, dying as the DO levels decrease below tolerance ranges.
- Decomposer populations increase due to the rapid growth of algae and aquatic plants. Because some individuals might get shaded out OR have short life spans OR because of the overcrowding and competition that happens in rapidly growing populations, plant/algae individuals die and become available material for decomposers.

(e) IDENTIFY ONE recommendation that you would make to local landowners within the drainage area of the enriched pond that would improve water quality by reducing the likelihood of eutrophication. JUSTIFY your recommendation's impact on water quality.

Maximum 2 points total: 1 point for IDENTIFYING a viable strategy or recommendation and 1 point for correctly JUSTIFYING how that recommendation improves water quality. Note: The explanation must describe/correlate to improvement in water quality. Students must link their identified recommendation with a justification that shows HOW the water quality will be improved:

Recommendation to Landowners	Associated Justification of How Water Quality Is Improved
Improve vegetation or address uses within riparian zone	The riparian zone, which is found immediately adjacent to water systems, is critical for protecting water. By maintaining or adding a vegetated barrier in this zone, water will be filtered and topsoil will be held in place, reducing the likelihood that high levels of nutrients or other pollutants will make their way into the water.
Change fertilizer type/ application schedule	By reducing the amount of fertilizer (inorganic is typically most harmful) used on land areas in the watershed, there will be less nutrient input and a lower likelihood of eutrophication. Organic fertilizers or slow-release fertilizers typically reduce the amount of nitrate and phosphate input just as well.
Sustainable agricultural practices	By practicing a range of sustainable agricultural practices such as intercropping, no-till farming, contour plowing, or crop rotation, there is a lower likelihood of having exposed fertilizer-rich soil moving into the adjacent water systems. Without this input, there will be fewer nutrient inputs and a lower risk of eutrophication.

Topic 1.8

Question 3

(a) *Maximum of 4 points.*

- *1 point for correctly identifying a biome that is consistent with the food web*
- *3 points (maximum) for correctly identifying three characteristics of a coniferous forest (1 point awarded for each correct characteristic)*

Identify a biome that would be consistent with the organisms shown in the food web and describe three characteristics of that biome.

A biome contains characteristic communities of plants and animals that result from and are adapted to its climate. A coniferous forest would be consistent with the biome shown. Coniferous forest vegetation is composed primarily of cone-bearing needle-leaved or scale-leaved evergreen trees found in areas that have long winters and moderate to high annual precipitation. Pines, spruces, and firs are the dominant trees in coniferous forests. The trees are similar in shape and height and often form a nearly uniform stand with a layer of low shrubs beneath, where the light levels are lower. Mosses and lichens cover the forest floor. The soil is usually acidic, with a compacted humus layer that may contain many types of fungi. The soil in a coniferous forest is usually low in mineral content, organic material, and invertebrates, e.g., earthworms. Mosquitoes, flies, and other insects are commonly found in a coniferous forest, but few cold-blooded vertebrates (e.g., snakes) are present in the more northern forests due to the cold air and soil temperatures. *Topic 1.2*

(b) *Maximum of 1 point for a correct description of how energy flows through trophic levels.*

In terrestrial and near-surface marine communities, energy flows from the sun to producers in the lowest trophic levels and then upward to higher trophic levels. Describe how energy decreases as it flows through ecosystems.

When energy moves between trophic levels, only ~10% of the energy is made available for the next higher level; the exception is the transition from the sun to producers, in which case only 1% of the energy is retained.

When a primary consumer, such as a deer, eats a plant, it gains energy from the plant, with that energy being used for growth, reproduction, and other biological processes, while some of it is lost as heat.

As we move up an energy pyramid or a trophic level, less and less of the original energy from the sun is available, roughly a 90% loss for each higher trophic level. In other words, ~10% of the previous trophic level's energy is available to the level immediately higher up—this phenomenon is known as the 10% Rule. *Topic 1.10*

(c) *Maximum of 1 point for a correct final answer and for showing all steps required to obtain that answer. All values for each trophic level must be correct.*

Suppose 1,000,000 joules of radiant energy from the sun reaches this biome in a defined time period. Draw a model (known as an energy pyramid) to show how the energy from the sun will ultimately be available to the wolf. Be sure to show how many joules of energy are available at each trophic level in your model.

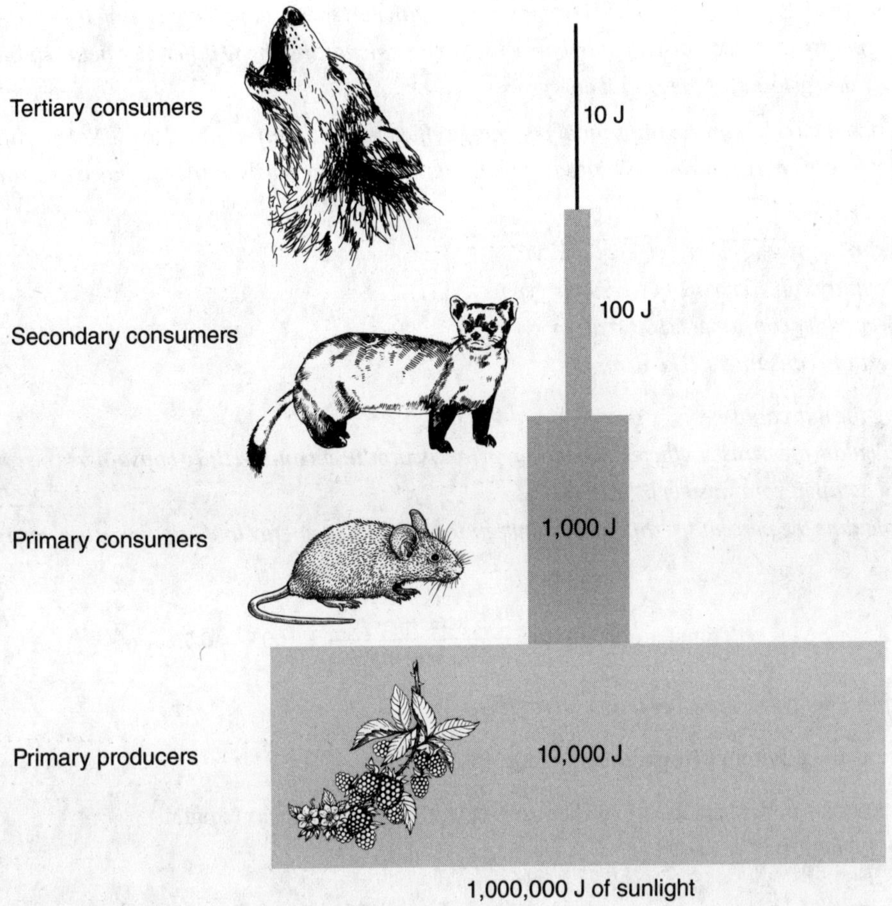

1,000,000 J of sunlight → 1% primary producers (plants) or 10,000 J → 10% primary consumer (rat) or 1,000 J → 10% secondary consumer (ringtail) or 100 J → 10% tertiary consumer (wolf) or 10 J. *Topic 1.10*

(d) *Maximum of 2 points.*

1 point for correct explanation of the First Law of Thermodynamics

1 point for correct explanation of the Second Law of Thermodynamics

Explain how energy loss through higher and higher trophic levels is consistent with the first and second laws of thermodynamics.

Energy transfers within food webs are governed by the first and second laws of thermodynamics. The first law relates to quantities of energy and states that energy can be transformed from one form to another, but it cannot be created or destroyed. This law suggests that all energy transfers, gains, and losses within a food web can be accounted for.

The second law relates to the quality of energy and states that whenever energy is transformed, some of it must be degraded into a less useful form. In ecosystems, the biggest losses occur during cellular respiration, usually in the form of heat. The second law explains why energy transfers are never 100% efficient. Ecological efficiency, which is the amount of energy transferred from one trophic level to the next, can range from 5–30% but averages ~10%. Because ecological efficiency is so low, each trophic level has a successively smaller energy pool from which it can withdraw energy, and it's the reason why food webs have no more than four to five trophic levels. Beyond that number, there is not enough energy available to sustain higher-order predators. *Topic 1.10*

(e) *Maximum of 2 points: 1 point for correct answer with correct setup for calculating transfer efficiency and 1 point for correct answer with correct setup for calculating the % difference between the theoretical energy loss and the actual energy loss.*

The 10% Rule is an approximation of the energy efficiency between trophic levels. Net productivity, or the rate of energy storage as biomass, by the various trophic levels in the food web diagram above are as follows:

- *Primary producers: 7,500 kcal/m^2/yr*
- *Primary consumers: 1,000 kcal/m^2/yr*
- *Secondary consumers: 100 kcal/m^2/yr*
- *Tertiary consumers: 5 kcal/m^2/yr*

Using the data above:

1. *calculate the transfer efficiency between primary producers and tertiary consumers (be sure to include the setup in your answer).*

2. *calculate the percentage difference between the theoretical energy loss and the actual energy loss.*

1.

$$\text{Transfer efficiency} = \frac{5\ \cancel{\text{kcal/m}^2\text{/yr}}}{7,500\ \cancel{\text{kcal/m}^2\text{/yr}}} \times 100\% = \mathbf{6.7\%}$$

1 point for correct answer with correct setup.

2. Expected difference (theoretical) using 10% Rule:

7,500 kcal/m^2/yr (plants) → 750 kcal/m^2/yr (rat) → 75 kcal/m^2/yr (ringtail) → **7.5 kcal/m^2/yr** (wolf)

$$\%\ \text{difference} = \frac{\text{actual} - \text{expected}}{\text{expected}} \times 100\%$$

$$= \frac{|5\ \text{kcal/m}^2\text{/yr} - 7.5\ \text{kcal/m}^2\text{/yr}|}{7.5\ \text{kcal/m}^2\text{/yr}} \times 100\% = \mathbf{33.3\%\ difference}$$

Topic 1.10

Index